Biodiversity: Sustaining Life

Biodiversity: Sustaining Life

Edited by **Jason Hendon**

R CALLISTO
REFERENCE

New York

Published by Callisto Reference,
106 Park Avenue, Suite 200,
New York, NY 10016, USA
www.callistoreference.com

Biodiversity: Sustaining Life
Edited by Jason Hendon

© 2015 Callisto Reference

International Standard Book Number: 978-1-63239-095-0 (Hardback)

Printed in the United States of America.

Contents

Preface

Biodiversity refers to the diversity of plant and other life forms in a biosphere. Biodiversity is often defined by biologists as the "totality of genes, species, and ecosystems of a region". Therefore traditionally, it can be identified at three levels - species diversity, ecosystem diversity and genetic diversity.

Extinction of species or other abiotic factors caused by geologically rapid changes in climate have been known to affect the rise or fall in biodiversity. However, with the evolution of man and the advancement in conditions of human life there has been a definite decrease in biodiversity and an accompanying loss of genetic diversity and the current rate and magnitude of extinctions are much higher than what was estimated. This decrease in biodiversity is primarily a result of human impacts, particularly destruction of environment.

While Darwinian theory of natural selection promotes the idea of "survival of the fittest" within a given species, however, each species depends on other species to ensure survival. Biodiversity hence plays an important role in increasing productivity of ecosystems where each species plays an equally important role in mutual survival and help create what is often called a balanced ecosystem. It is because of this reason that preservation of biodiversity is the need of the hour.

This book provides an overview of the evolution of biodiversity and how it has been affected by various factors at different stages of Earth's life. We hope that this book facilitates development in the field of science. I would like to thank our researchers and writers for their efforts and contributions. This book would not have been possible without your constant support and insights.

Editor

Edge-Interior Disparities in Tree Species and Structural Composition of the Kilengwe Forest in Morogoro Region, Tanzania

David Sylvester Kacholi

Department of Biological Sciences, Dar es Salaam University College of Education (DUCE), P.O. Box 2329, Dar es Salaam, Tanzania

Correspondence should be addressed to David Sylvester Kacholi; kacholi78@yahoo.com

Academic Editors: H. Ford and P. M. Vergara

A survey to determine the variation in species and structural composition of trees along the edge-interior gradient was done in the Kilengwe forest in Morogoro region, Tanzania. The forest was categorized into three habitats, namely, edge (0–100 m), intermediate (100–200 m), and interior (>200 m) depending on the distance from the forest margin. A total of six plots of 0.04 ha each were randomly placed in each of the habitats whereby all trees with DBH ≥ 10 cm were inventoried. A total of 67 species representing 26 families were recorded. Fabaceae was the most speciose and abundant family. *Brachystegia spiciformis* was the most abundant species. Of the recorded species, 10.45% were common in the three habitats while 8.95%, 13.43%, and 26.86% occurred exclusively to the edge, intermediate, and interior habitats, respectively. The forest interior was significantly rich in terms of species richness, diversity, density, and basal area than the edge and intermediate habitats. The edge had significantly higher number of stumps/ha. In summary, the results suggest that edge/intermediate and interior are contrasting habitats in terms of tree species richness, diversity, and structural composition. Moreover, the forest edge and intermediate habitats were found to be characterized by high anthropogenic activities compared to the forest interior habitat.

1. Introduction

Forest fragmentation and deforestation are among the critical environmental problems with possible implications on a global scale [1, 2] as human activities convert continuous forests into number of patches [3]. The tropical forests that encompass 6% of the world's land area and harbor at least 50% of the world's biodiversity are deforested and fragmented at an alarming rate [4, 5]. For instance, the global annual losses of 5.2 × 106 ha/year and 8.3 × 106 ha/year due to fragmentation and deforestation in the tropical forest area were recorded between 1990–2000 and 2000–2010, respectively [6]. In Tanzania, the annual loss of forest cover increased by 37% from the period of 1990–1995 (322,000 ha/year) to 2000–2010 (403,000 ha/year), which was mainly due to increased demand as the population grows [6, 7]. The forest fragmentation process disrupts structure and spatial continuity as it reduces original area, increases edge formation, and

isolates remaining forest patches [3, 8]. The formation of edges is recognized to be fundamental of ecological change as it involves alteration of microclimatic conditions, enhances invasion by exotic species, and increases the human pressure [9, 10]. Due to this, several organisms, for instance, mammals, amphibians, birds, and tree communities, have been reported to suffer significant changes in their local abundance and distribution in the forest fragments [11–14].

Various studies done in the tropical forests have reported negative impacts of forest fragmentation, mainly associated with the edge effects. These impacts include (1) reduction in recruitment rates of trees due to habitat desiccation and seedling damage by litter and tree falling near forest edges [11], (2) increasing sapling mortality rate by competition with lianas, vines, and ruderal species [15], and (3) increased mature tree mortality due to increased rates of uprooting and breakage near forest edges [2, 15, 16], which results in the decrease in canopy height [17, 18]. Due to the above explained effects and other edge-related processes, it is reasonable to expect that tree species in the forest edges will differ markedly from the forest interior in terms of species richness, diversity, and structure as well as ecological and taxonomical composition [12, 19, 20].

The Eastern Arc and Coastal forests (Kilengwe inclusive) of Tanzania are known to be centers of endemism, but the forests are facing danger of losing some of these due to increased anthropogenic activities as a result of population growth and fragmentation [21–23]. Thus, determining how tree species diversity and abundance vary within forests is a vital step in plant ecology and biodiversity conservation [9]. In Tanzania, no known study has examined the edge-interior differences in the tree species and structural composition in any forest. Thus, due to the existence of this knowledge gap, this study intends to (1) provide an understanding of the existing knowledge discrepancy by comparing the species richness, diversity, abundance, and basal area of tree communities along edge-interior gradient in the Kilengwe forest reserve and (2) quantify and compare impact of anthropogenic disturbances using number of stamps/ha observed in each established habitat. The findings from this study will contribute to the management of the forest reserve and other similar tropical forests in Tanzania and elsewhere.

2. Materials and Methods

2.1. Description of the Study Site. The Kilengwe Forest Reserve is located at latitude 7′ 29° South and longitude 37′ 32° East at an elevation of 182 to 228 m above mean sea level covering an area of 995 ha. The forest is in Kisaki Ward, Bwakira Chini Division in the Morogoro Rural District. The forest is surrounded by two villages, namely, Kilengwe and Zongomero, and is owned and managed by the local government. A number of seasonal streams that provide water to the local community for domestic purpose originate from this forest reserve. The climate of the region is oceanic due to nearness (about 200 km) to the Indian Ocean and the rainfall regime is bimodal. The long rains last from March to May, peaking in April while the short rains last from October to December. The mean annual rainfall and temperature in the Morogoro region are about 740 mm and 25.1°C, respectively. Agriculture is the most important socioeconomic activity done by locals living in the two adjacent villages. Other land use practices done by locals include livestock keeping, especially poultry, goats, and cows. Bee keeping and carpentry are done at small scale. Illegal logging activities were observed in the forest, especially close to the edges.

2.2. Data Collection. To characterize tree species and structural composition between edge (0–100 m), intermediate (100–200 m), and interior (>200 m) of the forest, six plots of 20 m × 20 m were randomly placed in each of the three established forest habitats. This design was due to the fact that the edge effects can penetrate to 100 m to the forest interior from the margin [2]. In each plot, all trees with diameter at breast height (DBH) ≥ 10 cm measured at 1.3 m above the ground were counted and identified and stem diameters were recorded from each plot in the habitats. Trees with multiple stems at 1.3 m height were treated as the single individual whereby the diameters of all stems were taken and averaged. If a tree had buttress or an abnormality at 1.3 m height, the diameter was measured just above the buttress where the stem assumes near cylindrical shape. The identification was done with the help of an expert from the forest department of the Morogoro region. Also, in each plot, tree stumps were counted and recorded. Each plot among the established habitats was considered as an independent sample.

2.3. Data Analysis. Species diversity was calculated using the Shannon-Wiener diversity index (H′) and Margalef index (D) while the equitability was determined by Pielou's evenness using the Species Richness and Diversity IV (SDR IV) software [24]. Species richness was expressed by the number of observed species in the forest while the first order Jackknife species richness estimator was used to estimate potential species richness in the three studied habitats [25]. The species accumulation curves were constructed for comparing the increase of the number of species with increasing sample area within the habitats. The forest structure was explained in terms of density (stems/ha) and basal area (m^2/ha). A multivariate agglomerative clustering technique using the Ward's group linkage based on the Bray-Curtis distance measure was performed using the Community Analysis Package 4 (CAP IV) software [24] to analyze species compositional similarities between the studied habitats. The significant differences in structural composition and stumps/ha between the three habitats were tested using the One Way Analysis of Variance (ANOVA) followed by the *post hoc* Tukey's Highly Significance Difference (Tukey's HSD) test at the 5% significance level using the QED statistical software [24]. Before undertaking ANOVA, all the data were tested for normality using the Shapiro-Wilk tests.

3. Results

3.1. Overall Floristic Composition and Abundance. A total of 199 stems belonging to 67 tree species and 26 families were recorded from an area of 0.72 ha. Family Fabaceae was the most dominant family with 21 species in the studied area of the forest, followed by Moraceae (5 species), and Sterculiaceae (4 species). The most abundant species were *Brachystegia spiciformis* with a relative abundance of 4.02%, followed by *Julbernardia globiflora, Burke Africana,* and *Synsepalum cerasiferum* with a relative abundance of 3.52% each. *B. spiciformis* and *J. globiflora* were the most frequent species in the studied area with a relative frequency of 3.73% each. Of the total observed species, 10.45% (7 species: *Acacia nigrescens, Albizia glaberrima, B. spiciformis, Cussonia spicata, S. cerasiferum, Vitex doniana,* and *Stereospermum kunthianum*) were common in all the three studied forest habitats while 8.95% (6 species), 13.43% (9 species), and 26.86% (18 species) occurred exclusively to the edge, intermediate, and interior habitats, respectively (see Table 1).

TABLE 1: The tree species distribution and abundance in the edge, intermediate, and interior of the forest.

Scientific names	Family	Edge	Intermediate	Interior	Total	Relative abundance
Brachystegia spiciformis Benth.	Fabaceae	2	3	3	8	4.02
Burkea Africana Hook. f.	Fabaceae	1	—	6	7	3.52
Jubernadia globiflora (Benth.) Troupin	Fabaceae	—	2	5	7	3.52
Synsepalum cerasiferum (Welw.) T.D.Penn.	Sapotaceae	2	1	4	7	3.52
Bauhinia petersiana Bolle	Fabaceae	1	—	5	6	3.02
Dombeya natalensis Sond.	Sterculiaceae	—	—	6	6	3.02
Ophrypetalum odoratum Diels.	Annonaceae	—	2	4	6	3.02
Acacia nigrescens Oliv.	Fabaceae	1	2	2	5	2.51
Diospyros squarrosa Klotzsch.	Ebenaceae	—	2	3	5	2.51
Ehretia amoena Klotzsch.	Boraginaceae	1	4	—	5	2.51
Stereospermum kunthianum Cham.	Bignoniaceae	2	2	1	5	2.51
Vitex doniana Sweet	Verbenaceae	1	2	2	5	2.51
Albizia glaberrima (Schum. and Thonn.) Benth.	Fabaceae	2	2	1	5	2.50
Dalbergia boehmii Taub.	Fabaceae	—	1	4	5	2.50
Albizia versicolor Welw. ex Oliver	Fabaceae	—	—	4	4	2.01
Cynometra uluguruensis Harms.	Fabaceae	1	—	3	4	2.01
Deinbollia borbonica Scheff.	Sapindaceae	3	—	1	4	2.01
Markhamia obtusifolia (Baker) Sprague.	Bignoniaceae	—	1	3	4	2.01
Myrianthus holstii Engl.	Moraceae	—	—	4	4	2.01
Oncoba spinosa Forssk.	Salicaceae	—	4	—	4	2.01
Sclerocarya birrea (A. Rich.) Hochst	Anacardiaceae	—	—	4	4	2.01
Strychnos spinosa Lam.	Loganiaceae	—	4	—	4	2.01
Terminalia sambesiaca Engl. and Diels.	Combretaceae	—	—	4	4	2.01
Trema orientalis (L.) Blume	Ulmaceae	2	—	2	4	2.01
Combretum molle R.Br. ex G.Don.	Combretaceae	3	—	—	3	1.51
Commiphora africana (A. Rich.) Endl.	Burseraceae	—	1	2	3	1.51
Cussonia spicata Thunb.	Araliaceae	1	1	1	3	1.51
Dalbergia melanoxylon Guill. and Perr.	Fabaceae	—	2	1	3	1.51
Dombeya rotundifolia (Hochst.) Planch.	Sterculiaceae	1	2	—	3	1.51
Khaya anthotheca (Welw.) C. DC.	Meliaceae	1	2	—	3	1.51
Sterculia quinqueloba (Garcke) K. Schum.	Sterculiaceae	1	—	2	3	1.51
Acacia polyacantha Wild.	Fabaceae	—	1	2	3	1.50
Acacia seyal Del.	Fabaceae	—	1	2	3	1.50
Albizia petersiana (Bolle) Oliv	Fabaceae	—	—	3	3	1.50
Acacia caffra Thunb. Wild.	Fabaceae	1	—	1	2	1.01
Allanblackia uluguruensis Engl.	Clusiaceae	1	1	—	2	1.01
Anthocleista grandiflora L.	Loganiaceae	1	—	1	2	1.01
Antiaris toxicaria Lesch.	Moraceae	—	—	2	2	1.01
Brachystegia temarindoides Benth.	Fabaceae	1	—	1	2	1.01
Breonadia salicina (Vahl) Happer and J.R.I. Wood	Rubiaceae	1	1	—	2	1.01
Bridelia micrantha (Hochst.) Baill.	Euphorbiaceae	—	2	—	2	1.01
Cassipourea mallosana Alston	Rhizophoraceae	2	—	—	2	1.01
Englerophytum natalense (Sond.) T.D. Penn.	Sapotaceae	—	—	2	2	1.01
Ficus exasperata Vahl.	Moraceae	—	1	1	2	1.01
Ficus lutea Vahl.	Moraceae	—	2	—	2	1.01
Grewia similis K. Schum.	Tiliaceae	1	—	1	2	1.01
Margaritaria discoidea (Baill.) G.L. Webster	Euphorbiaceae	—	1	1	2	1.01
Parkia filicoidea Welw.	Fabaceae	1	—	1	2	1.01
Annona senegalensis Pers.	Annonaceae	—	1	—	1	0.50
Cassia abbreviate Oliv.	Fabaceae	—	—	1	1	0.50
Combretum adenogonium Steud. ex Rich.	Combretaceae	—	1	—	1	0.50

TABLE 1: Continued.

Scientific names	Family	Edge	Intermediate	Interior	Total	Relative abundance
Commiphora eminii Engl.	Burseraceae	—	—	1	1	0.50
Cussonia zimmermannii Harms.	Araliaceae	—	—	1	1	0.50
Diplorhynchus condylocarpon (Muell. Arg.) Pichon	Apocynaceae	—	—	1	1	0.50
Erythrophleum suaveolens (Guill and Perr) Brennan	Fabaceae	—	1	—	1	0.50
Harrisonia abyssinica Oliv.	Simaroubaceae	—	—	1	1	0.50
Harungana madagascariensis Lam. ex Poiret	Clusiaceae	—	—	1	1	0.50
Lonchocarpus bussei Harms.	Fabaceae	1	—	—	1	0.50
Markhamia zanzibarica Bojer ex DC.	Bignoniaceae	1	—	—	1	0.50
Milicia excelsa (Welw.) C.C. Berg	Moraceae	—	1	—	1	0.50
Oxyanthus goetzei K. Schum	Rubiaceae	—	—	1	1	0.50
Pouteria altissima Baehni	Sapotaceae	—	—	1	1	0.50
Pseudolachnostylis maprouneifolia Pax	Phyllanthaceae	—	—	1	1	0.50
Pterocarpus tinctorius Welw.	Fabaceae	—	—	1	1	0.50
Scorodophloeus fischeri (Taub.) J. Leon	Fabaceae	1	—	—	1	0.50
Sterculia appendiculata K. Schum.	Sterculiaceae	1	—	—	1	0.50
Tabernaemontana pachysiphon Stapf	Apocynaceae	—	1	—	1	0.50
Total		39	55	105	199	100

3.2. Species Diversity, Similarity, and Species Accumulation Curves. The species richness, Shannon-Wiener diversity index (H'), Margalef index (D), Pielou's Evenness (E), and the first order Jackknife estimates of the studied habitats are shown in Table 2. The forest interior habitat was revealed to be the most diverse community with significantly higher species richness (47 species), Shannon-Wiener index (3.65), and Margalef index (9.86) compared to edge and intermediate habitats, which did not differ appreciably. Though evenness did not differ significantly between habitats, the edge community had relatively higher evenness than other habitats. Figure 1 shows the similarity between the three studied habitats. The distance correlation (Ward linkage using the Bray-Curtis measure) between the edge and intermediate is minimal, and this explains that edge and intermediate had a high similarity index (0.32). The forest interior habitat was less similar to edge and intermediate habitats (see Figure 1). The species accumulation curves (Figure 2) revealed an increasing trend as the number of sample areas increased in each studied habitat. The first order Jackknife species richness estimator projected higher species richness in each studied habitat than the observed ones (Table 2). Moreover, the rate of species increase per unit area (i.e., regression slope) was significantly higher in the interior habitat compared to edge and intermediate habitats, which did not differ substantially (Table 3).

3.3. Structural Composition. The densities of stems between the studied forest habitats varied significantly ($F(2, 15) = 15.18$, $P = 0.002$) ranging from 163 stems/ha (at the edge) to 442 stems/ha (at the interior) with an overall average of 278 stems/ha when all habitats are pooled together (Figure 3). *Post hoc* Tukey's HSD test ($P < 0.05$) confirmed the interior to have a significantly higher density than edge and intermediate, which did not differ appreciably ($P > 0.05$) in their tree densities. The basal areas differed significantly

between the studied habitats ($F(2, 15) = 5.98$, $P = 0.01$) ranging from 3.7 m^2/ha (at the edge) to 11.1 m^2/ha (at the interior) (Figure 4). *Post hoc* Tukey's HSD test ($P < 0.05$) revealed the interior of the forest to have a significantly higher basal area compared to the edge and intermediate, which also did not differ considerably ($P > 0.05$) in their basal areas. The species with the highest basal areas were *Dalbergia melanoxylon* and *J. globiflora* with a relative basal area of 8.10% and 6.00%, respectively. Moreover, number of stumps/ha among the habitats varied extensively ($F(2, 15) = 5.19$, $P = 0.02$) from 2 to 14 stumps/ha (Figure 5). *Post hoc* Tukey's HSD test ($P < 0.05$) revealed the forest edge (2 stumps/ha) to have a significantly lower number of stumps/ha than intermediate (8 stumps/ha) and interior (14 stumps/ha), which also differed significantly in their number of stumps/ha ($P < 0.05$).

4. Discussion

4.1. Overall Floristic Composition of the Forest. Tropical forests are recognized to harbor more than half of the global species richness, and they are often subjected to increasing anthropogenic pressure, which poses a great threat to existing biodiversity [26]. The Eastern Arc and Coastal forests of Tanzania (Kilengwe inclusive) are known to be rich in species diversity as they are located within the region of high rainfall and habitat heterogeneity [27]. This study has observed Fabaceae to be the most dominant family (with 21 species), which implies that the Fabaceae could be the most dominant tree family in the region. This finding is in line with the results of other authors [22, 28–30] who reported the same family to be the most dominant tree family in the Coastal and Eastern Arc forests of Tanzania. The family Fabaceae was reported to dominate by 25%–50% [28, 31, 32] and by 33% [22] in their studies. The present study revealed the Fabaceae to dominate by 31%. Moreover, like many other studies done

TABLE 2: Species richness, diversity indices, and evenness values of the three studied habitats.

Habitats	Observed species richness (±SE)	Jackknife 1 (±SE)	Shannon-Wiener (H′)	Margalef index (D)	Pielou's evenness (E)
Edge	29 ± 4[a]	51 ± 4[a]	3.28 ± 0.19[a]	7.64 ± 1.03[a]	0.97 ± 0.06
Intermediate	32 ± 2[a]	47 ± 2[a]	3.34 ± 0.05[a]	7.74 ± 0.40[a]	0.96 ± 0.02
Interior	47 ± 5[b]	68 ± 5[b]	3.65 ± 0.11[b]	9.86 ± 0.94[b]	0.95 ± 0.03

Note: values with different letters indicate significant differences between habitats (Tukey's HSD test, $P = 0.05$).

elsewhere [33–37] this study confirms that the Fabaceae is the speciose family in most of the lowland forests of Africa. Other recorded families in this study have similarly been reported by various authors [29, 30, 32] in their studies in the Eastern Arc and Coastal forests of Tanzania.

4.2. Species and Structural Composition along the Edge-Interior Gradient.

The study has registered a clear pattern of changes in the tree community whereby the tree species diversity, richness, density, and basal area increased towards the forest interior (Table 2, Figures 2 and 4). In all cases, the forest interior habitat had significantly higher values than the edge and intermediate habitats, which did not differ appreciably. The results signify that edge effects can penetrate to a distance of 200 m from the forest margin. Also, these findings suggest that the forest interior and the edge/intermediate are contrasting habitats. A forest community is said to be speciose if it has a Shannon-Wiener diversity index value ≥3.5 [38]. This confirms the observation of this study, which revealed significantly higher Shannon-Wiener diversity index value of 3.65 compared to edge (3.28) and intermediate (3.34) habitats. The presence of less species diversity, richness, density, and basal area at the edges and intermediate can be linked to the fact that these two habitats are easily accessible by local people while fetching for their basic needs like firewood, charcoal, building poles, and traditional medicines [22]. The observed illegal timber harvest contributes to the observed low values, especially on tree density and basal area on the edge/intermediate habitats as it is easier for the timber thieves to cut and carry logs to the truck during the night [22, 39]. The grazing and cattle trampling can also be the causal factor for the observed low values at edges as livestock were observed to reach the forest edges in the studied forest. The Luguru people prefer to use fire to prepare their farms before starting a new agricultural season and for hunting bush animals like *Hyrax pimbi*, *Sus scrofa*, and *Thryonomys* spp., which are used for nutritional purposes. The penetration of fire to forest edges could also account for the observed low species diversity and density values as it kills the fire-sensitive species and affects their regeneration [10, 22]. The periodic fires reduce canopy cover and drastically change vegetation structure and composition [15, 39]. Another observed important factor was the action of sporadic winds that cause great damage to tree communities and high mortality of canopy trees at the edges, which subsequently affect the composition and abundance. Moreover, the present findings are comparable to other works done in Brazil and other countries, in which the forest interior was found to be rich in species diversity and the penetration of edge effects was reported to vary from 15 m to 200 m [12, 40–43]. Distinct abiotic conditions [18, 40–42, 44],

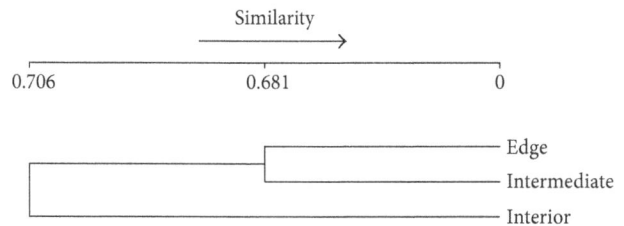

FIGURE 1: Similarity in species composition between the three studied habitats.

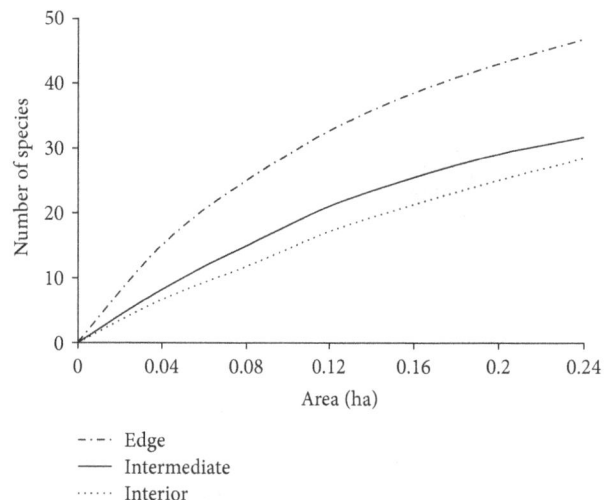

FIGURE 2: Species-area curves based on the cumulative sample area in the studied habitats.

vegetation structure [40], tree mortality rate [2], and high predation, loss of pollinators, and seed dispersers, at edges [11], were reported to be the common cause of differences in species composition between forest edge and interior in various studies done elsewhere.

The presence of significantly less tree density at the edges/intermediate habitats may also be the result of the interplay of factors of two kinds: firstly, those that reduce the possibility of seedling establishment as seedling is the first size class to be affected by edge effects due to its sensitivity to environmental changes and biotic interactions [9, 10], and secondly, those factors that increase seedling, sapling, and adult tree mortality rate [11]. Some processes related to edge effects that could have contributed to low tree density on the edge/intermediate habitats include; reduction in seedling recruitment at edges due to uprooting and breakage due to wind turbulence [2, 45], seedling damage caused by an increasing litter fall near forest edges [11], sapling mortality

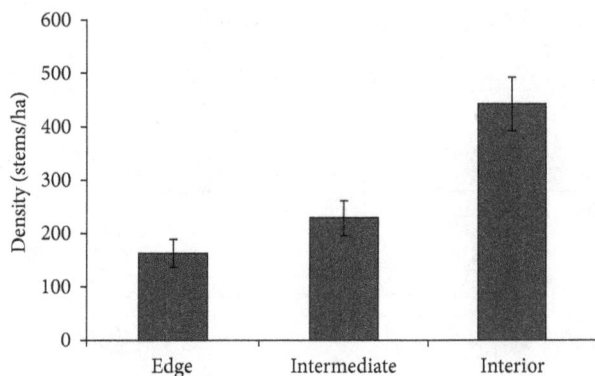

FIGURE 3: Mean tree density (±standard error) in the three studied forest habitats.

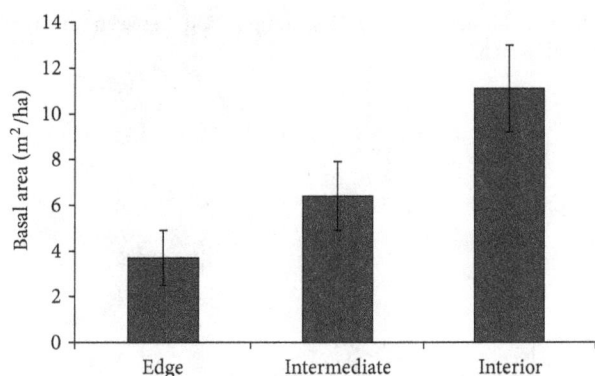

FIGURE 4: Mean basal area (±standard error) in the three studied forest habitats.

TABLE 3: Relationship between species richness and forest area.

Location	r^2-value	Regression equation
Edge	0.99	Species richness = 14.72 ln area − 2.15
Intermediate	0.98	Species richness = 16.77 ln area − 1.78
Interior	0.97	Species richness = 24.32 ln area − 0.88

by competition with lianas, vines, and invasive species [2, 46], and easy accessibility to edges by locals and their livestock [47]. Also, the low basal area at the edge and intermediate habitats could be the result of continuing anthropogenic activities especially illegal logging done by unfaithful locals. Various authors have indicated that the presence of harsh microenvironmental conditions such as air and soil temperatures, high light transmittance, lower relative humidity [17, 23, 48], increased wind forces [2], lower availability of soil carbon, total nitrogen and phosphorus [49], and lower soil and litter moisture [17, 49] near edges contributes to drastic changes in the abundance and distribution of trees in the forest.

4.3. Species Accumulation Curves. The species-area curves displayed an escalating trend, which suggest that increasing the sampling effort could have increased the species richness observed in each habitat. This is due to the fact that the larger

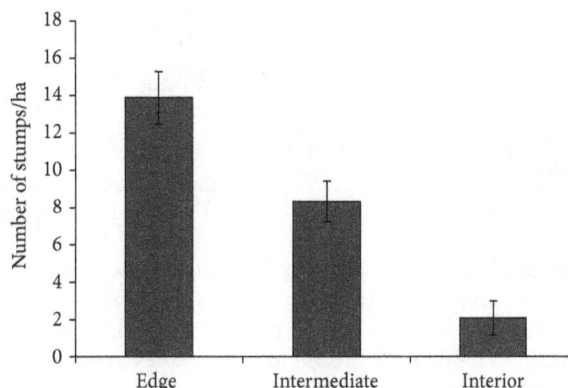

FIGURE 5: Mean stumps/ha (±standard error) in the three studied forest habitats.

the forest area sampled is the more environmentally heterogeneous the sampling area becomes and hence the higher the possibility of having many species is. The observed trend in both curves concurs with the first order Jackknife species richness estimator, which provides higher species richness estimates than the observed in each studied habitat. This observation basically informs us that the sample size used was not enough to capture all the species in the forest habitats, which implies that more plots will be required for future inventories in the habitats. Furthermore, the regression slopes have registered an increasing trend from the edge to the forest interior habitat (Table 3). This observation provides a clear indication that the interior habitat is rich and has a high recruitment rate per unit area than edge and intermediate habitats, which did not differ markedly. The observation can possibly be linked with ongoing anthropogenic disturbances observed at the edge and intermediate habitats (Figure 5) whereby the number of stumps/ha was significantly higher in these two habitats compared to the forest interior habitat.

5. Conclusion

In summary, the findings indicate that the edge/intermediate habitats are different from the forest interior in terms of tree species richness, diversity, and structural composition. The forest interior possesses higher species richness, diversity, tree density, and basal area while forest edges and intermediate habitats had impoverished assemblage of tree species, diversity, and density. Also, the edge and intermediate were characterized by high anthropogenic activities by having more stumps/ha than forest interior habitat. If the Kilengwe forest will continue being fragmented, there will be an increase of edge related habitats, which will cause structural and floristic composition changes due to increased edge effects and the forest will face a great threat of losing its original biota especially the rare species. The study recommends long-term research to study microenvironmental factors such as light availability, air and soil temperature, humidity, and nutrients along the edge-interior gradient in the forest in order to determine their influence on tree species richness, composition, and structure.

Conflict of Interests

The author declares that there is no conflict of interests regarding the publication of this paper.

Acknowledgments

The author is specially thankful to the Dar es Salaam University College of Education (DUCE) and the Deutscher Akademischer Austausch Dienst (DAAD) for funding this study. The thanks are also extended to the Morogoro Forest Department for permission to access the forest and providing technical and transport logistics to and from the study sites.

References

[1] B. L. Turner II and W. B. Meyer, "Land use and land cover in global environmental change: considerations for study," *International Social Science Journal*, vol. 43, no. 130, pp. 669–679, 1991.

[2] W. F. Laurance, L. V. Ferreira, J. M. Rankin-de Merona, and S. G. Laurance, "Rain forest fragmentation and the dynamics of Amazonian tree communities," *Ecology*, vol. 79, no. 6, pp. 2032–2040, 1998.

[3] L. Fahrig, "Effects of Habitat Fragmentation on Biodiversity," *Annual Review of Ecology, Evolution, and Systematics*, vol. 34, pp. 487–515, 2003.

[4] P. Ehrlich and A. Ehrlich, *Extinction*, Oxford University Press, Oxford, UK, 1981.

[5] S. L. Pimm, G. J. Russell, J. L. Gittleman, and T. M. Brooks, "The future of biodiversity," *Advances in Complex Systems*, vol. 1, p. 203, 1998.

[6] Food and Agriculture Organization of the United Nation (FAO), "Global forest resources assessment 2010," FAO paper 163, Rome, Italy, 2010.

[7] W. D. Newmark, *Conserving Biodiversity in East African Forests. A Study of the Eastern Arc Mountains*, Springer, New York, NY, USA, 2002.

[8] S. Walker, G. M. Rogers, W. G. Lee et al., "Consequences to threatened plants and insects of fragmentation of Southland floodplain forests," *Science for Conservation*, vol. 265, p. 86, 2006.

[9] C. Murcia, "Edge effects in fragmented forests: implications for conservation," *Trends in Ecology and Evolution*, vol. 10, no. 2, pp. 58–62, 1995.

[10] D. A. Saunders, R. J. Hobbs, and C. R. Margules, "Biological consequences of ecosystem fragmentation: a review," *Conservation Biology*, vol. 5, no. 1, pp. 18–32, 1991.

[11] J. Benitez-Malvido, "Impact of forest fragmentation on seedling abundance in a tropical rain forest," *Conservation Biology*, vol. 12, no. 2, pp. 380–389, 1998.

[12] M. A. Oliveira, A. S. Grillo, and M. Tabarelli, "Forest edge in the Brazilian Atlantic forest: drastic changes in tree species assemblages," *Oryx*, vol. 38, no. 4, pp. 389–394, 2004.

[13] I. M. Turner, "Species loss in fragments of tropical rain forest: a review of the evidence," *Journal of Applied Ecology*, vol. 33, no. 2, pp. 200–209, 1996.

[14] P. A. Zuidema, J. A. Sayer, and W. Dikman, "Forest fragmentation and biodiversity: the case for intermediated-sized conservation areas," *Environmental Conservation*, vol. 23, pp. 290–297, 1996.

[15] M. Tabarelli, J. M. Cardoso Da Silva, and C. Gascon, "Forest fragmentation, synergisms and the impoverishment of neotropical forests," *Biodiversity and Conservation*, vol. 13, no. 7, pp. 1419–1425, 2004.

[16] R. C. G. Mesquita, P. Delamônica, and W. F. Laurance, "Effect of surrounding vegetation on edge-related tree mortality in Amazonian forest fragments," *Biological Conservation*, vol. 91, no. 2-3, pp. 129–134, 1999.

[17] R. K. Didham and J. H. Lawton, "Edge structure determines the magnitude of changes in microclimate and vegetation structure in tropical forest fragments," *Biotropica*, vol. 31, no. 1, pp. 17–30, 1999.

[18] M. Oosterhoorn and M. Kappelle, "Vegetation structure and composition along an interior-edge-exterior gradient in a Costa Rican montane cloud forest," *Forest Ecology and Management*, vol. 126, no. 3, pp. 291–307, 2000.

[19] J. C. Jiquan Chen, J. F. Franklin, and T. A. Spies, "Vegetation responses to edge environments in old-growth Douglas-fir forests," *Ecological Applications*, vol. 2, no. 4, pp. 387–396, 1992.

[20] B. J. Fox, J. E. Taylor, M. D. Fox, and C. Williams, "Vegetation changes across edges of rainforest remnants," *Biological Conservation*, vol. 82, no. 1, pp. 1–13, 1997.

[21] R. P. C. Temu and S. M. Andrew, "Endemism of plants in the Uluguru Mountains, Morogoro, Tanzania," *Forest Ecology and Management*, vol. 255, no. 7, pp. 2858–2869, 2008.

[22] D. S. Kacholi, *Effects of habitat fragmentation on biodiversity of Uluguru Mountain forests in Morogoro region, Tanzania [Ph.D. thesis]*, Georg-August University Goettingen, Goettingen, Germany, 2013.

[23] W. D. Newmark, "Tanzanian forest edge microclimatic gradients: dynamic patterns," *Biotropica*, vol. 33, no. 1, pp. 2–11, 2001.

[24] R. M. Seaby and P. A. Henderson, *Species Diversity and Richness Version 4*, Pisces Conservation Ltd, Lymington, UK, 2006.

[25] A. N. Magurran, *Measuring Biological Diverisy*, Blackwell Science Ltd. A Blackwell Publishing Company, Melden, Mass, USA, 2004.

[26] N. Myers, R. A. Mittermeier, C. G. Mittermeier, G. A. B. da Fonseca, and J. Kent, "Biodiversity hotspots for conservation priorities," *Nature*, vol. 403, no. 6772, pp. 853–858, 2000.

[27] S. Madoffe, G. D. Hertel, P. Rodgers, B. O'Connell, and R. Killenga, "Monitoring the health of selected eastern arc forests in Tanzania," *African Journal of Ecology*, vol. 44, no. 2, pp. 171–177, 2006.

[28] N. D. Burgess and G. P. Clarke, *The Coastal Forests of Eastern Africa*, IUCN, Cambridge, UK, 2000.

[29] J. P. Mrema, *Conservation of Brachylaena huillensis O. Hoffm (Asteraceae) in Dindili Forest Reserve Morogoro [M.S. thesis]*, Addis Ababa University, 2006.

[30] S. P. Rwamugira, *Impact of mining on forest ecosystem and adjacent communities of Eastern Arc Mountains. A case study of Ruvu catchment forest reserve [M.S. dissertation]*, Sokoine University of Agriculture, 2008.

[31] N. D. Burgess and C. Muir, "Coastal forests of Eastern Africa: biodiversity and conservation," in *Proceedings of the Workshop held at the University of Dare s Salaam*, Society for Environmental Exploration/Royal Society for the Protection of Birds, UK, 1994.

[32] G. Eilu, D. L. N. Hafashimana, and J. M. Kasenene, "Density and species diversity of trees in four tropical forests of the Albertine rift, western Uganda," *Diversity and Distributions*, vol. 10, no. 4, pp. 303–312, 2004.

[33] F. Wittmann, J. Schöngart, J. C. Montero et al., "Tree species composition and diversity gradients in white-water forests across the Amazon Basin," *Journal of Biogeography*, vol. 33, no. 8, pp. 1334–1347, 2006.

[34] A. Gentry, "Changes in plant community diversity and floristic composition on environmental and geographical gradients," *Annals of Missouri Botanical Garden*, vol. 75, pp. 1–34, 1988.

[35] M. W. Cadotte, R. Franck, L. Reza, and J. Lovett-Doust, "Tree and shrub diversity and abundance in fragmented littoral forest of southeastern Madagascar," *Biodiversity and Conservation*, vol. 11, no. 8, pp. 1417–1436, 2002.

[36] P. Addo-Fordjour, S. Obeng, A. K. Anning, and M. G. Addo, "Floristic composition, structure and natural regeneration in a moist-semi deciduous forest following anthropogenic distur-bances and plant invasion," *International Journal of Biodiversity and Conservation*, vol. 1, no. 2, pp. 21–37, 2009.

[37] E. N. Mwavu, *Human impact, plant communities, diversity and regeneration in Budongo forest reserve, North-Western Uganda [Ph.D. thesis]*, University of the Witwatersrand, Johannesburg, 2007.

[38] M. Kent and P. Coker, *Vegetation Description and Analysis*, Belhaven Press, London, UK, 1992.

[39] S. B. Fontoura, G. Ganade, and J. Larocca, "Changes in plant community diversity and composition across an edge between Araucaria forest and pasture in South Brazil," *Revista Brasileira de Botanica*, vol. 29, no. 1, pp. 79–91, 2006.

[40] G. Williams-Linera, "Vegetation structure and environmental conditions of forest edges in Panama," *Journal of Ecology*, vol. 78, no. 2, pp. 356–373, 1990.

[41] G. R. Matlack, "Vegetation dynamics of the forest edge—trends in space and successional time," *Journal of Ecology*, vol. 82, no. 1, pp. 113–123, 1994.

[42] S. M. Gehlhausen, M. W. Schwartz, and C. K. Augspurger, "Vegetation and microclimatic edge effects in two mixed-mesophytic forest fragments," *Plant Ecology*, vol. 147, no. 1, pp. 21–35, 2000.

[43] O. Honnay, K. Verheyen, and M. Hermy, "Permeability of ancient forest edges for weedy plant species invasion," *Forest Ecology and Management*, vol. 161, no. 1–3, pp. 109–122, 2002.

[44] V. Kapos, "Effects of isolation on the water status of forest patches in the Brazilian Amazon," *Journal of Tropical Ecology*, vol. 5, no. 2, pp. 173–185, 1989.

[45] L. V. Ferreira and W. F. Laurance, "Effects of forest fragmen-tation on mortality and damage of selected trees in Central Amazonia," *Conservation Biology*, vol. 11, no. 3, pp. 797–801, 1997.

[46] W. F. Laurance, "Fragmentation and plant communities: syn-thesis and implications for landscape management," in *Lessons from Amazonia: The Ecology and Conservation of a Fragmented Forest*, Yale University Press, New Haven, Conn, USA, 2001.

[47] M. A. Cochrane and W. F. Laurance, "Fire as a large-scale edge effect in Amazonian forests," *Journal of Tropical Ecology*, vol. 18, no. 3, pp. 311–325, 2002.

[48] M. Yan, Z. Zhong, and J. Liu, "Habitat fragmentation impacts on biodiversity of evergreen broadleaved forests in Jinyun Mountains, China," *Frontiers of Biology in China*, vol. 2, no. 1, pp. 62–68, 2007.

[49] S. Jose, A. R. Gillespie, S. J. George, and B. M. Kumar, "Vegetation responses along edge-to-interior gradients in a high altitude tropical forest in peninsular India," *Forest Ecology and Management*, vol. 87, no. 1–3, pp. 51–62, 1996.

Floristic Composition, Structure, and Species Associations of Dry Miombo Woodland in Tanzania

Ezekiel Edward Mwakalukwa,[1,2] Henrik Meilby,[1] and Thorsten Treue[1]

[1] *Department of Food and Resource Economics, Faculty of Science, University of Copenhagen, Rolighedsvej 23, 1958 Frederiksberg C, Denmark*
[2] *Department of Forest Biology, Faculty of Forestry and Nature Conservation, Sokoine University of Agriculture, P.O. Box 3010, Chuo Kikuu, Morogoro, Tanzania*

Correspondence should be addressed to Ezekiel Edward Mwakalukwa; ezedwa@yahoo.com

Academic Editors: M. Drielsma, H. Ford, M. Tigabu, and A. Viña

For the majority of forest reserves in Tanzania, biodiversity is poorly documented. This study was conducted to assess species richness (woody species), diversity, and forest structure and to examine relationships between species occurrence and topographic and edaphic factors in the Gangalamtumba Village Land Forest Reserve, a dry Miombo woodland area in Tanzania. A total of 35 nested circular plots with radii of 5, 15, and 20 m were used to collect data on woody species and soil samples across the 6,065 ha community-managed forest reserve. Stumps were measured 20 cm above ground. A total of 88 species belonging to 29 families were identified. Generally forest structure parameters and diversity indices indicated the forest to be in a good condition and have high species richness and diversity. Vegetation analysis revealed four communities of which two were dominated by the family Caesalpiniaceae, indicating large variation of site conditions and possible disturbances in the study area. The high level of diversity of woody species and the high basal area and volume indicate that the forest is in good condition, but the effect of anthropogenic activities is evident and stresses the need for proper management to maintain or enhance the present species diversity.

1. Introduction

Miombo woodland is the most widespread and dominant dry forest formation in Eastern, Central, and Southern Africa. It is characterized by an abundance of tree species in the legume subfamily Caesalpinioideae, including the three dominant genera of *Brachystegia*, *Julbernardia*, and *Isoberlinia* [1, 2]. Covering an area of about 3.6 million km^2, miombo woodland supports the livelihoods of more than 100 million rural and urban dwellers by providing a wide range of products such as firewood, charcoal, timber, and forage and services such as soil conservation and water catchment [3–5]. However, due to the rapid population growth and the high level of poverty across the Miombo region, the human pressure on its woodlands has steadily increased over the last decades, leading to increasing deforestation and forest degradation [6–8].

The effects of increasing rates of deforestation and forest degradation on biodiversity in developing countries have been thoroughly studied [9–12]. Habitat loss due to deforestation reduces not only the number of species in the ecosystem but also the number and extent of places where species coexist. Activities such as charcoal production, firewood collection for subsistence use and for tobacco curing, conversion of woodlands to farmland, and seasonal forest fires are among the major drivers of deforestation and forest degradation in the Miombo region [13–17]. It is estimated that 1.4 million ha of woodlands is lost annually in the countries where Miombo woodlands dominate, leading to a loss of carbon stocks, biodiversity, and, through soil degradation, loss of plant nutrients [4, 5]. Syampungani et al. [5, p. 151] stated that "loss of biodiversity and extinction of most of the woodland resources are imminent if the current intensive exploitation of Miombo resources continues unchecked." More specifically, FAO (2000a, cited by Syampungani et al. [5]) reported that 191 tree species in the Miombo ecoregion are endangered due

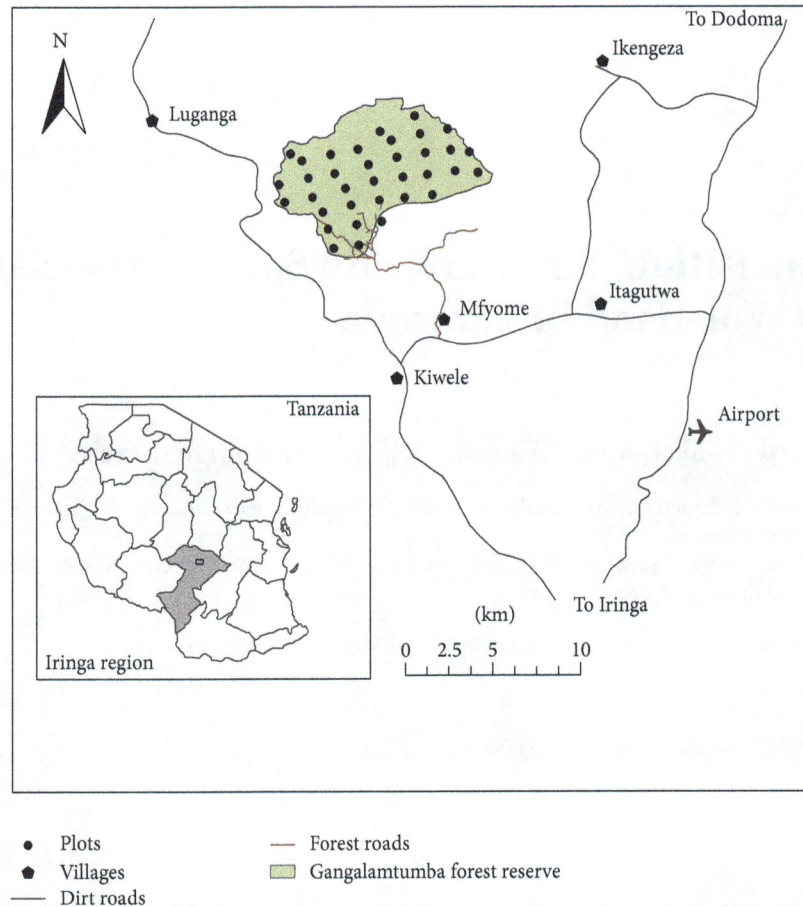

FIGURE 1: Map showing the location of the study area. The inserted map of Tanzania shows the location of the Iringa region.

to conversion of forest areas into agricultural lands or through charcoal production.

Gangalamtumba Village Land Forest Reserve (GVLFR) in Iringa rural district, Tanzania, which is owned and managed by the village of Mfyome, was established in 2002 under Tanzania's national participatory forest management programme and thus represents one of approximately 1500 Village Land Forest Reserves (covering some 2.4 million ha), the progressive establishment of which is intended to promote conservation of approximately 16.5 million ha of hitherto unreserved forest on general and village land [18–21]. Villages' control over Village Land Forest Reserves is conditional on their conservation/protection of these forests and the executive management is performed by an environmental/forest management committee whose members are directly elected for five-year terms by all members of the village above the age of 18 [18]. As such GVLFR is a typical example of an area which, at least until 2002, might have experienced loss of biodiversity due to increasing human activity, including charcoal production and extraction of wood for tobacco curing, giving cause for concern with respect to the maintenance of forest biodiversity [15]. The extent to which the activities have led to loss of biodiversity and deterioration of the plant community structure is so far unknown, but for the development of sustainable woodland management strategies

and for planning of future management and conservation, information on these issues is urgently needed.

Although many quantitative ecological studies have been undertaken in places where Miombo dominates, its extensiveness and the large between-site variation, which is caused by climatic and edaphic factors and anthropogenic activities, appear to warrant further case studies [20, 22–29]. Ecological case studies are particularly relevant when the information generated is required for sound decision making about forest management, conservation strategies, and determination of sustainable harvesting levels. Hence, the objectives of this study were (1) to provide a detailed assessment of the current standing stock, species diversity, richness, and structure and (2) to understand the relationship between species abundance and a range of environmental and topographic factors that shape plant communities and species associations in the GVLFR.

2. Materials and Methods

2.1. Study Site. Gangalamtumba Village Land Forest Reserve is located in central-southern Tanzania (7°35′S; 35°35′E), about 30 km north of Iringa town, the administrative capital of the Iringa region (Figure 1). The forest covers 6,065 hectares and is part of the Mfyome village area, which is located in the ward of Kiwele. The forest vegetation has been described as dry Miombo woodland, similar to the dry woodland type described from other countries such as Zimbabwe and Mozambique [1]. The forest is located in a relatively flat area at an elevation of 850–1,300 metres above sea level. The region is characterised by distinct wet and dry seasons with almost no rain in the four months of June-September and about 80% of the annual precipitation falling in December–March. Average rainfall data covering the last 50 years (1960–2010) were obtained from the meteorological station at Nduli airport, which is located about 30 km from the forest, and indicate that the area receives an average annual precipitation (mean ± standard error) of 617 ± 17 mm (448–1085 mm). The mean annual temperature is 19.8°C and the average relative humidities at 06.00 and 12.00 GMT are 53.9% and 51.4%, respectively.

The GVLFR is a production forest which is managed by the Mfyome village under a community-based forest management (CBFM) arrangement established in 2002; compare above. The primary economic activity in Mfyome is smallholder agriculture, and the main economic uses of the forest are production of timber, charcoal, and firewood [20]. The woodland is also used for grazing and is an important source of subsistence products such as firewood, construction materials, fruit, mushrooms, wild vegetables, and medicinal plants [18].

2.2. Vegetation Survey. The field survey was conducted in July and August 2009 and involved establishment of a total of 35 permanent, nested circular sample plots distributed across the entire forest. Plots were established along transect lines and the distance between plots was approximately 2 km (Figure 1). The radii of the nested circular plots were 5 m (0.0079 ha), 15 m (0.0707 ha), and 20 m (0.1257 ha). The following parameters were recorded within each of the 35 plots: within the 5 m radius all small trees and shrubs (<150 cm tall or ≥150 cm but <1 cm Dbh) were counted and their species were identified, and medium-size trees and shrubs (≥1 cm Dbh but <5 cm Dbh) were identified and measured with respect to diameter. Within 15 m radius, the species were identified and the diameter was measured for all large trees and shrubs with Dbh ≥5 cm. Within 20 m radius, all stumps of trees and shrubs were identified to species level and measured for diameter 20 cm above ground. Initial identification of species for both standing trees/shrubs and stumps relied on the knowledge of local botanists (using local vernacular species names and features such as color of the bark, smell, and leaves) and was later confirmed by botanists from Tanzania Forest Research Institute (TAFORI) based at Lushoto Silviculture Research Station. For species that were difficult to identify in the field, samples were taken to the herbarium at Lushoto for reidentification. Other measurements taken within the plots were geographical location (UTM coordinates) and elevation (m) using GPS and slope (%) using a Suunto clinometer.

2.3. Soil Sampling. Soil samples were collected from five points, which were located at the centre of the plot and 10 m from the centre in the four cardinal directions (North, East, South, and West). At each point two samples were taken, 0–15 cm and 15–30 cm below the surface. The five samples taken from each depth range were mixed in the field to obtain one composite sample per depth range and plot. Thus, 70 soil samples (35 from each depth range) were collected from the 35 plots. In addition, a soil core device with an inner diameter of 5 cm and a length of 5 cm was used for extracting soil bulk density samples from the centre of each plot and at each depth. Hence, a total of 70 samples were collected for bulk density determination.

2.4. Laboratory Analyses. In the laboratory all soil samples were ground and passed through a 2 mm sieve to remove stones and gravel. Fine and coarse roots were also removed. Subsequently, soil samples collected at 0–15 cm depth were analysed for soil pH, soil texture, cation-exchange capacity (CEC, cmol(+)/kg), available phosphorus (ppm), and exchangeable bases (Ca^{2+}, Mg^{2+}, and K^+, cmol(+)/kg). Samples from both depth ranges (0–15 and 15–30 cm) were analysed for percentages of organic carbon and total nitrogen. Standard methods for soil analysis were used in order to obtain estimates for each of the mentioned variables that can be compared with results reported in the literature [30–34]. Soil pH was determined electrometrically using 10 g of soil sample diluted in 25 mL distilled water, that is, using a 1 : 2.5 ratio of soil to water. Soil texture was determined by the hydrometer method and the textural classification was done by the use of the soil texture triangle [35].

The Bray 1 method was used for the determination of extractable P for acidic soils with pH less than 7 while the Olsen method was used for soils with pH above 7 (alkaline soils). The ammonium acetate method at pH 7 was used in determination of CEC, and by the use of an atomic absorption spectrophotometer in a UNICAM 919 AA Spectrometer, all exchangeable cations (Ca^{2+}, Mg^{2+}, and K^+) were determined. Subsamples were finely ground into powder form (<1 mm) in an agate mortar and analyzed for total percentages of organic C and N by dry combustion (Dumas method) in a Leco CNS 2000 analyzer [33]. Samples for bulk density estimation were oven-dried at 105°C to constant weight and the weight was recorded (accuracy 0.01 g). The volume was calculated from length and cross-sectional area of the soil core, and bulk density was determined as dry weight (g) per unit volume (cm^3). Most analyses were conducted at the Laboratory of Forest Biology, Sokoine University of Agriculture (SUA), but C and N analyses were conducted at the Soil Science Laboratory at the Department of Forest and Landscape (now Department of Geosciences and Natural Resource Management), University of Copenhagen, Denmark.

2.5. Data Analysis. Based on the data collected the following measures were analysed: species composition was expressed through species richness and diversity measures; forest structure was expressed through stem density, basal area and volume for plant communities, species groups, and diameter classes. Total species richness was computed as the total number of species across all 35 plots. Species diversity was computed using Shannon's and Simpson's Diversity Indices

[36]. The volume of stumps was calculated as cylinder volume while total volume for standing trees was calculated using a regression equation developed for GVLFR by the authors [37, 38]: $\ln(V) = -8.4554 + 2.3236 \times \ln(\text{Dbh})$ ($R^2 = 0.983$, RMSE = 0.248, Dbh range: 1.4–62 cm, $n = 104$), where V is volume (m^3/tree); Dbh is diameter at breast height (≥ 1 cm), RMSE is the residual standard error, R^2 is the coefficient of determination, n is the total sample size, and ln is the natural logarithm. The Importance Value Index (IVI) for each species in each plot was calculated as the sum of relative density and dominance (basal area) and expressed in percent [39]. Percentage base saturation (%BS) was determined as the ratio of total base cation concentration to CEC, while the C : N ratio was determined using the estimated elemental percentages of carbon and nitrogen [35].

Using IVI for each species, plots were classified by agglomerative hierarchical cluster analysis using Sørensen's distance measure and a group linkage method with flexible β of -0.50. The 35 plots were ordinated by nonmetric multidimensional scaling (NMS) using the PC_ORD software version 6.0 [40]. Topographic variables (elevation and slope) and edaphic variables (pH, bulk density, texture, extractable P, CEC, exchangeable base cations, %BS, C : N ratio, %C, and %N) were correlated with the NMS ordination axes. Indicator/dominant species in each cluster were determined using percentage indicator values (%IV) where values of 0 correspond to no indication and 100 is perfect indication [41, 42]. The first three to five names of these Indicator/dominant species with the highest percentage indicator values (%IV), constancy, and significant indicator values ($P < 0.05$) were used to assign names to the clusters/plant community types [29, 41, 42]. The Steinhaus (Sørensen/Czekanowski) coefficient was used to assess the similarity/dissimilarity of the species compositions of the plant communities [39].

3. Results

3.1. Species Richness. Including all size categories a total of 88 species (29 plant families) of standing trees and shrubs/small trees were identified in the GVLFR (Table 1). Trees contributed 60% (21 plant families) and shrubs 40% (15 plant families) of the species. For stumps, a total of 42 species (20 plant families) of trees and shrubs/small trees, with basal diameter ranging from 2 to 50 cm, were identified. For stumps, trees contributed 76% (13 plant families) of the species while shrubs contributed 24% (10 plant families). All species represented by stumps were also represented by standing trees/shrubs. In general, tree and shrub species from the family Caesalpiniaceae contributed most (13%) to the total number of species (standing individuals), followed by those from the families Mimosaceae (10%), Rubiaceae (10%), Fabaceae (9%), and Euphorbiaceae (9%) (Table 1). Among standing trees, the greatest number of species was found in the four plant families: Caesalpiniaceae (17%), Mimosaceae (15%), Fabaceae (13%), and Combretaceae (9%), while shrubs/small trees included most species from the families Rubiaceae (26%), Euphorbiaceae (17%), and Capparaceae (9%). For stumps, tree and shrub species from

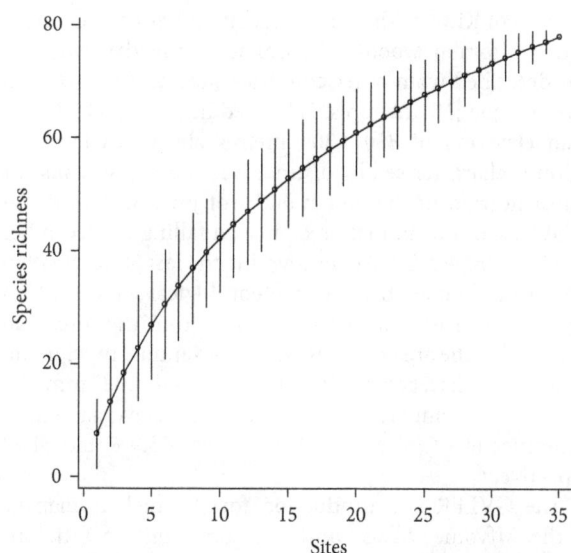

FIGURE 2: Species accumulation curve for large individuals (Dbh ≥ 5 cm) measured within circular plots with a radius of 15 m in the Gangalamtumba VLFR. Vertical lines indicate standard deviations (range 0–4.9).

the family Mimosaceae contributed most (17%) to the total number of species, followed by species from the families Caesalpiniaceae (14%), Combretaceae (10%), and Fabaceae (7%). Among species categorised as trees, the families Caesalpiniaceae (19%) and Mimosaceae (19%) contributed equal numbers of species, followed by Combretaceae (13%) and Fabaceae (9%). With respect to shrubs each of the families was represented by a single species (10%; see Table 1).

When considering different size categories and including both trees and shrubs (small sizes, Dbh < 5 cm and large sizes, Dbh ≥ 5 cm), a total of 78 species (28 families) were found among large sizes, with Caesalpiniaceae (13%), Mimosaceae (12%), and Fabaceae (10%) being the most species-rich plant families, while among small sizes, a total of 69 species (27 families) were observed, with Rubiaceae (13%), Caesalpiniaceae (12%), and Mimosaceae (10%) contributing the greatest number of species (Table 1). In general the average number of species per plot was found to be 14 species (range 5–24 species per plot).

The species accumulation curve (Figure 2) shows that the 35 sites/plots used in this study were sufficient to cover much (but not all) of the variation and species diversity of the study area. At 35 plots the graph has not yet reached its asymptotic level but is starting to converge, implying that any further increase of sample size would be expected to lead to inclusion of additional rare species. However, although the sample size was small (35 plots) and does not quite capture the full woody plant biodiversity of the reserve, the results are still useful for characterizing the tree/shrub species diversity and relationships between species and site.

3.2. Species Diversity. Shannon-Wiener diversity indices for large and small individuals were found to be 3.44 and 3.26, respectively, and the Simpson index for large individuals was

TABLE 1: Checklist of tree and shrub species recorded in Gangalamtumba VLFR showing frequency (%), density (mean ± SE), dispersion index (DI), and Importance Value Index (IVI), both for the current population of large individuals (plot size = 15 m radius; minimum Dbh = 5 cm) and for stumps (plot size = 20 m radius; minimum basal diameter = 2 cm); SH: shrub, ST: small tree, and T: tree.

S/no.	Vernacular/local name	Species/botanical name	Family	Habit/life form	Current population Frequency (%)	Current population Density stem/ha	Current population Basal area (m²/ha)	Current population IVI	DI	Stumps Frequency (%)	Stumps Density stem/ha	Stumps Basal area (m²/ha)	Stumps IVI
1	Mgiha	Dalbergia arbutifolia	Fabaceae, subfamily Papilionoiceae	ST	60	247 ± 69	1.36 ± 0.37	22.4	47.65	14	6.8 ± 4.1	0.06 ± 0.05	17.0
2	Mtono	Commiphora africana	Burseraceae	T	77	110 ± 23	1.59 ± 0.37	17.5	11.42	14	1.4 ± 0.6	0.01 ± 0.01	11.1
3	Mlama	Combretum molle G. Don	Combretaceae	T	74	116 ± 27	0.69 ± 0.15	13.2	15.23	34	6.8 ± 1.9	0.05 ± 0.02	30.7
4	Mkungugu	Acacia sp.	Mimosaceae	T	23	40 ± 22	1.02 ± 0.54	11.1	28.73	9	2.0 ± 1.4	0.01 ± 0.01	6.5
5	Mkwee	Brachystegia spiciformis Benth	Caesalpiniaceae	T	37	58 ± 20	1.18 ± 0.34	11.0	17.31	26	6.1 ± 2.4	0.20 ± 0.09	30.8
6	Mkalala	Albizia petersiana (Belle) Oliv.	Mimosaceae	T	40	80 ± 30	0.55 ± 0.20	8.5	27.11	6	3.6 ± 2.9	0.05 ± 0.04	9.4
7	Mlyasenga	Combretum zeyheri Sound	Combretaceae	T	43	60 ± 20	0.30 ± 0.11	7.4	15.97	20	2.0 ± 0.8	0.01 ± 0.01	4.7
8	Mkombivawo	Bauhinia petersiana	Caesalpiniaceae	SH	34	54 ± 21	0.29 ± 0.10	6.4	20.00				
9	Muguvani	Markhamia obtusifolia	Bignoniaceae	T	60	65 ± 13	0.38 ± 0.10	6.4	6.47	6	0.5 ± 0.3	0.0	2.3
10	Mdeke	Hymenodictyon parvifolium Oliv.	Rubiaceae	ST	31	50 ± 24	0.48 ± 0.24	5.8	27.49	3	0.2 ± 0.2	0.0	1.7
11	Mubwegele	Sclerocarrya birrea sbsp. birrea	Anacardiaceae	T	34	21 ± 6	0.61 ± 0.23	5.8	4.41	6	0.5 ± 0.3	0.01 ± 0.01	2.8
12	Mmemenamene	Margaritaria discoidea (Bail.)	Euphorbiaceae	SH	51	68 ± 24	0.32 ± 0.13	5.7	20.73	3	0.7 ± 0.7	0.0	0.6
13	Mkambala	Acacia mellifera (Vahl) Benth.	Mimosaceae	T	29	14 ± 5	0.49 ± 0.19	5.0	3.59	3	0.2 ± 0.2	0.0	0.8
14	Mugegere	Dichrostachys cinerea (L.) Wight & Arn	Mimosaceae	ST	31	53 ± 26	0.17 ± 0.09	4.7	30.86	29	5.9 ± 1.9	0.03 ± 0.02	17.8
15	Mbata	Acacia seyal Del. var. seyal	Mimosaceae	T	34	30 ± 10	0.19 ± 0.06	4.5	8.87	3	1.1 ± 1.1	0.01 ± 0.01	4.6
16	Mgulumo	Lannea schweinfurthii	Anacardiaceae	T	43	24 ± 7	0.32 ± 0.09	3.9	5.44	9	0.7 ± 0.4	0.02 ± 0.01	2.2
17	Muyombo	Brachystegia boehmii	Caesalpiniaceae	T	11	11 ± 7	0.19 ± 0.15	3.8	13.09	6	0.7 ± 0.5	0.01 ± 0.01	2.2
18	Mmulimuli	Cassia abbreviata	Caesalpiniaceae	T	34	31 ± 13	0.17 ± 0.07	3.6	13.53	9	1.6 ± 1.2	0.01 ± 0.01	4.9
19	Mdavi	Cordia sinensis Lam	Boraginaceae	ST	6	23 ± 23	0.11 ± 0.10	3.4	54.98	3	0.2 ± 0.2	0.0	0.6

TABLE 1: Continued.

S/no.	Vernacular/local name	Species/botanical name	Family	Habit/life form	Frequency (%)	Current population				Frequency (%)	Stumps		
						Density stem/ha	Basal area (m²/ha)	IVI	DI		Density stem/ha	Basal area (m²/ha)	IVI
20	Mdwendwe	Terminalia brownii	Combretaceae	T	17	14 ± 8	0.19 ± 0.09	3.2	11.29	3	0.5 ± 0.5	0.0	1.2
21	Mfilafila	Diplorynchus condylocarpon	Apocynaceae	T	31	25 ± 8	0.15 ± 0.05	3.0	6.46	6	0.5 ± 0.3	0.0	3.5
22	Mkwata	Codyla densiflora	Caesalpiniaceae	T	9	4 ± 2	0.33 ± 0.19	2.8	2.82	3	3.4 ± 3.4	0.04 ± 0.04	8.3
23	Mgunga	Acacia abyssinica (Hochst)	Mimosaceae	T	17	12 ± 7	0.14 ± 0.07	2.4	11.40	6	1.4 ± 1.2	0.01 ± 0.01	3.9
24	Mkole	Grewia bicolor	Tiliaceae	ST	26	21 ± 10	0.08 ± 0.03	2.1	10.78	6	0.5 ± 0.3	0.0	1.5
25	Mparapande	Strychnos potatorum L. f	Loganiaceae	T	23	23 ± 11	0.14 ± 0.09	2.0	13.31	6	0.7 ± 0.5	0.0	1.7
26	Kitimbwi	Ormocarpum kirkii	Fabaceae	SH	29	18 ± 6	0.07 ± 0.03	1.9	4.92				
27	Mtabagila	Lonchocarpus capassa	Fabaceae	T	26	11 ± 5	0.14 ± 0.06	1.7	5.50	3	0.2 ± 0.2	0.0	0.4
28	Muwotaponzi	Ozoroa insigns ssp. reticulata	Anacardiaceae	T	14	10 ± 5	0.10 ± 0.05	1.6	5.64	14	1.1 ± 0.5	0.01 ± 0.00	3.1
29	Msanzi	Gardenia resiniflua Hiern.	Rubiaceae	ST	9	13 ± 8	0.08 ± 0.05	1.6	12.29				
30	Mkoga	Vitex payos	Verbenaceae	ST	17	12 ± 5	0.11 ± 0.04	1.5	6.05	3	0.2 ± 0.2	0.0	0.2
31	Mpelemele	Grewia forbesii Haw. Ex Mast	Tiliaceae	ST	14	16 ± 9	0.05 ± 0.02	1.47	12.46				
32	Mnywenywee	Dalbergia boehmii	Fabaceae	T	14	11 ± 7	0.09 ± 0.07	1.5	12.21				
33	Mulagavega	Albizia amara (Roxb.) Boiv.	Mimosaceae	T	14	9 ± 5	0.04 ± 0.03	1.4	7.12				
34	Mkola	Afzelia quanzensis	Caesalpiniaceae	T	11	6 ± 3	0.15 ± 0.09	1.4	5.32				
35	Mugusi	Brachystegia manga	Caesalpiniaceae	T	20	11 ± 6	0.10 ± 0.07	1.3	8.24	6	1.8 ± 1.3	0.02 ± 0.02	2.5
36	Mwahama	Shrebera trichoclada	Oleaceae	T	17	11 ± 6	0.05 ± 0.02	1.2	6.97	3	0.9 ± 0.9	0.01 ± 0.01	1.5
37	Mpingo	Dalbergia melanoxylon	Fabaceae	T	11	11 ± 9	0.04 ± 0.03	1.2	16.55	3	1.4 ± 1.4	0.01 ± 0.01	2.6
38	Mubaya	Strychnos innocua	Loganiaceae	T	17	8 ± 4	0.11 ± 0.08	1.2	3.75	3	0.2 ± 0.2	0.01 ± 0.01	0.6
39	Msisina	Albizia harveyi Fourn.	Mimosaceae	T	14	5 ± 3	0.08 ± 0.05	1.1	3.59	3	0.5 ± 0.5	0.02 ± 0.02	1.5
40	Mlimbo	Euphorbia cuneata Valh.	Euphorbiaceae	ST	14	12 ± 6	0.04 ± 0.02	1.1	6.99				
41	Mwambi	Elaeodendron buchananii	Celasteraceae	T	3	8 ± 8	0.03 ± 0.03	1.0	20.00				
42	Mvembadanda	Xeroderris stuhlmannii	Fabaceae	T	14	4 ± 2	0.11 ± 0.05	0.9	1.91				
43	Mtelela	Brachystegia bussei Harms	Caesalpiniaceae	T	6	2 ± 1	0.12 ± 0.10	0.9	2.53				

TABLE 1: Continued.

S/no.	Vernacular/local name	Species/botanical name	Family	Habit/life form	Current population					Stumps			
					Frequency (%)	Density stem/ha	Basal area (m²/ha)	IVI	DI	Frequency (%)	Density stem/ha	Basal area (m²/ha)	IVI
44	Mteresi	Premna holstii Gürke	Verbenaceae	ST	9	8 ± 5	0.03 ± 0.02	0.9	8.99				
45	Mkongolo	Commiphora ugogensis	Burseraceae	T	6	1 ± 1	0.08 ± 0.06	0.8	0.97	3	0.2 ± 0.2	0.01 ± 0.01	0.9
46	Muwisa	Boscia angustifolia A. Rich var. angustifolia Canthium	Capparaceae	T	17	8 ± 4	0.04 ± 0.02	0.8	4.26				
47	Mnyalupuko	pseudoverticillatum Hien	Rubiaceae	ST	20	8 ± 4	0.04 ± 0.02	0.7	4.14				
48	Mnyaluhanga	Bridelia scleroneura Muell.Arg.	Euphorbiaceae	ST	14	6 ± 3	0.04 ± 0.02	0.7	3.41				
49	Mdide	Manilkara mochisia (Bak.) Dubard	Sapotaceae	T	9	5 ± 4	0.02 ± 0.01	0.7	7.93				
50	Muhemi	Erythrina abyssinica	Fabaceae	T	6	1 ± 1	0.07 ± 0.06	0.7	1.63				
51	Msambarawe	Vangueria infausta Burch. Ssp. Rotundata	Rubiaceae	ST	6	8 ± 5	0.02 ± 0.02	0.6	9.25				
52	Mhehefu	Allophylus ferrugineus Taub.	Sapindaceae	SH	9	5 ± 3	0.02 ± 0.01	0.5	5.40	3	1.4 ± 1.4	0.01 ± 0.01	1.4
53	Mnyeng'enyenge	Excoecaria bussei (Pax)	Euphorbiaceae	ST	11	4 ± 2	0.01 ± 0.01	0.5	3.21				
54	Muganga	Berchemia discolor	Rhamnaceae	T	6	4 ± 3	0.04 ± 0.03	0.4	7.17				
55	Mninga	Pterocarpus angolensis	Fabaceae	T	6	2 ± 1	0.05 ± 0.04	0.4	2.53	3	0.9 ± 0.9	0.06 ± 0.06	2.5
56	Mtanangwe	Ziziphus mucronata	Rhamnaceae	T	3	4 ± 4	0.03 ± 0.03	0.4	9.00				
57	Muhekele	Euclea divinorum Hiern	Ebenaceae	ST	3	3 ± 3	0.01 ± 0.01	0.4	8.00	3	0.2 ± 0.2	0.0	4.9
58	Mubumila	Cassipourea mollis (R.E.fr.) Alstom	Rhizophoraceae	T	9	4 ± 3	0.02 ± 0.01	0.4	5.26	3	0.2 ± 0.2	0.0	0.4
59	Mkwambe	Flueggea virosa Willd.	Euphorbiaceae	SH	6	3 ± 2	0.03 ± 0.03	0.3	4.06				
60	Kivanga	Zanha africana (Radlk.) Exell	Sapindaceae	T	9	2 ± 1	0.01 ± 0.01	0.2	2.12	3	0.2 ± 0.2	0.0	0.2
61	Mninga maji	Pterocarpus tinctorius	Fabaceae	T	3	0	0.02 ± 0.02	0.2	1.00				
62	Mtosi	Maerua triphylla A. Rich.	Capparaceae	ST	3	2 ± 2	0.01 ± 0.01	0.2	6.00				
63	Mugosa	Ehretia amoena	Boraginaceae	ST	9	1 ± 1	0.0	0.1	0.94				
64	Mnyongamembe	Steganotaenia araliacea	Apiaceae	T	6	1 ± 1	0.01	0.1	1.63				
65	Mvelevele	Vernonia amygdalina	Asteraceae	SH	3	2 ± 2	0.01 ± 0.01	0.1	4.00				
66	Mpongolo	Catunaregam spinosa (Thunb.)	Rubiaceae	ST	3	1 ± 1	0.01 ± 0.01	0.1	2.00				

TABLE 1: Continued.

S/no.	Vernacular/local name	Species/botanical name	Family	Habit/life form	Current population					Stumps			
					Frequency (%)	Density stem/ha	Basal area (m²/ha)	IVI	DI	Frequency (%)	Density stem/ha	Basal area (m²/ha)	IVI
67	Mboliboli	Acacia drepanolobium (Harms)	Mimosaceae	T	3	0	0.0	0.1	1.00				
68	Mwimakigulu	Maerua engolensis	Capparaceae	SH	3	1 ± 1	0.0	0.1	3.00				
69	Mnyali	Tamarindus indica	Caesalpiniaceae	T	6	1 ± 1	0.01 ± 0.01	0.1	0.97				
70	Mgombwani	Combretum aculeatum	Combretaceae	T	6	1 ± 1	0.0	0.1	0.97	3	0.2 ± 0.2	0.0	0.4
71	Msangala	Burkea africana Hook	Caesalpiniaceae	T	3	0	0.01 ± 0.01	0.1	1.00	3	0.2 ± 0.2	0.0	0.4
72	Mkokonza	Opilia amentacea Roxb.	Opiliaceae	SH	3	0	0.0	0.0	1.00	3	1.1 ± 1.1	0.0	4.2
73	Mfumbi	Kigelia africana	Bignoniaceae	T	3	0	0.0	0.0	1.00				
74	Mangali	Euphorbia candelabrum	Euphorbiaceae	T	3	0	0.0	0.0	1.00				
75	Mpululu	Terminalia sericea	Combretaceae	T	3	0	0.0	0.0	1.00				
76	Mdaha	Diospyros usambarensis F. White	Ebenaceae	T	3	0	0.0	0.0	1.00	3	0.5 ± 0.5	0.01 ± 0.01	1.5
77	Mtundwa	Ximenia caffra Sond	Olacaceae	ST	3	0	0.0	0.0	1.00				
78	Mwimaperu	Tarenna glaveolens (S. Moore) Brem	Rubiaceae	ST	+								
79	Mwesa	Boscia mossambicensis Kl.	Capparaceae	ST	+								
80	Musasamulo	Psychotria schumanniana	Rubiaceae	ST	+								
81	Muhanza	Senna singueana	Caesalpiniaceae	ST	+								
82	Mtundwahavi	Ximenia americana	Olacaceae	ST	+								
83	Mtingiligiti	Phyllanthus engleri Pax	Euphorbiaceae	T	+								
84	Msada	Vangueria madagascariensis Gmel.	Rubiaceae	ST	+								
85	Mkunungu	Zanthoxylum chalybeum	Rutaceae	T	+								
86	Lukali	Croton scheffleri Pax	Euphorbiaceae	SH	+								
87	Kilimandembwe	Gardenia ternifolia K. Schum. & Thonn.	Rubiaceae	ST	+								
88	Mpela/Mbuyu	Adansonia digitata	Bombacaceae	T	++								
	Total (all species)				1351	1521 ± 594	13.55 ± 5.52	200		294	60 ± 30	0.72 ± 0.49	200

+ indicates species identified among smaller individuals within 5 m radius plots (Dbh < 5 cm), and ++ indicates that a large individual was identified within a 15-radius plot but was not included in the analyses of stem density, basal area, and volume as it was considered an outlier due to its tremendous size (Dbh = 340 cm).

0.05 and that of small individuals was 0.06. The following species were observed to have the greatest contributions to the Shannon-Wiener diversity index of large individuals: *Dalbergia arbutifolia* (contributing 0.30), *Combretum molle* (0.20), *Commiphora africana* (0.19), *Albizia petersiana* (0.16), and *Margaritaria discoidea* (0.14), while for smaller ones the greatest contributions were found for *Brachystegia spiciformis* (0.26), *Dalbergia arbutifolia* (0.24), *Grewia forbesii* (0.23), *Margaritaria discoidea* (0.20), and *Dichrostachys cinerea* (0.16). In terms of frequency of occurrence for standing individuals (large sizes) *Commiphora africana* was the most frequent species (77% of plots), followed by *Combretum molle* (74%) and *Dalbergia arbutifolia* (60%), while for small sizes *Margaritaria discoidea* (66%), *Markhamia obtusifolia* (63%), and *Dalbergia arbutifolia* (57%) were the most frequent species. The Importance Value Index (IVI) for large individuals (Dbh $\geq$ 5 cm) shows that *Dalbergia arbutifolia* (22.4), *Commiphora africana* (17.5), and *Combretum molle* (13.2) were the most important species among standing individuals, while *Brachystegia spiciformis* (30.8), *Combretum molle* (30.7), and *Dichrostachys cinerea* (17.8) appeared to be the most important among harvested individuals (stumps). These species were also found to have higher frequencies than any other harvested species observed in the GVLFR (Table 1).

3.3. Stem Density.

The total mean stem density for large individuals with Dbh $\geq$5 cm was 1521 $\pm$ 594 stems/ha (Table 1) and that of small individuals with Dbh <5 cm (including individuals with Dbh <1 cm) was 14318 $\pm$ 6956 stems/ha. Among large individuals the most abundant species were *Dalbergia arbutifolia* (16.2% of 1521 stems/ha), *Combretum molle* (8%), *Commiphora africana* (7.2%), and *Albizia petersiana* (5.3%). Among small individuals, the most abundant species were *Brachystegia spiciformis* (13% of 14318 stems/ha) followed by *Dalbergia arbutifolia* (11%) and *Grewia forbesii* (10%). For stumps, the overall mean density was 60 $\pm$ 38 stems/ha, with *Combretum molle* (12%), *Dalbergia arbutifolia* (12%), *Brachystegia spiciformis* (10%), and *Dichrostachys cinerea* (10%) contributing the most (Table 1). Generally, the distribution of standing trees to size classes showed the usual reverse J shape, which was also approximately observed for stumps (Figure 3). However, for stumps the density of stems in the 1–10 cm diameter class was slightly lower than what would be expected if tree felling had been a random event.

3.4. Basal Area.

For the GVLFR as a whole the mean basal areas for large ($\geq$5 cm Dbh) and small individuals (<5 cm Dbh) were 13.55 $\pm$ 5.52 m^2/ha (Table 1, Figure 4) and 3.05 $\pm$ 0.02 m^2/ha, respectively. The species contributing most to the basal area of large individuals were *Commiphora africana* (12%), *Dalbergia arbutifolia* (10%), and *Brachystegia spiciformis* (9%), while those contributing most to the basal area of smaller individuals were *Dalbergia arbutifolia* (16%), *Grewia forbesii* (15%), and *Dichrostachys cinerea* (13%). The mean basal area for stumps was 0.72 m^2/ha with *Brachystegia spiciformis* contributing the greatest individual proportion (28%); 41 species made up the rest (Figure 4).

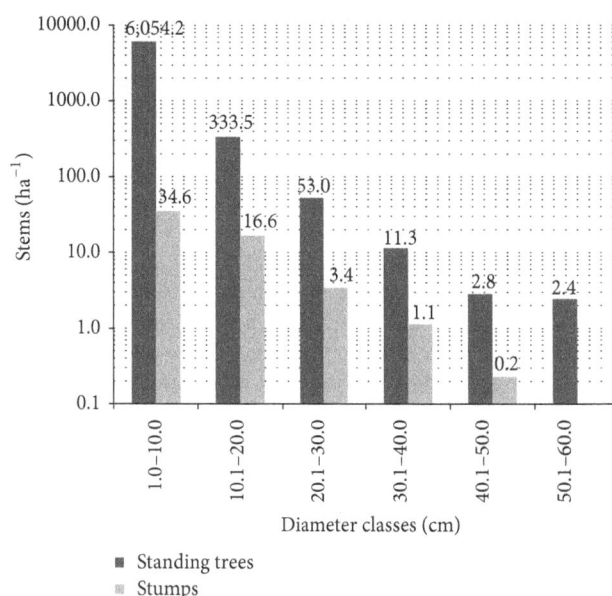

FIGURE 3: Density of standing trees $\geq$1 cm Dbh and stumps $\geq$1 cm by diameter class in Gangalamtumba VLFR (n = 35). NB: logarithmic scale on vertical axis.

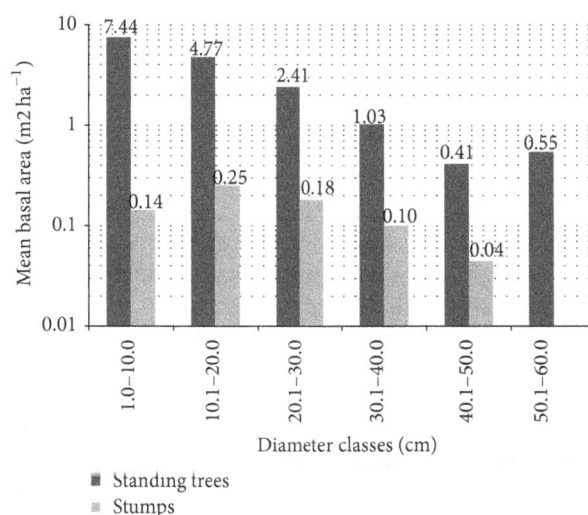

FIGURE 4: Distribution of basal area per hectare for standing trees $\geq$1 cm Dbh and stumps $\geq$1 cm by diameter classes in the Gangalamtumba VLFR (n = 35). NB: logarithmic scale on vertical axis.

3.5. Volume.

The mean volumes for large ($\geq$5 cm Dbh) and small individuals (<5 cm Dbh) were 92.17 $\pm$ 39.0 m^3/ha and 12.57 $\pm$ 6.35 m^3/ha, respectively (not shown in tables). The species contributing most to the volume of large individuals were *Commiphora africana* (12%), *Brachystegia spiciformis* (10%), *Acacia* sp. (9%), and *Dalbergia arbutifolia* (9%). For smaller individuals, the species that contributed most to volume were *Dalbergia arbutifolia* (16%), *Grewia forbesii* (14%), and *Dichrostachys cinerea* (13%). The mean remaining volume of stumps was found to be 0.15 $\pm$ 0.1 m^3/ha with *Brachystegia spiciformis* contributing the greatest individual percentage (28%); 41 species made up the rest.

3.6. Spatial Distribution. Dispersion indices (DI) are presented for individual species in Table 1. The dispersion index values range from 0.94, indicating almost complete spatial randomness or slight underdispersion, to 54.98, indicating considerable overdispersion, that is, a patchy or clustered distribution. Out of 77 species, excluding the single, very large *Adansonia digitata* that was considered an outlier, 64 species (83%), 9 species (12%), and 4 species (5%) were found to have DI > 1, DI = 1, and DI < 1, respectively, so the majority of species are characterised by a patchy distribution across the forest. The species with the lowest estimated DI was *Ehretia amoena* (0.94) while the highest DI was estimated for *Cordia sinensis* (54.98). The most abundant species, including *Dalbergia arbutifolia* (DI = 47.65), *Combretum molle* (DI = 15.23), and *Commiphora africana* (DI = 11.42), are strongly overdispersed.

3.7. Plant Communities and Species Association. Four plant communities were identified through cluster analysis based on the statistical significance (5%) of the observed maximum indicator values (Table 2). Only two of these plant communities were dominated by species from the family Caesalpiniaceae (Communities 1 and 24). The other two showed great variation/overlap between species of different plant families (Communities 3 and 5). The estimated Steinhaus similarity indices between pairs of plant communities varied from 32% for Communities 1 and 3 and Communities 5 and 24, to 37% for Communities 3 and 5 (not shown in tables).

The ordination diagram (Figure 5) shows that one topographic and six edaphic variables appear to be associated with the four plant communities and the species distribution in the study area. The strongest correlation with community composition was observed for elevation, followed by soil pH at 0–15 cm depth, base cation Ca^{2+} at 0–15 cm depth, percent base saturation (%BS) at 0–15 cm depth, percent clay at 0–15 cm depth, C : N ratio at 15–30 cm depth, and percent sand at 0–15 cm depth. The base cations magnesium (Mg^{2+}) and potassium (K^+) were not directly correlated with the community composition, but since they are included in the base saturation percentage they are indirectly related. Plant communities were ordered along the second ordination axis, which was positively correlated with pH15 ($r = 0.71$), Ca15 ($r = 0.69$), BS ($r = 0.65$), and Clay15 ($r = 0.52$) and negatively correlated with Sand15 ($r = -0.46$). Axis 1 of the ordination was positively correlated with Elev ($r = 0.85$) and CNrat30 ($r = 0.51$). As shown in Table 3 and indicated graphically in Figure 5 correlations between many of the edaphic variables are strong, particularly between pH15, Sand15, Clay15, Ca15, and BS, whereas correlation between topographic and most edaphic variables is small and nonsignificant. However, there is a significant negative correlation ($r = -0.42$) between Elev and Sand15, indicating that sandy soils are mostly found at lower elevations.

4. Discussion

4.1. Species Composition. The results reported in this study show that the composition of the vegetation types found in

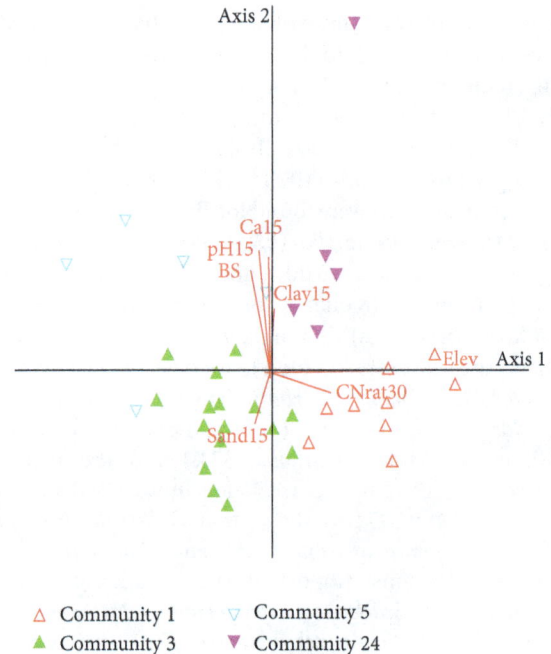

FIGURE 5: Compositional gradients and plant communities in NMS ordination of 35 vegetation plots for trees ≥5 cm in Gangalamtumba VLFR. Four plant communities are recognized (cf. Table 2). Community 1 = *Brachystegia spiciformis, Diplorynchus condylocarpon*, and *Lannea schweinfurthii* woodland; Community 3 = *Dalbergia arbutifolia, Commiphora africana*, and *Albizia petersiana* woodland; Community 5 = *Acacia* sp., *Acacia abyssinica*, and *Albizia amara* woodland; and Community 24 = *Bauhinia petersiana* and *Shrebera trichoclada* woodland. Elev: elevation (m); CNrat30: C : N ratio at 15–30 cm depth; Sand15: sand (%) at 0–15 cm; Clay15: clay (%) at 0–15 cm; Ca15 = Ca^{2+} (cmol(+)/kg) at 0–15 cm; pH15 = soil pH at 0–15 cm; BS: % base saturation at 0–15 cm.

the GVLFR, especially the dominance of species from the family Caesalpiniaceae, agreed well with previous descriptions and classifications of plant communities commonly found in miombo woodlands [1]. However, the observed dominance based on IVI of the genera *Dalbergia, Commiphora*, and *Combretum* contrasts with patterns usually considered common for miombo woodlands. The frequency of species in these genera was also high compared to other species observed in the GVLFR (Table 1). Similar deviations exist between the results obtained by Banda et al. [26], in the Katavi-Rukwa ecosystem, where they observed that *Terminalia* and *Combretum* were the dominant genera, and the findings of Giliba et al. [27] and Njana [43], who both noted the dominance of the two common miombo genera *Brachystegia* and *Julbernardia*. However, *Combretum* also occurred in their study areas. The results suggest that on a larger spatial scale the species diversity of miombo woodlands is very high and that the three common genera *Brachystegia, Julbernadia*, and *Isoberlinia* are not always dominant at the local scale.

The species richness observed in the GVLFR compares well with miombo community studies in other areas of dry miombo in Tanzania and elsewhere, which receive an average

TABLE 2: Plant communities and species associations in the dry Miombo woodlands of Gangalamtumba VLFR. Community numbers 1, 3, 5, and 24 represent the different plant associations: n_species: number of species (77 in total) with Dbh ≥ 5 cm; IV−: species indicator value; mean ± std. error; and BA is basal area.

Community	Associated species	Family	Habit	IV+ (%)	Constancy (%)	P value	Stems/ha	BA m²/ha	Mean Dbh
Brachystegia spiciformis—Diplorhynchus condylocarpon—Lannea schweinfurthii woodland 1 (n_species = 47, n_plots = 9) Site variables (mean ± SE): Elev = 1200 ± 20, pH15 = 6.44 ± 0.13, Ca15 = 6.04 ± 1.28, Sand15 = 72.1 ± 4.31, Clay15 = 13.7 ± 1.88, Silt15 = 14.2 ± 3.35, CNrat30 = 13.5 ± 0.5 and BS = 52.6 ± 6.5	Brachystegia spiciformis	Caesalpiniaceae	T	82.0	100	0.0002	164 ± 59	3.56 ± 0.77	13.8 ± 0.7
	Diplorynchus condylocarpon	Apocynaceae	T	75.9	89	0.0008	66 ± 18	0.44 ± 0.13	8.2 ± 0.5
	Lannea schweinfurthii	Anacardiaceae	T	71.9	100	0.0004	49 ± 14	0.72 ± 0.22	12.6 ± 0.9
	Boscia angustifolia	Capparaceae	T	53.6	56	0.0056	27 ± 12	0.14 ± 0.07	7.6 ± 0.6
	Euphorbia cuneata	Euphorbiaceae	ST	42.8	44	0.0200	42 ± 19	0.12 ± 0.06	6.5 ± 0.6
	Other species	—	—	—	—	—	1075 ± 574	10.07 ± 6.47	10.2 ± 0.3
	Subtotal						1423 ± 174	15.05 ± 1.78	10.9 ± 0.3
Dalbergia arbutifolia—Commiphora africana—Albizia petersiana woodland 3 (n_species = 55, n_plots = 16) Site variables (mean ± SE): Elev = 986 ± 19, pH15 = 6.48 ± 0.09, Ca15 = 8.56 ± 1.09, Sand15 = 76.7 ± 1.56, Clay15 = 13.5 ± 1.55, Silt15 = 9.7 ± 0.96, CNrat30 = 10.6 ± 0.3 and BS = 74.4 ± 4.4	Dalbergia arbutifolia	Fabaceae, subfamily Papilionoideae	ST	79.3	100	0.0002	500 ± 123	2.63 ± 0.64	7.7 ± 0.2
	Commiphora africana	Burseraceae	T	54.2	94	0.0026	187 ± 40	2.72 ± 0.64	11.9 ± 0.6
	Albizia petersiana	Mimosacea	T	53.6	75	0.0200	157 ± 57	1.03 ± 0.40	8.4 ± 0.2
	Sclerocaryya birrea	Anacardiaceae	T	50.4	63	0.0186	40 ± 11	1.28 ± 0.45	16.8 ± 1.7
	Other species	—	—	—	—	—	852 ± 449	6.69 ± 3.77	8.8 ± 0.3
	Subtotal						1735 ± 166	14.35 ± 0.71	9.2 ± 0.2
Acacia sp.—Acacia abyssinica—Albizia amara woodland 5 (n_species = 33, n_plots = 5) Site variables (mean ± SE): Elev = 986 ± 52, pH15 = 7.34 ± 0.23, Ca15 = 13.9 ± 3.3, Sand15 = 74.5 ± 6.78, Clay15 = 14.8 ± 4.7, Silt15 = 10.7 ± 2.6, CNrat30 = 10.6 ± 0.7 and BS = 96.3 ± 8.9	Acacia sp.	Mimosaceae	T	98.1	100	0.0002	269 ± 111	6.62 ± 2.86	17.2 ± 1.4
	Acacia abyssinica	Mimosaceae	T	59.4	80	0.0030	31 ± 9	0.60 ± 0.32	9.9 ± 0.8
	Albizia amara	Mimosaceae	T	55.2	60	0.0066	54 ± 30	0.23 ± 0.16	7.1 ± 0.8
	Grewia bicolor	Tiliaceae	ST	53.3	60	0.0106	113 ± 52	0.40 ± 0.17	7.0 ± 0.4
	Acacia seyal	Mimosaceae	T	43.3	60	0.0526	62 ± 34	0.34 ± 0.17	9.0 ± 0.5
	Other species	—	—	—	—	—	665 ± 533	5.52 ± 4.98	7.2 ± 0.4
	Subtotal						1194 ± 169	13.67 ± 3.47	10.1 ± 0.5
Bauhinia petersiana—Shrebera trichoclada woodland 24 (n_species = 35, n_plots = 5) Site variables (mean ± SE): Elev = 1175 ± 25, pH15 = 7.3 ± 0.4, Ca15 = 25.4 ± 6.9, Sand15 = 52.2 ± 6.8, Clay15 = 33.1 ± 2.8, Silt15 = 14.7 ± 4.4, CNrat30 = 9.61 ± 1.3 and BS = 104.1 ± 20.4	Bauhinia petersiana	Caesalpiniaceae	SH	55.1	80	0.0118	192 ± 121	1.11 ± 0.50	7.5 ± 0.3
	Sarebera trichoclada	Oleaceae	T	50.5	60	0.0110	57 ± 32	0.22 ± 0.13	8.0 ± 0.5
	Manilkara mochisia	Sapotaceae	T	37.6	40	0.0332	34 ± 27	0.14 ± 0.09	6.9 ± 1.0
	Dalbergia melanoxylon	Fabaceae	T	37.0	40	0.0380	68 ± 58	0.26 ± 0.22	6.6 ± 0.3
	Euretia amoena	Boraginaceae	ST	34.7	40	0.0344	6 ± 4	0.02 ± 0.01	7.0 ± 0.4
	Other species	—	—	—	—	—	982 ± 759	6.43 ± 4.79	9.9 ± 0.5
	Subtotal						1338 ± 137	8.17 ± 0.87	9.1 ± 0.4
	Total						1521 ± 594	13.55 ± 5.52	9.8 ± 0.1

Elev: elevation (m); CNrat30: C: N ratio at 15–30 cm depth; Sand15: sand (%) at 0–15 cm; Clay15: clay (%) at 0–15 cm; Silt15: silt (%) at 0–15 cm; Cal5: Ca^{2+} (cmol(+)/kg) at 0–15 cm; pH15: soil pH at 0–15 cm; BS: % base saturation at 0–15 cm.

TABLE 3: Pearson-correlation matrix for two topographic and eight edaphic variables[†] associated with the classified plant communities in the Gangalamtumba VLFR ($n = 35$). Applied significance levels: [*]5%, [**]1%, and [***]0.1%, [+]not significant.

	Elev	Slope %	pH15	Ca15	K15	N15	Sand15	Clay15	CNrat30
Slope %	0.19[+]								
pH15	−0.13[+]	−0.33[*]							
Ca15	0.02[+]	−0.20[+]	0.83[***]						
K15	−0.29[+]	0.00[+]	0.10[+]	0.12[+]					
N15	0.06[+]	0.36[*]	0.10[+]	0.41[**]	0.33[*]				
Sand15	−0.42[**]	0.03[+]	−0.53[***]	−0.74[***]	0.06[+]	−0.46[**]			
Clay15	0.31[+]	−0.12[+]	0.51[**]	0.76[***]	−0.01[+]	0.47[**]	−0.86[***]		
CNrat30	0.36[*]	0.10[+]	−0.39[*]	−0.51[**]	−0.15[+]	−0.19[+]	0.40[*]	−0.48[**]	
BS	−0.24[+]	−0.17[+]	0.84[***]	0.91[***]	0.15[+]	0.37[*]	−0.58[***]	0.60[***]	−0.51[**]

[†]Elev: elevation (m); pH15: soil pH at 0–15 cm; Ca15: Ca^{2+} (cmol(+)/kg) at 0–15 cm; K15: K^+ (cmol(+)/kg) at 0–15 cm; N15: % N concentration at 0–15 cm; Sand15: sand (%) at 0–15 cm; Clay15: clay (%) at 0–15 cm; CNrat30: C : N ratio at 15–30 cm depth; BS: % base saturation at 0–15 cm. t-value was calculated as $t = r\sqrt{(n-2)/(1-r^2)}$, where r is the Pearson-correlation coefficient and n is the total sample size.

TABLE 4: Species richness observed in other studies from dry miombo woodlands.

Plot size (ha)	Total number of plots	Total sample area (ha)	Total number of species	References
0.071	70	4.95	82	[43]
0.071	80	5.65	110	[27]
0.071	133	9.40	229	[26]
0.071	247	17.46	102	[44]
0.04	40	1.60	81	[45]
0.25	14	3.50	69	[25]
0.1	2	0.20	40	[28]
0.1	152	15.20	92 genera	[22]
0.071	35	2.47	88	This study

annual rainfall of 565–1500 mm (Table 4). Using plot sizes of 0.04–0.25 ha and sample sizes of 2–247 plots, a total number of species ranging from 40 (Dbh ≥ 10 cm) to 229 (Dbh ≥ 2 cm) have been reported from Miombo woodlands [22, 25–28, 43–45].

The high number of species reported by Banda et al. [26], Chidumayo [22], Isango [45], and Chamshama et al. [44] is likely to be a consequence of the spatial scale and coverage of these studies as they all cover large areas, include more than one site, and operate with large sample sizes. For example, Chidumayo [22] included species from both wet and dry Miombo areas in Zambia. The study by Sauer and Abdallah [15], who reported a total of 131 species occurring in the tobacco growing zone of Iringa rural district, included two forest reserves and three family-managed forests. Therefore, considering the climatic conditions in the GVLFR (617 mm of rainfall per year) and the sample size (35 plots), the species richness reported in this study can be ranked at least as high or higher than those found in the studies mentioned. The selective sampling approach, focusing on various microhabitat types, adopted by Banda et al. [26] may have contributed to the large number of species observed in their study area. Similarly, the higher number of species found by Giliba et al. [27] was probably due to the presence of riverine forest, offering site conditions favourable for many species.

The values of the Shannon-Wiener (H' = 3.44) and Simpson indices (D = 0.05) reported for large individuals in this study are within the range observed for most communities of particular life forms [36]. For example, H' usually does not exceed 5, although this maximum value varies depending on the type of the biological community sampled and the sampling approach applied (e.g., minimum diameter and size of sample units). A threshold value of 2 for H' has been mentioned as a minimum value, above which an ecosystem can be regarded as medium to highly diverse [27]. Chamshama et al. [44] reported three H' values of 3.1, 3.2, and 3.3 from Kitulangalo Miombo forest in Eastern Tanzania, while Njana [43] reported a H' value of 3.40 from dry Miombo forest in Western Tanzania. The studies by Sauer and Abdallah [15] and Giliba et al. [27] reported particularly high values of 3.46 and 4.27, respectively. This could be attributed to the very large sample sizes used in both studies and the presence of riverine forest where the chance of encountering many rare species is likely to be high. The forests examined in the mentioned studies receive an average annual rainfall of 700–1000 mm, compared to which the rainfall in the GVLFR is low (617 mm on average per year, c.f. above). The relatively high diversity found in GVLFR must therefore be attributed to other factors than climate. The soil analyses indicate that the soil fertility level in the GVLFR is relatively high, with considerable proportions of all 2 : 1 lattice clay minerals (114%) (i.e., illite, montmorillonite, and vermiculite) plus a good amount of organic matter, which may suggest that the soil has large reserves of important plant nutrients, for example, K, Mg, and Fe. The base saturation percentage (BS), which is normally used to indicate the soil fertility status (Landon, 1991), was also high, 76% on average, and with 86% of the samples (30 samples) with BS values >50% (fertile soil), and only 14% of the samples (5 samples) with BS below 50% (less fertile soil). As another indication of high soil fertility status the mean C : N ratio was 12 ± 0.43

(range 7–19) for top soil (0–15 cm) and 11 ± 0.36 (range 5–16) for subsoil (15–30 cm). In general, the values for both soil strata fall within the expected range of 8–17. Soil pH values (mean 6.7 ± 0.1, range 5.67–8.52) were also within the optimal range where all the nutrients that are most important for plant growth (Ca, Mg, K, and P) are available and accessible to plants [46]. In this study, 97% of the soil samples had pH below 8 (one sample had a pH of 8.52). Fisher and Binkley [47] noted that high pH is almost never a problem in forests as most trees do well across the full range of common pH values. Therefore, considering the low average annual rainfall compared to other forests, the relatively high fertility of the GVLFR may be the main factor creating an environment favourable to many species, thus leading to a quite high species diversity observed in this forest.

Based on the Importance Value Index *Brachystegia spiciformis* and *Combretum molle* appeared to be the most important species among harvested trees (stumps) in the study area. This result agreed very well with the large values of basal area and volume and the information obtained from local scouts and members of the village environmental committee, who said that the most frequently harvested species for charcoal making include all *Brachystegia* species found in the forest (i.e., *Brachystegia spiciformis*, *Brachystegia manga*, *Brachystegia boehmii*, and *Brachystegia bussei*). Among the *Combretum* species, the most important one (based on IVI), *Combretum molle*, is mostly used for house construction and firewood rather than for charcoal making. Apart from the *Brachystegia* species, the only other species mentioned as important for charcoal making was *Acacia mellifera*. Other species are only used in case they happen to fall when felling the preferred species during the preparation of charcoal kilns. With respect to stumps, *Brachystegia spiciformis*, *Brachystegia manga,* and *Combretum molle* are among the ten most abundant species in terms of density, basal area, and volume per hectare. Stumps of the other species mentioned are not commonly found, a fact which may be due to the small sample size being unable to capture sites where these species are harvested but may also be related to their limited distribution in the forest as indicated by the estimated dispersion indices (see Table 1).

4.2. Forest Structure. Species densities reported from other Miombo woodland areas in Tanzania are typically 348–1495 stems/ha for trees with Dbh >4 cm [24, 43, 45]. Compared with this the GVLFR is highly stocked as the estimated density is 2296 stems/ha for trees with Dbh >4 cm. The mean stem density of 1521 ± 594/ha for trees with Dbh ≥5 cm is also more than twice as large as values found in other studies [25–27, 44, 45, 48]. A similar pattern is seen when comparing the density of regeneration between the studies. The diameter distribution is characterised by a very clear trend of decreasing stem density with increasing diameter. The shape of the distribution is thus an inverted "J" (Figure 3), which is a common feature of natural forests with active regeneration and recruitment [49]. However, not many large trees were captured by the sample, whereas a considerable number of relatively large stumps were observed, suggesting that anthropogenic activities such as charcoal making may

have affected the structure and ecological balance of the forest (see Figure 3). Since the woodlands are often hit by fire which tends to kill the seedlings (especially the late fires) and thus only few seedlings can be expected to reach larger diameter classes, lack of mature trees and the resulting lack of seeds may eventually threaten the biodiversity of the forest.

The mean basal area of 13.55 ± 5.52 m²/ha observed in this study is slightly above the range of values typically reported from the miombo region, 7.5–12.6 m²/ha [26, 28, 43, 44, 50]. However, the basal areas reported by Isango [45] were a bit higher (15.04–15.63 m²/ha) than observed in this study.

The mean total volume in GVLFR was estimated at 92.17± 39.0 m³/ha for trees with Dbh ≥5 cm. Other studies in dry Miombo woodlands have reported mean volumes of 16.7–76.03 m³/ha [24, 44, 45, 50]. The current standing volume of GVLFR is thus slightly higher than values typically reported for forests in the Miombo ecoregion. A plausible reason for this may be that, compared to other forests, the GVLFR is still well stocked despite ongoing human activities such as charcoal making. The basal area observed for stumps was only 0.72 m²/ha so there is no indication of intensive extraction over many years. This corresponds well with the findings of Treue et al. [20] who conclude that the estimated 0.4 m³/ha annual extraction of woody biomass from GVLFR is considerably below its estimated annual increment of 1.5 m³/ha and that the local forest managers seem capable of regulating the harvesting activities, which, in addition to village members' collection for subsistence uses, also included commercial charcoal production by external companies that pay a fee per bag of charcoal extracted to the village government.

4.3. Plant Communities and Species Associations. Except for *Brachystegia spiciformis*, the majority of the *Brachystegia* and *Julbernardia* species that are common to miombo woodlands elsewhere are not among the species most commonly observed in this study. A similar pattern was observed by Banda et al. [26] in the Katavi-Rukwa ecosystem where the genera that are most common in miombo woodlands, *Brachystegia, Julbernardia,* and *Isoberlinia*, were not common in their study sites. Furthermore the four plant communities distinguished in this study are comparable with those reported by other studies in Tanzania, including the work by Munishi et al. [29] in the Miombo woodlands of Rukwa basin, Chunya district, Tanzania, and the study by Banda et al. [26] in the Katavi-Rukwa ecosystem.

Elevation was noted by Munishi et al. [29] to be the most important factor shaping the species communities in their study area. This study found a similar pattern with the highest coefficient of correlation observed for elevation, suggesting that topographical variation is among the strongest determinants of community composition in dry Miombo woodlands, hence influencing the spatial distribution of species strongly. However, in our study area edaphic factors also influenced the species distribution directly. Two plant communities occurred mainly on clayey soils with high pH at intermediate elevations, one on sandy soils with low pH at low elevations and one on sandy-clayey soils at higher elevations and with neutral soil pH (Figure 5). Specifically, it appeared that the *Acacia* woodland category (Community 5) grows at low elevations around temporary streams (or at least where the ground water table is relatively high), and the *Bauhinia* woodland category (Community 24) appears mostly in places with a relatively high percentage of clay (and therefore high values of pH, Ca, and BS). By contrast it appeared that the *Dalbergia* woodland category (Community 3) is located at more sandy sites (with low values of pH, Ca, and BS), and the *Brachystegia* woodland category (Community 1) seemed to be located mostly at higher elevations where the CN-ratio is quite high. Correlations between elevation and the basic soil variables pH, Ca, and percentages of sand and clay were low (c.f. Figure 5). Thus, the results indicate that the plant communities of the dry miombo woodlands in GVLFR are not only shaped by the topographic variation (elevation) and the groundwater level but also by basic soil characteristics. The results thus confirm findings from other studies detecting effects of elevation and soil characteristics on species composition [51, 52].

5. Conclusion

Considering that the sample size used in this study was smaller than samples used in other studies in dry miombo woodlands, the results show that the species diversity in GVLFR is relatively high compared to other forest reserves. The vegetation of GVLFR is characterised by high density, basal area, and volume, and despite the scarcity of large diameter trees this indicates that the forest is in a good condition. The effect of anthropogenic activities is nevertheless evident and stresses the need for proper management, especially for economically important species preferred for charcoal making, (e.g., *Brachystegia spiciformis*), if the current species diversity is to be maintained or enhanced. A repeated future study would be needed to assess whether the current community-based management regime yields this intended outcome, but the available information and analyses allow for some optimism in this respect.

Conflict of Interests

The authors declare that there is no conflict of interests regarding the publication of this paper.

Acknowledgments

The authors wish to thank people from Mfyome village for their invaluable assistance during field work: *Mzee* Iddi Kaheya, Damian Chotamasege, Augustino Mkwama, Liberatus Simime, and Mawazo Nyangwa. Furthermore, thanks are due to Venancia Mlelwa from Sokoine University and Mads Madsen Krag from University of Copenhagen for doing the laboratory work. The authors also wish to thank Goodluck Moshi, their driver, Marco Njana and Nassoro H. Magogo, the TAFORI botanist for assisting with species identification, and Klaus Dons for preparing a study site map. Financial support was provided by Danida through the ENRECA Project, "Participatory forest management for rural livelihoods, forest conservation and good governance in Tanzania" (no. 725).

References

[1] P. G. H. Frost, "The Ecology of Miombo Woodlands," in *The Miombo in Transition: Woodlands and Welfare in Africa*, B. Campbell, Ed., pp. 11–57, 1996.

[2] N. Burgess, J. Hales, E. Underwood et al., *Terrestrial Ecoregions of Africa and Madagascar: A Conservation Assessment*, Island Press, World Wildlife Fund, 2004.

[3] J. Clarke, W. Cavendish, and C. Coote, "Rural households and miombo woodlands: use, value and management," in *The Miombo in Transition: Woodlands and Welfare in Africa*, B. Campbell, Ed., pp. 101–135, 1996.

[4] B. M. Campbell, A. Angelsen, A. Cunningham, Y. Katerere, A. Sitoe, and S. Wunder, "Miombo woodlands—opportunities and barriers to sustainable forest management," https://energypedia.info/images/6/6d/Miombo2007.pdf.

[5] S. Syampungani, P. W. Chirwa, F. K. Akinnifesi, G. Sileshi, and O. C. Ajayi, "The Miombo woodlands at the cross roads: potential threats, sustainable livelihoods, policy gaps and challenges," *Natural Resources Forum*, vol. 33, no. 2, pp. 150–159, 2009.

[6] J. I. O. Abbot and K. Homewood, "A history of change: causes of miombo woodland decline in a protected area in Malawi," *Journal of Applied Ecology*, vol. 36, no. 3, pp. 422–433, 1999.

[7] E. J. Luoga, E. T. F. Witkowski, and K. Balkwill, "Land cover and use changes in relation to the institutional framework and tenure of land and resources in eastern Tanzania Miombo woodlands," *Environment, Development and Sustainability*, vol. 7, no. 1, pp. 71–93, 2005.

[8] FAO, *The State of Food and Agriculture 2010-2011*, Forest and Agriculture Organization, Rome, Italy, 2010.

[9] A. H. Gentry, "Tropical forest biodiversity: distributional patterns and their conservational significance," *Oikos*, vol. 63, no. 1, pp. 19–28, 1992.

[10] O. E. Sala, F. S. Chapin III, J. J. Armesto et al., "Global biodiversity scenarios for the year 2100," *Science*, vol. 287, no. 5459, pp. 1770–1774, 2000.

[11] H. J. Geist and E. F. Lambin, "Proximate causes and underlying driving forces of tropical deforestation," *BioScience*, vol. 52, no. 2, pp. 143–150, 2002.

[12] R. S. DeFries, "Tropical forests deforestation and ecosystems services Contributions of remote sensing," in *Geoinformatics For Tropical Ecosystems*, P. S. Roy, Ed., pp. 1–32, 2003.

[13] B. M. Campbell, P. Frost, and N. Bryon, "Miombo woodlands and their use: overview and key issues," in *The Miombo in Transition: Woodlands and Welfare in Africa*, B. Campbell, Ed., pp. 1–10, 2006.

[14] H. J. Geist, "Global assessment of deforestation related to tobacco farming," *Tobacco Control*, vol. 8, no. 1, pp. 18–28, 1999.

[15] J. Sauer and J. M. Abdallah, "Forest diversity, tobacco production and resource management in Tanzania," *Forest Policy and Economics*, vol. 9, no. 5, pp. 421–439, 2007.

[16] E. N. Chidumayo and K. J. Mbata, "Shifting cultivation, edible caterpillars and livelihoods in the Kopa area of northern Zambia," *Forests Trees and Livelihoods*, vol. 12, no. 3, pp. 175–193, 2002.

[17] E. N. Chidumayo and L. Kwibisa, "Effects of deforestation on grass biomass and soil nutrient status in miombo woodland, Zambia," *Agriculture, Ecosystems and Environment*, vol. 96, no. 1–3, pp. 97–105, 2003.

[18] J. F. Lund and T. Treue, "Are we getting there? Evidence of decentralized forest management from the Tanzanian miombo woodlands," *World Development*, vol. 36, no. 12, pp. 2780–2800, 2008.

[19] T. Blomley and S. Iddi, *Participatory Forest Management in Tanzania: 1993-2009—Lessons Learned and Experiences to Date*, Ministry of Natural Resources and Tourism, Forestry and Beekeeping Division, Dar es Salaam, Tanzania, 2009.

[20] T. Treue, Y. M. Ngaga, H. Meilby et al., "Does participatory forest management promote sustainable forest utilisation in Tanzania?" *International Forestry Review*, vol. 16, no. 1, pp. 1–16, 2014.

[21] Ministry of Natural Resources and Tourism, *Participatory Forest Management in Tanzania. Facts and Figures*, Forestry and Beekeeping Division, Dar es Salaam, Tanzania, 2008.

[22] E. N. Chidumayo, "Species structure in Zambian miombo woodland," *Journal of Tropical Ecology*, vol. 3, no. 2, pp. 109–118, 1987.

[23] J. D. Lowore, "Coppice regeneration in some miombo woodlands of Malawi," FRIM Report 99001, Forestry Research Institute of Malawi, 1999.

[24] E. J. Luoga, E. T. F. Witkowski, and K. Balkwill, "Harvested and standing wood stocks in protected and communal miombo woodlands of eastern Tanzania," *Forest Ecology and Management*, vol. 164, no. 1–3, pp. 15–30, 2002.

[25] M. Williams, C. M. Ryan, R. M. Rees, E. Sambane, J. Fernando, and J. Grace, "Carbon sequestration and biodiversity of regrowing miombo woodlands in Mozambique," *Forest Ecology and Management*, vol. 254, no. 2, pp. 145–155, 2008.

[26] T. Banda, N. Mwangulango, B. Meyer et al., "The woodland vegetation of the Katavi-Rukwa ecosystem in western Tanzania," *Forest Ecology and Management*, vol. 255, no. 8-9, pp. 3382–3395, 2008.

[27] R. A. Giliba, E. K. Boon, C. J. Kayombo, E. B. Musamba, A. M. Kashindye, and P. F. Shayo, "Species composition, richness and diversity in Miombo Woodland of Bereku Forest Reserve Tanzania," *Journal of Biodiversity*, vol. 2, no. 1, pp. 1–7, 2011.

[28] D. D. Shirima, P. K. T. Munishi, S. L. Lewis et al., "Carbon storage, structure and composition of miombo woodlands in Tanzania's Eastern Arc Mountains," *African Journal of Ecology*, vol. 49, no. 3, pp. 332–342, 2011.

[29] P. K. T. Munishi, R. A. P. C. Temu, and G. Soka, "Plant communities and tree species associations in a Miombo ecosystem in the Lake Rukwa basin, Southern Tanzania: implications for conservation," *Journal of Ecology and Natural Environment*, vol. 3, no. 2, pp. 63–71, 2011.

[30] J. Dewis and F. Freitas, "Physical analysis of soils," in *Physical and Chemical Methods of Soil and Water Analysis*, FAO Bulletin no. 10, pp. 51–57, 2nd edition, 1970.

[31] S. R. Olsen and L. E. Sommers, "Phosphorus," in *Methods of Soil Analysis, Part 2, Agronomy Monograph, No 9*, A. L. Page, R. H. Miller, and P. R. Keeney, Eds., pp. 403–430, American Society of Agronomy, Madson, Wis, USA, 1982.

[32] J. D. Rhoades, "Cation exchange capacity," in *Methods of Soil Analysis, Part 2, Agronomy Monograph No. 9*, A. L. Page, R. H. Miller, and P. R. Keeney, Eds., pp. 149–157, American Society of Agronomy, Madson, Wis, USA, 1982.

[33] I. Matejovic, "Determination of carbon, hydrogen, and nitrogen in soils by automated elemental analysis (dry combustion method)," *Communications in Soil Science and Plant Analysis*, vol. 24, no. 17-18, pp. 2213–2222, 1993.

[34] J. P. Møberg, *Soil Analysis Manual*, Revised Edition, Laboratory of Soil Sciences, Department of Soil Science, Sokoine University of Agriculture, Morogoro, Tanzania, 2001.

[35] N. C. Brady, *The Nature and Properties of Soils*, Macmillan, 8th edition, 1974.

[36] C. J. Krebs, *Ecological Methodology*, Hamper Collins Publishers, New York, NY, USA, 2nd edition, 1999.

[37] B. Husch, T. W. Beers, and J. A. Kershaw, *Forest Mensuration*, John Wiley & Sons, Hoboken, NJ, USA, 4th edition, 2003.

[38] E. E. Mwakalukwa, H. Meilby, and T. Treue, "Volume and aboveground biomass models for dry Miombo Woodland in Tanzania," *In press.*

[39] M. Kent, *Vegetation Description and Analysis, A Practical Approach*, Wiley-Blackwell, John Wiley & Sons, Hoboken, NJ, USA, 2nd edition, 2012.

[40] B. McCune and M. J. Mefford, *PC-ORD. Multivariate Analysis of Ecological Data*, Version 6, MjM Software, Gleneden Beach, Ore, USA, 2011.

[41] M. Dufrene and P. Legendre, "Species assemblages and indicator species: the need for a flexible asymmetrical approach," *Ecology Monography*, vol. 67, pp. 777–795, 1997.

[42] J. E. Peck, *Multivariate Analysis for Community Ecologists: Step-by-Step using PC-ORD*, MjM Software Design, Gleneden Beach, Ore, USA, 2010.

[43] M. A. Njana, *Arborescent species diversity and stocking in Miombo Woodland of Urumwa forest reserve and their contribution to Livelihoods, Tabora, Tanzania [M.S. thesis]*, Sokoine University of Agriculture, Morogoro, Tanzania, 2008.

[44] S. A. O. Chamshama, A. G. Mugasha, and E. Zahabu, "Stand biomass and volume estimation for Miombo woodlands at Kitulangalo, Morogoro, Tanzania," *Southern African Forestry Journal*, no. 200, pp. 59–70, 2004.

[45] J. Isango, "Stand structure and tree species composition of Tanzania miombo woodlands: a case study from miombo woodlands of community based forest management in Iringa district," in *MITMIOMBO—Management of Indigenous Tree Species For Ecosystem Restoration and Wood Production in Semi-Arid Miombo Woodlands in Eastern Africa*, Proceedings of the 1st MITMIOMBO Project Workshop held in Morogoro, Tanzania, 6th-12th February 2007. Working Papers of the Finnish Forest Research Institute no. 50, pp. 43–56, 2007.

[46] J. R. Landon, *Booker Tropical Soil Manual. A Handbook For Soil Survey and Agricultural Land Evaluation in the Tropics and Subtropics*, Longman Scientific & Technical and John Wiley & Sons, 1991.

[47] R. F. Fisher and D. Binkley, *Ecology and Management of Forest Soils*, John Wiley & Sons, 3rd edition, 2000.

[48] W. A. Rees, "Preliminary studies into bush utilization by Cattle in Zambia," *Journal of Applied Ecology*, vol. 11, pp. 207–214, 1974.

[49] S. M. Philip, *Measuring Trees and Forests*, CAB International, Wallingford, UK, 2nd edition, 1994.

[50] J. D. Lowore, P. G. Abbot, and M. Werren, "Stackwood volume estimations for miombo woodlands in Malawi," *Commonwealth Forestry Review*, vol. 73, no. 3, pp. 193–197, 1994.

[51] D. Kubota, T. Masunaga, Hermansah et al., "Soil environment and tree species diversity in tropical rain forest, West Sumatra, Indonesia," in *Soils of Tropical Forest Ecosystems. Characteristics, Ecology and Management*, A. Schulte and D. Ruhiyat, Eds., pp. 159–167, 1998.

[52] P. K. T. Munishi, T. H. Shear, T. Wentworth, and R. A. P. C. Temu, "Compositional gradients of plant communities in submontane rainforests of Eastern Tanzania," *Journal of Tropical Forest Science*, vol. 19, no. 1, pp. 35–45, 2007.

3

Is Cut-Flower Industry Promotion by the Government Negatively Affecting Pollinator Biodiversity and Environmental/Human Health in Uganda?

Bin Mushambanyi Théodore Munyuli[1,2,3,4]

[1] *Academic Affairs and Research Program, Cinquantenaire University (UNIC/Lwiro), D.S. Bukavu, South-Kivu Province, Democratic Republic of Congo*
[2] *Departments of Biology and Environment, National Center for Research in Natural Sciences (CRSN/Lwiro), D.S. Bukavu, South-Kivu Province, Democratic Republic of Congo*
[3] *Centre of Research for Health Promotion (CRPS), Department of Nutrition and Dietetics, Institute of Higher Education in Medical Techniques (ISTM/Bukavu), P.O. Box 3036, Bukavu, South-Kivu Province, Democratic Republic of Congo*
[4] *Department of Natural Resources and Environmental Economics, Faculty of Natural Resources and Environmental Sciences, Namasagsali Campus, Busitema University, P.O. Box 236, Tororo, Uganda*

Correspondence should be addressed to Bin Mushambanyi Théodore Munyuli; tmunyuli@gmail.com

Academic Editors: I. Bisht, H. Ford, R. Rico-Martinez, and P. K. S. Shin

A study was conducted from 2010 to 2012 around the flower growing areas in central Uganda to generate baseline information on the status of pollinators. Primary data were gathered using a questionnaire that aimed at determining farmers and flower farm officials' perceptions on the impact of activities carried out inside greenhouses on pollinators, human health, and on crop production in the surroundings. Results indicated that the quantity of pesticides and fertilizers applied daily varied among the different flower farms visited. Bee species richness and abundance varied significantly ($P < 0.01$) according to flower farm location, to the landscape vegetation type, and to field types found in the surrounding of flower farms. Bee richness found around flower farms varied in number from 20 to 40 species in total across seasons and years. Bee density increased significantly with the increase in flower density. Small-scale farmers were aware of the value and importance of pollination services in their farming business. There was no clear evidence of a direct effect of agrochemicals application on bee communities living in the surrounding habitats. There is a need for further research to be conducted on human health risks and for toxicological studies on soils, plants, flowers, and bees in the farm landscape.

1. Introduction

Due to government policy of enhancing crop productivity in response to population growth, agricultural modernization in many forms is increasing at high speed in Uganda. Uganda produces approximately 11.1 million tonnes of flowers and is the second largest in South Saharan Africa after Nigeria. Uganda is among the top 10 producing flowers in the world. The first rose farms in Uganda were planted in 1992 and since then, the country flower industry has grown gradually. The average exports of flowers increased from US$9.72 million in 1998 to US$29.45 million in 2009. About 95% of the production is exported and 5% is sold on local market or thrown away [1]. Uganda's floricultural sector has over the last 16 years emerged as an important nontraditional export earner, contributing over US$35 million in foreign exchange earnings in 2012 and directly employing over 6500–9000 people. The current government of Uganda's objective for the flower industry is to stimulate its rapid development because of its contribution to the diversification of the export base and rural development.

Is Cut-Flower Industry Promotion by the Government Negatively Affecting Pollinator Biodiversity and
Environmental/Human Health in Uganda?

25

Practically, cut-flower industry is growing in Uganda. Currently, there are many flower firms established in the country. However, various stakeholders from government departments/agencies, the academia, research organization, community based organizationss, and nongovernmental organizations are of the view that this industry is impacting the health of the environment and the health of human beings [1]. Biodiversity (such as bee biodiversity) is suspected to be affected (disappearing) in the surrounding of flower farms [1]. Therefore, there was a need to conduct a baseline study to gather information about the potential effect of flower farms on pollinators (bees).

A pollinator is the biotic agent (vector) that moves pollen from the male (anthers) of a flower to the female (stigma) of a flower [2]. Pollinators are a key component of global biodiversity that play an important functional role in most terrestrial ecosystems [2]. They represent a key ecosystem service that is vital to the maintenance of both wild plant communities and agricultural productivity [3–6]. Pollinators are critically important for the maintenance of the human agricultural enterprise, since they provide vital ecosystem services to crops and wild plants through their pollinating activities [7–9]. The most important pollinators in the world are bees. Bees are essential for healthy and diverse ecosystems through their pollinating activities. Approximately 80% of flowering plants depend on pollinators, mainly bees. However, bees constitute a fragile link to food production chains due to their vulnerability to various factors, mainly anthropogenic factors. Without pollinators (bees), ecosystem functioning, trophic cascades, and the survival and maintenance of genetic diversity of many wild plant populations would be at risk and economic yields of crops may suffer a drastic reduction [2, 10]. Hence, pollination is an essential step in the production of fruits and many vegetables [11]. An estimated 70% of world crops experience increases in size, quality, or stability because of pollinator services, benefiting 35% of the global food supply [7–10, 12, 13]. Animal pollination also contributes to the stability of food prices, food security, food diversity, and human nutrition [14–16].

The value of pollination to agricultural production worldwide is currently estimated to be worth €153 billion per year or approximately 39% of the world crop production value (€675 billion) from the total value of 46 insect pollinated direct crop species [14–16]. This value (€153 billion) is of more than €600 billion when added the economic benefit received from beekeeping products (sale of honey, propolis, etc.).

Though crop pollinators include a wide array of insects (e.g., beetles, butterflies, flies, etc.), bees are the most important and effective of these pollinators [14, 15]. As the world's primary pollinators, bees are a critically important functional group because, roughly 90% of the world's plant species are pollinated by animals and the main animal pollinators in most ecosystems are bees [15–17]. Although other taxa including butterflies, flies, beetles, wasps, bats, birds, lizards, and mammals can be important pollinators in certain habitats and for particular plant species, none achieve the numerical dominance as flower visitors worldwide as bees [18]. The likely reason for this is that unlike other taxa, bees are obligate florivores throughout their life cycle, with both adults and

larvae depending on floral products, primarily pollen and nectar. Bees (Hymenoptera: Apoidea) constitute an extremely species rich fauna, with an estimated 20,000–30,000 species worldwide [19–21] but approximately 3000 afrotropical species and only 700–1100 species recorded in Uganda so far [21–28].

The tropics are home to immense faunal and floral diversity and encompass much of the world's biodiversity hotspots including bees [19]. Much of the tropics exist as a mosaic of agricultural lands and forest patches, and these human-altered landscapes can have strong impacts on local bee biodiversity [21–28]. Tropical Sub-Saharan Africa is also characterized by strong ecological and agricultural dependencies on pollination [24, 25]. Hence, pollination shortage/decline is of great concern for food security in a continent where scientists are just beginning to understand how anthropogenic land-use impacts wild and managed pollinators. Since bees are very important in agricultural production, their status is therefore of great concern not only to the farmers but to any responsible government as well, as it has a direct impact on people's livelihoods and the economy. This concern can therefore be translated into developing management techniques for the conservation of effective native bee species.

Although bees provide enormous ecological and economic benefits to flowering plants, wildlife, and humans, they are, however, under increasing threat from anthropogenic factors. There is considerable evidence for the negative impacts of habitat alteration on pollinators in highly disturbed regions of the world [28, 29], particularly in Europe and North America [29]. Pollinator crisis exists even in tropical and subtropical areas, where natural habitat is well represented [28]. Studies indicate that native pollinator populations face many threats, and evidence of a global pollination crisis is steadily growing [28, 29]. Currently, there are sufficient scientific evidence of a sizeable decrease in the population and range of many pollinators such as bees, butterflies, moths, hummingbirds, and bats from most biomes of the globe [29, 30].

Pollinators are at risk from numerous threats and this, in turn, threatens the many benefits people and ecosystems derive from pollination services. Drivers (disturbance types) of pollinator loss/decline [31, 32] include seminatural and natural habitats loss/degradation (destruction) and fragmentation through intensive land-use, misuse/over-use of toxic pesticides in agriculture (agricultural chemicals), pathogens [31, 32], alien species, toxic effects of secondary compounds produced by genetically engineered plants [33], and climate change and the interactions between them [34–36]. The current challenge for the conservation of pollination services in rural landscapes is to better quantify the relative importance of a range of drivers (and pressures) and in particular their simultaneous synergistic effects in order to understand the magnitude of their impact, particularly if these are coupled with the clear ecological and economic risks associated with pollinator loss and crop yield failures [35, 36] such as agriculture intensification activities.

Agricultural intensification has got around 13 components of intensification [20–24]. However, the use of insecticides and fungicides is the component known to have consistent negative effects on biodiversity and ecosystem services delivery [35, 37]. Agriculture modernization or modern agricultural practices (agrochemical applications), landscape fragmentation, and habitat degradation have been identified as key drivers that negatively affect bee populations in rural landscapes by the elimination of resources needed for successful reproduction such as nesting sites and pollen and nectar sources [35]. In Uganda, agricultural intensification is taking place. An example of agricultural modernization in Uganda includes the upsurge of the floriculture industry. Flower farms with high agrochemical inputs are clear evidence of agricultural modernization. The negative effects of agrochemicals on biodiversity in farmlands are well documented [35–39]. As mentioned above, pesticides are considered a threat to pollinators [40, 41] although little is known about the potential impacts of their widespread use on pollination services in flower growing regions in Uganda.

The negative effects of agrochemicals (synthetic insecticides, botanical insecticides, miticides, acaricides, biologicals and natural enemies, fungicides, herbicides, seed dressing, adjuvant, nematicides, horticultural detergents, flower preservatives, plant growth regulators, foliar fertilizers, soil amendments, chelates, specialty fertilizers, grain storage insecticides, termiticides, rodenticides, etc.) on biodiversity in farmlands are well documented [35, 38, 39]. As mentioned above, pesticides are considered as a threat to pollinators [40, 41] although little is known about the potential impact of their widespread use on pollination services in habitats surrounding flower growing regions in Uganda. Since it is well established that agricultural chemicals can pose negative effects to biodiversity and to the environment of the areas where they have been applied [40, 41], loss of biodiversity around flowers is therefore expected in Uganda. However, assessing negative effects of agrochemicals applied by flower farms on biodiversity in the surrounding environment is very challenging since these agrochemicals are applied inside greenhouses and therefore expected to have almost no effect on the surrounding environment.

To our knowledge, there exists no study on the impact of inorganic fertilizer (NPK) application in greenhouses of flower farms on pollinators living in the surrounding habitats. Even when studies of the impact of fertilizers on biodiversity exist, they generally conclude the overall range of effects of inorganic fertilizers on species richness and abundance being arguably negligible or of little impact [42]. Contrastingly, Le Féon et al. [41] showed that increased nitrogen (and related inorganic fertilizer) input can cause a decline in floral resource diversity and abundance (nectariferous native plants) in European farmland habitats.

Overall, there is a need to determine the trends in response of biotic organisms (including bees) to inorganic fertilizations in and outside flower farms since they are agrochemicals and they cannot be without disturbance on surrounding local ecosystems. The best way to detect negative effects of pesticides application by the flower farm on biodiversity is to study responses of most sensitive biota to pesticides application regimes. Pollinators (bees) are the best candidates for such studies. They are good bioindicators of environmental health [42–44]. Since it is well established that agricultural chemicals can have negative effects on biodiversity and on the environment of the areas where they have been

applied, loss of biodiversity around flowers is expected [40]. However, assessing negative effects of agrochemicals applied by flower farms on biodiversity in the surrounding environment is very challenging since these agrochemicals are applied inside greenhouses and therefore expected to have almost no effect on the surrounding environment.

Local people's experiences and perceptions of the effects of rural development projects (e.g., flower farm industry) are often not reported or taken into account by decision-makers despite strong arguments that local opinions can help in building a policy enabling achieving a win-win conservation/development scenario to meet development targets, and that a community's willingness to become involved in decision-making for the establishment of a project in their village is closely linked to their past experiences and to their perception of the benefit. Much as gender-based differences exist in perceptions of problems, integrating understanding of people's perceptions with field observations of the functioning of environmental systems is critical for developing sustainable resource management activities. Previous work examining farmers' perceptions about importance of pollinators in crop production has shown that farmers often have acute and accurate awareness of problems, and they can propose effective interventions, even if they appear unable or unwilling to tackle them [24].

In the absence of relevant information, decision-makers may be obliged to formulate their policies based on perceptions and views of farmers that appear very relevant. Since historical data collections do not exist in Uganda, it is difficult to know which bee species has declined. Thus, farmers' surveys remain the only reliable source of information that can help to provide researchers with an idea or an indication of what might have happened in the environment few years ago. Using farmers' knowledge and perceptions about changes in bee populations over the last 5 to 50 years can help to understand what happened in the area several years ago and potential causes that led to such a situation.

Even though pollinator declines are a global biodiversity threat, drivers of pollination decline/loss in natural and in agricultural ecosystems of Uganda have not been taken into account by policy-makers, conservationists, and researchers, although Ugandan agriculture owe much of its production to services delivered by locally available diverse pollinator species [20–25]. The real magnitude of pollinator decline is not easy to determine, particularly in countries like Uganda where there has been no historical data collections.

Based on the above background, there was a need to assess bee activities in the surrounding environment where different flower farms have been established. Accurate measurements of population densities (visitations) and species richness of bees are essential for any meaningful assessment of decline [23–25]. For nonsocial bees, this can be done with direct counts of individuals and classical abundance measures. For social bees, however, the number of colonies rather than the number of individuals is the crucial parameter for conservation [36] and assessment of decline. Density estimates derived from direct counts of bees are tedious and can be unreliable because natural nests are hard to detect particularly when there is little time in the field. For rapid surveys, the number

Is Cut-Flower Industry Promotion by the Government Negatively Affecting Pollinator Biodiversity and Environmental/Human Health in Uganda?

27

of individuals can be used to give an indication on the potential richness of colonies in the landscape because the number of colonies in the wild can be sometimes very difficult to assess. The census of managed hives and detection of wild colonies of honeybees within a radius of 0.1–3 km from a given sampling location can exhaustively be surveyed but it needs personnel with knowledge of local beekeeping operations. Therefore, for a study of short period (<3 years) like the one presented here, the spatiotemporal assessment of visitations of different bee species to flowering plants in a given habitat can be a useful approach to detect historical or previous changes (decline, increase) in bee species and populations in relationship to farm management and local and landscape drivers such as pesticide application intensity, availability of floral resources, and so forth.

Currently, there exist no studies from Uganda addressing effects of multiple drivers on bee abundance and species richness. Information on how pollinators respond to different drivers may improve the understanding of the nature, causes, and consequences of declines in pollinator services at a local and national scale, as well as providing light on how to invest for the development of mitigation options to slow the decline of pollinators in Uganda.

As previously highlighted, in Uganda, there have been various protestations and complaints of farmers and various stakeholders from government agencies, academia, research organizations, civil society organisations, and nongovernmental organisations about the boom of flower farms [1]. There were strong views and suspicion regarding the negative unknown impact of flower farms on health of humans, environment, and biodiversity in areas where the flower farms have been established. Consequently, there was a need to check whether these claims/complaints had a scientific foundation, when regarding effects on sensitive taxa such as bees. A preliminary field visit was conducted by the researcher and from that visit, it was pointed out that pollinators (specifically bees) inhabiting the surrounding of flower growing areas were likely to be at risk of disappearing. To verify this suspicion, an in-depth field study was therefore needed to be conducted. This was found to be necessary to ascertain community complaints.

The general objective of this study was therefore to conduct a rapid assessment on the status of pollinators (bees) around flower farms in central Uganda and provide guidelines on the preparation of a monitoring plan for the pollinators in the flower growing areas around Lake Victoria.

The specific objectives were (i) to gather information on the agrochemicals used by flower farms and their potential impacts on pollinator bees inhabiting the surroundings of the flower farms; (ii) to measure pollinator activities and assess the richness, abundance, and diversity of bees in relationship to landscape and habitat types found in the surroundings of flower farms; (iii) to assess the level of knowledge of pollination service importance in crop production by small-scale farmers living in the surroundings of flower farms; (iv) to document perceptions of local people with regards to cause of pollinator decline in their villages, (v) to collect views and perceptions of farmers about benefits and negative effects of the flower farms established in their villages; (vi) to collect views of farmers on the ways to promote flower industry in a sustainable manner (floriculture industry that matches people's needs and desires, requirements for living in a clean non-polluted environment); (vii) to outline bee monitoring and conservation strategies in the flower farm growing regions.

2. Materials and Methods

2.1. Study Area Description, Visits to Flower Farms, and Dialogue with Production Managers. This study was conducted in the banana-coffee system of Lake Victoria Arc covering several districts of central Uganda. The study zone (latitude: 0.5°31'22''; longitude: 31°11'71''; altitude: 1080–1325 m) is characterized by ferris oils with high to medium fertility level and receives on average 1000–1800 mm of rainfall per annum on a bimodal pattern (rainy seasons: March–May, September–November; semidry to dry seasons: June–August, December–February) with $28.7 \pm 2.77°C$ and $68.65 \pm 8.91\%$ of mean annual temperature and relative humidity, respectively [22–26]. But the rainfall amounts and patterns are unpredictable. The study zone belongs to the Lake Victoria phytochorion [23–25] with shrubs of *Acacia* spp., legume trees, melliferous plant species, *Papyrus*, and palms ranging from 2 to 15 m high dominating the remnant secondary vegetation [20–22]. In this study region, coffee (*Coffea canephora*, Pierre ex Froehner) is the main cash crop and banana the main staple food crop. Several pollinator-dependent food and cash crops are grown in small-scale monoculture and/or polyculture fields that are integrated into this coffee-banana agroforestry system including home-gardens. There were no standard crops per study sites but most crops were found grown in almost all study sites. Crops grown as sole or in association with coffee and or banana include cassava (*Manihot esculentum* L.), sweetpotato (*Ipomoea batatus* L.), maize (*Zea mays* L.), beans (*Phaseolus vulgaris* L.), groundnut (*Arachis hypogea* L.), tomato (*Lycopersicon esculentum* L.), watermelon (*Citrullus lanatus* L.), pumpkin (*Cucurbita moschata* L.), cucumber (*Cucumis sativus* L.), melon (*Cucumis melo* L.), chilies (*Capsicum* spp.), and several other fruits, vegetables, and horticultural crops (cabbage, onion, etc., egg plants, sim-sim, etc.). The majority of these crops are grown in small-scale monoculture and or polyculture fields that are integrated into the coffee-banana agroforest production systems. The agroforestry system is also dominated by several native/indigenous fruit and agroforestry tree species. Banana-coffee agroforests and small-scale farms cover about 60% of the land, whereas mixed mosaic seminatural habitats cover approximately 40% of the farm-landscape studied. There exist in this study region some large monoculture plantations (sugar cane plantations, coffee plantations, tea plantations, etc.) and some flower farms companies.

Rural central Uganda is a mosaic landscape where "islands" of patches of natural habitats (forest fragments, forest reserves, wetlands, woodlands) and linear (eg., hedgerows) and nonlinear (fallow fields, grasslands, woodlots, cattle pastures, or rangelands) features of seminatural habitats [21] that serve as "field boundaries" of the variety of small-scale fields are found scattered within agricultural matrices.

Compared to other districts of the country, the study area (central Uganda) is also characterized by high demographic pressure, limited access to arable lands, continuous cultivation, and over-exploited lands under unrevised land policies [20]. All study sites had also some forest remnant tree species retained within them, ranging from 1 to 175 trees/ha found both in crop fields as well as inside remnant natural vegetations scattered inside the forest. Flower farms are located in four major zones of central Uganda: the first zone is Entebbe airport zone located at 60 km from Kampala city (example: Wagagai flower farm and Rosebud-II. Wagagai and Rosebud II are separated by a distance of 30 km). The second zone is in Mukono district located at about 250 km from Kampala city (example: Mairye estates), the third zone is Ntungamo zone located at about 500 km far away from Kampala city (example: Pear flower farm). In addition, mantel test ($P <$ 0.05) showed that there was no evidence of significant spatial autocorrelation between pollinator counts on transects within a landscape surrounding a flower farm. The distance between greenhouses and people's homes varied from a flower farm to another one: >20 m–100 m. In many cases, small-scale gardens are established closer (0–20 m) to greenhouses. Thus, field visits were conducted to make a rapid survey on the status of pollinators (bees) around the flower farms in the Lake Victoria shores. To be able to understand the actual activities and operations within the flower farm, four flower farms were visited out of a total of 12 existing and operating flower farms in central Uganda, given the budgetary and time constraints. The 4 selected flower farms included Fiduga (located at 15 km far from Kampala, Wakiso district), Pearl flower farm (located at about 400 km far from Kampala in Ntungamo district), Mairye estates (located at 30 km far from Kampala in Mukono district) and Wagagai flower famr (located at about 40 km far away from Kampala along Entebbe airport road). Overall, commercial floriculture is still a new industry in Uganda, dating back to only 1993. Cut-flowers, cut foliage and, to a lesser degree, pot plant cuttings are the main outputs. Cut-flowers include a variety of roses, chrysanthemum cuttings, carnations, and summer flowers. The four flowers grow different types of flowers and varieties (Table 1) and apply different levels of pesticides. The major flower varieties grown and exported from Kenya are roses, carnation, alstroemeria, lisianthus, statice, and cut foliage. Rose flower dominates.

Flower farm visits were conducted for two years (2010 to 2012). During a visit of each flower farm, discussions with production manager were engaged by the researchers. The discussion focused on collection of information likely to enlighten the potential negative effect of activities conducted at the flower farm on pollinators as well as collecting information likely to help generate information that is more rewarding to policy-makers. During the course of discussions, various types of information were collected, mainly, information about the type of agrochemicals (pesticides, fertilizers) used by the flower farms, the type of varieties grown, the type of production conducted (e.g., exporting flowers cuts or stems or roses), the total number of people employed (including the proportion of females employed), the total size of the flower farm (including the number of hectares in production), the number of years since the flower farm was established, the monthly total production exported, the gross income from sale of flower products, the costs of labor, the costs of other general inputs, costs of pesticides/fertilizers purchases, and information about measures taken to control runoff of chemicals into the surrounding environment. The researcher visited stores of agrochemicals to confirm trade names reported by the flower farm production managers. In addition, production managers were asked whether they understood pollination and if pollination by bees was an important factor in their production business.

2.2. Landscape/Habitat, Bee Biodiversity, and Floral Resources Surveys

2.2.1. Landscape/Habitat Surveys. After farmers' interviews, surveys of bees were conducted in different landscape/habitats in the immediate surroundings of the flower farms. For each flower farm, surveys were conducted in all directions considering the flower farm to be at the center. Two types of fields were found in the surroundings of flower farms: cultivated fields and noncultivated fields. The type of dominant landscape vegetations characterizing these cultivated and uncultivated fields were cropland and grassland vegetations. Cropland vegetation types that dominated cultivated fields were composed of a mixture of crops grown under various cropping/agroforestry systems. The grassland vegetation types that dominated the uncultivated fields were composed by a mosaic of seminatural habitats. The dominant habitat types in the croplands were different land-use types or crop associations (e.g., maize + beans + banana), whereas the dominant seminatural habitats in grasslands were pastures, fallows, woodlots, hedgerows, field margins, and so forth.

2.2.2. Transect Counts for Bee Surveys. Different beneficial insect taxa respond to agricultural practices at different spatial scales and often at multiple spatial scales [45]. Therefore, it was found important to assess bee diversity and visitations at different spatial scales by placing sampling transects at different distances from the flower farm. Placing sampling transects at different distances was sought to help in determining at which spatial scale chemical applications in the flower farms could affect significantly activities of bees in cultivated and noncultivated fields in the surrounding landscape habitats. The collection on bees was conducted for five consecutive rounds across dry and rainy seasons during two years (2010–2012) at regular periods (dates) of visits of the flowers farms. Hence, five transects were set from the edge of the flower farm into either croplands or grasslands or both. Transects were established, extending from the flower farm boundary into farmlands. Transects were separated by a distance of 3–5 km. They were set in north, east, west, and southern part of each flower farm. With this distance, bee samples were independents since the normal foraging distance of most bees is 2 km. Each transect measured a total length of 2.2 km. Bees were sampled on these transects (20 m-large × 2.2 km-long) at 0–10 m, 500 m, 1500 m, and 2000 m from the edge of the flower farm. Transects were set parallel to flower farms at the above mentioned distances. Plots were set

Is Cut-Flower Industry Promotion by the Government Negatively Affecting Pollinator Biodiversity and
Environmental/Human Health in Uganda?

29

TABLE 1: Example of type/category of flower varieties grown by different flower farms.

MAIRYE	PEARL and WAGAGAI	FIDUGA
Category of cut-flowers	*Intermediate Varieties*	*120 varieties of Crysantemum cuttings*
Sweet heart	Red Calypso (Red)	Eg. Ibis lime, Nimba, Copper, Maiko, Noa, Parrot
Super sweet heart	Akito (White)	City, Art Sunny, Ludmilla, Fire, Zembla, Avoriaz
Intermediate	Blushing Akito (Pink)	Grand Salmon, Zakumi, Goody, Fuego, Voyager, Lexy, Breezers,
	Jetset (Yellow)	Tory, Safin Purple, Bacardi Yellow, Managua,
Varieities are	Mylo (Pink)	Vivid Cream, Tuvalu Sunny, Kindly Salmon, Zembla Cream,
For Super sweet heart:	Inka (Bicolur Red/Yellow)	Champagne Pink, Ibis Pearl, Ping Pong, Browny,
Valentino	Tropizal Amazon (Orange)	Motown, Baracardi, Raisa, Verdy
Akito	Sonrisa (Yellow)	Rwins, Merlot, Art, Ping Pong, Kuga, Quinty
Banjo	Jupiter (Yellow)	Kindly, Ludmilla, Puma Sunny, Redstart
For sweet heart:		
Tropical Amazon	Wild Calypso (Hot Pink)	Leopard, Zembla Yellow, Grand Pink, Athos, Anastasia Pink,
Jet Set	Belle Rose (Pink)	Crystal, Streamer Splendid, Dance Sunny, Felling Green Dar
Red Ribbon		Handsome, Gabd Cherry, Froggy, Arctic Queen, Roma, Ping Pong,
Red Giant		Marabou, Fire, Parrot, Noa, Nimba, Kaitylnn, Tiger, Harley, Browny
		Cheops, Purple Rain, Sound, Delianne Yellow, Ritmo, Charming,
Intermediate varieties		Salinas, Jazzy, Husky, Oxana, Clearity, Lollipop Purple, Falcon
Blashing Akito		Joker, campus, Candor, le Mans, Planet, Lexy, Sarasarane Pink,
Lambada		Syrup, Moonlight, Lerbin, Minty, Anky, Edge, Katinka, Marabou, Marimo,
Frisco		Lollipop Yellow, Vogue, Energy, Magic, Teror, Ferry, Fuego, Central
		Leopard, Voyager, Swan, Puma, Vatican, Stromboli, Crystal Pink, Classy,
		Diva, Starling, Zodiac Lilac, Vulcano Dark, Spider White, Punch,
		Copper, Juicy, Chamapagne Ora, Boris Becker, Rocky, Pisang,
		Merlot, Rosas, Goody Orange, Supernova, Froggy, Gin Pink, Berdy, Pelican
		Eleonora Lilac, Vireo, Snowdown, Cheeks, Ritmo, Tigerrag, Ely, Charming
		Greenbird, Avoriaz, Swan, Motto, Katinka, Sound, Vivid, Vivid Cream
		Dance Salmon, Arctic Queen, Delianne White, Reagan White Eli, Balloon
		Tamarinda White, Tamarinda Violet, Margarita Dark Pin, Margarita Helena
		Hypnotica Red, Carolina Orange, Juventa, lrmon, Bromo, Margarita Lilac

mainly at different flower patches alongside transects. In this study, flower patches were isolated groups of blooming plants; the group was composed of various plants of same or various species associated in a given site. Flowers were composed of either natural vegetation or crop plants and in some cases both.

While walking along transect, at each flower patch, data was collected on bee populations using field observations, hand-nets, and transect-count methods. Observations were also made on other pollinators, such as butterflies, moth, and hover flies; although detailed data is not presented since the focus was on bees.

Within each plot, bee species within the plots were counted and their visitation intensity was measured and recorded on datasheets. Parallel to bee surveys, habitat/land-use variables were also recorded such as the percentage cover and number of seminatural habitats. When needed (in some few cases), nests of solitary bees were recorded using subjective (focused) searches in particular habitats (nesting habitats). The different nests were also located either by chance during random searches or by inspecting tree-holes/termite mounds located alongside established belt transects of width 20 m and length 1.1 km per transect.

Bee collections/censuses were conducted from morning to evening hours (07:00 h to 17:00 h, local time). Pesticides applications inside greenhouses are commonly conducted at different interval of times: 5:00 h–7:30 h, 7:30 h–10:30 h, 15:30 h–18:30 h. Therefore, some bee collections coincided with periods of pesticides application.

2.2.3. Censuses of Bee-Visitations. Visitations are part of four measures of pollination services delivery by pollinators to plants/crops: (a) pollination rate (number of pollinators visiting flowers/min), (b) proportion of visited flowers (number of flowers visited by pollinators/total number of flowers in a plot), (c) flower handling time (seconds) or the mean time pollinators spent on individual flowers in a plot, and (d) pollination efficiency (% fruit set after single visit by a p). These four parameters of pollination services can be used to forecast high reproductive success of plants/crops after visitation of crops/plants as compared to those that were not visited. Thus, pollinator visitations (number of bee individuals landing on flower reproductive parts and moving/collecting floral resources such as pollen/nectar for at least 0.1–1 minute before flying away) to plants/crops species in flower patches were conducted from morning to evening hours each sampling day.

All floral visitors, pollinators and nonpollinators to male and female flowers and inflorescences of all flowering plant species were collected at the time of their greatest activity. The pollinator censuses were conducted for 2 full years (2010–2012 and for each year, data was collected during four consecutive sampling rounds (R1: December–February, R2: March–May, R3: June–August, R4: September–November)). The pollinator censuses were conducted within permanent plots (20 m × 100 m) that were placed in a systematic way (along line transects of 20 m × 1100 m) and separated by 10 m. Pollination information was basically collected on transects. Overall, data recorded was based on an observable number of bees per flower patch per 25 m^2 (5 m × 5 m) from each sampling unit (plot). The data recorded and stored in datasheets was expressed as number of bee-visits per bee species per 100–500 flowers/15–20 min observations per flower patch of 25 m^2 (5 m × 5 m).

While walking along transect, censuses of bees were performed during 15–30 min at each flowering patch met along the transect. In each census, several floral stems with 100–5000 flowers were observed. A total of 10 to 50 flower patches were met and censured each study visit. Minimum observation distance from the flowers was 2 m. In each census, the number and identity of visitors were recorded, including the number of flowers visited, the visitation intensity of each bee species, and the behavior of these visitors on the flowers. The abundance, diversity, and composition of the flower visitor assemblage of the focal populations were determined by counting the number of insects visiting the flowers by means of point-centered 30 min surveys. Sampling took place during full bloom of plants. Thus, fully blooming inflorescences/flowers were monitored during the entire foraging period of the day including the period of peak insect visitation activity (9:00 h–15:00 h).

Flower-visiting insects were identified, and buzz-pollinating ability (based on observation of sonication and pollen release by some bees such as *Xylocopa*) was recorded for each visit whenever possible. During bee counts, local temperature and wind speed were recorded at the start of each transect as these can affect pollinator behavior, although counting was conducted only at temperatures above 18°C to eliminate the impact of low temperatures on counts. Censuses were conducted along the day from 6:30 h to 17:30 h local time. Each study visit, surveys were carried out across the five transect.

Duration of flower visits to individual inflorescences/flowers of each plant species was recorded for several individual bees and rounded up to full minutes. A flower visit was defined as the period between the first landing on the inflorescence and final departure (irrespective of short hovering flights for scent transfer). More specifically, a visit was defined to have occurred when the visitor's body contacted the reproductive organs (stigma or anther) of any available fresh flower. The bee observations were conducted each day that weather conditions allowed pollinator activities. The order of observation of plots alongside transect was random, each plot being observed only once per day. During each observation period on flowering patches in sampling plots,

the number and identity of flower visitors to flowers or inflorescences (depending on the species) of all species blooming in the plots were noted.

The identification of some flower visitors was done in the field with the experience of the researcher. For species that could not be identified in the field during foraging observations, the visitor was given a morpho-species name and immediately after finishing foraging observation data collection on datasheets, a handnet was used to collect specimens. Voucher specimens of visiting bees were collected from flowering plants on separate moments to avoid disturbance to pollinator activity. The specimens were saved in alcohol 70% and later the identification confirmed in the laboratory using available collection of bees of Uganda. The bees were identified using the author's reference collection of bees that is located at Makerere University zoology museum.

2.2.4. Assessment of Floral Resources. To account for the floral abundance of the plant species, after each observation period, each species' number of open attraction units to bees which could be flowers or inflorescences was counted. The data recorded was expressed as number of fresh flowers (all plant species combined) per flower patch per 25 m^2. In addition, the number of flowering plant species was recorded per sampling plot alongside transect walks.

2.3. Field Surveys and Farmer's Interviews on Pollination Knowledge and on Causes of Pollinators Decline in Relationship to Activities Carried Out at Flower Farms. In the surrounding of each flower farm, four directions were followed. In each direction, 2 villages were selected in the north, south, east, and western directions of the flower farm. In each direction, the first village was selected between 0 m and 1 km far from the edge of the flower farm; the second village was selected between 2 and 3 km far from the flower farm. These villages were selected to cover the range of space at which a farmer could smell and could not smell the scent of the chemicals from the flower farms. In each village 20 people (10 men and 10 women) were randomly selected while walking alongside main trucks in the village. In total, 160 people were interviewed from 4 villages from the four directions in the surrounding of the 4 flower farms, making them 640 people interviewed in total from the surroundings of the four flower farms. Farmers interviewed were those met in their gardens busy farming. During interviews, after finishing the dialogue with farmers, the researcher conducted field inspection under their guidance. Field visit and inspection were conducted to enable scientists/researchers to verify whatever farmers were reporting when interested in gathering information about their ecological knowledge of pollination process.

Primary data were collected by administering a questionnaire to production managers of different flower farms and to small-scale farmers living in the surroundings of the flower farms. In addition, a separate datasheet was used to gather data on bee species diversity and populations. The questionnaire captured information on agrochemicals used, bee diversity, landscape/terrain, weather conditions, especially at

Is Cut-Flower Industry Promotion by the Government Negatively Affecting Pollinator Biodiversity and
Environmental/Human Health in Uganda?

31

the time of bee surveys, farmers, and flower official perceptions and the impact of their activities on pollinators and crops.

More specifically, the effects of farm management practices (agrochemical pesticides utilization intensity), landscape vegetation types, distance from the flower farm, and the flower farm location on bee communities foraging in environments surrounding flower farms were investigated. The investigation aimed at identifying factors that could help in understanding the potential role of agrochemical activities carried out inside flower farms on bee biodiversity living in the adjacent habitats. In addition, farmers' surveys were conducted in order to get an idea of the potential causes of decline of bees in the villages including the impact of flower farm activities (agrochemical applications).

The questionnaire was also administered to assess if farmers understood the meaning of pollinators, pollination, and pollination service and its importance in crop production activity. Villages where flower farm activities were not expected to impact agricultural production were identified (villages located at more than 5 km far away from the edge of the flower farms) and sampled to ascertain the differences in bee populations and species richness. Only farmers found doing some activities in their gardens were interviewed. Efforts were made to be gender sensitive during the course of interview. However, women were frequently found in the gardens than men.

The questionnaire consisted of a mixture of open and closed ended questions. The questionnaire was pretested by the researcher a few weeks prior to field surveys. Semistructured interviews that had a number of predefined starting questions were used. Thus, farmers were interviewed using a pretested structured questionnaire. The questionnaire was piloted by the researcher one week prior to the actual survey and the necessary corrections were made. Interviews were largely conducted in the local languages (Luganda, Runyakore) with translation into English whenever necessary.

Data collection related to interviews at the different sites took place from 9:00 am onwards, because most women started their agricultural activities by 6:00 am and were completed by midday. Prior to data collection, the researchers solicited informed consent from participants. The questionnaires took approximately 20–30 minutes to administer to each individual respondent. The questionnaire was filled using face to face interviews. Interviews were conducted either at the farmer's home or in the field, where such fields were within 0.5–1 km from a farmer's homestead and the farmer was willing to be interviewed on site (field or garden). Once conversation on a topic was initiated, it was allowed to roam freely until exhausted at which point a new topic was begun. It was ensured that interviewees had the opportunity to ask the researchers questions at the end.

Notes were taken on individuals' responses. The researcher visited every respondent's crop field in order to verify some of their responses. The questionnaire submitted to small-scale farmers focused mainly on crop production and the relevance of pollinators in crop production. The survey questionnaire comprised two main parts. The first section sought general sociodemographic information about respondents, including age, gender, household income, gender labour in crop production, marital status, number of children, and formal education levels. The second section gathered information relating to respondents' knowledge of crop pollination, pollinator types, perception of the importance of pollinators to crop yield, potential causes of decline of pollinators in the village, and potential role played by activities (application of agrochemicals) carried out by the flower farm located in the village. Specifically, farmers were asked to (i) describe/define their understanding of pollination, (ii) to name, identify, and differentiate between wild bees and honeybees and other insect pollinators they knew and indicate the area where they sleep (nesting site), (iii) mention the role of bees and other pollinators in crop fruit/seed set, (iv) to explain the importance of pollinators in their farming business, (v) to comment on the effects of pesticides application on wild bees and other pollinators, (vi) to list (and justify why they think so) the potential causes of decline/loss of pollinators in their village including describing the role played by activities carried out at the flower farm such as intensive application of agrochemicals, (vii) to comment on the linkage between crop yields reduction and bee decline in the village, (viii) to list advantages (benefits) they get by living near the flower farms, (ix) to propose sustainable solutions to resolve their conflicts with flower farms. Photographs of different bee species and different other pollinator species were presented to respondents to help in identification of different species of bees visiting crop flowers.

2.4. Data Analysis

2.4.1. Cumulative Analysis of Agrochemicals Application. From the raw data obtained during discussions with the flower farm managers, the total amount of agrochemicals applied on a daily basis was used to calculate the amount of agrochemicals used per annum in relationship to total amount of water used. Later on, the annual amount of pesticides used was cumulated based on the number of years since the flower has been in production. The cumulative analysis was undertaken using Microsoft Excel 2007.

2.4.2. Bee Visitation and Floral Resources Raw Data Files Pooling and Organizations. Data on bee counts and visitations that were recorded per each sampling plot were summed to obtain the total number of bee-visits and bee species per transect each study round. Similarly, the total number of fresh flowers and number of flowering plant species per transect was obtained by summing values obtained per sampling plot ($20 m \times 100 m$).

2.4.3. Variation in Bee Abundance, Visitation, Species Richness, and in Flower Density between Surrounding Habitats of the Different Flower Farms. Cross-tabulations with selected variables (number of species and individuals, visitation frequency) were undertaken using pivot table in Microsoft Excel 2007, to verify anomalies and correct errors in raw data files before data analysis.

En ce qui concerne the diversity and numbers of bees, the abundance of flower visitors was estimated by standardizing the number of visits per time unit (expressed as visits per population per hour/flower patch). Flower visitor diversity was assessed by calculating species richness and evenness. Richness was calculated as the number of flower-visiting species found visiting flowers in flowering patches. Diversity of bee communities between the four flower farms was estimated using the Shannon-Wiener's diversity index (H') and the similarity in bee communities among the four flower farms was estimated by the Sorensen similarity index according to Magurran [46] using raw data collected across four rounds of data collection over two years (2010–2012).

2.4.4. Effects of Landscape/Habitat Types on Bee Abundance and Species Richness.

All variables were tested for normality and the strongly skewed variables were transformed prior to analyses if necessary to meet the assumption of normality and homogeneity of variances. Therefore, the percentage cover of flowering plants was arcsine-square-root (+0.5) transformed and number of species or counts of bee individuals and bee-visitations counts were $\ln(x + 1)$ transformed. The differences in number of bee individuals and species and in number of fresh flowers between the 4 flower farms were tested with general linear model (GLM) analysis of variance (ANOVA) in Minitab release version 165. GLM analyses were fitted with pesticides application intensity (very high: 4, high: 3, medium: 2, and low: 1) landscape vegetation types, flower farms, and number of transects as treatment factors (predictors) and the abundance, visitation frequency, species richness of bees, and number of fresh flowers as the response variables. Where GLM test indicated significant differences, posthoc Tukey's test was used for means separations.

2.4.5. Farmers' Surveys.

The survey data were entered into a spreadsheet and checked prior to analysis. Cross-tabulation with selected variables, percentages, and means were undertaken using pivot table in Microsoft Excel 2007. Percentages were based on either the total number of respondents or total responses, details of which are provided in the respective text or tables. Chi-square test was used to determine association between variables such as to determine the effects of farmers' sociodemographic profiles on their knowledge of pollinators, pollinator unfriendly farming practices, farmers' perceptions of crop yield reduction, and causes of decline of pollinators in their villages. During interviews of farmers about potential negative effects of agrochemicals applied by the flower farms on crop yields, decline of bees, and environmental pollution, most often a farmer could give more than one justification/statement (opinion). Therefore, both the summary of the statements made by farmers and the full statements were presented in separate tables/figures. The frequency of occurrence of the statements from the entire population interviewed was calculated and presented in respective tables.

3. Results

3.1. Agronomic and Socioeconomic Characteristics of Flower Farms Studied.

The different farms studied were established with clear production objectives. For example, Mairye and Pearl flower farms were found to be specialized in the production of cut flowers and roses whereas Fiduga flower farm was found to be specialized in the production and export of *Crysantemum* cuttings (stems). Hence, different companies grow different types of flower varieties (see Table 1). The land under production played a big role in the production potential for each farm.

The number of employed people was 800, 485, and 255, respectively, at Mairye, Fiduga, and Pearl flower farms. Across flower farms, the proportion of employed females oscillated between 55 and 65%. The monthly total production for export varied from one flower farm to another: 4 million cut-flowers for Mairye, 36 million cuttings for Fiduga, and 2 million cut flowers (including roses) for Pearl. Consequently, the declared total annual income obtained from sales of flower cuttings was US\$90,000, US\$47,000, and US\$7000 for Mairye, Fiduga, and Pearl flowers farms, respectively. The monthly cost for pesticides/fertilizers purchase oscillated between US\$10,000 and US\$53,000 across flower farms.

There was a positive correlation between the monthly total income obtained per flower farm after sales of flower cuttings and (i) the cost of general inputs ($y = 0.2667x + 61421$, $R^2 = 0.4714$, $P < 0.05$), (ii) cost of labor ($y = -0.7502x + 100423$; $R^2 = 0.4749$, $P < 0.05$), (iii) the number of employed females by the flower farm ($y = 51.036x + 59526$; $R^2 = 0.2035$; $P < 0.05$), and (iv) total size of the flower farm in production ($y = 3833.3x + 21000$, $R^2 = 0.306$, $P < 0.05$). On the other hand, there was no significant ($P > 0.05$) correlation between the monthly total income obtained per flower farm after sales of flower cuttings and (i) the number of employed males ($y = -61.626x + 89046$; $R^2 = 0.0966$, $P < 0.05$), (ii) the total number of workers ($y = 13.314x + 67832$; $R^2 = 0.023$, $P > 0.05$), (iii) the total farm size land ($y = 1684.2x + 37053$, $R^2 = 0.0312$, $P > 0.05$), and (iv) the cost pesticides/fertilizers ($y = 0.1344x + 67049$; $R^2 = 0.0159$, $P > 0.05$). This last result indicated that it may not be necessarily rational for a flower farm to over spend on agrochemicals (pesticides/fertilizers) that are dangerous to the health of human beings and to the environment. In other words, it is still possible to have a flower company reaching high profitability while spending little on toxic pesticides. Different options of buying and using nontoxic and effective pesticides can still be adopted by a flower company and obtain good profitability.

3.2. Variation among Flower Farms in the Application of Agrochemicals (Pesticides, Fertilizers) inside Greenhouses.

During field visits, information on agrochemicals used by the different flower farms was collected. A checklist of different fertilizers and pesticides used and dosages are given in Tables 3 and 4, respectively. Long-term use of fertilizers for agricultural purposes has been an issue of concern to researchers. The presence of metals in some agricultural fertilizers raised fears that continued application of fertilizers may lead to accumulation of these metals to toxic levels in the soil for living organisms including plants. Prolonged fertilizers use is also a concern in sites where cut-flower industries are established in that it may affect neighboring small-scale lands.

Is Cut-Flower Industry Promotion by the Government Negatively Affecting Pollinator Biodiversity and
Environmental/Human Health in Uganda?

33

TABLE 2: Types of fertilizers and pesticides used by different companies.

Fertilizers applied	Quantity mixed in 1000–2000 liters of water
MAIRYE flower farm	
Macroelements	
Calcium nitrate	600 g
Potassium nitrate	350 g
Magnesium sulfate	350 g
Ammonium phosphate	150 g
Monopotassium phosphate	150 g
Microelements	
Iron	10 g
Zinc	5 g
Manganese	5 g
Boran	5 g
PEARL flower farm	
Macroelements	
Calcium nitrate	86 Kg
Potassium nitrate	77 Kg
Magnesium nitrate	27 Kg
Monopotassium phosphate	30 Kg
Urea	29 Kg
Microfeed	2.2 Kg
Librel	3.4 Kg
Trait-elements	
Nitric acid	15 Kg
Phosphorc acid	12 Kg
Zinc sulphate	0.2 Kg
Borate	0.2 Kg
FIDUGA flower farm	
Macroelements	
Lime	25 Kg
TSP (triple super phosphate)	2.5 Kg
Calcium Nitrate	3.5 Kg
N-P-K (12-10-18)	3.5 Kg
Magnesium sulfate	150 Kg

Type of pesticides applied	Quantity mixed per liter of water
MAIRYE flower farm	
Fungicides	
Ortiva	1 g/liter
Nimrod	2.5 g/liter
Meltatox	2.5 g/liter
Previcur	2 g/liter
Ridonil	2.5 g/liter
Daconil	2 g/liter
Ravaral	2 mL/liter

TABLE 2: Continued.

Insecticide	
Methomex	1 g/liter
Confidor	1.5 g/liter
Duduthrin	0.8 g/liter
Othane	1 g/liter
Miticide/Nematicides	
Nemacin	1.1 mL/liter
Vydate	1.5 mL/liter
Herbicide	
Lalad Master	2 mL/liter
Round-up	3 mL/liter
PEARL flower farm	
Insecticides	
Meltaton	2.5 mL/liter
Milraz	2 mL/liter
Equation	0.4 mL/liter
Teldor	1 g/liter
Roural	1.5 mL/liter
Score	0.8 mL/liter
Previcur	1.5 mL/liter
Acaricide-Miticide	
Silwet	0.5 mL/liter
Fedion	2 mL/liter
Floxelle	0.7 mL/liter
Nissuron	0.6 mL/liter

In these flower farms, spraying of pesticide is done manually by male workers provided with almost no means of protection apart from gumboot. There was a high variability in quantity of pesticides and fertilizers applied daily among the different flower farms visited (Tables 3 and 4). In the pesticide class, fungicides, miticide-nematicide, insecticides, and herbicides were applied by the different flower farms at different levels.

Types (herbicides, insecticides nematicides, miticides, and fungicides) and number of pesticide applications on flowers were retained as a surrogate variable for agricultural practice intensity inside the flower farm. Indeed, this variable reflected both the amount of inputs in field greenhouse (to increase flower productivity) and disturbance of environment and human health caused by each spray session by farming pumps or spray machines. Overall, the intensity of pesticide application (that combines both the types and number of pesticides applied per month) was considered as a proxy of flower production practice intensity. This surrogate has a potential indirect impact on pollinator communities living in habitats surrounding flower farms. The intensity of pesticide application by the flower farm was measured at four levels (very high: 4, high: 3, medium: 2, and low: 1). Based on quantity reported by the production mangers and based on field experiences (observations, field impressions), Fiduga flower farm was classified as the flower farm with *very high* level of

TABLE 3: General linear model (GLM) for exploring the effects of the flower farm type (location), the intensity of agrochemical (fertilizers/pesticides) applications per flower farm, landscape vegetation types, and the distance at which the sampling transect was set (spatial scale) in explaining the patterns of variability in bee biodiversity inhabiting habitats surrounding major flower farms found in central Uganda.

Responses	Fixed factors	df	F value	P value
Bee species richness	Intensity of agrochemical applications	(3,33)	1.93	0.142
	Flower farm name (location)	(3,33)	2.94	0.048
	Landscape vegetation types	(1,33)	4.30	**0.047**
	Transects	(4,33)	3.88	**0.012**
Bee abundances	Intensity of agrochemical applications	(3,33)	9.881	**0.000**
	Flower farm name (location)	(3,33)	13.94	**0.000**
	Landscape vegetation types	(1,33)	2.361	0.135
	Transects	(4,33)	4.07	**0.009**
Bee visitation frequency	Intensity of agrochemical applications	(3,33)	8.02	**0.000**
	Flower farm name (location)	(3,33)	11.12	**0.000**
	Landscape vegetation types	(1,33)	0.09	0.773
	Transects	(4,33)	4.42	**0.006**
Blooming plants abundance	Intensity of agrochemical applications	(3,33)	0.64	0.595
	Flower farm name (location)	(3,33)	1.69	0.190
	Landscape vegetation types	(1,33)	5.27	**0.029**
	Transects	(4,33)	13.80	**0.000**
Blooming plant species richness	Intensity of agrochemical applications	(3,33)	0.711	0.598
	Flower farm name (location)	(3,33)	0.778	0.541
	Landscape vegetation types	(1,33)	2.415	**0.018**
	Transects	(4,33)	3.331	**0.0016**
Nests and nesting sites density (abundance) on transects	Intensity of agrochemical applications	(3,33)	2.743	0.033
	Flower farm name (location)	(3,33)	2.568	0.044
	Landscape vegetation types	(1,33)	2.796	**0.027**
	Transects	(4,33)	23.12	**0.000**

intensity of pesticide application; and Rosebud-II was classified as with *high* level of intensity of pesticide application whereas Mairye was classified as flower farm with *medium* level of intensity of pesticide applications. The flower farm that was classified as with *low* level of intensity of pesticide applications was Pearl flower farm.

It was not clear whether the quantity of pesticides had an effect on pollinators living in the surrounding habitats, since most of the agrochemicals were applied inside greenhouses of the flower farms. However, daily application of pesticides cannot be without effect on the surrounding environment with its living organisms. In the short term, the effects may not be visible or perceptible. However, in the long-term, the effects may be visible/perceptible given the fact that different types of pesticides have different characteristics of persistence in the environment. The more pesticides were applied from one focal point, the more they would accumulate in the environment with consequences of disturbing/disorganizing natural and ecological systems.

The results indicated that the amount of agrochemicals spent so far per flower varied with the number of years since the flower farm was established (Figure 2). Among the 3 flower farms for which data was available, Mairye is likely to have applied 250,000 kg (m.a) of fungicides; 1200000 kg (m.a.) of insecticides; 65000 kg of nematicide-maticide; 35000 kg of herbicides; and approximately 200,000,000 liters of water for a total land in production of 18 ha. This amount of pesticide application with high environmental persistence cannot be with any consequence to the local environment. There is a need to choose to apply less toxic pesticides by the flower firms.

3.3. Precautionary Measures Taken by Owners of Flower Farms for Containing Chemical Runoff from Flowers into the Surrounding Environment. When asked about the strategies/measures taken by the flower farm to reduce negative effects of chemicals runoff into the environment, all production managers said they recycled and controlled well the quality of water before releasing it into the environment. Across all flower farms, greenhouses were set in such a way as to be always open to the outside; and this was suspected to have a great influence on pollinators and movement of pesticides into farmland habitats.

Is Cut-Flower Industry Promotion by the Government Negatively Affecting Pollinator Biodiversity and
Environmental/Human Health in Uganda?

35

(a)

(b)

(c)

(d)

FIGURE 1: Plate showing greenhouses, cows grazing near the greenhouse, and crops grown near greenhouses and near water reservoirs at
Pearl flower farm in Ntungsamo district, Western Uganda.

Production managers interviewed said management of
agrochemicals and effluent from flower farms were a major
concern by their companies. They also said that they were
aware that pesticides, fertilisers, and herbicides can pollute
river, lake, and wetland systems as a result of poor management of effluent from the flower farms and this constitutes
a threat to aquatic life like fish and human health. Hence,
they had to take measures that minimise soil and water pollution, such as constructing lagoons and planting papyrus to
perform water purification (artificial wetlands). For example,
production managers said that the water recycling system has
enabled them to reduce water needs from 50,000 liters of
water/day/ha to 13–20 liters/day/ha. However, during office
discussion with the production managers, a question was
asked about their perceptions/views in response to farmers'
complaints of chemicals sprayed in the flower farms affecting
their crops/livestock and their own health in the village
nearby, some production managers said that they have never
received complaints from the nearby communities, others
said they control perfectly chemical runoff, and therefore
accusations of farmers living nearby were not correct.

*3.4. Effects of the Types of Habitats Found in the Surrounding
Landscape of Flower Farms on Bee Abundance and Species
Richness and on Availability of Floral Resources.* Bee nests
density, species richness, abundance, and visitation frequency
to blooming plants in landscapes found in the villages surrounding the flower farms varied significantly (GLM test, $P <
0.05$) across transects and locations of the flower farms. They

also varied by the intensity of agrochemicals (pesticides, fertilizers) applications by the flower farms (Table 3) since flower
farms that used more agrochemicals were also involved in
regular (frequent) throwing (dumping) of agricultural wastes
(agrochemical wastes) in the grasslands/croplands in the
villages nearby the flower farms. However, the richness and
the abundance of blooming plant species were not significantly ($P < 0.05$) affected by the intensity of application of
agrochemicals (pesticides) nor by the flower farm location.
But they varied significantly ($P < 0.05$, GLM test) across transects and according to landscape types/vegetation type. This
result indicated that agrochemicals application during flower
production process did not affect the richness of blooming
plants in the neighborhood; few plant species managed to get
adapted to such environment. Adapted plants were seen to be
abundantly in bloom around the neighborhood of the flower
farm (Table 3) even when there were few bees visiting such
blossoms. Also, there was a significant positive correlation
($r = 0.56$, $P < 0.05$) between the intensity of pesticides application by the different flower farms and the monthly net
income from sales of cuttings.

Farmers living in the surrounding of different flower
farms are engaged in the cultivation of different types of crop
species in association. The average number of pollinator-
dependent crop species inventoried during the study survey
was 5.1 ± 0.9 (Fiduga), 3.1 ± 0.6 (Mairye), 2.3 ± 0.9 (Pearl),
and 4.6 ± 0.85 (Rosebud = Wagagai). The number of non-
pollinator-dependent crop species grown was 3.3 ± 0.35
(Fiduga), 1.9 ± 0.7 (Mairye), 2.2 ± 0.45 (Pearl), and 2.3 ± 1.12
(Rosebud) (Table 4).

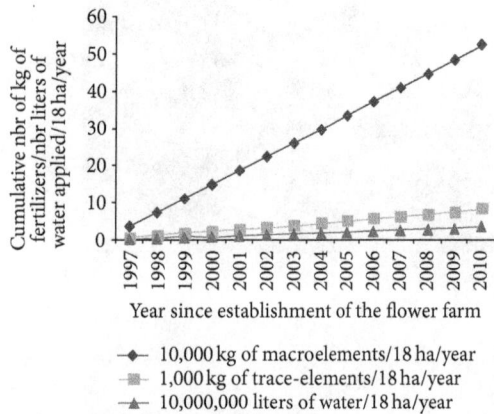

(a)

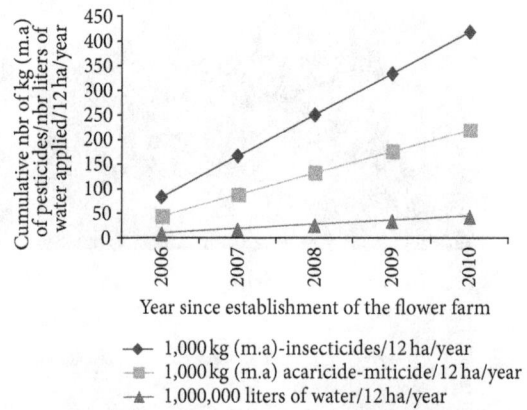

(d)

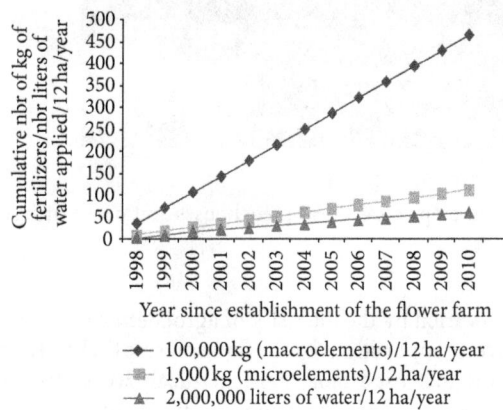

(e)

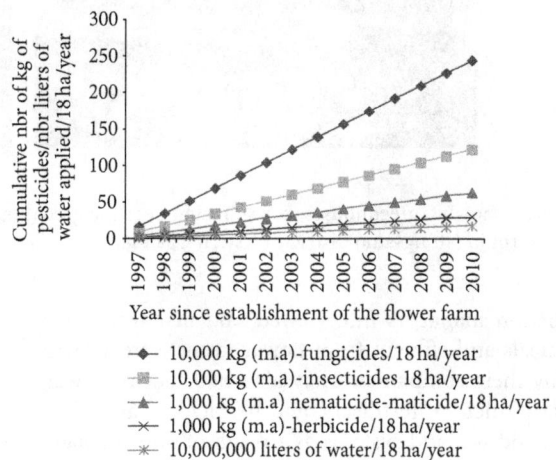

(b)

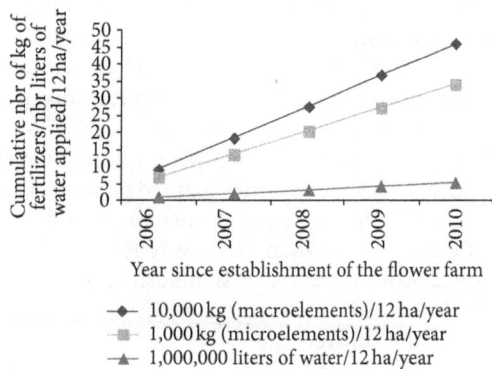

(c)

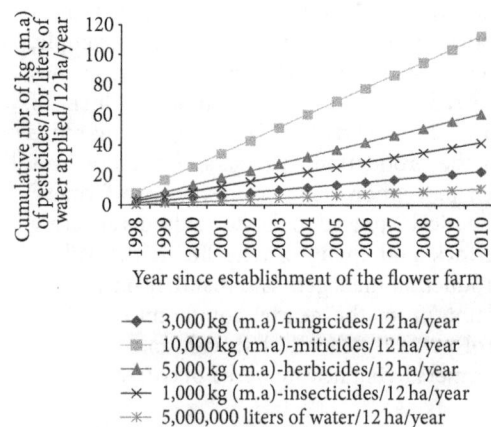

(f)

FIGURE 2: Long term application of (i) fertilizers ((a) = Mairye, (e) = Pearl, (c) = Rosebud-II) and (ii) pesticides ((d) = Mairye, (b) = Pearl, (f) = Rosebud-II) at different flower farms in Uganda. Table indicating the amount of each group of chemicals used at each farm is provided in the final year in Table 2.

Is Cut-Flower Industry Promotion by the Government Negatively Affecting Pollinator Biodiversity and
Environmental/Human Health in Uganda?

37

TABLE 4: Common crop associations practiced by farmers interviewed from 2010 to 2012.

Different types of crops grown in association by farmers	
Frequency of crop associations in non-pollinator-dependent crops	Freq. (%)
Cassava-Banana	14.06
Cassava-Banana-Yams-Taro	1.56
Banana-Maize-Millet	6.25
Banana-Sweetpotato-Cassava-Maize	3.13
Banana-Sweetpotato-Maize	7.81
Cassava-Maize-Sweetpotato-Banana	10.94
Cassava-Sweetpotato-Banana-Maize	3.13
Maize-Cassava-Sweetpotato-Banana	17.10
Sweetpotato-Banana	14.06
Sweetpotato-Banana-Cassava	9.38
Sweetpotato-Banana-Maize-Cabbage	10.94
Sweetpotato-Maize-Cassava	1.56
Frequeny of crop associations in pollinator-dependent crops	
Egg plant-Mango-Beans-Groundnut-Pepper-Pumpkin	3.03
Avocado-Mango-Pumpkin-Beans-Coffee	15.15
Avocado-Pumpkin-Beans-Tomato-Coffee	12.12
Beans-Egg plants-Tomato	3.03
Beans-Avocado-Mango	22.73
Beans-Groundnut-Pumpkin	6.06
Beans-Papaw-Watermelon-Pumpkin-Egg plant-Tomato	6.06
Beans-Tomato-Avocado-Papaya	6.06
Citrus-Papaw-Beans-Cowpea-Avocado-Mango	1.52
Coffee-Avocado-Passion fruit-Vanilla	1.52
Coffee-Tomato-Beans-Papaw-Vanilla	9.09
Pumpkin-Beans-Watermelon-Egg plant	7.58

In the surrounding of each flowering farm, there were significant ($P < 0.05$) differences in species richness, abundance, visitation frequency, and number of fresh flowers per transect at different distances (0 m, 10 m, 50 m, 500 m, 1500 m) far away from the flower farm into farmland habitats. For example, for Pearl flower farm, there were significant differences between the 5 distances (0 m, 10 m, 50 m, 500 m, 1500 m, 2000 m) in species richness (GLM: $F_{1,4} = 3.14$, $P < 0.001$), abundance (GLM: $F_{1,4} = 7.85$, $P < 0.001$), bee visitation frequency (GLM: $F_{1,4} = 9.43$, $P < 0.001$), and abundance of fresh flowers (GLM: $F_{1,4} = 12.76$, $P < 0.001$) (Figure 3). Similar trends in the results were observed at Mairye, Rosebud, and Fiduga flower farms (Figure 3).

In the surrounding of each flower farm and across sampling rounds (R1, R2, R3, R4), there were significant differences in the species richness and abundance of bees. In fact, for Pearl flower farm, there were significant differences in species richness of blooming plants in croplands (GLM: $F_{1,3} = 3.14$, $P < 0.001$) as well as in farmland habitats (GLM: $F_{1,3} = 7.14$, $P < 0.001$). Results of similar trends were observed for the percentage cover of mass blooming plants/ transects in both croplands (GLM: $F_{1,3} = 12.3$, $P < 0.001$) and grassland habitats (GLM: $F_{1,4} = 9.13$, $P < 0.001$; Figure 4).

The abundance of bees varied significantly ($P < 0.05$, χ^2 test) across the different land-use types and seminatural habitats encountered on transects during transect counts of bees across the surroundings of the four flower farms (Table 5). The most visited land-use types were the banana + bean + cassava, followed by young fallows, field margins, and coffee + banana + cassava + beans + fruit trees + agroforestry trees. The most visited seminatural habitat among those encountered during transect counts of bees in grassland and rangeland habitats was "shrubby fallow", followed by unfenced grazing plot followed by pad-docking fenced grazing plot with live fence, unreclaimed papyrus swampy habitat, and hedgerow (Table 5). Different bee species made visits to blooming plants in these different habitats/land-uses at different periods of the day; most frequently, they made intense visits from 10:30 h to 15:30 h. The spray of chemicals in flower farms is generally conducted between 6:30 h and 8:30 h or between 15:30 h and 18:30 h.

Different bee species were recorded significantly (χ^2 3 df = 7.87, $P < 0.001$) in the different cropland and grassland habitats in the surrounding of the flower farms. They occurred with different abundance. In total: 37, 26, 45, and 33 bee species were recorded in the surrounding of respectively, Fiduga, Mairye, Pearl, and Wagagai (Rosebud-II) flower farms. Landscape habitats surrounding Pearl flower supported more diverse bee communities (mean ± SE *of H'* = 2.68 ± 0.13) than Mairye (mean ± SE *of H'* = 2.53 ± 0.13), Fiduga (mean ± SE *of H'* = 2.17±0.11), and Wagagai (mean ± SE *of H'* = 1.91±0.09) flower farms (GLM-ANOVA: $F_{(4,37)}$ = 2.55, $P = 0.041$). However, there were no significant differences ($\chi^2 = 3.24$, $P = 0.198$, df = 3) among flower farms in the average similarity index values of shared bee species. In other words, bee communities from these flowers were statistically similar in species composition. Common bee species were frequently recorded to be abundant on flowers than specialist bees. In the tropics, the structure of most bee communities is not different from that of other insects. There is always 1–5% of dominant bee species and 90–95% of species that are rare or appear as singletons or doubletons.

There was a significance (χ^2 test, $P < 0.05$) in relative abundance of different bee species in the surrounding of different flower farms. The most abundant bee species in the surrounding habitats (croplands, grasslands) were *Apis mellifera adansonii* (14.08%), followed by *Meliponula ferruginea* (10.90%) and *Nomia brevipes* (9.03%). The most abundant species in the farmland around Mairye were *Apis mellifera adansonii* (23.3%), followed by *Meliponula ferruginea* (19.84%) and *Apis mellifera scutellata* (8.67%). The most abundant species in the surrounding of Pearl flower farm were *Apis mellifera adansonii* (23.67%), followed by *Apis mellifera scutellata* (10.64%) and *Ceratina tanganyicensis* (8.58%).

TABLE 5: Relative abundance of bees (%) per habitat types encountered during transect counts across the surroundings of the 4 flower farms investigated from 2010 to 2012 in Uganda.

(a)

Field types	Land-use types encountered during bee transect counts in croplands	Relative abundance (%)
Cultivated fields	Abandoned bushy gardens	0.1
Cultivated fields	Agroforestry trees + Banana + Beans + Maize + Fruit trees	3.3
Cultivated fields	Banana + Bean + Cassava crops	29.7
Cultivated fields	Banana + Beans + Sweetpotato + Agroforestry trees + Fruit trees	4.8
Cultivated fields	Banana + Cassava + Beans + Maize	0.9
Cultivated fields	Banana + Cassava + Beans + Sweetpotato + Maize + Mango + Avocado	3.3
Cultivated fields	Banana + Cassava + Groundnut + Fruit trees	0.8
Cultivated fields	Banana + Coffee + Beans + Maize + Mango + Avocado	1.7
Cultivated fields	Banana + Egg plant + Tomato + Bean + Sweetpotato	1.6
Cultivated fields	Banana + Groundnut + Casssava + Maize + Agroforestry trees	2.8
Cultivated fields	Banana + Maize + Beans + Sweetpotato + Fruit trees	2.2
Cultivated fields	Coffee + Banana + Cassava + Beans + Fruit trees	2.0
Cultivated fields	Coffee-Banana-Cassava-Beans–Fruit trees-Agroforestry trees	7.1
Cultivated fields	Coffee + Banana + Cassava + Groundnut + Maize + Fruit trees	0.5
Cultivated fields	Fruits orchards + Homegardens	1.2
Cultivated fields	Communal grazing fields	1.3
Cultivated fields	Field margins	9.3
Cultivated fields	Hedgerows (established in farmlands)	6.8
Cultivated fields	Swampy crops (Egg plant + Pumpkin + Watermelon + Tomato, etc.)	6.4
Cultivated fields	Herbaceous young fallows	15.8

Chi-square test: $\chi^2 = 179.61$, $N = 100$, DF = 19, $P < 0.0001$.

(b)

Field types	Semi-natural habitat types encountered during transect-counts of bees in grasslands/rangelands	Relative abundance (%)
Uncultivated fields	Edge of natural wetlands (various plant species mixed)	0.22
Uncultivated fields	Forest fallow (*Vernonia* sp.)	1.70
Uncultivated fields	Forest fallows	1.00
Uncultivated fields	Forested-swampy habitats	2.21
Uncultivated fields	Grassy fallows	1.50
Uncultivated fields	Hedgerows (established rangelands)	4.42
Uncultivated fields	Herbaceous/shrubby grasslands	1.43
Uncultivated fields	Lake Victoria-edge vegetation (various species)	5.88
Uncultivated fields	*Lantana camara-Erlangeya tomentosa* old fallows	0.83
Uncultivated fields	*Lantana camara-Erythrina* sp. hedges on roadsides	10.07
Uncultivated fields	Natural forest remnant patches	0.44
Uncultivated fields	Old fallows (*Lantana camara-Cassia* sp.-*Maesa lancelota*)	1.12
Uncultivated fields	Paddocking fenced grazing plots (with live fences)	12.64
Uncultivated fields	Reclaimed wetlands (with crops + natural vegetations)	0.99
Uncultivated fields	Shrubby old fallows	25.29
Uncultivated fields	Swampy habitat vegetation (various plant species mixed)	3.66
Uncultivated fields	Unfenced grazing plots	19.97
Uncultivated fields	Unreclaimed-papyrus swampy habitats	4.60
Uncultivated fields	Woodlands (*Euphorbia* sp.)	0.15
Uncultivated fields	Woodlots (*Eucalyptus*-Pine)	0.46
Uncultivated fields	Woodlots (native and exotic species mixed)	1.42

Chi-square test: $\chi^2 = 195.43$, $N = 100$, DF = 20, $P < 0.0001$.

Is Cut-Flower Industry Promotion by the Government Negatively Affecting Pollinator Biodiversity and
Environmental/Human Health in Uganda?

39

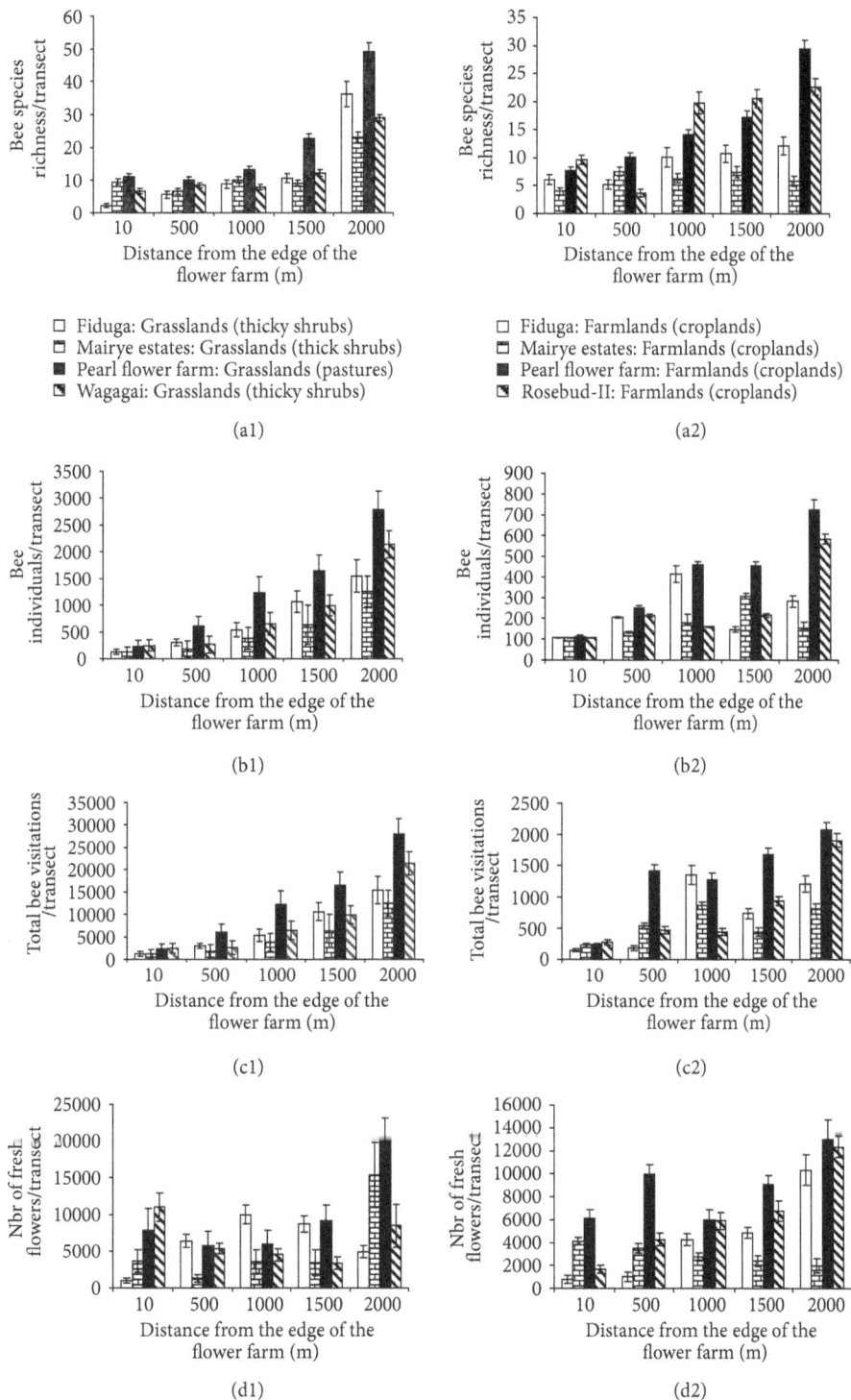

FIGURE 3: (i) Variability in bee species richness (a1), bee abundance (b2), total visitation frequency (c1) and in number of fresh flowers (d1) countable per transect with the distance way from the edge of the flower farm into grassland habitat; (ii) variability in bee species richness (a2), bee abundance (b2), total visitation frequency (c2) and in Number of fresh flowers (d2) countable per transect with the distance way from the edge of the flower farm into farmland (cropland) habitat. [The data is the total from 10 focal observation plots (25 m^2) per transect (20 m × 1 km). For each sampling plot, data recorded was then number of bee-visits (number of bee coming and landing on the flowers)/20 min observations/50–500 flower patches/25 m^2). The flower patch was composed of diverse flowering plants.] Concerning the pattern (trend) in the raw data collected, it was observed that across the five distances (10, 500, 1000, 1500, 2000 m), there were significant differences (Tukey test at $P < 0.05$) in the number of bee individuals/species richness/transect and in the number of total bee visits/number of fresh flowers/transect. Either in cropland or in grassland habitats, within each distance, overlapping error bars indicate no significant differences (Tukey test at $P > 0.05$) between consecutive flower farms; Error bars are ± SE.

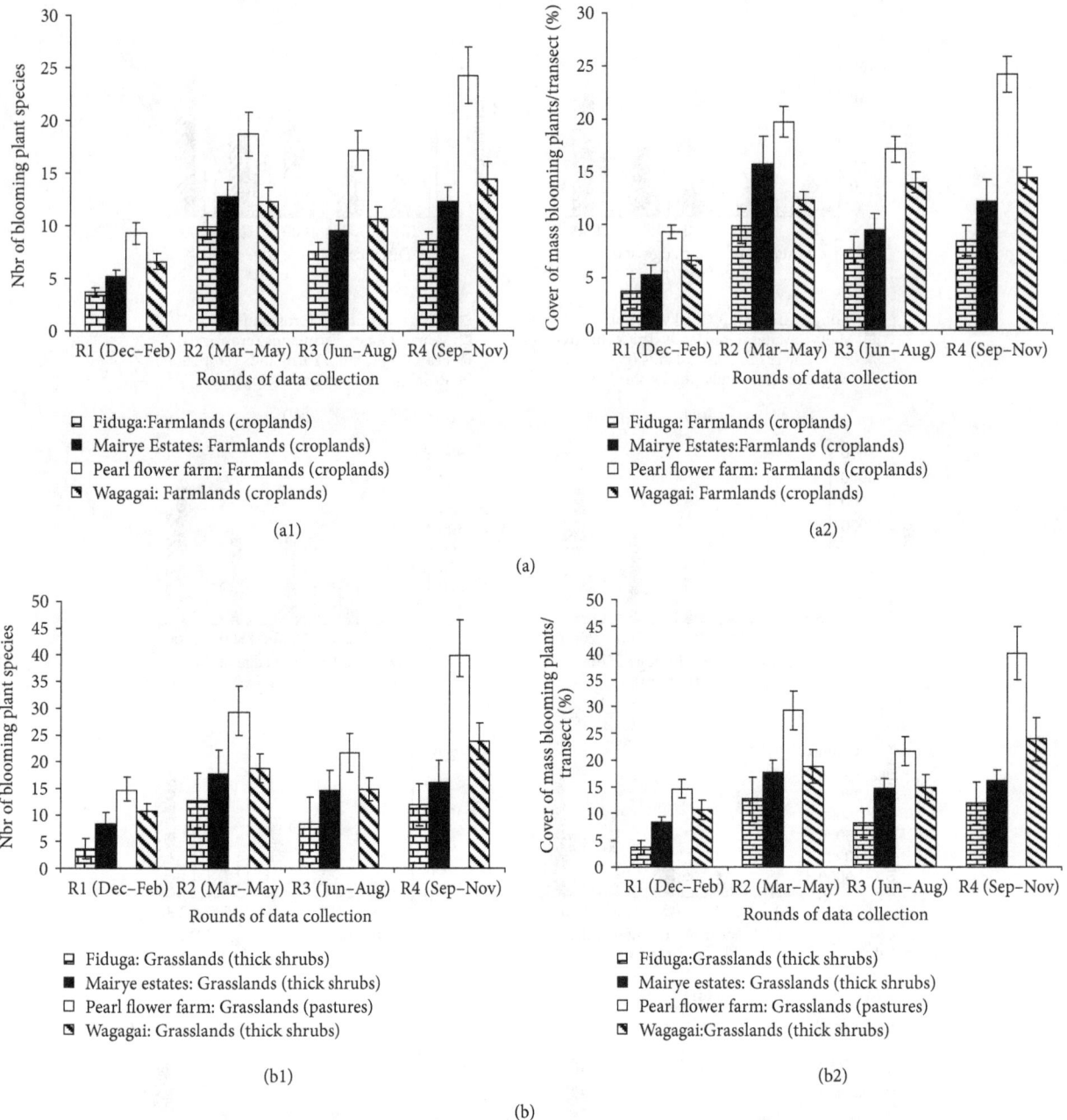

(a1)

(a2)

(a)

(b1)

(b2)

(b)

FIGURE 4: (i) Variability in the richness of blooming plant species (a1) and in the abundance of blooming plants (a2) in cropland habitats (a) during each of the four rounds of data collection conducted across the four flower farms (data are means of five transects); (ii) variability in the richness of blooming plant species (a1) and in the abundance of blooming plants (a2) in grassland habitats (b) during each of the four rounds of data collection conducted across the four flower farms (data are means of five transects) during four rounds of data collection from 2010 to 2012. Concerning the pattern (trend) in the raw data collected, it was observed that across the four sampling rounds (R1 to R4), there were significant differences (Tukey test at $P < 0.05$) in the number of blooming plant species and in the % cover of mass blooming plants/transect. Either in cropland or in grassland habitats, within each sampling round, overlapping error bars indicate no significant (Tukey test at $P > 0.05$) differences between consecutives flower farms; Error bars are ± SE.

The most abundant species in the surrounding habitats of Wagagai (Rosebud-II) were *Apis mellifera adnasonii* (37.25%), followed by Apis *mellifera scutelatta* (15.7%) and *Halictus orientalis* (9.59%) (Table 6).

Overall, high bee species richness was associated with different habitats (land-uses) found around Pearl flower farm,

probably because the flower farm was young (recently established). There may be little accumulation in the environment of chemicals applied at Pearl flower farm to affect bee populations. The majority of bee species recorded were characterized by different ecological requirements. They belonged to different functional groups, but on overall most species

Is Cut-Flower Industry Promotion by the Government Negatively Affecting Pollinator Biodiversity and
Environmental/Human Health in Uganda?

41

TABLE 6: List of bee species recorded in different habitats around flower farms from 2010 to 2012 (data include relative abundance of the species over the five rounds of data collection conducted on five transect from 2010 to 2012. The abundance is the average visitation frequency (number of individual bee-visits counted on flowers patches met while walking along transect). The data are means of data collected in transects set in farmlands and in grasslands.

Name of the flower farm	Bee species recorded around	Relative abundance (%) (n = 3638 individuals)
Fiduga	*Apis mellifera adansonii* (Linnaeus, 1758)	14.08
Fiduga	*Meliponula ferruginea* (Lepeletier, 1836)	10.90
Fiduga	*Nomia brevipes* (Friese, 1914)	9.03
Fiduga	*Lassioglossum kampalense* (Cockerell, 1945)	6.92
Fiduga	*Scrapter armatipes* (Friese, 1913)	5.62
Fiduga	*Ceratina rufigastra* (Cockerell, 1937)	5.13
Fiduga	*Halictus jucundus* (Smith, 1853)	4.98
Fiduga	*Apis mellifera scutellata* (Latreille, 1804)	3.69
Fiduga	*Ceratina whiteheadi* (Eardley and Daly, 2007)	3.47
Fiduga	*Pseudoheriades moricei* (Friese, 1897)	3.19
Fiduga	*Patellapis flavorufa* (Cockerell, 1937)	3.11
Fiduga	*Megachile torrida* (Smith, 1853)	2.64
Fiduga	*Megachile rufipennis* (Fabricius, 1793)	2.48
Fiduga	*Anthophora armata* (Friese, 1905)	2.30
Fiduga	*Xylocopa inconstans* (Smith, 1874)	2.09
Fiduga	*Ctenoplectra ugandica* (Cockerell, 1944)	2.03
Fiduga	*Scrapter flavipes* (Friese, 1925)	1.95
Fiduga	*Scrapter flavostictus* (Cockerell, 1934)	1.55
Fiduga	*Braunsapis angolensis* (Cockerell, 1933)	1.49
Fiduga	*Megachile rufiventris* (Guérin-Méneville, 1834)	1.37
Fiduga	*Xylocopa caffra* (Linnaeus, 1767)	1.37
Fiduga	*Ceratina tanganyicensis* (Strand, 1911)	1.19
Fiduga	*Amegilla calens* (Lepeletier, 1841)	1.18
Fiduga	*Nomia bouyssoui* (Vachal, 1903)	1.14
Fiduga	*Patellapis neavei* (Cockerell, 1946)	1.12
Fiduga	*Amegilla velutina* (Friese, 1909)	1.02
Fiduga	*Tetraloniella braunsiana* (Friese, 1905)	0.77
Fiduga	*Nomia atripes* (Friese, 1909)	0.68
Fiduga	*Tetralonia macrognatha* (Gerstäcker, 1870)	0.60
Fiduga	*Tetralonia caudata* (Friese, 1905)	0.59
Fiduga	*Lipotriches dentipes* (Friese, 1930)	0.57
Fiduga	*Megachile acraensis* (Friese, 1903)	0.52
Fiduga	*Lasioglossum ugandicum* (Cockerell, 1937)	0.43
Fiduga	*Patellapis vittata* (Smith, 1853)	0.42
Fiduga	*Lithurgus rufipes* (Smith, 1853)	0.39
Name of the flower farm	Bee species recorded around	Relative abundance (%) (n = 4492 individuals)
Mairye Estates	*Apis mellifera adansonii* (Linnaeus, 1758)	22.30
Mairye Estates	*Meliponula ferruginea* (Lepeletier, 1836)	19.84
Mairye Estates	*Apis mellifera scutellata* (Latreille, 1804)	8.67
Mairye Estates	*Halictus jucundus* (Smith, 1853)	4.44
Mairye Estates	*Ceratina braunsi* (Eardley and Daly, 2007)	4.25
Mairye Estates	*Amegilla calens* (Lepeletier, 1841)	4.17
Mairye Estates	*Patellapis terminalis* (Smith, 1853)	4.17
Mairye Estates	*Patellapis schultzei* (Friese, 1909)	4.12
Mairye Estates	*Lithurgus pullatus* (Vachal, 1903)	3.75
Mairye Estates	*Braunsapis angolensis* (Cockerell, 1933)	3.24

TABLE 6: Continued.

Mairye Estates	*Nomia bouyssoui* (Vachal, 1903)	3.21
Mairye Estates	*Lassioglossum kampalense* (Cockerell, 1945)	2.60
Mairye Estates	*Nomia rozeni* (Pauly, 2000)	1.79
Mairye Estates	*Xylocopa caffra* (Linnaeus, 1767)	1.57
Mairye Estates	*Tetraloniella junodi* (Friese, 1909)	1.35
Mairye Estates	*Allodape friesei* (Strand, 1915)	1.28
Mairye Estates	*Nomia lutea* (Warncke, 1976)	1.23
Mairye Estates	*Scrapter algoensis* (Friese, 1925)	1.20
Mairye Estates	*Megachile hopilitis* (Vachal, 1903)	1.01
Mairye Estates	*Tetralonia caudata* (Friese, 1905)	1.01
Mairye Estates	*Stenoheriades braunsi* (Cockerell, 1932)	0.91
Mairye Estates	*Megachile torrida* (Smith, 1853)	0.78
Mairye Estates	*Meliponula nebulata* (Smith, 1854)	0.71
Mairye Estates	*Megachile rufipennis* (Fabricius, 1793)	0.79
Mairye Estates	*Ctenoplectra armata* (Magretti, 1895)	0.64
Mairye Estates	*Amegilla acraensis* (Fabricius, 1793)	0.61
Mairye Estates	*Thyreus bouyssoui* (Vachal, 1903)	0.59

Name of the flower farm	Bee species recorded around	Relative abundance (%) (n = 12454 individuals)
Pearl	*Apis mellifera adansonii* (Linnaeus, 1758)	23.67
Pearl	*Apis mellifera scutellata* (Latreille, 1804)	10.64
Pearl	*Ceratina tanganyicensis* (Strand, 1911)	8.58
Pearl	*Ceratina rufigastra* (Cockerell, 1937)	4.12
Pearl	*Nomia senticosa* (Vachal, 1897)	5.75
Pearl	*Nomia theryi* (Gribodo, 1894)	5.43
Pearl	*Ceratina lineola* (Vachal, 1903)	3.33
Pearl	*Halictus jucundus* (Smith, 1853)	2.57
Pearl	*Halictus niveocinctulus* (Cockerell, 1940)	2.24
Pearl	*Amegilla velutina* (Friese, 1909)	2.20
Pearl	*Lasioglossum aethiopicum* (Cameron, 1905)	2.00
Pearl	*Meliponula ferruginea* (Lepeletier, 1836)	1.80
Pearl	*Braunsapis angolensis* (Cockerell, 1933)	1.79
Pearl	*Anthophora armata* (Friese, 1905)	1.73
Pearl	*Megachile rufiventris* (Guérin-Méneville, 1834)	1.68
Pearl	*Nomia bouyssoui* (Vachal, 1903)	1.55
Pearl	*Braunsapis facialis* (Gerstäcker, 1857)	1.39
Pearl	*Megachile junodi* (Friese, 1904)	1.18
Pearl	*Patellapis vittata* (Smith, 1853)	1.18
Pearl	*Megachile felina* (Gerstäcker, 1857)	1.15
Pearl	*Megachile eurymera* (Smith, 1864)	1.03
Pearl	*Lipotriches digitata* (Friese, 1909)	1.02
Pearl	*Andrena africana* (Friese, 1909)	1.01
Pearl	*Patellapis disposita* (Cameron, 1905)	0.99
Pearl	*Allodape stellarum* (Cockerell, 1916)	0.96
Pearl	*Afromelecta bicuspis* (Stadelmann, 1898)	0.95
Pearl	*Megachile niveofasciata* (Friese, 1904)	0.79
Pearl	*Amegilla acraensis* (Fabricius, 1793)	0.77
Pearl	*Thyreus somalicus* (Strand, 1911)	0.73
Pearl	*Megachile rufipennis* (Fabricius, 1793)	0.70
Pearl	*Lithurgus rufipes* (Smith, 1853)	0.67
Pearl	*Macrogalea candida* (Smith, 1879)	0.63
Pearl	*Megachile nasalis* (Smith, 1879)	0.63

Is Cut-Flower Industry Promotion by the Government Negatively Affecting Pollinator Biodiversity and
Environmental/Human Health in Uganda?

43

TABLE 6: Continued.

Pearl	*Halictus obscurifrons* (Cockerell, 1945)	0.59
Pearl	*Lipotriches tanganyicensis* (Strand, 1913)	0.59
Pearl	*Melitta whiteheadi* (Eardley, 2006)	0.58
Pearl	*Plebeiella lendliana* (Friese, 1900)	0.51
Pearl	*Melitta arrogans* (Smith, 1879)	0.31
Pearl	*Amegilla niveata* (Friese, 1905)	0.27
Pearl	*Melitta danae* (Eardley, 2006)	0.27
Pearl	*Megachile natalica* (Cockerell, 1920)	0.26
Pearl	*Amegilla rufipes* (Lepeletier, 1841)	0.26
Pearl	*Plebeina hildebrandti* (Friese, 1900)	0.23
Pearl	*Allodape friesei* (Strand, 1915)	0.23
Pearl	*Ctenoplectra terminalis* (Smith, 1879)	0.20
Pearl	*Ceratina viridis* (Guérin-Méneville, 1844)	0.18
Pearl	*Nomia rozeni* (Pauly, 2000)	0.18
Pearl	*Tetralonia macrognatha* (Gerstäcker, 1870)	0.18
Pearl	*Xylocopa caffra* (Linnaeus, 1767)	0.18
Pearl	*Megachile torrida* (Smith, 1853)	0.11
Name of the flower farm	Bee species recorded around	Relative abundance (%) (n = 11038 individuals)
Wagagai	*Apis mellifera adansonii* (Linnaeus, 1758)	37.25
Wagagai	*Apis mellifera scutellata* (Latreille, 1804)	15.76
Wagagai	*Halictus orientalis* (Lepeletier, 1841)	9.59
Wagagai	*Meliponula ferruginea* (Lepeletier, 1836)	6.63
Wagagai	*Hypotrigona gribodoi* (Magretti, 1884)	8.23
Wagagai	*Braunsapis bouyssoui* (Vachal, 1903)	3.04
Wagagai	*Ceratina tanganyicensis* (Strand, 1911)	2.63
Wagagai	*Nomia atripes* (Friese, 1909)	1.89
Wagagai	*Allodape friesei* (Strand, 1915)	1.88
Wagagai	*Allodape ceratinoides* (Gribodo, 1884)	1.81
Wagagai	*Allodapula acutigera* (Cockerell, 1936)	1.78
Wagagai	*Braunsapis angolensis* (Cockerell, 1933)	1.70
Wagagai	*Nomia bouyssoui* (Vachal, 1903)	1.51
Wagagai	*Tetralonia caudata* (Friese, 1905)	0.84
Wagagai	*Megachile eurymera* (Smith, 1864)	0.74
Wagagai	*Patellapis disposita* (Cameron, 1905)	0.64
Wagagai	*Megachile rufipes* (Fabricius, 1781)	0.63
Wagagai	*Scrapter whiteheadi* (Eardley, 1996)	0.64
Wagagai	*Thyreus neavei* (Cockerell, 1933)	0.55
Wagagai	*Macrogalea candida* (Smith, 1879)	0.54
Wagagai	*Lasioglossum ugandicum* (Cockerell, 1937)	0.45
Wagagai	*Amegilla calens* (Lepeletier, 1841)	0.40
Wagagai	*Tetralonia macrognatha* (Gerstäcker, 1870)	0.31
Wagagai	*Megachile rufipennis* (Fabricius, 1793)	0.30
Wagagai	*Xylocopa inconstans* (Smith, 1874)	0.25

recorded were solitary, polylectic, multivoltine, and ground nesting bees. However, high population density was observed in the less rich functional groups of species: social bees (Apini, Meliponini). In addition, a high number of nesting sites and nests was counted for various solitary bee species in the landscapes. On average, nest density (10.98 to 183.91 nests/transect) was high in rangelands/pasturelands than in agroforestry landscapes around Pearl flower farm. Few (2.5 to 5.89/transect) stingless bee nests were counted in croplands, indicating the fact that most managed and wild bee species found in the surrounding of flower farms used natural and seminatural habitats as preferential nesting sites (reservoirs). However, the different bee species used different foraging habitat types. While walking in croplands, "banana + beans-cassava" and "young fallows" were found to harbor a high number of bee foragers, whereas frequency of visitations by

TABLE 7: Social, economic, environmental, and pollination attitudes of farmers towards activities of bees on flowers of their gardens.

Statement no.	Question: What do you think bees are doing while visiting crop flowers?	Freq. (%) of the statements among interviewed people
1	Bees come and urinate on flowers of my crops; that urine makes fruits/seeds to come, a blessing from God carried out by bees	4.6
2	Bees come to facilitate marriage of my crops	0.8
3	Bees and other pollinators facilitate the marriage of flowers	2.3
4	Bees are just visiting my crops to get nectar only	1.5
5	Bees bring wild pollen ("Ngwaso"), drop them onto flowers to enable flower to open and give fruits/seeds	3.8
6	Bees come to drink water and collect other foods I do not know	11.5
7	Bees come to my crops to do the natural job of fertilizing my crops as God created them for that job	5.4
8	Bees come, bust, and later fertilize the flowers of my crops	3.1
9	Bees comes to take pollen/nectar from crops to their hives because they are pests (thieves), and of what they take, my crop miss it	0.8
10	Bees fall on flower of my crops, eating from there and later I get fruits/seeds	1.68
11	Bees fertilize my crops by transporting pollen between male and female flowers	3.1
12	Bees fertilize my crops, but wind can also fertilize my crops if bees do not come, so I am not bothered	5.4
13	Bees get nectar from my crops, they help us to get better yield and help themselves	2.3
14	Bees just come to drink juice from my crops but they do nothing good or bad on my crops	2.3
15	Bees just take pollen ("ebivu") from flowers, go away, and disseminate these pollen on other plants in forests	8.5
16	Bees must visit flowers of my crops to get good yield (seeds/fruits): rules of Nature and God	0.8
17	Bees pollinate on my crop for good growth and good seeds/fruit	1.5
18	Bees provide a two-way benefit (collect nectar for their hives and bring pollen to my crops)	0.8
19	I do not know (I cannot tell)	1.5
20	I do not know but was told in a workshop that bees have to pollinate my crops for good yield	1.5
21	I see bees visiting flowers of my crops but I do not know what they are after on my crops	2.3
22	I still do not know/understand what bees bring to my flowers to enable me to get better fruits/seeds	0.8
	If bees fall on flowers of my crops, I will see fruits/seeds coming in 1 to 3 weeks	14.6
	Just playing (they do nothing valuable)	3.8
	They pollinate most of the crops I grow	5.4
	They come to fertilize crop flowers	3.8

Chi-square test for difference in knowledge of what bees are after on crop flowers: $\chi^2 = 85.92$, $N = 93.8$, DF = 26, $P < 0.0001$.

individual bees belonging to different bee species was intense in old and bushy fallows and in hedgerows. This indicated that the conservation of bees in the flower producing zones has to involve the conservation of seminatural habitats (hedgerows, fallows) in the surrounding of flower farms. It may be relevant to say here that both flower farm managers and small-scale farms should be sensitized about the value of conserving seminatural habitats for the maintenance of pollinators in the habitats surrounding their flower farms.

3.5. Farmers' Surveys Results

3.5.1. Characteristics of Interviewed Farmers. Several farmers from Baganda, Bakiga, and Banyankore tribes were interviewed. Across flower farm location, the majority of respondents were females (62%), aged between 35–60 years. The main lucrative activity of these farmers was crop production, although in Ntungamo, farmers interviewed were cattle keepers (cattle keeping being the main lucrative activity and crop production being the secondary subsistence activity). The majority of small-scale farmers interviewed had a total land allocated to crop production of 0.1 to 10 ha maximum. The majority of these farmers hired or paid 1 to 2 workers and this result indicated that they had almost no labour cost. Interviewed farmers did grow various crop species in association. Most frequently, it was common to find a mixture of pollinator-dependent crops with non-pollinator-dependent crops (Table 7).

3.5.2. Farmers' Knowledge of Pollination, Pollinator Groups, Pollination Processes, and Value of Pollinating Services to Their Crops. The percentage of farmers understanding the word pollination (those knowing different pollinators of their crops) increased and was significantly ($P < 0.05$) positively related to (a) the education level of the farmer (number

Is Cut-Flower Industry Promotion by the Government Negatively Affecting Pollinator Biodiversity and Environmental/Human Health in Uganda?

45

TABLE 8: Perception of the importance of pollinators by bees in farming business.

Statement no.	Question: Why do you think crop pollination by bees is important in your farming business?	Freq. (%)
1	Bees are important for honey production and for pollinating flowers of my crops	1.48
2	I know bees collect only their nectar but I am not sure if they do anything beneficial to my crops	1.48
3	I know no bee-visits, no yields. Please teach me how to make a home for them near my crops	0.74
4	I was told by extension workers and neighbours to care for bees because if no bee-visits, no fruit/seeds I will harvest and sell, yet I need money to pay school fees	8.15
5	If bees fall on flowers of my crops (avocado), I see a fruit coming; cropping is my business and bees help me free	0.74
6	Bee visitation to coffee has no value; they add nothing, whether they come or not, my coffee will set fruits	0.74
7	Bees are important because if no bee-visits, no yields I get at all	7.41
8	Bees are important to get honey and also if no bee-visits, no fruit/seed I get from most of my crops	8.15
9	Bees are important because I have seen that no bee-visits, no fruits on all crops I grow	2.22
10	Bees are important and visit my commercial and vegetable crops; but I ask myself a question, where do they sleep?	2.96
11	Bees are important for my pulse crops, I need to know how to build a small house for them so that they stay near	0.74
12	When building a house for wild bees, how should it look? Like a beehive or like a rat hole?	0.74
13	Bees are important because all flowering plants/crops on this earth want bee-visits to perpetuate well	2.96
14	Bees are important for honey production and for pollinating flowers of crops I feed on and make money from	0.74
15	Bees are important for honey production and for pollinating flowers of my crops, especially crops wanted at market	1.48
16	Bees are important for pollinating our crops and they provide honey, wax, propolis, candle	5.19
17	Bees are important to help make my crops grow and get better yield;they do free work	2.22
18	Bees get nectar from my crops, but they bring manure from the hive to make turn into seeds/fruits	5.19
19	By field experience, I know no bee-visits, all flowers will die off, no fruits/seeds will come out	1.48
20	Either bees visit my beans/peas or not, I will still get my 5 to 12 pods per plant (bees have no value to my crops)	0.74
21	Either bees visit my beans or not, I will still get my 5 to 12 pods per plant (bees have no value to my crops)	1.48
22	Every time bees fall on flowers of my crops (avocado, mango), few days later, I see a beautiful fruit/seed/pod coming; no hunger at home in the future	0.74
23	If there are few or no visits of my crop flowers, I get no or poor yield	0.74
24	For most fruit crops (mango, avocado, papaw) I grow, they have to get pollinated by bees to get sweet fruits and high yields competitive in the markets	0.74
25	From my observations if no bees visit flowers in the village abundantly, no pods on my bean plants	0.74
26	I know without bee-visitations to flowers I can not harvest. However, I do not know where they come from or sleep; with bee-visits, I get fruits most wanted at market by my customers	2.96
27	I know, no bees, no harvest. However, I do not know how to rear and make them many in my crops	2.22
28	I am not sure if bees are important in crop production business (for my crops)	11.11
29	I know from my grandparents that no bees (pollinators) visits, no single mango fruit I will get acceptable at market	0.74
30	I know they are important but can not explain why it is only after their visit that I see fruits/seeds coming?	2.96
31	I see bees visit my coffee flowers but I do not know for which purpose apart from feeding there	0.74
32	I see bees visit my avocado/coffee flowers but I donot know what do they bring to my flowers to get exciting fruits/seeds	2.22
33	I was told by my neighbours by extension service agents that if no bee-visits to my crops, no yield at all but I can not explain why	0.74
34	If bees fall on flowers of my crops (avocado), I see a fruit coming later and that fruit is wanted at market and if children feed on it, they grow well because it is rich in vitamins, hence no stunting of children in my family	1.48
35	If no bee-visits, no yields; however, I do not know where they come from (sleep)	0.74
36	Important because all the crops I grow need bee-visitations, even flowers of my cassava	5.19
37	My crops have to get high bee-visitations, otherwise they will not yield properly well	2.96
38	No mango fruits or bean seeds I get if no abundant and diverse bees flying around in the village, if you feed on fruits not visited by bees, you fall ill	1.48

TABLE 8: Continued.

Statement no.	Question: Why do you think crop pollination by bees is important in your farming business?	Freq. (%)
39	The more bees fall on my (avocado/watermelon/pumpkin/tomato) flowers, the more fruits I get in my bag Fruits/seeds that were obtained after bee-visits last longer and are attractive to customers at local market	1.48
40	The more bees fall on my crop flowers, the more fruits/seeds I get	1.48
41	Without bees, no farmer can harvest in our village	0.74
42	Yes, if no bee-visits happen/occur to flowers during the blooming season, no yields I will get	0.74

Chi-square test for difference in knowledge of what bees are after on crop flowers: $\chi^2 = 103.49$, $N = 100$, DF = 41, $P < 0.0001$.

of years schooling), (b) the age of the respondent, (c) the number of years the farmer has been growing pollinator-dependent crops among those interviewed, (e) the total land allocated for crop production by a farmer, and (f) the proportion of rich farmers in the village (community) among those interviewed (Figure 5). Surprisingly, there was a negative relationship between the number of rich farmers and knowledge about pollination (Figure 5(f)).

When asked the question, do you know or understand what we mean by pollination?, the percentage of farmers saying they understand what pollination means was of 80% against 20% who said they did not understand or know what pollination means (χ^2 1 df = 36.56, $P < 0.0001$). When asked to name 1 to 6 species of pollinators they knew and saw visiting the flower of their crops, approximately 5%, 37%, 38%, 14%, 4%, and 2% of farmers interviewed declared knowing (were able to name), respectively, at least 0, 1, 2, 3, 4, and 6 bee species/groups (χ^2 1 df = 85.129, $P < 0.0001$). But, when asked to describe the types of pollinator/bee groups (species) they see visiting flowers of their crops, interviewed farmers had significantly (χ^2 test, $P < 0.05$) correct knowledge of more than 2 pollinator groups. Farmers (14.1–19.3%) knew honeybees, Xylocopa ("Civuvumira" in local language: Luganda) and stingless bees ("Kadoma" in local language: Luganda) as the frequent flower bee species/groups of their crops (Figure 6).

There were significantly ($P < 0.05$, χ^2 test) different farmers' perceptions on roles played by bees in crop flowers. When asked about what they think bees are doing on flowers of their crops, farmers provided different responses. Most frequent statement (14.6%) from farmers was that they believed that if "bees fall on flowers on their crops, they will see fruits/seeds coming in 1 to 3 weeks." Other farmers (11.5%) believed that "bees come to drink water and collect other foods on their crops." In 0.8% of frequency of statements, some farmers believed that "bees come to facilitate marriage of their crops." However, some farmers believed that bees were just playing with flowers of their crops and doing nothing valuable for their crops (Table 7). When asked the question: "is crop pollination by bees important in your crop production?", 73.9%, 9.9%, and 16.2% (χ^2 2 df = 74.81, $P < 0.001$) of farmers declared, respectively, that they believe (i) bees are important, (ii) they are not important, and (iii) they are not sure if bees are important in their crop production activities. More frequently, farmers reported that they "think crop pollination by bees is important in their farming business because they frequently believed (8.15%) that "with bees, they will get honey and they are convinced that if no bee-visits, there is no yield from their crops" (Table 8). Farmers who grow vegetables and fruits had higher understanding of pollination than those who grow legume, cereals, root, and tubers. Other farmers said bees contribute little and for them they are aware that bees that visit their crops come from community hives and or from surrounding forests and lake edges/wetlands; but for them they are convinced that "if no bees visiting crop flowers, wind & other insects will still pollinate and they will still harvest something."

When asked "how much do you think bee-visitations to flowers of your crops contribute to crop yields?", approximately 23 to 28% of farmers perceived significantly (χ^2 test, $P < 0.05$) that bees contributed, respectively, to half (41–50%) or to third (26–40%) of yield increase of crop yields in their villages (Figure 7). In fact, there were significant differences (χ^2 test, $P < 0.05$) between the average pollination experimental data [21, 22] and the farmers' perceived contribution of bees to yield of different crops such as beans, citrus, coffee, cowpea, mangoes, passion fruit, and pepper (Table 10). In most cases, farmers guessed little value as compared to the pollination experimentally derived data [21–23]. For other crops (avocado, egg plant, watermelon, tomato, etc.) farmers perceived the value of the contribution of bees to yield that was statistically (χ^2 test, $P > 0.05$) similar to the one that was derived empirically after conducting field pollination experiments (Table 9) by the author.

3.5.3. Farmers' Knowledge and Perception of Drivers of Bees in Villages Immediately Surrounding Flower Companies. When farmers were asked to explain where do they think bees are abundant between the edge of the flower farm (0.01–0.2 km) and far away (>2 km) in the village, most (90%) farmers believed that bees should be abundant in their villages since the "fumes" or chemicals sprayed daily inside greenhouses of the flower farms will not reach at such distance (>2 km) (Table 10). Some villagers frequently stated that bees were many in the village (>2 km far away from flower farm) because "bees cannot survive where they spray daily toxic chemicals." However, the answer "I do not know, I am not sure, I cannot tell" was frequently given by farmers (18.3% of frequency) and this indicated that some farmers were not good naturalists or had almost no interest in understanding the work of bees in their farming business. Different justifications (reasons) for getting higher yields far away from

Is Cut-Flower Industry Promotion by the Government Negatively Affecting Pollinator Biodiversity and
Environmental/Human Health in Uganda?

47

TABLE 9: Farmers' perception of the importance and value of pollinating services delivered by bees to their crops.

Crop categories	Crops reported	Yield parameters	YNFBV (Mean ± SD)	YHBV (Mean ± SD)	A = Fruit/Seed set (%)	B = Fruit set (%) Experimental data (2007)	Statistics for difference in fruit set between A and B	
							χ^2 (df = 1)	P
Fruits	Avocado	Nbr fruits/branch	4.037 ± 3.76	9.49 ± 6.14	57.45	82.67	4.53931	0.033
Pulses	Beans	Nbr pods/plant	1.33 ± 1.14	6.05 ± 4.53	77.95	27.11	24.6022	<0.001
Fruits	Citrus-lemon	Nbr fruits/branch	0.71 ± 0.29	4.31 ± 0.82	83.43	24.65	31.9679	<0.001
Fruits	Citrus-orange	Nbr fruits/branch	2.45 ± 0.57	8.39 ± 1.88	70.47	32.23	14.2385	<0.001
Industrial	Coffea arabica	Nbr berry-fruits/branch	21.84 ± 3.28	53.62 ± 7.35	59.26	62.12	0.067389	0.795
Industrial	Coffee robusta	Nbr bags of coffee beans/ha	4.16 ± 0.99	8.62 ± 1.79	51.75	68.71	2.387869	0.122
Industrial	Coffee robusta	Nbr berry-fruits/branch	11.65 ± 6.10	25.23 ± 4.52	53.86	78.32	4.52634	0.033
Pulses	Cowpea	Nbr pods/plant	1.61 ± 0.45	6.31 ± 0.82	74.51	45.43	7.05058	0.008
Vegetables	Egg plant	Nbr fruits/plant	1.46 ± 1.21	3.62 ± 0.63	59.73	41.65	3.22437	0.073
Pulses	Groundnut	Nbr pods/plant	2.34 ± 0.70	8.61 ± 1.65	72.85	68.55	0.130764	0.718
Fruits	Mangos	Nbr sacs of 100 Kg/tree	8.53 ± 1.76	12.77 ± 2.66	33.21	83.87	21.9204	<0.001
Fruits	Mangos	Nbr Kg fruits/branch	3.71 ± 3.48	9.276 ± 3.76	60.05	29.34	10.5504	<0.001
Fruits	Papaya	Nbr fruits/plant	1.59 ± 0.44	5.25 ± 0.89	69.81	31.55	14.4419	<0.001
Fruits	Passion fruit	Nbr fruits/plant	8.57 ± 0.150	11.00 ± 1.31	22.10	78.98	32.0077	<0.001
Pulses	Pigeonpeas	Nbr pods/branch	2.13 ± 1.56	6.31 ± 1.05	66.28	65.72	0.002376	0.961
Pulses	Peas	Nbr pods/plant	1.13 ± 1.17	4.07 ± 0.73	72.21	65.41	0.335998	0.562
Vegetables	Pepper	Nbr fruits/branch	0.88 ± 0.19	2.75 ± 0.47	67.87	19.89	26.2315	<0.001
Fruits	"Prune"	Nbr fruits/branch	0.00 ± 0.00	3.20 ± 0.58	100.0	76.54	3.11755	0.077
Fruits	Pumpkin	Nbr fruits/plant	0.00 ± 0.00	3.92 ± 0.88	100.0	82.21	1.73692	0.188
Pulses	Simsim	Nbr pods/plant	0.84 ± 0.20	5.19 ± 1.12	84.1	71.67	0.991878	0.319
Industrial	Sunflower	Nbr seeds/head	290.5 ± 71.9	951.5 ± 186.7	69.47	69.77	0.000646	0.98
Vegetables	Tomato	Nbr fruits/plant	12.56 ± 10.3	22.06 ± 11.90	43.06	37.54	0.378045	0.539
Fruits	Watermelon	Nbr fruits/plant	0.00 ± 0.00	3.43 ± 0.69	100.0	79.43	2.35816	0.125
Fruits	Apple	Nbr fruits/branch	0.00 ± 0.00	6.20 ± 2.58	100.0	78.45	2.60242	0.107

Legend:
YNFBV: When no/few; YNFBV: yields when very few or no bees visit my crop flowers (bad yield).
YHBV: When I receive high bee-visits to crops; YHBV: yields if bees came with high visitation frequency to flowers of my crop.
A: (Fruit/seed set in %) = [(mean YHBV − mean YNFBV)/mean YHBV]*100.
B: (Fruit/seed set in %); this is data obtained after conducting pollination experiments.

the flower farm were given, but must frequently, farmers believed that higher yields can only be due to difference in field management systems, fertility levels, types of varieties grown, and to difference in bee-visitations (Table 11) because bee-visitations are almost absent near the flower farm where they apply toxic chemicals.

When asked if they have ever observed any changes (reduction, increase) in crop yields in the village over the last 5 to 20 years, and if yes, this may be due to what farmers perceived as significant (χ^2 test, $P < 0.001$) changes (reductions) in crop yields during the last 5–20 years for various reasons such as soil infertility of their lands (Table 11). The presence of the flower farm spraying chemical toxic pesticides nearby and environmental degradation were key reasons provided by farmers even when some farmers said that there has been little change (14.62%) while most farmers (20.10%) said they were not sure if change has ever happened or affected crop yields in their villages (Table 11).

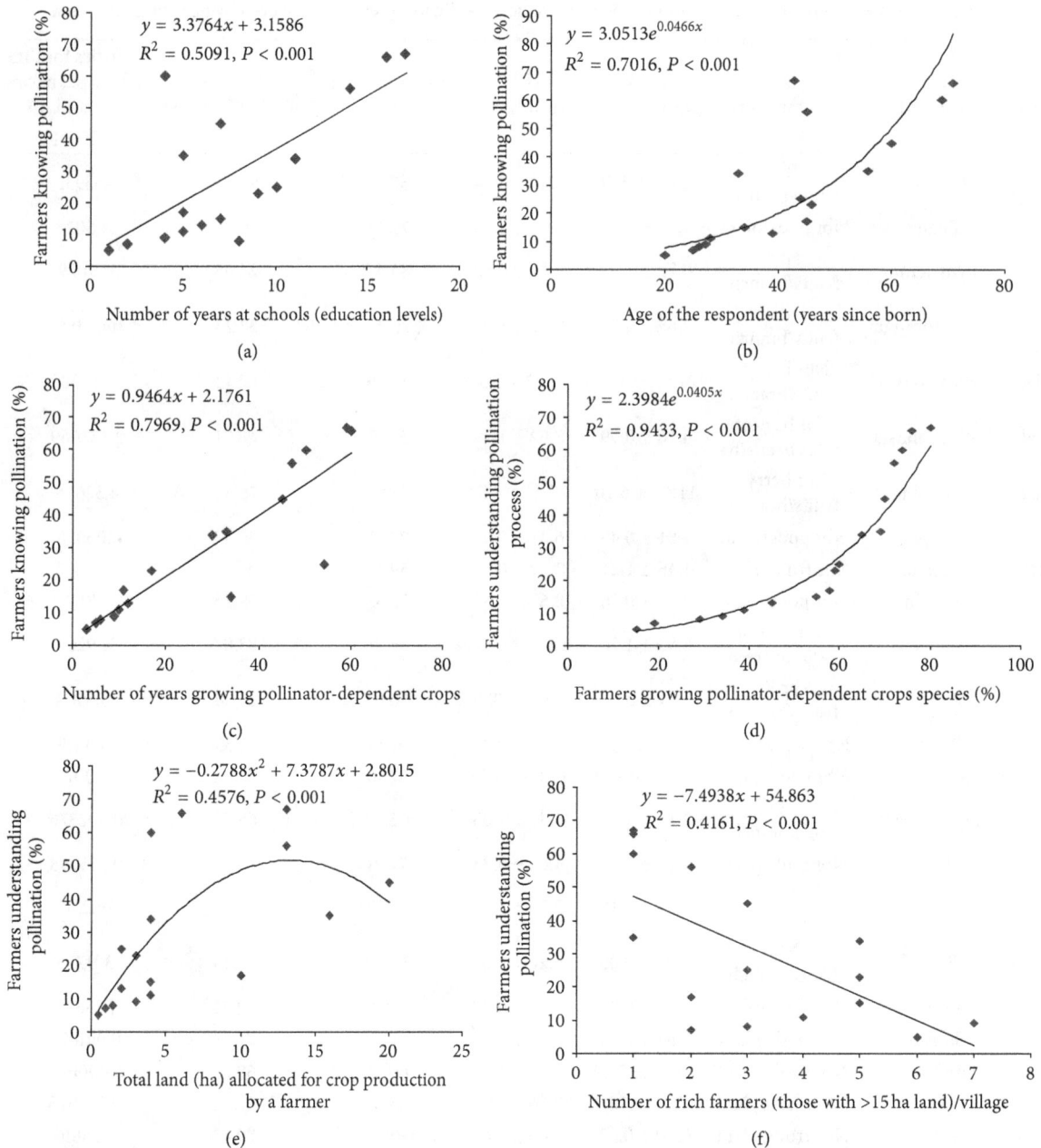

FIGURE 5: Trends in the influences of socioeconomic and demographic characteristics ((a) = level of education, (b) = age; (c) = crop production experience, (d) = type of crops grown by the farmers, (e) = land allocated to production of pollinator dependent crops, and (f) = level of richness of the farmer) of the respondents on the level of knowledge/understanding of the word pollination by farmers interviewed from different villages surrounding flower farms.

When farmers were asked if they have seen any change in the population density (abundance) of bees in the village, most farmers were not sure if there has been change. Farmers who believed in changes attributed that to declining in bee-keeping, "flower fumes", bad farming practices, or forest bush clear-cutting (Table 12). When asked if they felt that bee populations have been stable or changing (increasing/decreasing) over the last 5 to 15 years in their village landscapes, there were various answers (perceptions) among farmers. Most (45%) of farmers perceived that bee populations have been reducing (declining) seriously in their villages in the last 5 to 15 years.

Some farmers (32.1%) were not sure or had no idea (they did not know anything) whether changed has occurred or not. A few farmers (19.8%) said they perceived no changes while only 3.1% of these respondents felt that there has been an increase of bee populations in the village over the last 5 to 15 years. The difference in perceptions among the four categories of respondents was significant (χ^2 3 df = 38.367, $P < 0.001$).

Also, when asked if they felt that crop yields have been stable or changing (increasing/decreasing) over the last 5 to 15 years in their village landscapes, most (48.3%) of the farmers perceived that crop yield has been reduced by 10–50% in their

Is Cut-Flower Industry Promotion by the Government Negatively Affecting Pollinator Biodiversity and Environmental/Human Health in Uganda?

49

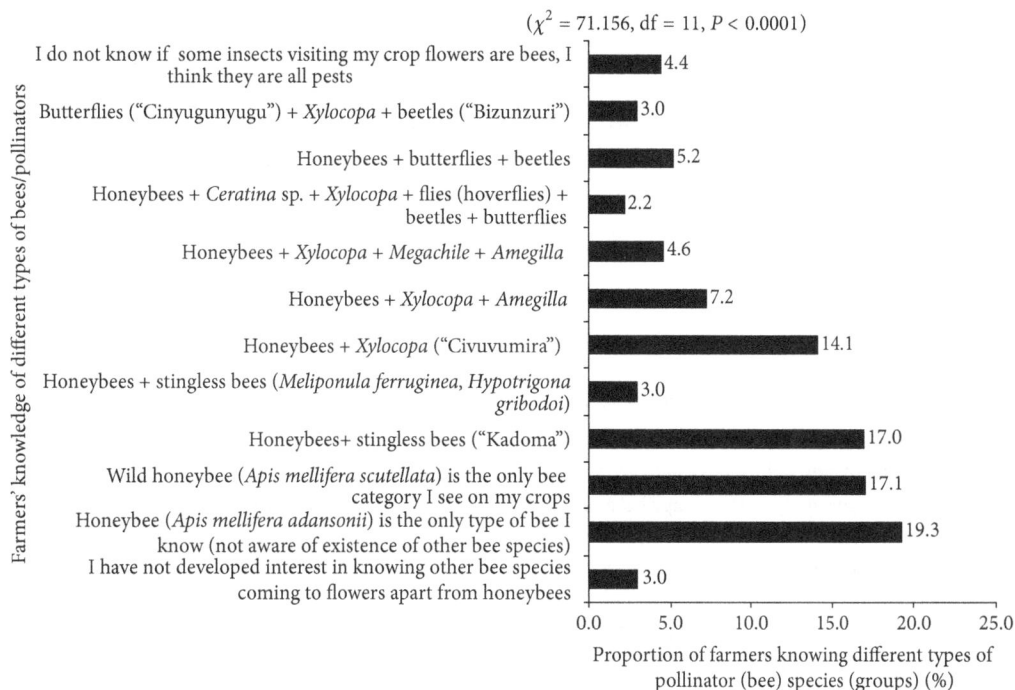

FIGURE 6: Type of species/functional group of pollinators farmers know pollinate their crops.

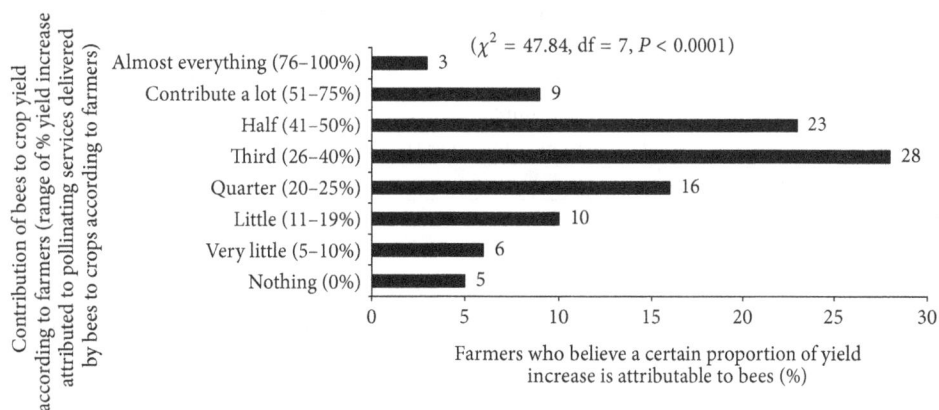

FIGURE 7: Farmers' perception of the general contribution of bees to crop yield.

villages in the last 5 to 15 years. Some farmers (17.8%) felt that crop yield has been stable. Another group of respondents (18%) perceived that they were not sure (they could not tell) if change has occurred or not. A small portion of respondents (6.1%) perceived that crop yield has been increasing slightly in the 10 to 35% proportions. Approximately 3.8% of respondents perceived that crop yield has increased in the 40–85% proportion while another small group of respondents (6%) perceived that crop has declined seriously by 50 to 85%. The difference in views/perceptions among the 6 groups of respondents interviewed was significant (χ^2 5 df = 7.87, P < 0.001). Overall, when asked if the currently observed changes (reductions) in crop yield are a consequence of reduction in bee populations in the villages, most farmers perceived that there has been a reduction in crop yield as a consequence of the decline in bee populations in the villages (Table 13).

When asked if they believed that if "crop yield reduction/ loss was linked to decline in bee populations in the village", there were different statements given by farmers (Table 14). Overall, 35.4% of respondents were of neutral views, 39.2% disagreed, and 25.45% agreed. The difference in views between the 3 categories of respondents (neutral, disagreeing, agreeing) was significant (χ^2 2 df = 3.056, P < 0.05). In fact, around 58% of farmers interviewed believed that growing crops far away (>2 km) from the flower farm was a better option to obtain higher yields even when some farmers (22.9%) believed that it should be similar yields. Onlya few farmers (1.5%) believed that the one growing crops at (0.1–0.3 km) the edge of the flower farm could get better yield. A small proportion (17.6%) of farmers stated that they were not sure (they did not know). The difference in views/perceptions of the 4 groups of respondents was statistically significant

TABLE 10: Perception of farmers about the abundance of bees between the rural landscape and the neighborhood of a given flower farm.

Statement no.	Question: Where do you think bees are abundant, near the flower farm (0.1–0.5 km) or further (>2 km) in the village?	Freq. (%)
	I have no idea but think bees should be many in the villages where there are abundant flowers for them	11.5
1	Bee populations should be high up-country than near the flower farm with toxic chemical sprays they use	2.3
2	Bees are many in the village (>2 km far from the flower farm) and no bees will stay near chemicals	4.6
3	Bees are many in the village (>2 km far from the flower farm), everywhere where wild/crop flowers are freshly/abundant	3.8
4	Bees are many in the village because we have abundant flowers there than near the flower farm with toxic fumes	2.3
5	Bees are many in the village compared to near (edge) the flower farm	0.8
6	Bees are many in the village (>2 km) because they spray daily near the farm and bees can not survive there	14.5
7	Bees are many in the villages (in every place where there are many flowers, bees will be there)	3.1
8	Bees are many in the villages because they have no enemy there, they are free to go everywhere in the village	2.3
9	Bees are many of course in the village (>2 km far from the flower farm)	8.4
10	Bees are no longer many in the village because the flower farm is killing them with the toxic sprays they make every day	1.5
11	Bees are not strong like humans, if they smell the fumes from the flower farm when foraging, they end up dying in their nests	0.8
12	Bees are obviously many in the village with fresh air, not polluted by toxic chemicals of flower farms	3.1
13	Bees are still many in the village compared to near the flower farm where nesting sites have been destroyed	5.3
14	Every bee flying for forage from the village to the edge of the flower farm will die there because of too many pesticides there	1.5
15	I am not sure (I cannot tell with confidence where exactly)	5.3
16	I am not sure if bees are many in the village (>2 km) than near the flower farm but I see few bees on gardens located in the proximity of flower farm (0.2 km)	3.8
17	I cannot tell exactly but I think they should be many in the village where healthy-non-polluted wild flowers are	3.1
18	I donot know; I cannot tell; I am not sure (I cannot even guess)	18.3
19	Obviously bees are many in the village than near the flower farm where they spray a lot	1.5
20	Pesticides have a long-persistence smell near the flower farm so bees will die there quickly	1.5

Chi-square test for differences in beliefs of farmers about area where bees are abundant: $\chi^2 = 93.1291$, $N = 100$, $DF = 20$, $P < 0.0001$.

(χ^2 3 df = 68.032, $P < 0.05$). Overall the majority of farmers doubted negative effects of the proximity to the flower farm on yields of their crops.

When asked to guess the place where someone can get better yield between near the flower farm and far away >2 km) from the flower farm, most farmers were not sure (10.9%), while others stated that there should be no difference in yield (11.6%). A small proportion of farmers felt that yield will be higher in the villages (>2 km far away) because the fumes of the flower farms do not reach over there (Table 13).

When farmers were asked to say when they started observing changes (in bee populations/crop yields) in their villages the majority of farmers (22.3–44%) believed there has been significant (χ^2 test, $P < 0.05$) changes (reductions) occurring in either crop yields or bee populations. However, some farmers (29-30%) believed that changes in bee populations started occurring some 5 to 10 years ago (Figure 8).

Several drivers of bees in farmlands have been identified worldwide. During discussion meetings, farmers were asked to name key factors likely leading to the current ongoing bee loss in their villages. These drivers were put together on a paper and farmers were again asked to rank them in terms of importance from the most important (dangerous) to the less

important factors that may be associated directly/indirectly with bee loss in the village. There were variations in assessment (ranking) of drivers by farmers from the four study districts (where the 4 flower farms are located). However, on average, all farmers ranked as significant ($P < 0.005$, Kruskal-Wallis test) primary drivers the following drivers: (i) fragmentation of national habitats such as forests/wetlands (average rank: 1.90 ± 0.13), (ii) forests/bush clear-cutting in the village (average rank: 1.26 ± 0.002), (iii) logging, charcoal burning, timber/poles/firewood collection (average rank: 1.26 ± 0.02), (iv) fires burning intensification (1.69 ± 0.04), and (v) local climate change such as rainfall pattern changes (1.84 ± 0.002).

Some worldwide documented key drivers of bee biodiversity loss (road construction up-country, land-use change, global climate change, environmental/degradation, grazing intensification, and agriculture modernization plan by the government) were perceived by local farmers as factors playing a tertiary role or as factors playing no role in the decline process or as factors having nothing to do with bee decline in the villages. Most farmers said they were not sure if these factors could affect bees even when a few of them believed that these factors could be associated to some extent with bee

Is Cut-Flower Industry Promotion by the Government Negatively Affecting Pollinator Biodiversity and
Environmental/Human Health in Uganda?

51

TABLE 11: Perceptions of farmers about changes (reduction/increase) in crop yields in the village in the last 5–20 years.

Statement no.	Question: Have you observed any changes (reduction, increase) in crop yields in the village over the last 5 to 20 years?	Freq. (%)
1	Yield of crop is also reducing in the last 10–20 years. I think it is because; soil is old/currently infertile in our area	5.38
2	Changes in crop yield may be due to infertility of the lands and to flowers fumes	0.77
3	I have not yet observed changes (reduction/increase) in my yields of crops	8.46
4	I have observed changes (reduction) in yields for the last 5–10 years	5.38
5	I have seen some change these days (reduction) in yields of many crops although I donot know the exact causes	8.46
6	No change, for me, yield is increasing because I live far from the flower farm and I manage well my gardens	1.54
7	No changes (reduction), instead I see increase because I do fertilization with cow dung to increase my yield	0.77
8	No changes (reduction) in yields apart from thieves in the villages	2.31
9	Not sure (I can not tell)	20.10
10	There has been change (reduction) and this may be linked to practice of bad farming methods	1.54
11	There has been changes in yield of many crop species for unknown reasons of changes in the global system	0.77
12	There have been drastic changes (reduction of 50 to 80% compared to previous harvests) in most of the crop yields I grow	3.08
13	There is a general decrease in yield of all crops we grow but I think it may be due to soil/land infertility	0.77
14	There is a change (reduction) and bad yields are due to soil infertility, excess sunshine/rains, and climate change	4.62
15	There is a change (reduction) in crop yields but do not know what it is due to exactly	3.08
16	Yes, there is a change (reduction) in yields over time but I can not explain clearly the changes are due to what exactly	0.77
17	There is a change in yields probably due to soil infertility because I have over-used the land to grown different crops over years	1.54
18	There is a reduction in yield due to over-use of soils, growing crops/bad varieties in a wrong season, no respect of growing seasons	3.85
19	There are changes (reduction) in yields probably due to soil infertility of my gardens and to fumes from flower farms	0.77
20	There is yield reduction because bees are disappearing in the village due to chemical sprays from the flower farm	6.15
21	There is very little change (decrease)/to not much changes in yield of my crops	14.62
22	Yield of crops has reduced in the village over the last 5–20 years but I do not know why	5.38

Chi-square test for difference in arguments (statements) frequency: $\chi^2 = 110.15$, $N = 100$, DF = 21, $P < 0.0001$.

disappearance in the villages (Table 16). There was a great variability in perception of farmers concerning the place (area) where to get better yield between the near and the far areas to the flower farm (Table 15).

Farmers were requested to relate the rate of disappearance of bees due to degradation of natural/seminatural habitats; and it was observed that only old farmers (aged 70–80 years) think/perceive that forests and other good seminatural habitats disappeared 10–50 years ago. Old farmers perceived also that in areas covered by forests, it is possible to observe visits of bees to crop flowers at the rate of 500 bee-visits/10 min/50 m^2 garden plot. Young people (20–35 years old) perceived that forests disappeared in the villages some 5–10 years ago and that bee-visits to crop do not exceed 20 bee-visits/10 min observations/50 m^2 garden plots (Figure 9) even in regions covered by forests or plenty of seminatural habitats.

3.5.4. Farmers' Attitudes towards Flower Industry Growth and Development in the Surrounding of Their Villages: Perceptions of the Negative Effects of Flower Industries on Environmental, Human, and Agricultural Health. When asked about the benefit of living near the flower farm, most (66%) respondents

significantly (χ^2 test, $P < 0.05$) declared (stressed) gaining no benefit (advantage) by having their homesteads established close to flower farms (Figure 10). However, the proportion of farmers declaring having no problem with the presence of the flower farm in their villages was of 86% against 14% of farmers who said they had a problem with the flower farm (χ^2 1 df = 51.46, $P < 0.001$). Those who had problems with the presence of the flower farm in their village raised several reasons for why they had problems with the presence of the flower farm in their village (Table 17).

When asked to propose durable solutions to the problems of the presence of the flower farms in their villages, more frequently (47.4%) farmers stated that they had no power to propose what should be done because when these cut-flower industries were established, politicians were consulted and they ended up allowing theses industries to settle in our villages without the consent of communities (Table 18). Only a small proportion of farmers frequently (5.9%) affirmed that if flower farms stop applying dangerous chemicals, then they can stay in the villages. Another group said the flower companies should use nontoxic dangerous chemicals (8.1%). Other farmers said they wanted the flower farms to pay all land titles from the surrounding villages (22%). Some other farmers

TABLE 12: Perception of changes in the populations of bees in the villages over the last 5–20 years.

Statmenet no.	Question: Have you observed any changes in bee populations (abundances) in the village over the last 5 to 20 years? It is because of what?	Freq. (%)
	Not sure if there has been strong changes in bees, but I watch few bee-visits (I see few bees feeding) in my farm these days	8.40
1	I have observed no reduction in bees in the village; there are enough hives, wild nest sites, and food plants for bees	5.34
2	I think bees are reducing in the village because of too much pollutants and toxic chemicals applied in the flower farm around there	3.05
3	Not sure if any change has occurred because I donot know how many bees were there before compared to now	6.11
4	Bee populations are declining over years because many termite mounds and tree/stamps where bees used to nest have been destroyed.	0.76
5	Bee populations are declining in the village over the years because beekeeping is also declining in the village (few hives in the village)	2.29
6	Beekeeping attracts all wild bees to hives and concentrates them where the hives are placed, so they will visit only that area, then we shall think bees have disappeared	3.05
7	Bees have been reducing in the village over the last 5–10 years, but I do not know why Maybe you can tell me more about it	0.76
8	Bees are increasing these days because we have planted many flowering trees and shrubs they are interested in	1.53
9	Bees are increasing; they are many in all bushes in the village, even in crops, I see many these days	1.53
10	Bees are no longer coming on my crops (beans) these days and I do not know why	0.76
11	Bees are reducing because many beehives/wild nests have been destroyed 5–10 years ago with crop production intensification due to too many children we are producing these days	0.76
12	Bees are reducing because the flower farms do not want bees, any insect around their flower farm is killed	0.76
13	Bees are reducing because the crop varieties we grow these days are noninteresting foods for bees anymore. No good scents of flowers to attract bees	0.76
14	Bees are reducing because may be these days we are growing crops that produce noninteresting/attractive flowers to bees	0.76
15	Bees are reducing because we have destroyed underground nests of wild bees and we are cutting trees to make charcoal	0.76
16	Bees have been reducing in the village over the last 5–10 years, but do not know why May be you can tell me more about it because you stay with people who know things than us	0.76
17	Bees have been decreasing because no abundant hives in the village these days	0.76
18	Bees have drastically reduced; they are few these days because farmers are cuttingoff all flowering herbs/weeds near fields (bad farming practices)	3.05
19	Bees have reduced because many of them are dying in the village after smelling the toxic fumes of the flower farm	3.05
20	Bees have reduced because the flower farm has killed them with their toxic sprays from the flower farm in the surroundings	0.76
21	Bees have reduced because the flower farm sprayed everywhere at night and their fumes chased away all bees	0.76
22	Bees have reduced ever since the flower farm came around here, no more bees come to my crop flowers	1.53
23	Bees number have drastically reduced (e.g., could count around 100 bees/ $\mathbf{m}^2$ in the bean field, these days I count less than 10/ $\mathbf{m}^2$)	3.05
24	Bees reduced because no abundant bees coming to the village these days compared to a previous time when I was young	0.76
25	Bees used to be many in the village (5–20 years ago), these days, they have reduced I do not know why	0.76
26	Changes (reductions) came with human population increasing (pressure), destroying nesting sites (habitats) of bees to grow crops	0.76
27	Drastic reduction in bees occurred because people have cleared all bushes and forests for livestock/crop productions intensification	3.05
28	I am not sure if bees have reduced/increased in the village, I have no interest in bees	2.29
29	I am not sure, I have never paid attention (no curiosity) to observe change in bee populations	1.53
30	I cannot explain (do not know) why bee populations are reducing in the village these days	0.76
31	I do not know, I think bees will be every where because there are tasty/fresh flowers in the village	0.76

Is Cut-Flower Industry Promotion by the Government Negatively Affecting Pollinator Biodiversity and
Environmental/Human Health in Uganda?

53

TABLE 12: Continued.

Statmenet no.	Question: Have you observed any changes in bee populations (abundances) in the village over the last 5 to 20 years? It is because of what?	Freq. (%)
32	I have observed no changes in bee populations in the village since my childhood up to now, it is the same thing I see	3.82
33	I have never paid attention (no curiosity) to observe changes in bee populations	4.58
34	I have no idea if bees have reduced (declined) or not; I will start making my observations from today in many gardens	6.87
35	I have observed no changes in bee numbers these days; in fact, there are many coming to visit flowers as usual	6.11
36	I see no change in bees numbers in the village because we have even received beehives from NAADS; numbers are increasing these days	0.76
37	I used to see 2-3 colonies moving (flying around/month) in the village, these days I see one colony flying-out around our villages once a year	0.76
38	I used to see bees hovering cups, saucepans left outside, these days, no bees come to visit my cups left outside, only flies and ants	0.76
39	I used to see many bees flying around/feeding on flowers in my fields; but since the flower farm came, I see few or no bees at all	1.53

Chi-square test for differences in explanations about causes of changes in bee numbers: $\chi^2 = 72.07$, $N = 87$, DF $= 39$, $P = 0.001$.

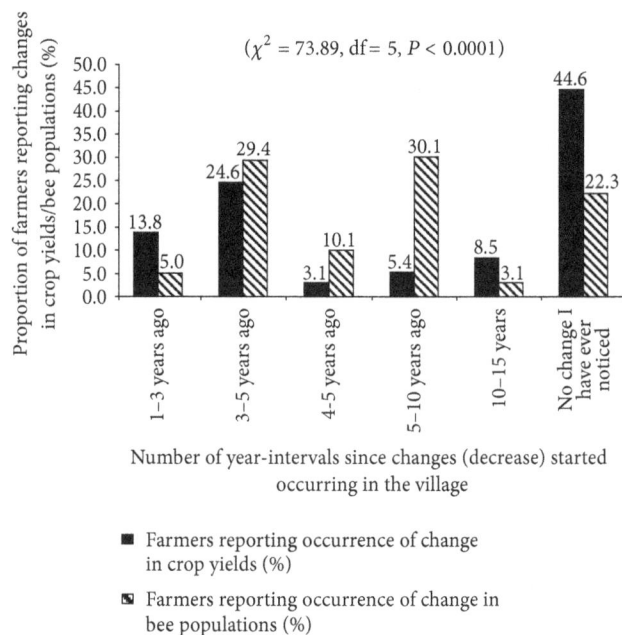

FIGURE 8: Perception of time since change (decrease in crop yields and bee populations) started occurring in the village.

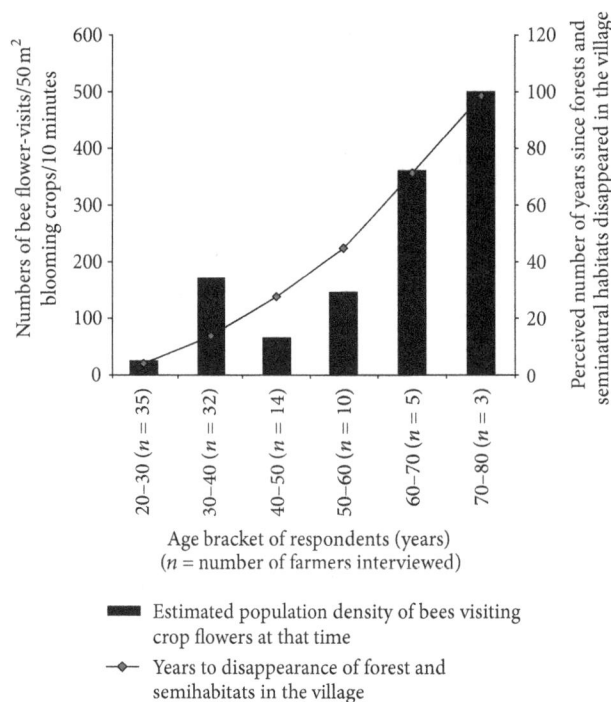

FIGURE 9: Farmers' perception/estimation of periods of disappearance of bee population visiting crop flowers in relationship to periods since disappearance of natural habitats (forests)/seminatural habitats in the village due to various drivers (grazing, fires, crop cultivation intensity, and demographic pressures).

perceived that leaders of flower companies should organize regular meetings with the surrounding communities (5.2%). Some farmers requested that these flower companies should be requested to go away (14%). Another small group of farmers said that flower companies should be allowed to continue their business (14%). The difference in opinions among the respondent groups was significant (χ^2 5 df = 118.9, $P < 0.001$; Table 18).

Overall, with regard to sustainable development of the flower industry in Uganda, farmers who were living within less than 1 km of the flower farm's boundaries indicated that they had a problem with the existence and operations of cut-flower forms in their village. Various reasons were provided

for this dissatisfaction. More specifically, the following reasons were given among others: (i) rotting of fruits before fruit maturity from most cultivated fruit tree species such as avocado, mango, pawpaw, and jackfruit; (ii) for farmers practicing animal rearing, farmers accused the flower farms of being responsible for abortion when cattles drank from a water stream near the flower farms and yet was declared free of effluents; (iii) low yield of fish from fish ponds established

TABLE 13: Farmers' perceptions of the changes in crop yield that have occurred in the last 5 to 20 years.

Statement no.	Question: Have you observed any changes (reduction, increase) in crop yields in the village over the last 5 to 20 years?	Freq. (%)
1	Yield of crop is also reducing in the last 10–20 years, we think it is because soil is old/infertile in our area	5.38
2	However, changes in yield may be due to infertility of the lands and to flowers fumes	0.77
3	I have not yet observed changes (reduction/increase) in my yields of my crops	8.46
4	I have observed changes (reduction) in yields for the last 5–10 years	5.38
5	I have seen these days some change (reduction) in yields of many crops although I donot know the exact causes	8.46
6	No change, for me, yield is increasing because I live far from the flower farm and I manage well my gardens	1.54
7	No changes (reduction), instead I see an increase because I do fertilization with cow dung to increase my yield	0.77
8	No changes (reduction) in yields apart from thieves	2.31
9	Not sure (I cannot tell)	20.10
10	There have been a change (reduction) and this may be linked to practice of bad farming methods	1.54
11	There have been changes in yield of many crop species for unknown changes in the global system	0.77
12	There has been drastic changes (reduction of 50 to 80% compared to previous harvests) in most of the crop yields I grow	3.08
13	There is a general decrease in yield of all crops we grow but I think it may be due to soil/land infertility	0.77
14	There is a change (reduction) and bad yields are due to soil infertility, excess sunshine/rains, and climate change	4.62
15	There is a change (reduction) in crop yields but do not know what it is due to exactly	3.08
16	There is a change (reduction) in yields over time but we can't explain clearly the changes are due to what exactly	0.77
17	There is change in yields probably due to soil infertility because I have over-used the land to grow different crops over the years	1.54
18	There is areduction in yield due to over-use of soils, growing crops/bad varieties in a wrong season	3.85
19	There are changes (reduction) in yields probably due to soil infertility of my gardens and to fumes from flower farms	0.77
20	There is yield reduction because bees are disappearing in the village due to chemical sprays from the flower farm	6.15
21	There has been very little change (decrease)/to not much changes in yield of my crops	14.62
22	Yield of crops has reduced in the village over the last 5–10 years but I do not know why	5.38

Chi-square test for difference in arguments (statements) frequency: $\chi^2 = 110.15$, $N = 100$, DF $= 21$, $P < 0.0001$.

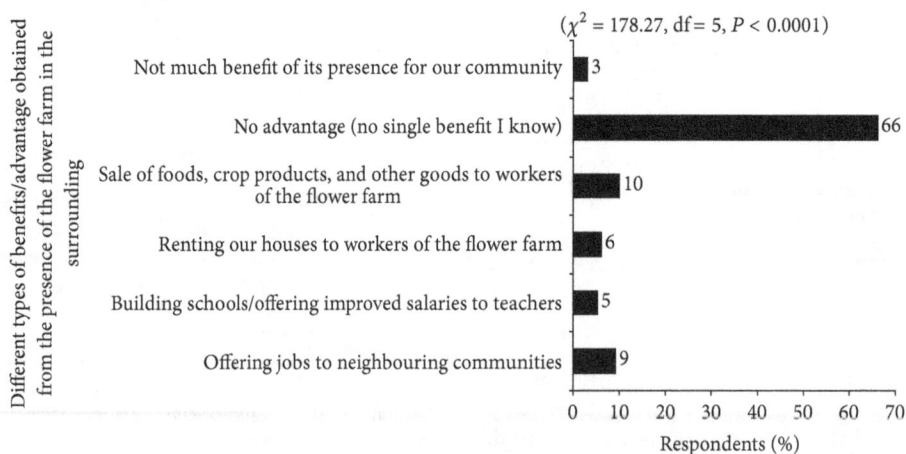

FIGURE 10: Perception of farmers about the benefits (advantages) they get from the presence of the flower farm in their village.

in the village and those around Lake Victoria resulted in fish dying and found floating on the lake in areas surrounding the flower farms (no evidence was gathered by the researchers on this); and (iv) low honey yield due to death of bees and they attributed this to agrochemicals used by the flower farms.

Farmers said they see no added advantage/benefit of the flower farm being in their locality, whereas some said that flower farms were offering jobs, building schools for their children, and improving road infrastructure (Figure 8). Farmers requested that the government should relocate

Is Cut-Flower Industry Promotion by the Government Negatively Affecting Pollinator Biodiversity and Environmental/Human Health in Uganda?

55

TABLE 14: Views/beliefs of farmers' about crop yield reduction being linked or not to decline in bee-visitations to crop flowers in villages.

St. no	Question: Do you think (agree/disagree) that changes (reductions) in crop yields are associated with changes (decline/loss) in bee-visitations? Why not?	Freq. (%)
1	I am not sure and do not know, someone should explain to me why there is too much yield reduction these days in our village	2.3
2	I am not sure (I have no idea at all and cannot even speculate or guess)	8.5
3	Because most of our grown crops need bees, if bees are few in the village, then low yield obviously is due to that	14.6
4	Change (reduction) in crop yields over time is due to the fact that FIDUGA dumped toxic chemicals around our villages	1.5
5	Change (reduction) in yields can only be related to thieves of my harvests	3.8
6	Changes in yield may be due to land infertility/degradation and fumes from the flower farms	0.8
7	Changes in yields are due to lack (reduction) in bee population in the village because all crops I grow need bees to fertilize them	1.5
8	Changes in yields are due to lack (reduction) in bee population in the village because all crops I grow need bees to fertilize them	0.8
9	Currently observed yield loss (reduction) in many crops is due to pests/diseases emergence but not at all to decline in bees	1.5
10	I am not sure if change in yield can be associated with decline in bees, I need someone to explain to me how	8.5
11	I can't relate yield reduction to loss of bees because in the village, their habitats (nests) are still many	3.1
12	I do not think low yield is linked to reduction because theses days yield reduction is generalized in all crops even in those that don't need pollinators to yield better	5.4
13	I don't agree/believe that low yield is due to lack (low) of numbers of bees in the village	2.3
14	I don't think the changes (reductions) in yield are due to few bees currently seen in the villages	0.8
15	I have no idea if they have changed or not	11.5
16	No, I think yield reduction is due to drought and climate change, but not to low bee-visitations or lack of bees in the village	1.5
17	Not sure (I donot believe low yield is due to absence or low bee-visitations to my crops)	4.6
18	Reduction in coffee/beans yields may be due to climate change and to diseases/pests emergency	3.1
19	The change (reduction) in yield is due to change in rainfall pattern prevalence infertile soils, but not due to bees reduction in the village	3.1
20	The change (reduction) in yields is only associated with low rainfall, these days, we receive little rainfall that can help get good harvests	3.8
21	The change (reduction) in yield is due to chemical sprays of the flower farm that have killed bees and then we get little yield	5.4
22	There have been changes in yields of many crops but I don't believe low yields is due to lack of bees	2.3
23	There is little harvest these days because there is not much rainfall coming these days	1.5
24	Yield reduction is due to interacting factors (including chemical, poor farming methods) leading to reduced bee-visitations to flowers of our crops	6.2
25	Yield reduction is due to the fact that NARO (agricultural research institute) has released crop varieties that are very susceptible or contain pests/disease germs in the village	1.5

Chi-square test for difference in statement frequency: $\chi^2 = 74.2604$, $N = 100$, DF = 24, $P < 0.0001$.

flower farms in areas that are not close to villages or to crop production and animal rearing. They alternatively suggested that flower farms should be encouraged to use nontoxic agrochemicals (if they exist on the market) if they have to stay for a longer period in their village (Figure 9).

When interviewing the four agronomic managers of the flower farms about pollination, most of them were aware of the word pollination and affirmed that they did not need pollinators to produce flowers inside their greenhouses. They also confirmed that as opposed to farmers' accusations they do not experience any problem of runoff of water or of waste chemicals into the environment. These agronomic managers

suggested controlling well their wastes by setting some mechanisms such as pumping back the water and recycling/reusing the water from the flower farm. They also suggested not to dump wastes at night in villages (as claimed by farmers). However, field observations showed that infrastructures to control runoff in the environment were old and poorly constructed/managed. During field visits, it was easy to see waste running into water crossing neighboring villages. In other words, agronomic managers were at a defensive front but actually the complaints of farmers may be correct. In addition, while sampling bees, several dumping points of agrochemical wastes were observed in the landscape (grasslands

Table 15: Perception of the place (area) where to get better yield between the near and the far areas to the flower farm.

Statement No.	Question: Who can get higher yields between the one growing close to the flower farm compared to the one growing crops far away (>2 km) and why? (all growing/environmental conditions are standards and similar)	Frequency (%)
1	I am not sure (I do not know)	10.9
2	I am not sure if the distance has an effect on yield but difference in yields can be due to varieties, age of plants, and fertility conditions	2.3
3	I know, I will get higher yield at >2 Km far from the flower than growing near it, but I still cannot explain why	2.3
4	No difference in yield and if there exist differences, it depends on good care and management of crops and fields	3.9
5	The same yield everywhere (I see no reason for difference in yields), there should be no difference in yields	10.8
6	The yield is high in the village because of difference in fertility levels of fields	1.6
7	There should be no difference in yields but difference can be due to difference in fertility levels of the fields	11.6
8	Definitively, mangos/avocado will yield more in the village because bees have been killed near the flower farm	7.8
9	Digging at the edge of the flower farm with toxic fumes, you must get no (low) yields	2.3
10	Even when the level of fertility of the soil is the same, yield will still differ (being high in the village) but I do not know why?	1.6
11	Growing crops near flower farm, no yield because no bees; but in the village there are some bees, flies, and wasps to pollinate my crops	1.6
12	Higher yields are in the village because we pose no harm to bees in the village but the flower spray alot to chase them away	1.6
13	Higher yields are in the village because soils are fertile there and people care more for their crops up there	0.8
14	I am not sure if someone who grows coffee faraway from the flower farm will harvest more than me	4.7
15	I can harvest more in the village than near the flower farms, I am sure	0.8
16	I think it should be the same yield if all variables and growing conditions (rainfall, management, bee-visits, soil, and fertility levels) remain the same over time	0.8
17	I think the one who grows far away will get higher yield even if I donot why	3.1
18	I will get the same yields if we receive the same quantity of rain	4.7
19	If everything is constant at all farms, the yield would be the same, I see nothing bringing a difference	0.8
20	No difference in yields, but if it exists, it can only be due to bad weather, infertility of soils, management techniques, and planting bad varieties	1.6
21	Similar yields if we manage fields equally	2.3
22	The yields will be different since we have different levels of weed management system/soil fertility/bee-visitations and thus bee-visitations will be different and consequently yield different	3.9
23	Higher yields are in the village due to different levels of fertility, management, and bee-abundances compared to flower farm edge where bees are continuously killed	3.1
24	There will be higher yields in the village because there are many bees to enable getting better yields than near the flower farm	0.8
25	Yields near the flower farm cannot be similar to that in the village (>2 km) because growing conditions are not the same	3.9
26	Yields will be higher in the village because of high fertility levels in the village	1.6
27	Yields will be higher in the village probably due to difference in environmental conditions and not due to difference in bees	1.6
28	Yields will be high in the village because there are many bees at >2 km far from the flower farm	7.0
29	Yields are high in the village (>2 km) because polluted water from the flower farm is not running up there to kill crops and bees are not drinking polluted water here	0.8

Is Cut-Flower Industry Promotion by the Government Negatively Affecting Pollinator Biodiversity and Environmental/Human Health in Uganda?

57

TABLE 16: Farmers' perception of key drivers (factors) of the ongoing decline (loss/erosion) in wild and managed pollinators in villages in relationship to multiple drivers and pressures that act simultaneously/synergistically on pollinator communities in villages (farmers' rating of key drivers of bee diversity in their village). Among various interacting drivers (types of disturbances) of wild bee abundance and species richness, the most important factors likely being responsible for ongoing decline (loss/erosion) in wild and managed pollinators in the villages are ranked by order of importance.

Type of drivers/pressures/disturbances of pollinators	Wagagai	Mairye	Fiduga	Pearl	Overall mean	Kruskal-Wallis test		
	Mean ± SE	Mean ± SE	Mean ± SE	Mean ± SE	Mean ± SE	DF	H (adj)	P
General land-use changes/shifts pattern	4.91 ± 0.05	5.29 ± 0.06	4.01 ± 0.09	3.85 ± 0.05	4.51 ± 0.06	3	81.71	<0.0001
Fragmentation of natural habitats (forests, wetlands)	2.93 ± 0.03	1.38 ± 0.03	1.29 ± 0.05	1.99 ± 0.43	1.90 ± 0.13	3	88.78	<0.0001
Loss/degradation of semi-natural habitats (hedges, grasslands)	4.24 ± 0.05	4.12 ± 0.19	2.97 ± 0.07	3.08 ± 0.06	3.60 ± 0.09	3	79.51	<0.0001
Forest/bushes clear-cutting in the village	1.03 ± 0.01	1.98 ± 0.03	0.99 ± 0.03	1.05 ± 0.01	1.26 ± 0.02	3	62.44	<0.0001
Logging, charcoal burning, timber/poles/firewood collection	1.07 ± 0.02	1.08 ± 0.01	0.95 ± 0.03	1.93 ± 0.03	1.26 ± 0.02	3	65.13	<0.0001
Forest habitats conversion into grazing fields/crop fields	3.35 ± 0.06	1.12 ± 0.02	3.5 ± 0.10	5.09 ± 0.01	3.27 ± 0.05	3	88.82	<0.0001
Over-harvesting of specific resources in the forests	4.26 ± 0.09	1.07 ± 0.01	4.76 ± 0.13	5.18 ± 0.04	3.82 ± 0.07	3	76.48	<0.0001
Agriculture modernization plan by the government	5.14 ± 0.02	4.98 ± 0.03	5.31 ± 0.04	5.13 ± 0.08	5.14 ± 0.04	3	16.27	0.001
Cultivation intensification (cropping intensification)	4.50 ± 0.07	5.22 ± 0.06	4.08 ± 0.01	4.27 ± 0.06	4.52 ± 0.05	3	63.92	<0.0001
Indigenous/native trees cutting/degradation in villages	3.24 ± 0.04	3.31 ± 0.04	1.11 ± 0.01	2.10 ± 0.01	2.44 ± 0.03	3	88.92	<0.0001
Intensification of farm management practices	5.33 ± 0.05	5.49 ± 0.07	5.28 ± 0.04	4.17 ± 0.06	5.06 ± 0.05	3	89.07	<0.0001
Increasing application of pesticides/other agrochemicals	3.12 ± 0.06	1.25 ± 0.04	4.12 ± 0.02	3.21 ± 0.05	2.92 ± 0.04	3	74.56	<0.0001
Grazing intensification	5.38 ± 0.06	3.35 ± 0.05	5.22 ± 0.04	4.96 ± 0.04	4.73 ± 0.05	3	73.38	<0.0001
Fires burning intensification	1.27 ± 0.02	2.05 ± 0.03	1.29 ± 0.06	2.16 ± 0.04	1.69 ± 0.04	3	80.11	<0.0001
General environmental pollution/degradation	5.25 ± 0.03	5.24 ± 0.04	4.32 ± 0.04	5.22 ± 0.04	5.01 ± 0.04	3	61.61	<0.0001
Decreased/degradation of floral/nesting resources diversity	4.01 ± 0.02	1.34 ± 0.05	3.08 ± 0.01	3.34 ± 0.05	2.98 ± 0.03	3	91.04	<0.0001
Alien invasive species spread in the village	4.92 ± 0.03	5.17 ± 0.06	4.24 + 0.05	5.08 ± 0.02	4.85 ± 0.04	3	64.23	<0.0001
Cultivation of exotic high yielding varieties/genotypes	3.02 ± 0.04	1.24 ± 0.03	3.54 ± 0.09	5.20 ± 0.04	3.25 ± 0.04	3	90.85	<0.0001
Bee diseases/parasites and predation	4.92 ± 0.02	1.39 ± 0.06	4.09 ± 0.08	5.56 ± 0.08	3.99 ± 0.06	3	94.22	<0.0001
Increasing urbanization of rural towns	4.12 ± 0.04	4.06 ± 0.03	3.94 ± 0.06	5.38 ± 0.06	4.38 ± 0.05	3	83.85	<0.0001
Road construction	5.19 ± 0.04	5.08 ± 0.03	5.15 ± 0.06	4.98 ± 0.04	5.13 ± 0.05	3	6.89	0.075
Demographic pressures (human population increase, refugees migration)	4.45 ± 0.05	4.10 ± 0.01	4.16 ± 0.10	4.31 ± 0.04	4.26 ± 0.05	3	14.67	0.02
Natural calamities (landslide, river floods)	4.91 ± 0.05	1.93 ± 0.03	4.12 ± 0.09	4.06 ± 0.08	3.75 ± 0.06	3	76.73	<0.0001
Research and tourism (wildlife visits) intensification	5.11 ± 0.06	0.98 ± 0.01	5.15 ± 0.06	4.48 ± 0.09	3.93 ± 0.06	3	67.15	<0.0001
Local climate change (rainfall pattern changes)	2.18 ± 0.03	1.08 ± 0.02	2.14 ± 0.06	1.96 ± 0.02	1.84 ± 0.03	3	60.58	<0.0001
Regional and global climate changes	5.21 ± 0.04	4.99 ± 0.04	3.75 ± 0.07	4.34 ± 0.10	4.58 ± 0.06	3	77.20	<0.0001

Ranks: 1: very important (or primary causal) agent; 2: important causal agent; 3: associate (secondary) important causal agent, 4: playing tertiary role; 5: play very little or no role in the decline process; 6: has nothing to do with bee decline (not sure this can affect bees).

TABLE 17: Perceived problems of having a flower farm in the village.

Statement no.	Question: Do you have any problem with the presence of the flower farm in the village?	Frequency (%) of reasons/statements
1	I doubt if a single animal/insect will survive where that chemically-enriched waste soil has been dumped by flower farm	2.31
2	My animals have been seek ever since the farm came around	0.77
3	Papaw/avocado/pumpkin/mango flowers are aborting; even the few that formed fruits started rotting on the tree before the fruit is mature	1.55
4	People working for a long time (2–5 years) get contaminated by chemicals and eventually died when they came out from work	1.55
5	Ever since this flower farm come around, we are experiencing yield loss and abnormal (reduction) crop production	1.54
6	A flower farm should never be placed near people because the chemical sprays travel towards our village (>3 Km) up to inside our homestead and children get respiratory diseases	0.77
7	All our crops these days seek and can't breathe properly because of chemicals they apply there	0.77
8	Avocado, mango, papaw, and pumpkin are ripening/falling down earlier and are not tasty, ever since the flower farm came around	0.77
9	Avocado/mangos don't look good anymore since the establishment of the flower farm, may be because of the pesticides they spray which are poisons	1.54
10	Bees are moving away from the flower farm to our villages these days, crops near flower farms suffer	0.77
11	Cassava is stunt; other fruits are rotting these days due to chemicals; it used not be like that	0.77
12	Chemical dumped/deposited, at flower farm edge, during night hours, has a strong smell for >6 months	0.77
13	Chemical of the flower farm is dangerous to humans/livestock/wildlife because can't guarantee keeping polluted water there everyday	0.77
14	Chemical sprays from the flower farm smell alot in our village and they kill flowers of our crops	0.77
15	Chemicals sprayed daily from the flower farm have reduced the bee population in our village	1.54
16	Chemicals sprayed leave the flower farm and move in the air and we smell them too much; my kids fall sick regularly	0.77
17	Emerging pests and diseases in many crops may be due to fume from the flower farm	0.77
18	Every time the river/swamp water-enriched with water from the flower farm crosses my gardens, I harvestno fruits (egg plants/tomato)	3.80
19	Few crops are not very affected by these chemicals, but for most crops their yields are reducing dramatically	0.77
20	Flower farm dump regularly chemical wastes into the rivers around our villages and this kills fish	1.54
21	Crop flowers are quickly aborting and few formed fruits (mangos, avocado, papaw) do not taste well anymore	0.77
22	From most crops, few flowers are turning into fruits/seeds these days ever since the flower farm was established around	0.77
23	Fruit rotting problem started some 3 to 5 years ago; these days, no more good fruits; most of them are rotting	0.77
24	Fumes sprayed in the flower farm kill bees and other animals in the nature around because they smell strongly	0.77
25	I am convinced that bees disappearing in the village is due to toxic pesticides applied in the flower farm	3.08
26	I am not affected in my farming business by any activities around the flower farm	0.77
27	I am now abandoning growing fruit tree crops because they will not give good fruits or if a fruit comes, it is of low quality	0.77
28	I believe that the flower farm chemicals must have contributed to reduced yields and to poor yield performance of crops	0.77
29	I do not think chemicals sprayed in the flower farm are dangerous to our crops or to bees	0.77
30	I have no problem with the flower farm, either they are there or not I benefit or lose nothing	5.38
31	I have no problem, either they are there or not, it does not matter to me	0.77
32	I regularly see dead fish on the edge of the river that cross the flower farm into our village	3.08
33	I regularly see vehicles moving at night from the flower farm dumping smelly chemicals near our village	2.31

Is Cut-Flower Industry Promotion by the Government Negatively Affecting Pollinator Biodiversity and
Environmental/Human Health in Uganda?

59

TABLE 17: Continued.

Statement no.	Question: Do you have any problem with the presence of the flower farm in the village?	Frequency (%) of reasons/statements
34	I see no problem may be because I am a migrant worker, but one from this village has always complained about bees declining (disappearance)	2.31
35	I strongly believe that the fume of pesticides applied in the flower is responsible for currently observed flower abortion in most fruit crops in the village	0.77
36	I suspect the flower farm is dangerous to people and animals because of the nature of smelling the toxic chemicals they apply	0.77
37	I used to see many snakes/soil ant colonies around the village, these days I rarely see them passing around	0.77
38	Insect pests chased near the flower farm came in my (>1000 m) garden to eat/destroy my crops	0.77
39	Many children who went to work at the flower farm got exposed to these pesticides and died	0.77
40	Most of the mango/avocado fruits grown near the flower farm are no longer tasty or juicy	2.31
41	Never place flower farms near people because they don't need bees but farmers need bee'-visits to crops	0.77
42	No activity from the flower farm affects my crops; egg plants and tomato grow here in the village	0.77
43	No one has ever complained about chemicals applied there to kill his crops	0.77
44	No problem because I was informed that they have constructed several dams for recycling their water and for containing chemical wastes there	0.77
45	Papaw/avocado are no longer sweet, too many fruit diseases emerging these days; tree crops are currently yielding low	2.31
46	People living near the flower farm breathe polluted air with dangerous chemicals, fall sick frequently, and many die	0.77
47	People working at the flower farm become tiny and die after 3–5 years of working there	0.77
48	People fear to use the stream water because they believe it is full off chemical wastes dumped regularly by the flower farm	3.08
49	People living at the edge of the flower farm can smell the fumes (pesticides) scent from the flower farm daily	2.31
50	People who grow mango/avocado/papaw near (at the edge of) the flower farm experience low yields these days;they are complaining	2.31
51	People who work at the flower farm all get sick because of chemical breathing and many die after 2–5 years working	0.77
52	People who work in the flower farm start falling sick 3–5 years later	1.54
53	Pesticides applied in the flower farm are decolorating (changing the colour) our crops in the village	0.77
54	People who work in these flower farms complain all time about their health status andlow salaries (not paid in time)	3.08
55	Pumpkin and watermelon flowers do not turn into fruits anymore near the flower farm, but in the village (>2 km) fruits are many	0.77
56	Since the farm was established here, we experience rotting of fruits (avocado, papaw) and it may be due to chemicals	1.54
57	When smelling chemicals up to 2-3 Km here in the village, I lose my appétite and my children fall sick all the time	0.77
58	So far no complaint apart from the risk of our cows drinking the river water-enriched with chemical water from the flower farm	0.77
59	The flower farm is dumping the crop residues full of chemicals in our village and they smell a lot for 6 months	0.77
60	The flower farm pollute river/wetland waters crossing our village	1.54
61	The flower farm pour chemically-enriched water into the river that people use as drinking water and we drink that water	0.77
62	The flower farm spray daily massive toxic chemical that flies up to 2–4 km in the village	0.77
63	The fumes from the flower farm smell badly everyday their are applied; and we fall seek after being exposed for long time	0.77
64	The rate of abortion of my cows is very high these days and I think it is due to drinking polluted water from the flower farm	0.77
65	The rate of abortion of my cows is very high these days and I think it is due to drinking polluted water from the flower farm	1.54

Table 17: Continued.

Statement no.	Question: Do you have any problem with the presence of the flower farm in the village?	Frequency (%) of reasons/statements
66	The recycling of water around the flower farm is not properly done/controlled, there is still leakage of their water into the river	0.77
67	The reduction in crop yields in the village may be due to the fact that bees smell pesticides from the flower and die	1.54
68	The soil from the waste dumping point can smell a lot for more than 5 months	2.31
69	The yield of fruits (mango/avocado/passion fruits) has reduced to 15 to 35% since the flower farm came around	0.77
70	There are rumors that chemicals applied spoil our crops, but I am not sure if it is linked to seasonal/climate changes	2.31
71	There are several dumping holes of chemical wastes near our gardens	1.54
72	There are no more bees and honey production is absent in the village ever since the flower farm is around	0.77
73	There are several dumping points of chemically-enriched soils in our village	1.54
74	There was a problem of water leakage from the flower farm into the river, but they claimed to have controlled it (properly closed it), it is not correct	0.77
75	These days I see many fish dying/stunting because the water I use in my 12 fish ponds cross the flower farm	1.54
76	These days, there is too much water in the swamp because the flower sends a lot there daily	0.77
77	These people claim recycling/reutilizing water 2 times only, after that water, where do they put it? Not in the river?	0.77
78	These days, there is too much crop failure due to pests/diseases/chemicals applied in the flower farm	1.54
79	They kill bees with chemicals in the village because they don't want any single bee to land on the flower farm	0.77
80	Water from the flower farm is currently still leaking into the river and both human beings and cows drink that water and fall sick and die.	1.54
81	We meet several hundreds of dead bees/snakes/fish/lizard in the fields a day after they have sprayed	0.77
82	We are all migrant and care little about effects of pesticides on crops grown here	0.77
83	We regularly see several thousands fish dying near the flower farm where they eject water into Lake Victoria	0.77
84	Whenever they are at spraying the flower farm, the fumes comes into the air and reaches our houses and make us fall sick; including our wives and children	0.77
85	Yields of my crops are reducing because I think the flower farm killed all the bees in the village that pollinate my crops	0.77

Chi-square test for difference in frequency of statements: $\chi^2 = 59.98$, $N = 100$, DF = 84, $P = 0.983$.

4. Discussion

In this study, it was hypothesized that (i) pesticide applications inside greenhouses of the different flower farms, had no negative effects on bees inhabiting/foraging in the surrounding habitats; (ii) there was no difference in bee species and visitation frequencies among different types of landscape vegetations, fields, and habitats (land-uses) found in the surrounding of the flower farms; (iii) farmers were not aware of the importance of pollination in their farming business, (iv) small-scale farmers living in the surrounding of flower habitats that may be used by various bee species as refugia/nesting sites or as foraging habitats). Also, several other pollinator groups (Table 19) were identified during the course of this study although it was difficult to ascertain their role in the reproduction success of plants/crops visited.

farmers benefited a lot from the presence of the flower farm in their villages.

The results indicated (i) a lack of scientific evidence for direct effects of pesticides applied inside greenhouses on bee biodiversity although bee species richness and visitation intensities tended to increase with distance from the flower farm into farmlands; (ii) that bee species and visitations varied according to flower farm location, to the distance from the edge of the flower farm into farmland, to field types, landscape vegetations, and to habitat types found in the surroundings of the flower farms; (iii) that more than 60% of small-scale farmers were aware of the value and importance of pollination in their farming business and believed that causes of bee decline were pesticides applied by flower farm and natural habitat degradation (forest clear cut); and (iv) that the majority of farmers declared benefiting nothing from the presence of the flower farm in their village and requested that flower farm should be removed from their villages since it is

Is Cut-Flower Industry Promotion by the Government Negatively Affecting Pollinator Biodiversity and
Environmental/Human Health in Uganda?

61

TABLE 18: Opinion of farmers about sustainable solutions to solve the conflicts created by the establishment of the flower farm (farmers' sustainable and win-win solution) to the development of the flower industry in the villages of Uganda.

Statement no.	Question: Which sustainable solution do you propose to resolve your conflicts with the flower farm?	Frequency (%) of statements by farmers interviewed
1	Destroy or move away the flower farm where there are no people	5.9
2	I do not know what should be done because they have consulted politicians to come here	3.0
3	I do not know what should be done, but I am worried that chemicals will continue to kill/finish people in the village	2.2
4	I want them to leave our village	2.2
5	I want them to pay (buy) my land and I will leave the place, no other solutions	1.5
6	I want to go away to farm and leave this land for them	1.5
7	Isolate or remove the flower farm from our village	5.2
8	Maintain the flower farm around because we get clean water/electricity and charge our phones there	0.7
9	Need regular meeting with them to tell us that chemicals applied are not dangerous to living organisms	2.2
10	No problem, no need to disturb them in the village; they provide employment opportunity to communities	5.2
11	Sensitize people about the bad and the good things from the flower farm	2.2
12	Since the flower farm was established here, we are worried and we keep speculating about the negative effects on our crops and health	1.5
13	The flower farm should leave the village and go to a place where people are not	1.5
14	They can stay there because their activities do not affect me at all	8.1
15	They need to show us evidence that chemicals applied are not dangerous to crops/livestock/wildlife/fish	1.5
16	They should stop application of dangerous pesticides to humans and crops	5.9
17	They should use no toxic chemicals on bees/snakes/fish/lizard	2.2
18	I don't know what should be done (I have no power to propose what should be done)	47.4

Chi-square test for difference in frequency of statements: $\chi^2 = 347.129$, $N = 99.9$, DF $= 17$, $P < 0.0001$.
The frequency is the number of times the statement was given by people interviewed; as one interviewed person could offer several statements for each question asked by the researcher.

polluting their environment as well endangering the health of people working there.

4.1. Declining Status of Bee Diversity and Numbers in Flower Farm Zones and in Ugandan Farmlands.
Worldwide pollinators (bees) decline has been declared. However, in the African context and particularly in Uganda, there is no historical data collection to verify decline in pollinator species. It is therefore difficult to accurately say that pollinators have ever declined or not in Uganda. In the absence of historical databases, the only source of information is through asking small-scale farmers (local residents), particularly those who know pollination and pollinators and who care and believe pollination is an important production factor in their farming business. Reading from the memory of small-scale farmers, it appears that there has been some decline in bees in central Uganda, although it is difficult to say exactly at which rate bees did decline compared to current inventories or to identify the exact causes.

Although it was perceived in this study that bees have been declining around flower growing regions, there exist no or very few long-term data on bee populations; and this makes their conservation status difficult to assess in rural landscapes of Uganda. It is difficult to know which species are clearly declining given the absence of long-term monitoring data in Uganda. In addition, recent studies [19–23] indicated that bee communities from farmlands of central Uganda were characterized by a high rate of rare and unique taxa, and this indicated that the Apoidea communities from Uganda farmlands were at high risk of extinction. Obviously the current population of bees cannot be the same as to what happened some 10–50 years ago.

4.2. Influence of Habitat Types on Bee Richness and Abundance.
Bee communities living in the surrounding of the flower farms can show a stronger response to changes in habitat structural properties (vegetation composition) and to landscape composition [31]. Although bee community composition differed between habitats/land-uses within study sites (flower farms), only bee species richness showed a consistent difference between habitats across the studied sites (flower farms). The degree to which bee community composition differed between habitats within sites may be related to landscape variables. In fact, it was found in this study that the difference in bee species richness and abundance between habitats was strongly related to the difference in farm-land-scape vegetation types (croplands, grasslands). However, the strong influence of landscape vegetation on bee species richness and

TABLE 19: Preliminary list of other pollinators recorded in farmland/grassland habitats surrounding the four flower farms during the course of the study.

Order	Family	Subfamily	Species, morphospecies, and doubtful identification	Abundance*
Coleoptera	Chrysomelidae		*Eurythenes* sp.	1.3
Coleoptera	Melolonthinae		*Peritrichia albovillosa* (Schein)	0.6
Coleoptera	Nitidulidae		*Meligethes variabilis* (Reitter)	2.2
Coleoptera	Scarabaeidae		*Rhabdotis* sp.?	2.5
Diptera	Agromyzida	Agromizinae	*Melanagromyza* sp.	6.3
Diptera	Calliphoridae	Calliphorinae	*Hemigymnochaete varia*	5.3
Diptera	Calliphoridae	Rhiniinae	*Rhinia apicalis*?	3.0
Diptera	Calliphoridae	Chrysomyinae	*Chrysomya chloropyga*	15.4
Diptera	Lauxaniidae	Lauxaniinae	*Lauxania* sp.	10.6
Diptera	Muscidae	Muscinae	*Musca* sp.	47.8
Diptera	Sarcophagidae	Sarcophaginae	*Sarcophaga inaequalis*	50.9
Diptera	Sarcophagidae	Miltogramminae	*Hoplacephala tesselata*	8.5
Diptera	Syrphidae	Syrphinae	*Allobaccha* sp.	51.3
Diptera	Syrphidae	Syrphinae	*Betasryphus adligatus*?	79.6
Diptera	Syrphidae	Syrphinae	*Allograpta nasuta*?	116.7
Diptera	Bibionidae		*Bibio turneri* (Edwards)	16.2
Diptera	Bombyliidae	Bombyliinae	*Bombylisoma* sp.	82.8
Diptera	Calliphoridae	Chrysomyinae	*Chrysomya regalis* (Desvoidy)	41.1
Diptera	Calliphoridae	Rhiniinae	*Rhyncomya cassatis* (Walker)	67.0
Diptera	Muscidae	Phaoniinae	*Atherigona addita* (Malloch)	7.9
Diptera	Muscidae	Muscinae	*Musca lusoria* (Wiedemann)	27.6
Diptera	Mycetophilidae	Mycetophilinae	*Phronia* sp.	27.3
Diptera	Sarcophagidae	Miltogramminae	*Pterella* sp.	52.7
Diptera	Syrphidae		*Eristalinus laetus* (Wiedemann)	38.6
Diptera	Syrphidae		*Phytomia natalensis* (Macquat)	76.4
Diptera	Syrphidae		*Senaspis haemorrhoa* (Gerstacker)	58.8
Diptera	Syrphidae	Milesiinae	*Phytomia incisa* (Wiedemann)	118.3
Diptera	Syrphidae		*Metasyrphus* sp.	151.8
Diptera	Tachinidae	Goniinae	*Aplomya* sp.	156.0
Diptera	Tephritidae		*Neoceratitis* sp.	86.3
Hymenoptera	Formicidae	Formicinae	*Acantholepis* sp.	3.5
Hymenoptera	Formicidae	Formicinae	*Anoplolepis* sp.	5.3
Hymenoptera	Formicidae	Formicinae	*Camponotus* sp.	8.9
Lepidoptera	Nymphalidae	Nymphalinae	*Hypolimnas misippus* (Linnaeus)	0.7
Lepidoptera	Nymphalidae	Nymphalinae	*Junonia hierta cebrene* (Trimen)	1.0
Lepidoptera	Nymphalidae		*Byblia ilithyia* (Drury)	0.6
Lepidoptera	Nymphalidae	Nymphalinae	*Vanessa cardui cardui* (Linnaeus)	0.5
Lepidoptera	Hesperiidae	Coeliadinae	*Coeliades anchise anchise* (Gerstaecker)	0.4
Lepidoptera	Hesperiidae	Pyrginae	*Celaenorrhinus proxima proxima* (Mabille)	0.7
Lepidoptera	Hesperiidae	Pyrginae	*Spialia dromus* (Plötz)	0.4
Lepidoptera	Hesperridae	Hesperiinae	*Acleros mackenii* (Trimen)	0.4
Lepidoptera	Hesperridae	Hesperiinae	*Gegenes hottentota* (Latreille)	0.6
Lepidoptera	Lycaenidae	Polyommatinae	*Anthene amarach* (Guerin)	0.9
Lepidoptera	Lycaenidae	Polyommatinae	*Anthene butleri* (Oberthur)	0.3
Lepidoptera	Lycaenidae	Polyommatinae	*Anthene larydas* (Cramer)	0.2
Lepidoptera	Lycaenidae	Polyommatinae	*Anthene lunulata* (Trimen)	0.4
Lepidoptera	Lycaenidae	Polyommatinae	*Cacyreus palemon palemon* (Stoll)	0.6

Is Cut-Flower Industry Promotion by the Government Negatively Affecting Pollinator Biodiversity and
Environmental/Human Health in Uganda?

63

TABLE 19: Continued.

Order	Family	Subfamily	Species, morphospecies, and doubtful identification	Abundance[*]
Lepidoptera	Lycaenidae	Polyommatinae	*Cupidopsis cissus* (Godart)	1.1
Lepidoptera	Lycaenidae	Polyommatinae	*Cupidopsis jobates* (Hopffer)	3.2
Lepidoptera	Lycaenidae	Polyommatinae	*Euchrysops osiris osiris* (Hopffer)	10.8
Lepidoptera	Lycaenidae	Polyommatinae	*Leptotes pirithous* (Linnaeus)	2.4
Lepidoptera	Lycaenidae	Polyommatinae	*Zizeeria knysna* (Trimen)	1.3
Lepidoptera	Lycaenidae	Polyommatinae	*Zizula hylax* (Fabricius)	0.5
Lepidoptera	Lycaenidae	Polyommatinae	*Zizina antanossa* (Mabille)	0.5
Lepidoptera	Lycaenidae		*Leptosia nupta nupta* (Butler)	0.4
Lepidoptera	Lycaenidae	Theclinae	*Hypolycaena hatita ugandae* (Sharpe)	0.3
Lepidoptera	Lycaenidae	Theclinae	*Iolaus poultoni* (Riley)	0.3
Lepidoptera	Nymphalidae	Danainae	*Amarius ochlea* (Boisduval)	0.4
Lepidoptera	Nymphalidae	Danainae	*Danaus chrysippus* L.	0.3
Lepidoptera	Nymphalidae	Limenitinae	*Byblia anvatara acheloia* (Wallengren)	0.2
Lepidoptera	Nymphalidae	Limenitinae	*Pseudacraea boisduvali* (Doubleday)	0.4
Lepidoptera	Nymphalidae	Nymphalinae	*Junonia stygia gregorii* (Butler)	6.8
Lepidoptera	Nymphalidae	Nymphalinae	*Junonia terea elgiva* (Hewitson)	0.2
Lepidoptera	Nymphalidae	Satyrinae	*Bicyclus safitza safitza* (Hewitson)	2.0
Lepidoptera	Nymphalidae	Satyrinae	*Ypthima antennata antennata* (van Son)	0.3
Lepidoptera	Nymphalidae		*Hamanumida daedulus*	0.2
Lepidoptera	Nymphalidae		*Junonia eonone eonone* (Linnaeus)	0.3
Lepidoptera	Nymphalidae	Acraeinae	*Acraea zetes Hewitson*	0.2
Lepidoptera	Nymphalidae	Acraeinae	*Acraea acerata* (Hewitson)	0.4
Lepidoptera	Nymphalidae	Acraeinae	*Acraea acrita* (Hewitson)	0.4
Lepidoptera	Nymphalidae	Acraeinae	*Acraea aganice* (Butler)	0.2
Lepidoptera	Nymphalidae	Acraeinae	*Acraea cabira* (Hopffer)	0.8
Lepidoptera	Nymphalidae	Acraeinae	*Acraea eponina* (Cramer)	0.3
Lepidoptera	Nymphalidae	Acraeinae	*Acraea neobule* (Doubleday)	0.9
Lepidoptera	Nymphalidae	Acraeinae	*Acraea zonata* (Hewitson)	0.3
Lepidoptera	Papilionidae		*Papilio demodocus* (Esper)	0.7
Lepidoptera	Pieridae	Pierinae	*Belenois thysa* (Hopffer)	0.6
Lepidoptera	Pieridae	Pierinae	*Colotis antevippe zera* (Lucas)	0.2
Lepidoptera	Pieridae	Pierinae	*Leptosia alcesta* (Bernardi)	0.8
Lepidoptera	Pieridae	Pierinae	*Mylothris agathina agathina* (Cramer)	0.8
Lepidoptera	Pieridae	Pierinae	*Nepheronia argia* (Suffert)	0.4
Lepidoptera	Pieridae	Pierinae	*Pontia helice johnstoni* (Crowley)	0.5
Lepidoptera	Pieridae	Coliadinae	*Catopsilia florella* (Fabricius)	0.6
Lepidoptera	Pieridae	Coliadinae	*Colotis danae* (Fabricius)	0.3
Lepidoptera	Pieridae	Coliadinae	*Eurema brigitta* (Stoll)	0.7
Lepidoptera	Pieridae	Coliadinae	*Eurema hecabe solifera* (Linnaeus)	0.5
Lepidoptera	Sphingidae		*Agrius convolvuli* (Linnaeus)	0.7
Lepidoptera	Sphingidae		*Hippotion celerio* (Linnaeus)	0.6
Lepidoptera	Sphingidae		*Leucostrophus hirundo* (Gerstaecker)	0.4
Lepidoptera	Sphingidae		*Macroglossum trochilus* (Hubner)?	0.3

[*]Total visitations: average numbers of individuals counted collecting floral resources (nectar/pollen) on flower patches/180 min/24 h (night or day) across the
4 flower farms during faunistic surveys of 2010–2012.

abundance is likely to be caused by differences in quality and availability of floral resources and nesting sites across flower farms visited. Seminatural habitats like fallows had the greatest species richness and abundance in both croplands and grasslands; and this information indicated that increasing proportion cover of fallow resources in flower growing regions can benefit species diversity at both habitat and landscape scale levels through their ability to support a greater number of flowering species and nesting resources per unit area than other seminatural habitats/land-uses. Overall, a growing body of work has shown a positive relationship between crop yield per unit area, the diversity and abundance of insect pollinators, and the amount of native vegetation in the area around the crop. When temporally and spatially integrated into the agricultural matrix, natural environments benefit pollinators (bees) and become important sources of a diversity of pollinator species, probably because they provide good nesting sites and food resources [20–25].

4.3. Drivers of Bees in the Surrounding Habitats of Flower Farms. Two processes globally threatening natural ecosystems, biodiversity, and services delivered are changes in land-use and deforestation [47]. Bees are ecologically and functionally diverse, foraging on a broad array of floral forms, nesting in a wide variety of substrates [19], and responding to both nesting and foraging resources at a wide range of spatial scales. Bees require multiple resources to complete their life cycle, including pollen, nectar, nest substrates, and nest-building materials. These resources are often gathered from different locations, making bees reliant on multiple, "partial habitats." This might make bees vulnerable to disturbance insofar as they would be negatively affected by the loss of any of these habitats. Thus, rural landscapes that offer heterogeneous nesting and floral resources, in both cultivated and noncrop areas provide a good habitat for a diverse suite of bees. Also, the main cause of bee decline in the UK, USA, and Western Europe is widely agreed to be the agricultural intensification that took place in the 20th century [28, 47].

When asked about the causes of bee biodiversity decline in the village, most participants linked it to various factors. The vast majority of households interviewed were aware of significant changes in the local bee populations, most of which they interpreted as a product of several interacting anthropogenic factors. Among threats to pollinators identified by farmers included deforestation, habitat fragmentation, fires-slash and burn agriculture, use of agrochemicals, charcoal production, pests and diseases, poor honey harvesting methods, and climate change. However, nearly half (50%) of the interviewees believed that few factors played key a role in the decline of wild bees, including pesticide application and habitat alteration (fragmentation/degradation) as primary causes of pollinator declines. Most frequently, interviewees often pointed out forest clear cut and agricultural chemicals as key drivers of pollinator decline in their villages. Several other farmers commented with details on how pesticides used in the flower farm affect bees and their crops. Very few farmers assigned the problems of bees decline to climate change. Two to three households noted the importance of bee diseases

in the decline processes. Overall, this study surveys agrees with previous findings [35] about the role played by habitat alteration and pesticide applications as key drivers of bee decline worldwide. Pollinator unfriendly farming practices such as pesticides application, forest clearance, and annual fire burning are well acknowledged by farmers (from various tribes) as important factors accelerating bee decline in villages in Uganda [22]. However, such disturbances and related drivers (pressures) and associated loss of bees will have severe impacts on crop production and the livelihood of rural people in the nearest future, if nothing is done by the government (policy-makers).

Bees have been shown to increase plant seed set, improve fruit quality, and improve fruit quantity. Given their importance for floral biodiversity and food production, the status of bees is subject to increasing concern, with evidence of declines found in some species/groups from the local to the national scale. Several factors are recognized as potential drivers of bees decline [40]. The use of pesticides in agriculture, parasites, habitat loss, and fragmentation have all been linked to pollinator decline. As mentioned above, agricultural intensification, of which an increased use of pesticides is an integral part, is often argued to be a key driver of pollinator decline [36, 40, 48] worldwide. It is well documented that agricultural intensification (pesticides and fertilizers application intensification) has negative effects on species richness, abundance, and diversity of wild bees [37, 38, 41]. The application of pesticides quickly affects bees, particularly susceptible species. The loss or reduction in abundance of a species can subsequently indirectly affect other trophic levels, within the ecological network of which they are a part [40]. Pesticides are likely to affect whole plant-pollinator interactions and communities [40]. Pesticide application is an anthropogenic disturbance that can promote pollinator species extinction and induce a sudden collapse in pollination networks [49, 50]; and such situation can pose serious threats leading to high levels of crop production failure in rural landscapes.

For example, while studying the effects of farmland management (agriculture intensification) and environmental factors at local and landscape scales on bees in central Hungary, it was realized that increasing the amount of fertilizer decreased total and small bee abundance and increased the coefficient of variation of total bee abundance [50]. The study also realized that insecticide use had a significant negative effect on total and small bee species richness and on large bee abundance, although the percentage cover of seminatural habitats in a 500 m radius did not influence bee species richness and abundance. It was concluded that the intensification of farmland management poses a threat to bee diversity and thus may reduce pollination services [50]. The reduction of the amounts of fertilizer and a cessation of insecticide application were advocated for to lead to high bee species richness and abundance and ensure the pollination of wild plants and flowering crops [50]. The disappearance of bees has great a impact on the livelihood and food security of the population. It also has an impact on the national and regional economy for development.

Is Cut-Flower Industry Promotion by the Government Negatively Affecting Pollinator Biodiversity and
Environmental/Human Health in Uganda?

65

4.4. Role Played by Pesticides Applied inside Flower Farms on the Decline in Pollinators Living in the Surrounding Landscapes in Uganda. During field visits, information on agrochemicals used by the different flower farms was collected. A checklist of different fertilizers and pesticides used and dosages is given in Tables 4 and 5. Among the different agrochemicals used, the effects of fertilizers on pollinators are not clearly documented; however, the effect of pesticides on pollinators is largely documented in the literature [35, 40]. Herbicide applications are known to diminish food resources in fields and in field margins and fallows. Insecticides are expected to directly affect bees with lethal or sublethal effects. Fungicides can also be harmful to bees sometimes through synergic mechanisms with insecticides [41].

To our knowledge, there exists not one study of the impact of inorganic fertilizer (NPK) application in greenhouses of flower farms on pollinators living in the surrounding habitats. This is the first study examining such effect. Even when studies of the impact of fertilizers on biodiversity exist, they generally conclude that the overall range of effects of inorganic fertilizers on species richness and abundance is arguably negligible or of little impact [51]. For example, a recent study was conducted in UK farmlands to identify the impacts of farm management practices (cattle grazing, inorganic fertilizer applications) on species richness, abundance, and community composition and on functional groups of diverse invertebrate taxa (beetles, plant hoppers, true bugs, butterflies, bumblebees, and spiders) and plants (forb species richness, grass species richness, and sward architecture). The study concluded that the effects of inorganic fertilizer applications were generally weak, both across trophic levels and within the phytophagous feeding guilds of invertebrate groups studied [51]. Even bumblebees showed no strong response (reduction, increase) to inorganic fertilizer applications [51]. However, plant species richness and abundance increased with intensity of fertilizer application [51]. Contrastingly, another study showed that increased nitrogen (and related inorganic fertilizes) input can cause a decline in floral resource diversity and abundance (nectariferous native plants) in European farmland habitats [41]. Overall, there is a need to determine the trends in the response of bees to inorganic fertilizations in and outside flower farms since they are agrochemicals and they cannot be without disturbance of surrounding local ecosystems.

All insecticide types used by the flower farms are well documented to be toxic to bees and to other beneficial insects [20–26]. With the exception of fungicides to which negative effects on bees are less documented, most of the insecticides (including miticides/nematicides) and herbicides used by the flowers are of broad spectrum action and are very toxic to bees [35, 52]. Most of these insecticides are on the "red list of the World Health Organisation (WHO) classification of pesticides" with negative effects on beneficial organisms. Some other pesticides used are not on the list of dangerous pesticides by the WHO (because they are commercialized with different names) but are highly toxic to bees. It was commonly found that flower farms used WHO highly (Ia and Ib) and moderately (II) hazardous pesticides. Pesticides like Dithane, Gramaxon, Round-up, Decis, Copper, Pyrethrins,

Score, Previcur, Confidor Cypermethrin, Ambush, Larvin, Decis (Deltametrin), Rogor (Dimethoate), Dimethoate, Mancozeb, Malathion, Ambda-Cyhalothrin, and so forth have been classified by the WHO as highly (Class-I) to moderately (Class-II) hazardous. There is a need for flower farms to adopt the use of pesticides that are slightly hazardous to the environment or pesticides unlikely to present acute hazard in normal use following WHO guidelines [53]. Flower farms should adopt using approved pesticides with lowest residual toxicity, which should be selected and sprayed when bees are not actively foraging, during the late evening times, at night, or in very early morning hours.

Currently, a mechanism by which agrochemicals applied inside flower farms may affect bee biodiversity inhabiting habitats in the surrounding of flower farms remains largely unknown. No previous studies have compared pesticide toxicity in wild bees in Uganda [23]. However, laboratory studies suggest that bees feeding on contaminated pollen/nectar at high concentrations are likely to encounter/experience lethal effects in the wild [21]. In Uganda, there exist several other causes of pollinators decline including pests and diseases and climate change, but the application of agrochemicals weighs heavy compared to other causes (drivers). Responses of farmers interviewed about the negative effects of agrochemical applied by the flower farms were almost similarly immediate with a large proportion of households responding that these chemicals directly killed bees, indirectly disturbing the reproduction process of crops and directly damaging the quality of fruits they grow. However, it was difficult to verify such assumptions by farmers.

Pesticide application causes a lot of damage to wildlife and ecosystem biodiversity [54]. Pesticides have a major impact on both honeybee and wild bee populations; and most pesticides used in agriculture are toxic to bees [52]. For example, the annual external cost of honeybee and pollination losses from pesticide applications in the USA was estimated to be of US$409.8 million in 2002 [52, 54]. Later, it was estimated that the annual agricultural losses due to the reduction in pollination services caused by pesticides may be as high as $4 billion a year in the USA [52].

The government of Uganda should regulate importation of pesticides, mainly an officer should be trained and attached to flower farms to control the quality of chemicals imported, particularly verifying the name of active ingredients and commercial names and the degree of toxicity of these chemicals to living organisms (including bees), the length of residual toxic effects in hours or days.

Most of the chemical pesticides used by flower farms can easily diffuse into the environment. These chemicals pesticides are also of high persistence in the environment and highly toxic to several member species of Apoidea communities. Insecticides applied by the flower farms can affect the foraging and homing behaviour of wild bees at sublethal to lethal concentrations with acute effects [40]. Once the air is polluted by these chemicals, most bee species foraging in the surrounding environment cannot survive.

Much as there was no direct evidence of the impact of chemicals applied on bees, it was realized that bee species richness increased with distance from the edge of the flower farm into farmlands/grasslands, and this indicated that something was happening around flowers that may be one of the causes of low bee species richness and abundance in the proximity of different flower farms. Bees are highly sensitive to many insecticides [35] like those applied by the different flower farms in Uganda. Honeybees, and likely other bees as well, have relatively few detoxification genes, which increases their susceptibility to pesticides. Native bees nesting near crops can experience more exposure since they forage frequently on floral resources nearby their nesting sites in the surrounding of fields; thus, there is a need for growers to reduce or avoid spraying pesticides during periods of bee activities.

Although little is known about the effects on wild bee communities of typical insect pest management programs used pre- and postcrop blooms, native bees exhibit a wide range of responses to pesticide applications to crops, but most bees are very sensitive to toxic insect pest management practices [55]. Recent studies show that wild bee communities are negatively affected by increasingly intensive chemical pest management activities in crop fields. It is likely that a significant link between typical crop pesticide applications and bee abundance and species richness in the field exist [55] in most cropping systems of the world. The study also demonstrated that both solitary and social foraging bee species declined with increasing exposure to applied pesticides in the field [55]. Insecticide application before crop bloom has ceased and that is likely to have a greater impact on social species than on the solitary [55].

Much as it is not yet clear how pesticide application affects pollinator community structures and how these alterations subsequently affect pollination service delivery to crops, it remains a major threat to crop production in Uganda, particularly in regions where flower farms have been established.

Bees are highly sensitive to many insecticides [35] like those applied by the different flower farms in Uganda. All over the world, honeys have been found to be positively sensitive [56] to various types of pesticides (herbicides acaricide, fungicide, insecticide, nematicide, molluscicide, etc.). Although honeybees (Apis mellifera L.) are among the most important pollinators in natural and agricultural settings, they are like other beneficial insects, very sensitive to insecticides (carbamates, nicotinoids, organochlorines, organophosphates, pyrethroids and miscellaneous, etc.). Honeybees, and likely other bees as well, have relatively few detoxification genes, which increases their susceptibility to pesticides. Native bees nesting near crops can experience more exposure since they forage at times of day and times of year. In Uganda, there exist several other causes of pollinators decline including pests and diseases [57] and climate change, but application of agrochemicals weight is heavy compared to other causes (drivers). There is therefore a need for small-scale growers to reduce or avoid spraying pesticides on their crops during periods of bee activities on flowers.

While studying the effects of insecticide Phosmet application on solitary bee foraging and nesting in orchards in the USA, no difference was found in the number of native bee visits to several species of plants flowering in and near orchards immediately before and 1 day after spraying [58].

Conversely, the nesting studies using the semidomesticated alfalfa leaf cutting bee, Megachile rotundata (F.), showed strong significant declines in the number of adult males, nesting females, and progeny production subsequent to spraying at distances of up to 160 m from sprayed orchards where the bees were presumably foraging [58]. This result from the USA indicated that pesticide application does not only directly affect foraging activities of bees in the field sprayed but the effect can go far away from the central point of spraying in any areas where bees are apt to forage from. Hence, currently, much as there was no evidence for direct effects of pesticides applied in greenhouses on bees foraging in the habitats surrounding flowers farms, it is believed pesticides applied inside greenhouses could have broad effects beyond greenhouses.

Despite the fact that only limited research has been done on the topic, studies consistently find pesticide residues in air. Pesticide sprays can directly hit nontarget vegetation or can drift or volatilize from the treated area and contaminate air, soil, and nontargeted organisms such as flowering plants. Some pesticide drift occurs during every application, even from ground equipment. Drift can account for a loss of 2 to 25% of the chemical being applied, and this can spread over a distance of a few meters to several hundred kms. Practically, various types of pesticides can volatilize (i.e., they can evaporate from soil and foliage, move away from the application, and contaminate the environment). As much as 80–90% of an applied pesticide can be volatilized within a few days of application. Many pesticides used by the flower farms in central Uganda are of long-lived pesticides; they can be omnipresent in the air (atmosphere) throughout the year. In fact, most of the flower farms visited had their greenhouses open on at least two sides (Figure 1). Application of pesticides was conducted early morning and later in the evening at a time when local wind direction was high and since these greenhouses had two sides open, wind could circulate inside and carry droplets of pesticides outside the greenhouses into nearby farmland habitats. In addition, small-scale farmers located within 10 m to 1000 m far away from the flower farm said they could smell pesticide applications and knew when flower farms were applying their toxic fumes. Also, this study found that bee species richness and population density increased with distance away from the flower farm. High bee species richness and population density were found at beyond 1.5–2 km far away from the flower farms. These observations combined made to believe that pesticides applied inside flower farm were contributing to reduction of bee biodiversity, particularly in habitats immediately surrounding them.

Bees are highly sensitive to pesticides application although the dose may vary from a product to another one. At the current doses of application of pesticides inside the flower farms, it is possible that a high proportion is driven by the air runoff in the farmlands. However, it is not clear at which doses they reach farmland habitats. Bees are sensitive to pesticide applications at any amount. It is likely that most bees flying around the flower farms can die in the next 30 to 60 min after being exposed to these chemicals. To illustrate this assumption, an example of a scientifically proven study can be used. The impact of bifenthrin on honeybees, Apis mellifera

Is Cut-Flower Industry Promotion by the Government Negatively Affecting Pollinator Biodiversity and
Environmental/Human Health in Uganda?

67

(Hymenoptera: Apidae) was evaluated in both laboratory and semifield assays to determine killing time after 15, 30, and 60 min of honeybee exposure. Bifenthrin was applied at 9.7 mL/liter, 19.5 mL/liter, and 29.5 mL/liter of water to common landscape vegetation, *Melampodium paludosum* Melanie (show star) and *Duranta erecta* L. (golden dewdrop), a water control, was also used [59]. The results indicated that both dose and exposure time significantly affected honeybee mortality. The application dose of 35 mg/mL resulted in 100% bee mortality at all time intervals. Bee mortality was significantly higher at 29.5 mL/liter compared to the mortality at 19.5 mL/liter and 9.7 mL/liter application rates after 24 h exposure to the treated vegetation [59].

4.5. Potential Causes of Reduction in Bee Numbers and Species along Distance from the Flower Farms into Farmlands. In this study, it was observed a reduction in bee numbers and species at the edge of flower farms compared to far away from the flower farms. However, it was not clear if chemicals applied inside greenhouses could be the cause. Overall, there is no clear evidence of the decline in bee biodiversity due to activities within the flower farm but the fact that bee abundance and species richness tended to increase at distances beyond 2 km way from the flower into farmlands indicated that something was happening, forcing some sensitive bee species to find refuge far away from the flower farm.

Although there was no direct evidence for the flower farm activities to cause the reduction in bee species and numbers nearby the flower farm, it is believed that agrochemicals used inside flowers may be responsible for such reduction. Various scenarios (speculations) of some activities happening inside glasshouses may help in explaining possible connections with bee reduction in nearby farmland habitats.

(i) During field surveys, it was observed that the flower farmers did not have good control measures of the runoff of water from inside glasshouses to outside in the surrounding nature. In most cases, chemicals were not fully contained because dead fish and frogs were frequently observed in most wetlands and aquatic zones that receive water from flower farms. As evidence, farmers reported that "they cannot see any snakes, lizards or frogs moving around in the bush ever since the flower farms came around in their village." Also, containing runoff of chemical in the soil, air, and water was difficult in most flower farms. Chemicals in water could affect water bodies and this was visible by animals and fish seen dying along field transects. It was difficult to confirm the existence of chemical accumulation in soil of the surrounding farms; however, it is recommended that further investigations focus on quantifying activities of earthworms in soils around flower farms to determine if they are limited compared to soils sampled from farms located far away. The aim would be to identify if ground-nesting bees are not threatened by chemical accumulation in the soils of small-scale farms and of natural/seminatural vegetation found in the surrounding of flower farms. It is therefore recommended that good methods to contain runoff of water from flowers are set, tested, and evaluated.

(ii) Air is a vector of most dangerous chemical pollutants for bees. Chemicals from industrial factories can be driven in the air and dropped on the vegetation. Thereafter, bees that are busy collecting floral resources can ingest these pollutants. Similarly, a ground nesting bee will fail to make a choice of place as nesting site if felt it is polluted. During the time of surveys, it was difficult to detect pollution of the air by chemicals from flower farms. However, the distance at which these chemicals (commonly called by farmers "fumes") could move was high. During an early morning survey, the researchers were found near inhabitant houses located at 1000 m far away from the flower farm while chemicals were being sprayed in the flower farm (pearl flower farm) and 30 minutes after ending, the smell of the "fumes" could be felt high and strong. The "fumes" were driven by the air and farmers interviewed indicated that they have to run away from their houses for at least 2 hours when the flower is spraying.

The "walls" of the flower farms were not closed to contain the "fumes" inside glasshouses. Consequently, after spraying, under windy conditions, the "fumes" could move in the air and be felt by communities living even at 2500 m far away from the flower farm. These chemicals moving in the air, polluting the nearby environment could affect bees that could inhale them and die in their hives. Consequently, few bees could be observed around flower farms. It is therefore recommended that toxicological studies are conducted in all environments nearby flower farms to determine and identify bee toxic components in the air in relationship to the type of agrochemical used in flower farms. It is important to identify most hazardous chemicals among those preferred and imported by flower farms.

(iii) During farms' survey, most farmers indicated some flower farms (eg., Fiduga) were dumping, waste of the agrochemicals they use at the edge of their flower farms nearby villages. Dumping points were counted at the density of 20 per hectare by the researchers during the surveys. These dumping materials were not covered by soils.

(iv) In addition, some farmers living near Mairye estates were observed collecting and using waste of agrochemical and soil wastes soils as fertilizers in their gardens. Agrochemicals and soil wastes collected and used in gardens as fertilizers are actually mixtures of fertilizers and pesticides used inside the glasshouses. Hence, some farmers ignorantly contributed to the spread of chemicals into the farmlands and it was difficult to observe a bee foraging nearby here these waste chemicals have been put in the gardens. Bee nesting is difficult in such environments.

All these above described activities carried out inside glasshouses could lead pesticides outside. Other causes of reduction of bees around flower farms are human activities (of small-scale farmers) that are carried out near flower farms. In their farming process, small-scale farmers may be engaged in farming practices that are not pollinator-friendly or in activities that can negatively affect pollinators. During field work, the researcher inventoried a certain number of bad agricultural practices: bush burning, seminatural habitats clearing, over-cultivation, and herbicides and pesticides application by certain farmers. However, these observations were not generalized. In the surrounding of Fiduga flower farms, some farmers were observing that spraying chemicals and such activities could contribute to the reduction in bee numbers and species at the edge of flower farms. It is recommended that future studies are conducted to explore and determine the role played and the contribution of human activities (small-scale farmers' activities) in the reduction of bees in the landscapes immediately surrounding flower farms compared to the role played by agrochemicals from the flower farms. Such further research should be conducted to understand and describe how activities carried out inside greenhouses can affect bees living in adjacent habitats. Such research should identify and clearly point out factors involved in causing the reduction in bees.

4.6. Potential Consequences and Implications of Loss/Decline in Bees on Agricultural Production, the Economy, and Local Community Livelihoods. Evaluating the potential risks related to the loss of pollination services in Uganda is not easy. However, the risks posed by failure of pollination services delivery are many including depletion of biodiversity. The loss of pollinators might also precipitate shift in vegetation communities. Any decline in pollination services could lead to cascading effects on ecosystems and an overall loss of biodiversity and ecosystem services (such as agricultural production). Mutualistic relationships are most directly affected since the loss of individual pollinator species will lead to the extinction of any codependent plant species (and vice versa). Ecosystems possess some robustness to the loss of individual species since multiple pollinators can pollinate most plants, each with somewhat different effectiveness or responses to environmental change. However, the loss of particular pollinator species and the impoverishment of pollinator diversity diminish the resilience of ecosystems to change, which are subsequently less able to provide services to humans. Many pollinators are also important food sources for higher animals, so their loss may threaten birds, bats, and other small mammals. As individual species are lost from an ecosystem, the functional redundancy that diverse ecosystems generally display is reduced and resilience to change also tends to decline.

The consequences of continuing decline/loss of pollinators in rural landscapes of Uganda are many. If bee species richness was 3 to 5 times higher at 2 km distance away from the flower farm than at the edge of the farm, it means that with intensification of chemical applications, a high number of bee species will continue to get lost. During field work, it was observed that the few species that could forage in the 500–1000 m were generalist bee species. Specialized bee species were recorded at 1500–2000 m. This means that chemical applications by the different flowers at the current rate of application may erode specialized bee species and leave few common ones to occur. Unfortunately, 70 to 89% of crop species grown in Uganda require pollination by bees, and most often specialized bee species [22–25]. Practically, the majority of flowering plants and crops grown in Uganda are pollinated by several pollinator species, but about 45 percent of grown crops depend on a single taxon for pollination. The disappearance of specialist bee species may increase pollination limitation of many crops (e.g., *Cucurbits, vanilla*) that require these specialist bee species (*Megachile, Patellapis,*

Lipotriches, etc.). Different bee species have got different levels of sensitivity to pesticides applications; once these specialized bees are completely eroded, the remaining common species may not be enough to deliver pollination services of high quantity to different crops and wild plant species that may require these services to set fruits/seeds and for their perpetuation in the landscape. Thus, generalized bees decline in Uganda is likely to lead to high food insecurity in the country. The country will also lose a great deal of income. For example, Uganda earns approximately US$230-240 million per annum from coffee exports, yet 60% of the total value from coffee export is attributable to pollinating services delivered by native bees [20]. Uganda cannot afford losing its pollinators because most of the crops grown are pollinator-dependent. Pollinators contribute up to 67–75% of the annual income of small-scale farmers in Uganda. Total loss/decline of pollinators in Uganda may affect drastically livelihoods of small-scale farmers especially in the rural areas and could lead to high food insecurity. Therefore, the conservation and protection of pollinators by the government of Uganda is a national duty/obligation if the current government is interested in achieving and maintaining the stability of food security of its local population. Agricultural modernisation is good but at what cost?

Any deterioration of pollination services will also have an impact on the rural livelihoods of many rural communities in Uganda, particularly communities growing crops that rely on the specialist bee species. Quite simply, "poor people, with restricted access to resources and with lower integration into the cash economy, are less able to substitute human and physical capital and have less purchasing power, and are therefore particularly, and most directly, dependent on ecosystem services." Such reliance upon ecosystem services makes the poorest more vulnerable to any change in those services. Thus, a functioning pollination process is life-threateningly important to the rural poor who rely on animal-pollinated food crops and medicines for their subsistence.

4.7. Farmers' Knowledge of Pollination, Perceptions, and Attitudes towards Negative Effects of Agrochemicals Applied by Flower Farms on Pollinators, Crop Yields, and Human and Environment Health. Concerning the opinions of local community and agrochemical use and waste disposal by flower farms, in this study, it was realized that 80% of farmers interviewed knew pollination and were aware of the value of pollination in their farming business. These farmers were willing to cooperate in the restoration and maintenance of pollinators in their villages. These results indicate that a high number of farmers in Uganda understood pollination and were willing to care for pollinator conservation in farmlands. Similarly, Kasina et al. [8] conducted a survey in Kakamga region (western part of Kenya) on farmers' knowledge of pollinators and pollination services conservation. The study concluded about 50% of 354 farmers interviewed were aware of pollination and could relate it to crop productivity [18], and after having been introduced to the importance of bee pollination in crop production, they showed a willingness to conserve bees for such purposes [18].

Is Cut-Flower Industry Promotion by the Government Negatively Affecting Pollinator Biodiversity and
Environmental/Human Health in Uganda?

69

Local communities accused the flower firms of being responsible for the abortion of flowers of their crops in most cultivated fruit tree species (Avocado, Mango, Papaw) in their village. Overall, it was not clear if farmers were realistic or not since it was difficult to get a scientific explanation of most observations/perceptions by farmers. Farmers also accused flower farms of being responsible for the current high rate of fruit abortion of cucurbit species. Other farmers said that even when the fruit was formed, it could not reach mature stage or if it matures, it was not tasty at all. Many deep investigations were not carried out to provide solid scientific explanation of the farmers' accusations. However, according to field observations, these farmers' accusations were relatively consistent because during field work, the researcher was given a chance to observe some avocado flowers and fruits that prematurely fell on the ground. From a pathological point of view, symptoms observed could not help in thinking of a pathogen as a causal agent. For example, most avocado fruits observed presented some cracking probably due to environmental stresses (e.g., "heat"). The presence of certain chemical components (at high levels beyond the standards) in the air (atmosphere) can also bring similar environmental stresses on plants through interaction with the physiology mechanisms of the plant.

Farmers perceived that current low honey yield received these days is due to death of bees by chemical pesticides applied at the flower farms. This accusation is consistent because the researcher frequently observed in villages neighboring flower farms several dumping points of agrochemical wastes that have been either thrown on the vegetation or buried out in the soil. Honeybee workers collect nectar and pollen from various plants. During foraging activities, bees were observed feeding on flowers parts with a high level presence of toxic chemical compounds; in most cases, they may die while flying towards their hives or die inside hives. Ingestions of contaminated flower parts may have contributed to the reduction in honeybee population in the village; thereby leading to low honey yield despite the presence of abundant floral resources in the surrounding.

Farmers accused flower farms of being responsible for the current high rate of cow abortion and of low yield of fish from ponds they had established in their village. For such an accusation, it was difficult to offer a scientific explanation. It is difficult to link cow abortion with a particular chemical used in greenhouses. However, scientifically, animals drinking chemically polluted water can have their metabolism systems disorganized due to the presence of certain chemical components in their bodies.

Farmers also accused flower farmers of contributing to rise-up of illness and death of children working at the flower farm. For this accusation, it was difficult to offer a scientific explanation but the researcher had interactions with medical personnel working in the surrounding villages who reported that patients working from flower farms were the most difficult patients they received, in most cases, these patients suffered from low blood pressures, blood infection, septicemia, and so forth.

Most farmers responded that they receive no advantage (no single benefit) from living near the flower farm. This

perception was correct because when looking at the origin of most workers, very few (<10%) of them were from the surrounding with the exception of Mairye estates that employed approximately 20% of local people. In most cases, companies recruited workers from other districts and countries.

Most farmers did not propose a sustainable solution for the future of the flower industry in Uganda, not because that they were ignorant of what should be done, but in most cases, farmers remained hypocritical on this issue. In fact, some farmers disclosed and that their opinion did not matter because they knew big politicians in the country shared actions in most flower companies such as a political decision cannot be implemented. Farmers said that even if they recommend flowering farms to use nontoxic dangerous chemicals, it will not happen because somebody will make the decision for it not to be implemented.

Overall comments/observations given by farmers may reflect a certain reality that require further investigations to come up with a scientific explanation of how the flower farm affects livelihoods, environment, and ecosystem in the surroundings in Uganda.

As in the case of Uganda, Kenya's cut-flower industry has been praised as an economic success as it contributed an annual average of US$141 million foreign exchange (7% of Kenyan export value) over the period of 1996–2005 and about US$352 million in 2005 alone. The industry also provides employment, income, and infrastructure such as schools and hospitals for a large population around Lake Naivasha. On the other hand, the commercial farms have been blamed for causing a drop in the lake level, polluting the lake, and for possibly affecting the lake's biodiversity [60].

There are many human costs of cut flowers in Uganda. Health risks with flower industries are high. The majority (75%) of flower workers are women, most of them employed as temporary workers at low pay and in harsh working conditions. Many workers are being forced to enter greenhouses right after fumigation, at least 24 hours between the time flowers are sprayed and the time workers reenter the area. Because they are not consumable, flowers have escaped much of the antipesticide pressure that has begun to reform the way fruit and vegetables are grown. Because cut-flowers have no value if they are not attractive, chemicals are also used to preserve their beauty. Flower workers are exposed to various different pesticides, 10–20% of which are banned in the international trade market because they are highly toxic or carcinogenic. Because pesticides are used primarily in enclosed greenhouses with often high temperatures, they are more easily inhaled or absorbed through the skin. In Uganda, most flower workers exhibit at least one symptom of pesticide poisoning. Flower workers are constantly exposed to high levels of extremely toxic chemicals, and many report serious health effects, including skin lesions and allergies, respiratory problems, incidences of people fainting, headaches, eye problems, and chronic asthma.

Briefly, flower farm industry in Uganda is associated with several undeclared socioeconomic and environmental problems. During farmers' surveys, some farmers indicated that the health status of children working in the flower farms, fruit (papaw, avocado, and mango) abortion, and abortion of their cows could be associated with chemicals applied in the flower farms. As it was found in this study, the flower industry in Kenya is known to suffer from major social and environmental problems [61]. Several negative issues associated with the flower industry in Kenya include (i) sexual harassment; (ii) forced pregnancy testing/sterilization as condition of employment; (iii) severe occupational health and safety deficiencies, including long working hours, lack of training on safe chemical use, forced overtime, and lack of appropriate safety equipment; (iv) use of toxic pesticides and fungicides causing health problems including skin rashes, respiratory problems, eye problems, and miscarriages; (v) use of child labor; and (vi) lack of recognition of labor rights [61].

Practically, the main environmental problems [61–64] associated with flower farms in Uganda and in East Africa are related to the use of water and agrochemicals (pesticides). The need for water is very high in the flower industry. In Kenya, one hectare of chrysanthemums uses 150.000 litres of water per week; unsustainable cultivation led to sinking ground water levels and dry rivers so that the supply of drinking water is no longer guaranteed in neighboring villages.

Pesticides are also used in extreme amounts including sometimes dangerous and prohibited substances. About half of the women working in the flower industry in Kenya have showed symptoms of poisoning from pesticides after working there for 3–5 years. Similarly, acute pesticide poisonings were reported by small-scale farmers living around flower farms in central Uganda, although it was not possible for this study to verify these accusations made by farmers. Kenya principal flower exporters spray about 200 kg/hectare/year. This is double to almost the similar amount of pesticides used in most flower farms visited during the course of this study conducted in central Uganda.

5. Conclusion

The aim of this study was to conduct a rapid assessment on the status of pollinators around flower farms in central Uganda. This study aimed also at developing an understanding of the different perspectives of local people and see how to propose such information for integration in the decision-making by policy makers. Findings of the study provide insights into the status of pollinators around flower growing areas in Uganda. Although there is a lack of scientific evidence for direct effects of agrochemicals applied inside greenhouses on bee biodiversity, bee species richness and visitation frequencies tended to increase with distance from the flower farm into farmlands. The degree to which bee community composition differed between habitats within sites may be related to landscape variables. In fact, it was found in this study that the difference in bee species richness and abundance between habitats was strongly related to the difference in farm-landscape vegetation types (croplands, grasslands). However, the strong influence of landscape vegetation on bee species richness and abundance is likely to be caused by differences in quality and availability of floral resources and nesting sites across flower farms visited.

Seminatural habitats like fallows had the greatest species richness and abundance in both croplands and grasslands; and this information indicated that increasing proportion cover of fallow resources in flower growing regions can benefit species diversity at both habitat and landscape scale levels through their ability to support a greater number of flowering species and nesting resources per unit area than other seminatural habitats/land-uses. Understanding local perceptions of pollination importance and pollinators decline can be critical for gaining positive attitudes and support from communities for sustainable conservation and restoration of pollination services in rural landscapes, as well as improving livelihoods of local communities.

There are several negative and positive economic, social, and environmental effects associated with flower production in Uganda. Economically, cut flowers are an important export for Uganda, contributing significantly to both the local and national economies. The industry generates several billion Uganda shillings annually. Thus, the cut-flower industry has helped to reduce poverty in Uganda. The flower farms have provided job opportunities, which translate into better living standards. Some of the largest farms employ more than 5,000 workers each, many of them women. Households that are involved in working in cut-flower industries can now afford education and enjoy good health and balanced meals, among other essentials. People living near to the flower farms tend to be relatively well off, since the incidence of poverty there is between 10% and 29% compared to a national level that is well above 47%. However, there have been some reports of labour problems in the cut-flower industry. The sector relies heavily on temporary workers (seasonal or casual), which affects income and job insecurity. Casual labour rises during the annual production peaks. Anecdotal evidence suggests that women workers face problems such as sexual harassment, compulsory overtime, and job insecurity. Given the nature of the accusations very few women make public complaints. Several individuals, environmental bodies and NGOs have expressed concern not only about the health hazards involved in flower production but also about the environmental problems caused by the flower industry. Water crossing flower farms are always polluted, which endangers the health of cattles and human beings using that water at the lower level. A rising population also presents major environmental challenges; the other main pressures on Lake Victoria include water abstractions, agrochemicals and sewage pollution, destruction of papyrus, riparian habitat, over-fishing, soil erosion, and so forth. For example, these days, there has been a massive death of fish on the shore of Lake Victoria and this is attributed to poisoning by the agrochemicals emitted by the flower farms. In some cases (Wagagai flower farm), the flower farms rely on the lake for irrigation, and pipes run straight from the lake into the greenhouses.

This study revealed that more than 79% of small-scale farmers were aware of the value and importance of pollination services in their farming business. Local people tended to be most aware of potential negative effects of chemical applied by flower farms than by chemicals they apply themselves on their plots. Crop yield reductions (fruit abortion) were frequently mentioned possibly because fruit trees were the

Is Cut-Flower Industry Promotion by the Government Negatively Affecting Pollinator Biodiversity and Environmental/Human Health in Uganda?

71

main source of income of small-scale farmers interviewed. Although it was perceived in this study that pollinators have been declining around flower growing areas there exist no or limited long-term data on pollinator populations; and this makes their conservation status difficult to assess in rural landscapes in Uganda. It is difficult to know which species are clearly declining given the absence of long-term monitoring data in Uganda. Thus, maintaining diverse, healthy, and abundant communities of wild pollinators within farmland presents a challenge to farmers, natural resource managers, and policy-makers. However, conserving pollinator-supporting habitats within farmlands can clearly bring benefits to both agriculture and conservation. Overall, the conservation of pollinators in farmlands of Uganda, particularly in flower growing regions, should be monitored. Also, there is a need to define protocol for conducting faunistic surveys, research, and monitoring of pollinators (bees) in natural and agricultural landscapes in Uganda.

Many farmers interviewed were not open enough to suggest durable solutions, but the understanding of desires and preferences of local people is essential for building support for development projects in a given area. Therefore, education for and dialogue with communities are thus crucial since they can all strongly influence communities preferences, perceptions, and priorities. It is important to build community involvement in the development process to ensure the success of projects. Policy-makers should focus on encouraging dialogue between flower farm managers and local communities for enabling the coexistence of the flower farms in the villages. Local communities living in the surrounding need to be involved/informed about risks and benefits (goods and services) associated with the presence of flower farm in their villages. Conservation managers and policy-makers need to balance the demands of public health protection (for humans, crops, and the environment), agricultural production intensification, and the long-term conservation of bee biodiversity in rural landscapes.

Pesticides are often taken as the first line of defense against pests, yet they also impact at least two of the key ecosystem services that sustain crop yields: natural pest control and pollination [62]. Therefore, there is a need to develop pesticide application legislations that are pollinator-friendly in Uganda. There is also a need to develop, put in place, and enforce laws to control or regulate the use of pesticides to protect workers, environment, and inhabitants in areas where cut-flower industries are established. Greater respect for farmers' rights is key to reducing adverse pesticide impacts in flower growing regions in Uganda. Regulations concerning types of pesticides to be used by flower farms do not need to be flexible, but should be compatible with local demand for a healthy environment. Environmental/agricultural policies that authorize the abandonment of use of toxic pesticides are advocated for here, as part of a win-win solution to softly mitigate negative effects caused by flower farm industry in Uganda. It is also recommended to policy makers to commission studies that investigate and find evidence for all accusations made by farmers about the negative effects of agrochemicals applied on human, plant, and environmental healths.

NEMA (national environmental management agency) is the government body charged with environmental management, issuance of license, and monitoring of activities that are detrimental to the environment. It seems NEMA was not involved in the investment of flower firms in Uganda. However, NEMA should be visible at the feasibility study (to carry out its EIA = environmental impact assessment), and in the documentation of the flora/fauna prior to any investment in the country. Also, NEMA should be present during the implementation and routinely monitor the environment around the flower farms and related investments/projects. The agro-inputs department of the Ministry of Agriculture, Animal Industries, and Fisheries need to have a strong presence on the ground in vetting chemicals imported, in monitoring their correct usage, and in monitoring negative impacts on humans, environment, and biodiversity.

In this study it was observed that both the species richness and bee populations were lower near flower farms (despite availability of abundant and diverse floral resources) but increased with increasing distance away from the flower farms. However, it was not clear if chemicals applied inside greenhouses could be the cause. Although, there was no direct evidence that flower farm activities cause the reduction in bee species and numbers nearby flower farms, it is believed that agrochemicals used inside flowers may be responsible for such a reduction. To be more conclusive on this topic, it is therefore recommended that future research (using primary data) is carried out to experimentally test and establish whether or not the decline in bee species richness and numbers is due to the use of agrochemicals by flower farms. It may useful that toxicological studies are conducted in all environments nearby flower farms to determine and identify bee toxic components in the air in relationship to the type of agrochemicals used in flower farms. This may help in determining pesticides directly and indirectly involved in acute bee poisoning.

Flower farms have sprung up in many developing countries to provide a wide range of flowers year-round in western countries. They are a valuable source of foreign income and employment, but may have costs to the environment and local community. This work provides information on some of these potential impacts. To our knowledge, this is the first rigorous study providing useful scientific contribution in the area of pollination services delivery in farm landscapes with large flower farms in Sub-Saharan Africa and in the world. This is the first excellent review of how government involvment, antrophogenic stress, use of pesticides, among other factors are affecting biodiversity of pollinators in Uganda. The present study can serve as a valuable guide to natural resource managers and policy makers in Uganda, as conserving pollinator-supporting habitats within farmlands can clearly bring benefits to both agriculture and conservation.

Overall, a remarkably high diversity of bees was observed in this study; the diversity and numbers of bees were observed to be higher further away from the flower farms and in sites where there was more seminatural vegetation. Also, it was observed in this study that local farmers mostly had a good understanding of the importance of bees as pollinators of their crops. They also mostly felt negatively about the flower farms and believed that the farm's high use of pesticides has an impact on bee numbers and hence on the productivity or quality of their crops, and potentially on human health. It was not clear if flower farms were directly responsible for all these accusations by farmers, but it is believed that that there should be further investigation of the flower farms' impacts. It was observed that over time, there was an increase in the amounts of agrochemicals used by the flower farms. An interesting finding was the lack of relationship between farm income and amount of agrochemicals used. This result suggested that the use of chemicals could be reduced without loss of income (and as costs are reduced gross and net income should be increased). It was also interesting to find out that at least one flower farm was showing some environmental responsibility, by taking some care to reduce use of pesticides and also to recycle water and so forth.

It is therefore strongly recommended to conduct toxicological studies (to determine side-effects) on different bee species collecting floral resources or residing in farmlands where flower firms are established. The aim of such study will be to assess the exposure of bees to pesticide residues [62–64] from greenhouses. There is a need for monitoring programs to establish spatial and temporal trends in the abundance and bioavailability of pesticide residues for bees and other insects. Biomonitoring of organisms that accumulate contaminants in their tissues can be used to assess the relative quality of environments, including the presence, levels, and changes in contaminants [56].

Pollen loads/living bees have been suggested to be the best matrix for assessing/detecting the presence of pesticide residues (frequencies, concentrations) in samples from in the environment. Toxicological studies may help in yielding more significant results on influences of pesticides application inside greenhouses by flower firms on health of human communities. To provide more accurate results, there is also a need to conduct analysis of water quality (pond/river/stream flowing through/from flower farms to farm landscape), soil, and resources produced by bees (honey and wax) to check for traces of pesticides in these substances. Stingless bees and honeybee may be useful as environmental indicators for assessing the presence of environmental pollution [61] with heavy metals (important and potentially harmful pollutants such as Co, Cd, Cr, Cu, Fe, Mn, Ni, Pb, and Zn). The link between floriculture industry with the environment and health is poorly documented in Uganda. Management of agrochemicals and effluent from flower farms need to be revised by all flower firms. It is therefore recommended that good methods to contain runoff of water from flowers are set, revised, tested, and evaluated by all flower farmers to reduce pollution of the environment from pesticides and fertilizer runoff.

Long-term effects of regular exposure to pesticides can cause erosion of pollinators. Some pesticides can be persistent in the soils, air, and in plant parts. These can become toxic later to certain organisms in contact with the environment where they are deposited. It is important to determine the levels of certain pesticides (eg., organochlorine) in the local environment (air, soil, water, bee food plants and flower parts, air, etc.) compared to published lethal doses to bees. Explorative studies are recommended to determine the role played and contribution of human activities (small-scale farmers' activities) in the reduction of bees in the landscapes immediately surrounding flower farms compared to role played by agrochemicals from the flower farms. There is also a need to conduct vast awareness campaigns among small-scale farmers about some of their pollinator's unfriendly [65] farming practices. There is a need to conduct further deep research on issues of health risks of workers of cut-flower farms.

Ethical Approval

Ethical issues (including plagiarism, informed consent, misconduct, data fabrication and/or falsification, double publication, and redundancy, etc.) have been completely observed by the author.

Conflict of Interests

The author declares that he has no conflict of interests.

Acknowledgments

Early seed financial supports (2%) for conducting preliminary surveys (for 3 days) were obtained under a consultancy commissioned by the Economic Policy Research Centre (EPRC) and supported by the National Environment Management Authority (NEMA) Trade and Biodiversity Initiative of the Government of Uganda through a grant offered by the European Union and the United Nations Environment Programme (UNEP) to the Government of Uganda. EMRC-DRC funded 98% of the activities (major activities for 2 years: 2010–1012) related to this study. The author is very grateful to both sources of funding. The author is also very grateful to production managers of the different flower farms visited for allocating some of their busy time to answer the questionnaires submitted to them. The author is very grateful to small-scale farmers living in the surrounding of different flower farms who accepted freely to answer the questions. The author is grateful to Ms. Madina (research assistant) who assisted him with local translation of the questionnaire in local language in the early stages of this work. The author thanks Mr. Ziwa (EPRC driver) for driving safely and bringing us in time to the field and for the kind patience during field data collection by the researcher with his assistant. The author is grateful to the Director of EPRC. The warm and kind collaboration of other staff at EPRC is much acknowledged, mainly from Francis Mwaura (EPRC) and Francis Ogwal (NEMA). The author is also very grateful to various scientific help/support received from Dr. Hafashimana David, Dr. Talwana Hubert, Dr. Michael Otim, Dr. Anne Akol, Dr. Chris Bakuneta, Dr. Robert Kityo, Professor John Tabuti, Professor Samuel Kyamanywa, Professor Kansiime, and Professor Nyeko Philippe (Makerere University-Uganda).

Is Cut-Flower Industry Promotion by the Government Negatively Affecting Pollinator Biodiversity and Environmental/Human Health in Uganda?

73

References

[1] Economic Policy Research Center (EPRC), "Integrated Assessment of Trade-Related Policies on Biological Diversity in the Agricultural Sector in Uganda. The potential impacts of the EU-ACP Economic Partnership Agreement. A case study in the horticulture sector. Ministry of Water and Environment, Republic of Uganda, National Environment Management Authority (NEMA), United Nations Environment Programme (UNEP), Kampala, Uganda," 2009.

[2] G. Allen-Wardell, P. Bernhardt, R. Bitner et al., "The potential consequences of pollinator declines on the conservation of biodiversity and stability of food crop yields," *Conservation Biology*, vol. 12, no. 1, pp. 8–17, 1998.

[3] C. D. Eardley, M. Gikungu, and M. P. Schwarz, "Bee conservation in Sub-Saharan Africa and Madagascar: diversity, status and threats," *Apidologie*, vol. 40, no. 3, pp. 355–366, 2009.

[4] R. Winfree, N. M. Williams, H. Gaines, J. S. Ascher, and C. Kremen, "Wild bee pollinators provide the majority of crop visitation across land-use gradients in New Jersey and Pennsylvania, USA," *Journal of Applied Ecology*, vol. 45, no. 3, pp. 793–802, 2008.

[5] P. Neumann and N. L. Carreck, "Honey bee colony losses," *Journal of Apicultural Research*, vol. 49, no. 1, pp. 1–6, 2010.

[6] A.-M. Klein, B. E. Vaissière, J. H. Cane et al., "Importance of pollinators in changing landscapes for world crops," *Proceedings of the Royal Society B: Biological Sciences*, vol. 274, no. 1608, pp. 303–313, 2007.

[7] D. Goulso, "Conserving wild bees for crop pollination," *Food, Agriculture & Environment*, vol. 1, no. 1, pp. 142–144, 2003.

[8] O. MacIas-Macias, J. Chuc, P. Ancona-Xiu, O. Cauich, and J. J. G. Quezada-Euán, "Contribution of native bees and Africanized honey bees (Hymenoptera: Apoidea) to Solanaceae crop pollination in tropical México," *Journal of Applied Entomology*, vol. 133, no. 6, pp. 456–465, 2009.

[9] M. G. Park, M. C. Orr, and B. N. Danforth, "The role of native bees in apple pollination," *New York Fruit Quarterly*, vol. 18, no. 1, pp. 1–24, 2010.

[10] C. Kremen, N. M. Williams, M. A. Aizen et al., "Pollination and other ecosystem services produced by mobile organisms: a conceptual framework for the effects of land-use change," *Ecology Letters*, vol. 10, no. 4, pp. 299–314, 2007.

[11] L. A. Garibaldi, M. A. Aizen, S. A. Cunningham, and A. M. Klein, "Pollinator shortage and global crop yield," *Communicative and Integrative Biology*, vol. 2, no. 1, pp. 1–3, 2009.

[12] D. R. Campbell, "Pollinator shifts and the origin and loss of plant species," *Annals of the Missouri Botanical Garden*, vol. 95, no. 2, pp. 264–274, 2008.

[13] M. A. Aizen, L. A. Garibaldi, S. A. Cunningham, and A. M. Klein, "How much does agriculture depend on pollinators? Lessons from long-term trends in crop production," *Annals of botany*, vol. 103, no. 9, pp. 1579–1588, 2009.

[14] A. G. S. Cuthbertson and M. A. Brown, "Issues affecting British honey bee biodiversity and the need for conservation of this important ecological component," *International Journal of Environmental Science and Technology*, vol. 6, no. 4, pp. 695–699, 2009.

[15] R. Winfree, "The conservation and restoration of wild bees," *Annals of the New York Academy of Sciences*, vol. 1195, pp. 169–197, 2010.

[16] N. Gallai, J.-M. Salles, J. Settele, and B. E. Vaissière, "Economic valuation of the vulnerability of world agriculture confronted with pollinator decline," *Ecological Economics*, vol. 68, no. 3, pp. 810–821, 2009.

[17] M. A. Aizen and L. D. Harder, "The global stock of domesticated honey bees is growing slower than agricultural demand for pollination," *Current Biology*, vol. 19, no. 11, pp. 915–918, 2009.

[18] M. Kasina, M. Kraemer, C. Martius, and D. Wittmann, "Farmers' knowledge of bees and their natural history in Kakamega district, Kenya," *Journal of Apicultural Research*, vol. 48, no. 2, pp. 126–133, 2009.

[19] M. J. F. Brown and R. J. Paxton, "The conservation of bees: a global perspective," *Apidologie*, vol. 40, no. 3, pp. 410–416, 2009.

[20] T. M. B. Munyuli, S. G. Potts, and P. Nyeko, "Patterns of bee biodiversity and conservation values of localities with contrasting mini-ecological structures found within, nearby and between agricultural mosaic landscapes from central Uganda," *Life Sciences International Journal*, vol. 2, no. 4, pp. 866–914, 2008.

[21] T. M. B. Munyuli, "The status of pollinators around flower growing zones in Uganda," Economic policy Research Center (EPRC) and National Environmental management authority (NEMA).

[22] T. Munyuli, "Factors governing flower visitation patterns and quality of pollination services delivered by social and solitary bee species to coffee in central Uganda," *African Journal of Ecology*, vol. 49, no. 4, pp. 501–509, 2011.

[23] T. M. B. Munyuli, "Farmers' perceptions of pollinators importance in coffee production in Uganda," *Agricultural Sciences*, vol. 2, pp. 318–333, 2011.

[24] T. M. B. Munyuli, "Pollinator biodiversity in Uganda and in sub-Saharan Africa: landscape and habitat management strategies for its conservation," *International Journal of Biodiversity and Conservation*, vol. 3, pp. 551–609, 2011.

[25] T. Munyuli, "Assessment of indicator species of butterfly assemblages in coffee-banana farming system in central Uganda," *African Journal of Ecology*, vol. 50, no. 1, pp. 77–89, 2012.

[26] T. M. B. Munyuli, "Diversity of life-history traits, functional groups and indicator species of bee communities from farmlands of central Uganda," *Jordan Journal of Biological Sciences*, vol. 5, pp. 1–14, 2011.

[27] T. M. B. Munyuli, "Micro, local, landscape and regional drivers of bee biodiversity and pollination services delivery to coffee (*Coffea canephora*) in Uganda," *International Journal of Biodiversity, Ecosystem Services and Management*, vol. 8, pp. 190–203, 2012.

[28] T. M. B. Munyuli, "Is pan-trapping the most reliable sampling method for measuring and monitoring bee biodiversity in agroforestry systems in sub-Saharan Africa?" *International Journal of Tropical Insect Science*, vol. 33, no. 1, p. 1437, 2013.

[29] T. M. B. Munyuli, P. Nyeko, S. G. Potts, P. Atkinson, D. Pomeroy, and J. Vickery, "Patterns of bee diversity in mosaic agricultural landscapes of central Uganda: implication of pollination services for conservation for food security," *Journal of Insect Conservation*, vol. 17, pp. 79–93, 2013.

[30] T. M. B. Munyuli, "Climatic, regional land-use intensity, landscape, and local variables predicting best the occurrence and distribution of bee community diversity in various farmland habitats in Uganda," *Psyche*, vol. 2013, Article ID 564528, 38 pages, 2013.

[31] T. M. B. Munyuli, "Trends in responses to drivers by different bee ecological traits and functional groups in agricultural landscapes in Uganda," *Trends in Entomology*, vol. 9, no. 9, pp. 1–23, 2013.

[32] J. C. Biesmeijer, S. P. M. Roberts, M. Reemer et al., "Parallel declines in pollinators and insect-pollinated plants in Britain and the Netherlands," *Science*, vol. 313, no. 5785, pp. 351–354, 2006.

[33] J. C. Grixti, L. T. Wong, S. A. Cameron, and C. Favret, "Decline of bumble bees (Bombus) in the North American Midwest," *Biological Conservation*, vol. 142, no. 1, pp. 75–84, 2009.

[34] A. Holzschuh, I. Steffan-Dewenter, and T. Tscharntke, "How do landscape composition and configuration, organic farming and fallow strips affect the diversity of bees, wasps and their parasitoids?" *Journal of Animal Ecology*, vol. 79, no. 2, pp. 491–500, 2010.

[35] L. A. Morandin and M. L. Winston, "Wild bee abundance and seed production in conventional, organic, and genetically modified canola," *Ecological Applications*, vol. 15, no. 3, pp. 871–881, 2005.

[36] K. Goka, "Introduction to the special feature for ecological risk assessment of introduced bumblebees: status of the European bumblebee, *Bombus terrestris*, in Japan as a beneficial pollinator and an invasive alien species," *Applied Entomology and Zoology*, vol. 45, no. 1, pp. 1–6, 2010.

[37] R. Jaffé, V. Dietemann, M. H. Allsopp et al., "Estimating the density of honeybee colonies across their natural range to fill the gap in pollinator decline censuses," *Conservation Biology*, vol. 24, no. 2, pp. 583–593, 2010.

[38] S. Maini, P. Medrzycki, and C. Porrini, "The puzzle of honey bee losses: a brief review," *Bulletin of Insectology*, vol. 63, no. 1, pp. 153–160, 2010.

[39] S. G. Potts, J. C. Biesmeijer, C. Kremen, P. Neumann, O. Schweiger, and W. E. Kunin, "Global pollinator declines: trends, impacts and drivers," *Trends in Ecology and Evolution*, vol. 25, no. 6, pp. 345–353, 2010.

[40] L. A. Garibaldi, I. Steffan-Dewenter, R. Winfree et al., "Wild pollinators enhance fruit set of crops worldwide, regardless of honey-bee abundance," *Science*, vol. 339, no. 6127, pp. 1608–1611, 2013.

[41] V. Le Féon, A. Schermann-Legionnet, Y. Delettre et al., "Intensification of agriculture, landscape composition and wild bee communities: a large scale study in four European countries," *Agriculture, Ecosystems and Environment*, vol. 137, no. 1-2, pp. 143–150, 2010.

[42] C. A. Brittain, M. Vighi, R. Bommarco, J. Settele, and S. G. Potts, "Impacts of a pesticide on pollinator species richness at different spatial scales," *Basic and Applied Ecology*, vol. 11, no. 2, pp. 106–115, 2010.

[43] P. G. Kevan, "Pollinators as bioindicators of the state of the environment: species, activity and diversity," *Agriculture, Ecosystems and Environment*, vol. 74, no. 1–3, pp. 373–393, 1999.

[44] B. A. Woodcock, S. G. Potts, T. Tscheulin et al., "Responses of invertebrate trophic level, feeding guild and body size to the management of improved grassland field margins," *Journal of Applied Ecology*, vol. 46, no. 4, pp. 920–929, 2009.

[45] K. W. Richards and P. G. Kevan, "Aspects of bee biodiversity, crop pollination, and conservation in Canada," in *Pollinating Bees—The Conservation Link Between Agriculture and Nature*, P. Kevan and V. L. Imperatriz Fonseca, Eds., pp. 77–94, Ministry of Environment-Brasília, 2002.

[46] A. E. Magurran, *Measuring Biology Diversity*, Blackwell, Oxford, UK, 2010.

[47] C. Porrini, A. G. Sabatini, S. Girotti et al., "The death of honey bees and environmental pollution by pesticides: the honey bees as biological indicators," *Bulletin of Insectology*, vol. 56, no. 1, pp. 147–152, 2003.

[48] D. Gabriel, S. M. Sait, J. A. Hodgson, U. Schmutz, W. E. Kunin, and T. G. Benton, "Scale matters: the impact of organic farming on biodiversity at different spatial scales," *Ecology Letters*, vol. 13, no. 7, pp. 858–869, 2010.

[49] C. N. Kaiser-Bunbury, S. Muff, J. Memmott, C. B. Müller, and A. Caflisch, "The robustness of pollination networks to the loss of species and interactions: a quantitative approach incorporating pollinator behaviour," *Ecology Letters*, vol. 13, no. 4, pp. 442–452, 2010.

[50] A. Kovács-Hostyánszki, P. Batáry, and A. Báldi, "Local and landscape effects on bee communities of Hungarian winter cereal fields," *Agricultural and Forest Entomology*, vol. 13, pp. 59–66, 2011.

[51] D. Pimentel, "Environmental and economic costs of the application of pesticides primarily in the United States," *Environment, Development and Sustainability*, vol. 7, no. 2, pp. 229–252, 2005.

[52] S. Naidoo, L. London, A. Burdorf, R. N. Naidoo, and H. Kromhout, "Agricultural activities, pesticide use and occupational hazards among women working in small scale farming in Northern Kwazulu-Natal, South Africa," *International Journal of Occupational and Environmental Health*, vol. 14, no. 3, pp. 218–224, 2008.

[53] E. M. Tegtemeier and D. Duffy, "External cost of agriculture production in the United States of America," *International Journal of Agriculture Sustainability*, vol. 2, no. 1, pp. 1–20, 2004.

[54] J. K. Tuell and R. Isaacs, "Community and species-specific responses of wild bees to insect pest control programs applied to a pollinator-dependent crop," *Journal of Economic Entomology*, vol. 103, no. 3, pp. 668–675, 2010.

[55] M.-P. Chauzat, A.-C. Martel, N. Cougoule et al., "An assessment of honeybee colony matrices, *Apis mellifera* (Hymenoptera: Apidae) to monitor pesticide presence in continental France," *Environmental Toxicology and Chemistry*, vol. 30, no. 1, pp. 103–111, 2011.

[56] A. Alix and G. Lewis, "Guidance for the assessment of risks to bees from the use of plant protection products under the framework of Council Directive 91/414 and Regulation 1107/2009," *EPPO Bulletin*, vol. 40, no. 2, pp. 196–203, 2010.

[57] D. G. Alston, V. J. Tepedino, B. A. Bradley, T. R. Toler, T. L. Griswold, and S. M. Messinger, "Effects of the insecticide phosmet on solitary bee foraging and nesting in orchards of Capitol Reef National Park, Utah," *Environmental Entomology*, vol. 36, no. 4, pp. 811–816, 2007.

[58] R. Kajobe, G. Marris, G. Budge et al., "First molecular detection of a viral pathogen in Ugandan honey bees," *Journal of Invertebrate Pathology*, vol. 104, no. 2, pp. 153–156, 2010.

[59] M. M. Mekonnen, A. Y. Hoekstra, and R. Becht, "Mitigating the water footprint of export cut flowers from the lake Naivasha Basin, Kenya," *Water Resources Management*, vol. 26, pp. 3725–3742, 2010.

[60] W. A. Qualls, R.-D. Xue, and H. Zhong, "Impact of bifenthrin on honeybees and *Culex quinquefasciatus*," *Journal of the American Mosquito Control Association*, vol. 26, no. 2, pp. 223–225, 2010.

[61] D. Holt and A. Watson, "Exploring the dilemma of local sourcing versus international development—the case of the flower industry," *Business Strategy and the Environment*, vol. 17, no. 5, pp. 318–329, 2008.

[62] H. van der Valk, I. Koomen, R. C. F. Nocelli et al., *Aspects Determining the Risk of Pesticides to Wild Bees: Risk Profiles for Focal Crops on Three Continents*, FAO, Rome, Italy, 2013.

Is Cut-Flower Industry Promotion by the Government Negatively Affecting Pollinator Biodiversity and
Environmental/Human Health in Uganda?

75

[63] D. Van Engelsdorp, N. Speybroeck, J. D. Evans et al., "Weighing
risk factors associated with bee colony collapse disorder by
classification and regression tree analysis," *Journal of Economic
Entomology*, vol. 103, no. 5, pp. 1517–1523, 2010.

[64] M. N. Rashed, M. T. A. El-Haty, and S. M. Mohamed, "Bee
honey as environmental indicator for pollution with heavy
metals," *Toxicological and Environmental Chemistry*, vol. 91, no.
3, pp. 389–403, 2009.

[65] T. M. B. Munyuli, "Influence of functional traits on foraging
behaviour and pollination efficiency of wild social and solitary
bees visiting coffee (*canephora*) flowers in Uganda," *Grana*. In
press.

4

Vegetation Recovery in Response to the Exclusion of Grazing by Sika Deer (*Cervus nippon*) in Seminatural Grassland on Mt. Kushigata, Japan

Takuo Nagaike,[1] Eiji Ohkubo,[2] and Kazuhiro Hirose[3]

[1] *Yamanashi Forest Research Institute, Saisyoji 2290-1, Fujikawa, Yamanashi 400-0502, Japan*
[2] *Yamanashi Gakuin Junior College, Sakaori 2-4-5, Kofu, Yamanashi 400-8575, Japan*
[3] *Minami-Alps City Office, Ogasawara 376, Minami-Alps, Yamanashi 400-0306, Japan*

Correspondence should be addressed to Takuo Nagaike; nagaike-zty@pref.yamanashi.lg.jp

Academic Editors: H. Ford, P. K. S. Shin, P. M. Vergara, and J.-t. Zhang

We examined the recovery of vegetation in seminatural grassland in central Japan after eliminating grazing by sika deer (*Cervus nippon*) by fencing. By 2012, after 5 years of fencing for exclusion of sika deer, the species composition of quadrats within the enclosure reverted to the original species composition in 1981, not browsed by sika deer. Conversely, outside the fence was different from the baseline quadrats in 1981. *Iris sanguinea*, a prominent flower in the area, recovered within the enclosure, while it continued to decrease with grazing outside the fence. Nevertheless, the *I. sanguinea* cover had not recovered to the 1981 levels in the enclosure. Fencing can effectively restore vegetation as the species composition within the enclosure gradually reverts to the original vegetation. Preventing grazing in intensively grazed seminatural grassland might lead to different successional pathways. Since *I. sanguinea* did not recover fully within the enclosure and the species composition differed slightly from the original vegetation, this suggests that the vegetation within the enclosure will change to an alternative state. Therefore, different management is needed to promote the correct succession pathways for ecological restoration, perhaps by enhancing the colonization of target species, to prevent restored sites from giving rise to alternative states.

1. Introduction

Traditionally, seminatural grasslands in Japan have been managed for grazing cattle and harvesting agricultural materials (e.g., fertilizer; [1]). However, most of these have been abandoned with changes in lifestyle and farming methods [2], as in other countries (e.g., [3]). In Japan, forests cover 78% of the land area [4] and the mild, humid climate promotes the growth of forests [5]. Consequently, abandoned grasslands are invaded by trees and are important especially for early successional species and species favoring open habitats (e.g., [6]). So, such seminatural grassland has a high priority to conserve biological diversity.

Recently, the population of sika deer (*Cervus nippon*) in Japan has increased sharply because of the declining numbers of hunters and their aging, extinction of predator, and so on [7, 8]. The effects of deer on natural grassland vegetation have

been reported worldwide (e.g., [9]). In Japan, deer have had serious effects on natural forests [10, 11], plantations [12, 13], and grasslands [14, 15]. Since sika deer prefer open habitats as foraging sites [16], patchy grasslands surrounded by forests are used heavily by sika deer. Consequently, the plant species composition in the grasslands has been altered markedly by grazing sika deer.

On Mt. Kushigata (2053 m a.s.l. at the summit) in central Japan, seminatural grasslands are distributed patchily with plantations and fragmented natural forests [17]. In the grassland, the flowers of *Iris sanguinea* and other meadow species (e.g., *Veronicastrum japonicum*) had been renowned. However, those flowers have been decreasing since 2000 because of grazing by sika deer. As countermeasures for the grazing, fences were set in the grassland in 2007. Fencing is effective for recovering from herbivore grazing [18, 19] and promotes tree regeneration [20, 21]. The effects of grazing exclusion

Vegetation Recovery in Response to the Exclusion of Grazing by Sika Deer (Cervus nippon) in Seminatural
Grassland on Mt. Kushigata, Japan

77

from seminatural grasslands on species richness [22] and spatial patterns [23] have been studied. However, undesirable results, in which execution at ecological restoration might lead to alternative states [24], can occasionally occur when fencing is used as a tool for conservation in overgrazed environments [25]. Therefore, restoration efforts should be checked by comparing the results with the restoration target [3, 4].

This paper presents the vegetation recovery pattern in seminatural grassland in central Japan after preventing grazing with fencing and addresses the following question: can *I. sanguinea* and a vegetation community recover to the original situation and community before they were affected by sika deer?

2. Methods

2.1. Study Site. The study was conducted on Mt. Kushigata (2053 m a.s.l. at the summit), Yamanashi Prefecture, in the cool-temperate zone of central Japan (35°35'N, 138°22'E). The mean annual precipitation and temperature at the nearest meteorological station (Oizumi, 867 m a.s.l.) are approximately 1140 mm and 10.7°C, respectively. The estimated mean annual temperature at the summit of Mt. Kushigata is about 3.5°C. Snow cover is usually less than 1 m.

On Mt. Kushigata, seminatural grasslands are distributed patchily with plantations and fragmented natural forests [17]. The origin of the grassland is not clear, but elevations in this region around 2000 m are typically dominated by subalpine coniferous forests composed of *Abies* and *Tsuga* species. The grassland was probably the result of human activity, such as mowing or burning. We studied the Hadakayama area of Mt. Kushigata. Hadakayama means "naked mountain," which indicates that this area has not been covered by forest for a long time. The grasslands in the Hadakayama area are renowned for the flowers of *I. sanguinea*. Over a 10-year period, however, the number of *I. sanguinea* flowers has decreased, possibly as a result of the natural succession from grassland to the typical subalpine coniferous forest. Grazing by sika deer was another potential reason. Therefore, fences to exclude sika deer were erected to prevent grazing in 2007.

2.2. Field Study. In July 1981, the science club of Koma High School studied the vegetation in the Hadakayama area [6]. They set 32 1 × 1 m quadrats in the Hadakayama area typically dominated by *I. sanguinea* and applied the standard Braun-Blanquet scale: +, sparse cover; (1) cover <5%; (2) cover 5–25%; (3) cover 25–50%; (4) cover 50–75%; (5) cover 75–100%. The species composition in 1981 is considered the original vegetation not affected by grazing by sika deer.

In October 2007, we established a 20 × 20 m plot and fenced half of it (10 × 20 m) to protect it from deer grazing. The quadrats set by Koma High School were 20 m from the plot. In June 2008, we established 20 1 × 1 m quadrats, 10 inside and 10 outside the fence, and conducted annual surveys from 2008 to 2012 using the Braun-Blanquet method. We counted the number of individuals of *I. sanguinea* in each quadrat every year.

2.3. Analysis. Koma High School [6] did not list all rare species (i.e., species with low coverage in the quadrats studied). Therefore, we analyzed the species with coverage with a score >1 on the Braun-Blanquet scale.

We used nonmetric multidimensional scaling (NMS; [27]) to provide an ecologically interpretable quantification of the compositional differences among original vegetation (1981) and quadrats inside and outside the fence (2008–2012). NMS applied Sørensen's similarity index to calculate a distance matrix. We used the species cover in each quadrat for NMS after transforming the Braun-Blanquet scale quantitatively. The data transformed to cover values (the midpoints of the cover intervals for each score) were used; that is, scores of 1 to 5 were converted to the values 2.5%, 12.5%, 37.5%, 62.5%, and 87.5%, respectively [26]. NMS was performed using PC-ORD [27].

To show the recovery of *I. sanguinea*, we compared the number of individuals and cover.

3. Results

The recovery of *I. sanguinea* inside the fence was good, while outside the fence, it decreased continuously with grazing (Figure 1). The number of flowering *I. sanguinea* also increased inside the fence to 25 in 2010, 198 in 2011, and 307 in 2012, while no flowers occurred outside the fence (unpublished data, Committee of Conservation of *I. sanguinea* at Mt. Kushigata). Nevertheless, the *I. sanguinea* cover had not recovered fully compared to 1981 (Figure 2).

The changes in the number of species inside and outside the fence showed contrasting trends (Figure 3). The number of species increased inside the fence, but not outside it. The species composition differed markedly inside and outside the fence (Figure 4). In 2008, most of the quadrats were located in the upper left position in the NMS diagram. Then, the species composition of the quadrats within the fence shifted to the upper right position in the diagram, where the species composition in 1981 was located (i.e., not browsed by sika deer). Conversely, the species composition of the quadrats outside the fence was shifted to the lower right. Species that occurred in more quadrats inside the fence were *Dianthus superbus* var. *longicalycinus*, *Phedimus aizoon* var. *floribundus*, *Serratula coronata*, and *Chamerion angustifolium* as well as *I. sanguinea* (Table 1). *Angelica pubescens*, *Veronicastrum japonicum*, and *I. sanguinea*, which were categorized by tall herbs, were only dominant inside the fence. *Brachypodium sylvaticum* and *Ranunculus japonicas* were less dominant before grazing but were dominant after exclusion of sika deer. Outside the fence, *Artemisia princeps* initially dominated, and subsequently the graminoids *Stipa pekinensis* and *B. sylvaticum*, which appeared to be unpalatable, dominated.

4. Disscusion

Exclusion of sika deer by fencing was successfully recovering the cover and number of individuals of *I. sanguinea* because those outside the fence were continuously low. Thus, effects of grazing by sika deer were continuously severe. In

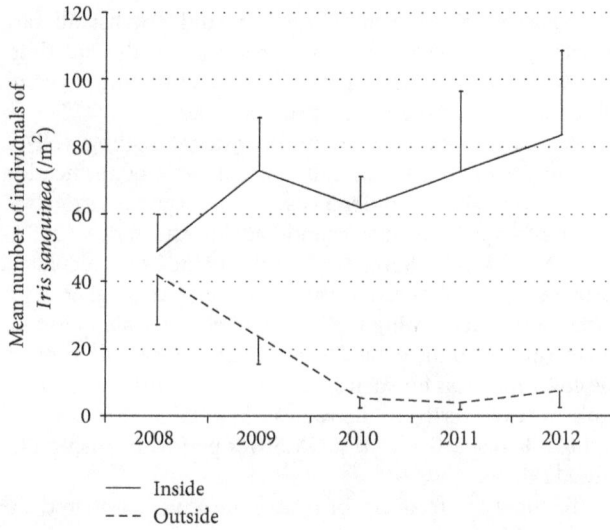

FIGURE 1: Changes of mean number of individuals of *Iris sanguinea* in each quadrat. Vertical bars showed standard deviation.

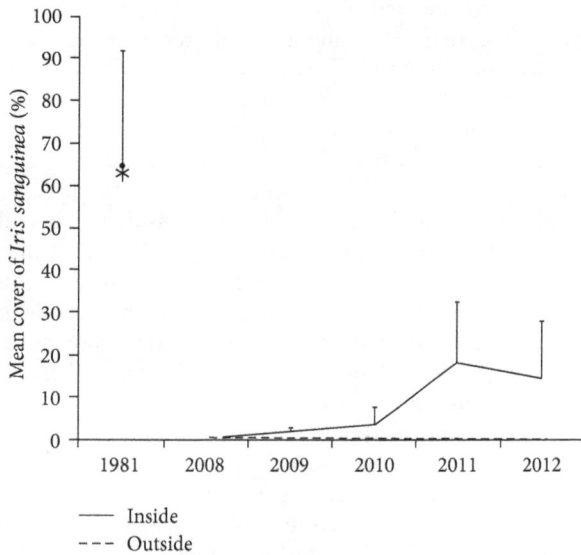

FIGURE 2: Changes of mean cover of *Iris sanguinea* in each quadrat. Vertical bars showed standard deviation.

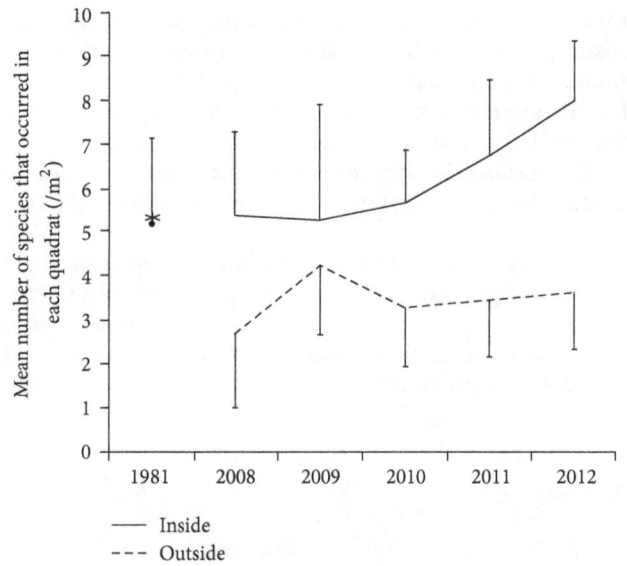

FIGURE 3: Changes of mean number of species that occurred in each quadrat. Vertical bars showed standard deviation.

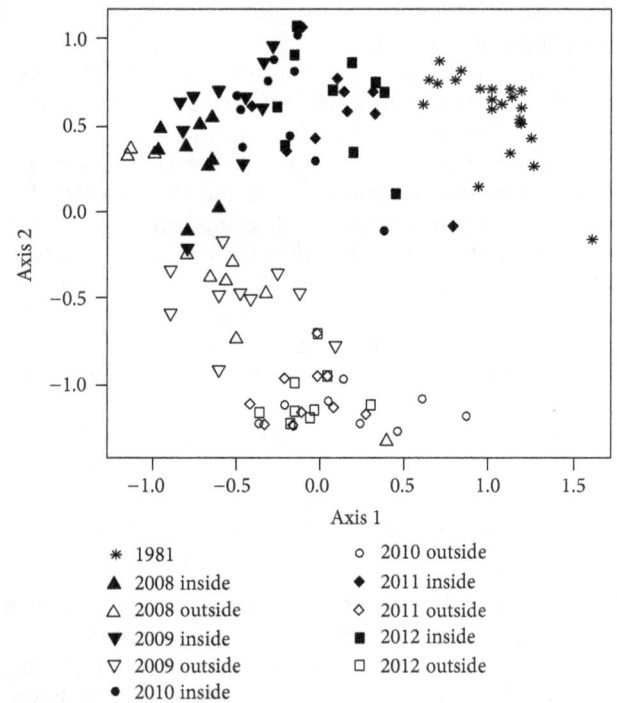

FIGURE 4: Results of nonmetric multidimensional scaling (NMS) for species that occurred in each quadrat.

the seminatural grassland of northern Japan, *Iris setosa* dominated the most at sites grazed by horses, which improved the surface soil characteristics [28]. In our study, sika deer grazed on *I. sanguinea* directly, causing serious damage. Thus, to conserve *I. sanguinea*, fencing seems necessary under the present circumstances. Since Tamura [29] showed that vegetation recovery, particularly tall herb species, was poor when fencing was delayed, it will be impossible for *I. sanguinea* to recover fully even if fences are erected now. However, *I. sanguinea* would not recover fully compared to situation before grazing despite preventing grazing.

Fencing appeared to be effective at restoring the vegetation as the species composition in the fence gradually reverted to the original vegetation. Herbivores often, but not always, increase plant diversity in grasslands [30]. Outside the fence, however, since the species composition was altered and the number of species was low, intense grazing pressure by sika deer likely existed.

Rooney and Dress [31] showed that species with relatively lower abundance were more likely to be missing due to

Vegetation Recovery in Response to the Exclusion of Grazing by Sika Deer (Cervus nippon) in Seminatural
Grassland on Mt. Kushigata, Japan

79

TABLE 1: Species list in each quadrat. Figures in the table show number of quadrats in species occurred. The total number of quadrats is 36 in "before grazing," 10 in "inside the fence," and 10 in "outside the fence," respectively.

Species	Before grazing	Inside the fence					Outside the fence				
	1981	2008	2009	2010	2011	2012	2008	2009	2010	2011	2012
Adenophora remotiflora	1										
Adenophora triphylla	3										
Agrostis clavata									1		
Anaphalis margaritacea	2										
Angelica pubescens	2	4	3	2	3	5					
Aquilegia buergeriana		1	1			1			1		1
Arabis hirsuta		1						1			
Arenaria serpyllifolia											1
Artemisia princeps	2	10	10	8	7	8	9	9			
Astilbe microphylla	6										
Brachypodium sylvaticum		3	2	4	4	3	5	7	8	10	10
Campanula punctata	2							1			
Chamerion angustifolium	6					1					
Cirsium gratiosum								1	1	1	1
Clinopodium chinense subsp. *grandiflorum*		5					2				
Dianthus superbus var. *longicalycinus*				1	3	2					
Dryopteris expansa	4										
Filipendula multijuga				1		1					
Fragaria nipponica		4	4	2	3	4		3	1	2	2
Geranium onoei	14		1	2	5	9	1				
Gymnadenia conopsea				1							
Hakonechloa macra	2										
Iris sanguinea	23		6	9	10	10		1			
Ixeridium dentatum	1										
Jacobaea cannabifolia	4		2	4	4	4	2				
Ligularia dentata	1								1	1	2
Malus toringo							1	1			
Moehringia lateriflora		3	2				1				
Oxalis corniculata								1	2	3	3
Phedimus aizoon var. *floribundus*	1				1	2					
Picris hieracioides	1			1	2			1			
Polygonatum odoratum	3										
Polygonum cuspidatum	1			1	1	3					
Potentilla freyniana		4	5	1		2					
Ranunculus japonicus	1	10	8	8	8	8	2	5	2	4	3
Scabiosa japonica	8		1	2	2	1					
Serratula coronata	19	1			2	2					
Solidago virgaurea	4										
Stipa pekinensis		1					4	8	10	10	10
Tephroseris flammea	3				1			2	6	4	3
Veronicastrum japonicum	12	7	8	10	9	10		1			
Viola acuminata					2	3					

browsing than more abundant species. Actually, species with lower abundance in 1981 (e.g., *Polygonatum odoratum*, *Ixeridium dentatum*, and *Hakonechloa macra*) were not recovered even after fencing. By grazing, tall-growing herbs were reduced and lower-growing species were increased [32] and grazing-resisted species were shorter in height than grazing-susceptible species [33]. Thus, tall herb species were tending to grazing and hard to recover after grazing.

5. Conclusions

Preventing grazing after intensive grazing of seminatural grassland might result in different successional pathways being followed and the species composition is slightly different from the original vegetation. This suggests that the vegetation inside the fence will change to an alternative state [24]. Galvánek and Lepš [3] showed that the species composition of the restored plots after the reintroduction of mowing was still far from the target composition. Therefore, different management methods are needed to ensure the correct succession pathways are followed for ecological restoration and to enhance colonization of the target species [34], rather than the restored site resulting in an alternative state [24]. Thus, other methods to restore the vegetation (e.g., removing of unpalatable recalcitrant species [35]) inside of the fence would be necessary. Moreover, as Wright et al. [36] suggested, complete removal of ungulates may be required for recovery in heavily browsed forest understory vegetation in New Zealand. Hence, the control of sika deer population should be required.

Conflict of Interests

The authors declare that there is no conflict of interests regarding the publication of this paper.

Acknowledgments

The authors thank members of the Committee of Conservation of *I. sanguinea* on Mt. Kushigata and Dr. Masako Kubo (Shimane University) and Dr. Hayato Iijima (Yamanashi Forest Research Institute) for helping with the field study and Ms. Atsuko Hayashi (Forest and Forest Products Research Institute) for finding the references.

References

[1] T. Suka, "A history of semi-natural grasslands and its causation to distribution of grassland species," *Grassland Science*, vol. 56, pp. 225–230, 2010 (Japanese).

[2] M. Kubo, T. Kobayashi, M. Kitahara, and A. Hayashi, "Seasonal fluctuations in butterflies and nectar resources in a semi-natural grassland near Mt. Fuji, central Japan," *Biodiversity and Conservation*, vol. 18, no. 1, pp. 229–246, 2009.

[3] D. Galvánek and J. Lepš, "How do management and restoration needs of mountain grasslands depend on moisture regime? Experimental study from north-western Slovakia (Western Carpathians)," *Applied Vegetation Science*, vol. 12, no. 3, pp. 273–282, 2009.

[4] T. Nagaike, T. Yoshida, H. Miguchi, T. Nakashizuka, and T. Kamitani, "Rehabilitation for species enrichment in abandoned coppice forests in Japan," in *Restoration of Boreal and Temperate Forests*, J. A. Stanturf and P. Madsen, Eds., pp. 371–381, CRC Press, Boca Raton, Fla, USA, 2005.

[5] T. Kira, "Forest ecosystems of east and southeast Asia in a global perspective," *Ecological Research*, vol. 6, no. 2, pp. 185–200, 1991.

[6] Koma High School, "*Iris sanguinea* on Mt. Kushigata," 1986.

[7] N. Agetsuma, "Ecological function losses caused by monotonous land use induce crop raiding by wildlife on the island of Yakushima, southern Japan," *Ecological Research*, vol. 22, no. 3, pp. 390–402, 2007.

[8] H. Iijima, T. Nagaike, and T. Honda, "Estimation of deer population dynamics using a bayesian state-space model with multiple abundance indices," *Journal of Wildlife Management*, vol. 77, pp. 1038–1047, 2013.

[9] S. Takatsuki, "Effects of sika deer on vegetation in Japan: a review," *Biological Conservation*, vol. 142, no. 9, pp. 1922–1929, 2009.

[10] T. Takeuchi, T. Kobayashi, and M. Nashimoto, "Altitudinal differences in bark stripping by sika deer in the subalpine coniferous forest of Mt. Fuji," *Forest Ecology and Management*, vol. 261, no. 11, pp. 2089–2095, 2011.

[11] R. Tsujino and T. Yumoto, "Effects of sika deer on tree seedlings in a warm temperate forest on Yakushima Island, Japan," *Ecological Research*, vol. 19, no. 3, pp. 291–300, 2004.

[12] Z. Jiang, H. Ueda, M. Kitahara, and H. Imaki, "Bark stripping by sika deer on veitch fir related to stand age, bark nutrition, and season in northern Mount Fuji district, central Japan," *Journal of Forest Research*, vol. 10, no. 5, pp. 359–365, 2005.

[13] T. Nagaike and A. Hayashi, "Bark-stripping by Sika deer (*Cervus nippon*) in *Larix kaempferi* plantations in central Japan," *Forest Ecology and Management*, vol. 175, no. 1–3, pp. 563–572, 2003.

[14] T. Kamei, K.-I. Takeda, K. Koh, S. Izumiyama, O. Watanabe, and K. Ohshima, "Seasonal pasture utilization by wild sika deer (*Cervus nippon*) in a sown grassland," *Grassland Science*, vol. 56, no. 2, pp. 65–70, 2010.

[15] T. Nagaike, "Effects of browsing by sika deer (*Cervus nippon*) on subalpine vegetation at Mt. Kita, central Japan," *Ecological Research*, vol. 27, no. 2, pp. 467–473, 2012.

[16] S. Takatsuki, "Edge effects created by clear-cutting on habitat use by sika deer on Mt. Goyo, northern Honshu, Japan," *Ecological Research*, vol. 4, no. 3, pp. 287–295, 1989.

[17] T. Nagaike, "Effects of altitudinal gradient on species composition of naturally regenerated trees in *Larix kaempferi* plantations in central Japan," *Journal of Forest Research*, vol. 15, no. 1, pp. 65–70, 2010.

[18] J. Beguin, D. Pothier, and S. D. Côté, "Deer browsing and soil disturbance induce cascading effects on plant communities: a multilevel path analysis," *Ecological Applications*, vol. 21, no. 2, pp. 439–451, 2011.

[19] M. Suzuki, "Succession of abandoned coppice woodlands weakens tolerance of ground-layer vegetation to ungulate herbivory: a test involving a field experiment," *Forest Ecology and Management*, vol. 289, pp. 318–324, 2013.

[20] H. Itô and T. Hino, "How do deer affect tree seedlings on a dwarf bamboo-dominated forest floor?" *Ecological Research*, vol. 20, no. 2, pp. 121–128, 2005.

[21] P. M. Perrin, F. J. G. Mitchell, and D. L. Kelly, "Long-term deer exclusion in yew-wood and oakwood habitats in southwest Ireland: changes in ground flora and species diversity," *Forest Ecology and Management*, vol. 262, no. 12, pp. 2328–2337, 2011.

[22] N. L. Schultz, J. W. Morgan, and I. D. Lunt, "Effects of grazing exclusion on plant species richness and phytomass accumulation vary across a regional productivity gradient," *Journal of Vegetation Science*, vol. 22, no. 1, pp. 130–142, 2011.

[23] C. Deléglise, G. Loucougaray, and D. Alard, "Spatial patterns of species and plant traits in response to 20 years of grazing exclusion in subalpine grassland communities," *Journal of Vegetation Science*, vol. 22, no. 3, pp. 402–413, 2011.

Vegetation Recovery in Response to the Exclusion of Grazing by Sika Deer (Cervus nippon) in Seminatural
Grassland on Mt. Kushigata, Japan

81

[24] K. N. Suding, K. L. Gross, and G. R. Houseman, "Alternative states and positive feedbacks in restoration ecology," *Trends in Ecology and Evolution*, vol. 19, no. 1, pp. 46–53, 2004.

[25] V. Aschero and D. García, "The fencing paradigm in woodland conservation: consequences for recruitment of a semi-arid tree," *Applied Vegetation Science*, vol. 15, pp. 307–317, 2012.

[26] A. Shoji, H. Hayashi, K. Kohyama, and H. Sasaki, "Effects of horse grazing on plant species richness and abundance of Iris setosa in a boreal semi-natural grassland, Japan," *Grassland Science*, vol. 57, no. 1, pp. 1–8, 2011.

[27] B. McCune and M. J. Mefford, "PC-ORD. Multivariate analysis of ecological data," ver. 4, MjM Software Design, Gleneden Beach, Ore, USA, 1999.

[28] J. Lepš and V. Hadincová, "How reliable are our vegetation analyses?" *Journal of Vegetation Science*, vol. 3, pp. 119–124, 1992.

[29] A. Tamura, "Effect of time lag of establishment of deer-proof fences on the recovery of perennial herbs in a cool temperate deciduous forest diminished by sika deer browsing in the Tanzawa Mountains, central Japan," *Japanese Journal of Conservation Ecology*, vol. 15, no. 2, pp. 255–264, 2010 (Japanese).

[30] H. Olff and M. E. Ritchie, "Effects of herbivores on grassland plant diversity," *Trends in Ecology and Evolution*, vol. 13, no. 7, pp. 261–265, 1998.

[31] T. P. Rooney and W. J. Dress, "Species loss over sixty-six years in the ground-layer vegetation of heart's content, and old-growth forest in Pennsylvania, USA," *Natural Areas Journal*, vol. 17, no. 4, pp. 297–305, 1997.

[32] K. J. Kirby, "The impact of deer on the ground flora of British broadleaved woodland," *Forestry*, vol. 74, no. 3, pp. 219–229, 2001.

[33] S. Lengyel, K. Varga, B. Kosztyi et al., "Grassland restoration to conserve landscape-level biodiversity: a synthesis of early results from a large-scale project," *Applied Vegetation Science*, vol. 15, no. 2, pp. 264–276, 2012.

[34] S. Díaz, I. Noy-Meir, and M. Cabido, "Can grazing response of herbaceous plants be predicted from simple vegetative traits?" *Journal of Applied Ecology*, vol. 38, no. 3, pp. 497–508, 2001.

[35] T. Nuttle, T. E. Ristau, and A. A. Royo, "Long-term biological legacies of herbivore density in a landscape-scale experiment: forest understoreys reflect past deer density treatments for at least 20 years," *Journal of Ecology*, vol. 102, no. 1, pp. 221–228, 2014.

[36] D. M. Wright, A. J. Tanentzap, O. Flores et al., "Impacts of culling and exclusion of browsers on vegetation recovery across New Zealand forests," *Biological Conservation*, vol. 153, pp. 64–71, 2012.

5

Diversity, Uses, and Threats in the Ghodaghodi Lake Complex, a Ramsar Site in Western Lowland Nepal

Pramod Lamsal,[1] Krishna Prasad Pant,[2] Lalit Kumar,[3] and Kishor Atreya[2]

[1] *Himalayan Geo-En. Pvt. Ltd., 133 Mokshya Marga, Kathmandu 4, Nepal*
[2] *Kathmandu University, P.O. Box 6250, Dhulikhel, Nepal*
[3] *University of New England, Armidale, NSW 2351, Australia*

Correspondence should be addressed to Pramod Lamsal; pramod lamsal@yahoo.com

Academic Editors: A. R. Atangana, I. Bisht, A. Chistoserdov, P. De los Ríos Escalante, and P. M. Vergara

This study documents aquatic and terrestrial/riparian biodiversity in an anthropogenically disturbed Ramsar site, the Ghodaghodi Lake complex, in the Western Nepal surveyed during the summer season (March-April) of 2007. The study site comprises three major interconnected lakes: Ghodaghodi (138 ha), Nakharodi (70 ha), and Bainshwa (10 ha). Five transect lines for aquatic macrophytes and three transect lines and 37 sampling plots were laid to sample terrestrial/riparian plants, birds, and animals. Five sample plots were established for fish and aquatic bird. A total of 45 species of aquatic macrophytes, 54 species of terrestrial/riparian vegetation, 19 fish species, 41 bird species, 17 mammals (endangered and vulnerable), and 5 reptiles (critically endangered, vulnerable, and near threatened) were recorded at the lake complex. Local people have used most of the aquatic and terrestrial plants for different purposes while many of the potential medicinal plant species were still untapped. Persistent anthropogenic threats, like excessive harvesting and poaching, habitat destruction—population pressure, forest fragmentation, siltation, fertilizer and pesticide seepage, water pollution, overgrazing, and unmanaged irrigation system found over the lake complex, endangered the existing biodiversity. The suggested remedial measures are further exploration of medicinal potential, prioritization of in situ biodiversity conservation strategies, and implementation of awareness program at local level against anthropogenic threats.

1. Introduction

Wetlands are defined as lands transitional between terrestrial and aquatic eco-systems where the water table is usually at or near the surface or the land is covered by shallow water [1] and as in such state, supports considerable amount of biological diversity of the earth. The values of wetlands are receiving global attention due to their high contribution to human kind. But it is also a most delicate and threatened habitat as compared to other types because of the close interaction with us. Wetlands occupy approximately 5% (743,756 ha) of the total area of Nepal including high altitude glacial lakes, hot springs, ponds, oxbow lakes, river flood plains, swamps, and marshes and are critical habitats for many plant and animal species. MFSC [2] mentioned that the biogeographic location of Nepal and its complex mountain systems have given rise to a proportionately high level of biodiversity relative to its total area (i.e., 147,181 sq. km). Dugan [3] reported that wetlands occupy only about 4 to 6% of the Earth's surface but provide habitat for about 20% of the world's species. Although Nepal represents only 0.1% of the global terrestrial surface, it houses 0.3% of the world's species including 4.5% of the mammalian species, 4.2% of the butterfly species, 2.7% of the flowering plant species, 2.2% of the freshwater fish species, 1.6% of the reptile species, and 1% of the amphibian species [2]. A. R. Joshi and D. P. Joshi [4] reported that 25% of Nepal's estimated 7,000 vascular plant species are wholly or partially wetland dependent. Though wetlands form only 5% of Nepal's area, a total of 193 bird species (22% of the total recorded in the country) fully depend on it [5]. Of the 35 globally threatened species recorded in Nepal, 15 (43%) are wetland birds. In addition a total of 12 (50%) out of 24 near-threatened species inhabit wetlands [6]. According to IUCN [7], there are 10 species of amphibians, 1 reptile species, and 8 species of flowering plants endemic to Nepal's wetlands.

Medicinal plants have contributed significantly to the livelihood of Nepalese people [8] and their use is prevalent in different parts of the country [9, 10]. In Nepal, traditional use of plant resources for medicinal purpose has a long history and is gaining popularity due to a lack of side effects, easy availability at affordable prices; and in many circumstances, it is the only source of health care to the poor communities [11]. Around 6,500 plant species are used for medicinal purposes in Asia [12] while at least 1,600 to 1,900 plant species are used traditionally for medicinal practices in Nepal [13, 14]. The government has also recognized the importance of proper conservation of medicinal plant diversity and the issue is highlighted in the national level Nepal Biodiversity Strategy [2]. It is estimated that only 15–20% of the population of Nepal, who live in and around urban areas, have access to modern medicinal facilities, whereas the rest depend on traditional medicines [8].

Biodiversity loss has become a major issue over the last few decades and its protection has emerged as a main agenda within national nature conservation policies, international conventions, conservation targets, and political programmes [15]. Freshwater biodiversity has been threatened by a number of major impacts such as overexploitation, water pollution, and flow modification including water abstraction, destruction, or degradation of habitat as well as invasion by alien species [16]. The diversity and distribution of wetland flora and fauna are affected by the changes in the water chemistry [17]. Schuyt [18] reported various threats to the major wetlands of Africa: reclamation for agriculture and settlement expansion at Nakivubo wetland, Uganda; reduction of water level by overabstraction within catchment and degradation of the catchment itself due to overpopulation, overexploitation of wetland resource, soil erosion and siltation, pollution from the use of agrochemicals, and invasion by water hyacinth at Lake Chilwa wetland, Malawi; reduced water flow by overabstraction, aquatic weed infestation, overuse of wetland resource, pollution, and deforestation at Zambezi Basin wetland, South Africa. Five threats have been documented by Dudgeon et al. [19] including overexploitation of the natural resources, water pollution from the siltation and agricultural wastes, flow modification, habitat degradation, and exotic species invasion, whose combined and interactive influences are the causes of declining populations and shrinking global freshwater biodiversity and could be compared with the existing condition of Ghodaghodi Lake complex. Siwakoti and Karki [20] highlighted the heavy dependency of the poor people on wetlands as a major issue for wetland conservation in Nepal. The authors were of the view that the involvement of wetland-dependent communities in the decision-making process could assist sustainable resource conservation efforts and help them benefit in lieu of conservation. The effective systems of resource management can ensure that biological resources not only survive, but increase as well while they are being used by the people, thus providing the foundation for sustainable development [21]. Protection of freshwater biodiversity is perhaps the ultimate conservation challenge in the current global scenario [19]. Since wetlands are considered as a common property resource, the conservation of this ecosystem is an uphill task unless the principal stakeholders are involved in the process.

Ghodaghodi Lake complex is the largest interconnected natural lake system in the plain land of Nepal and was designated a Ramsar site in 2003 due to its high biodiversity value. Though it is a very important lake ecosystem, only a few studies [7, 22] are found on the biodiversity and threats, while medicinal uses of plant species are hardly tapped by the scientific research. IUCN [7] carried out a full inventory of the Lake Complex spread over six months (Nov.–May, 1997/98) while Kafle [22] studied certain lake diversity focussing only on Ghodaghodi Lake in summer (Jan.-Feb.) and winter (Sept.-Oct.) of 2005. This study investigates the terrestrial/riparian and aquatic biodiversity of the lake complex, existing and potential utility including the medicinal values of the plant species, while at the same time identifies the anthropogenic threats to the lake biodiversity. The results of this research are intended to inform managers and decision makers about the importance of this lake ecosystem and how planning decisions could impact the lake's biodiversity.

2. Materials and Methods

2.1. Study Area. The Ghodaghodi Lake complex (28°41′17″ N, 80° 56′47″ E) lies in the Kailali district of far Western Terai in Nepal (Figure 1) and is 205 m above mean sea level. This wetland covers approximately 2500 ha, 14 large and small oxbow lakes with associated marshes, swamps, streams, springs, seasonal marshy grasslands, and artificial wetlands (canals, irrigated fields, ponds, etc.) surrounded by tropical deciduous mixed *Shorea robusta* forest in the lower slopes of Siwalik hills. Major lakes of the complex include Ghodaghodi (138 ha), Nakharodi (70 ha), and Baishawa (10 ha) [20].

The lake system is connected with extensive forests along the Siwalik (Churia) Hills to the north and falls between two of Terai's protected areas—the Royal Bardia National Park and the Royal Suklaphanta Wildlife Reserve—and functions as an important corridor for the movement of wildlife. The lake complex is characterised by three types of wetland habitats: (i) riverine including perennial rivers and river flood plain; (ii) lacustrine including oxbow lakes and ponds; and (iii) palustrine including marshes and swamps. Marshy areas on the fringes of the lakes are subject to periodic inundation. Only Ghodaghodi and Nakharodi Lakes are perennial while Bainshwa is seasonal and turns marshy during the dry season. Similarly, the wetland forest complex has three types of forest habitats: *Shorea robusta* forest, *Terminalia alata* forest, and mixed deciduous riverine forest.

The area has a tropical monsoon climate. The average annual rainfall ranges between 1630 mm (recorded at Tikapur, 35 km to the southeast of the lake) and 1705 mm (recorded at Dhangadhi) where about 80–85% of total rainfall occurs during the monsoon period (mid-June to late September). The average monthly maximum temperature ranges from 21°C to 38°C and minimum 6°C to 25°C (at Tikapur). At Dhangadhi, the maximum temperature ranges from 22°C to 37°C and minimum from 5°C to 25°C. The soil at the bed and surrounding area of the lake complex was

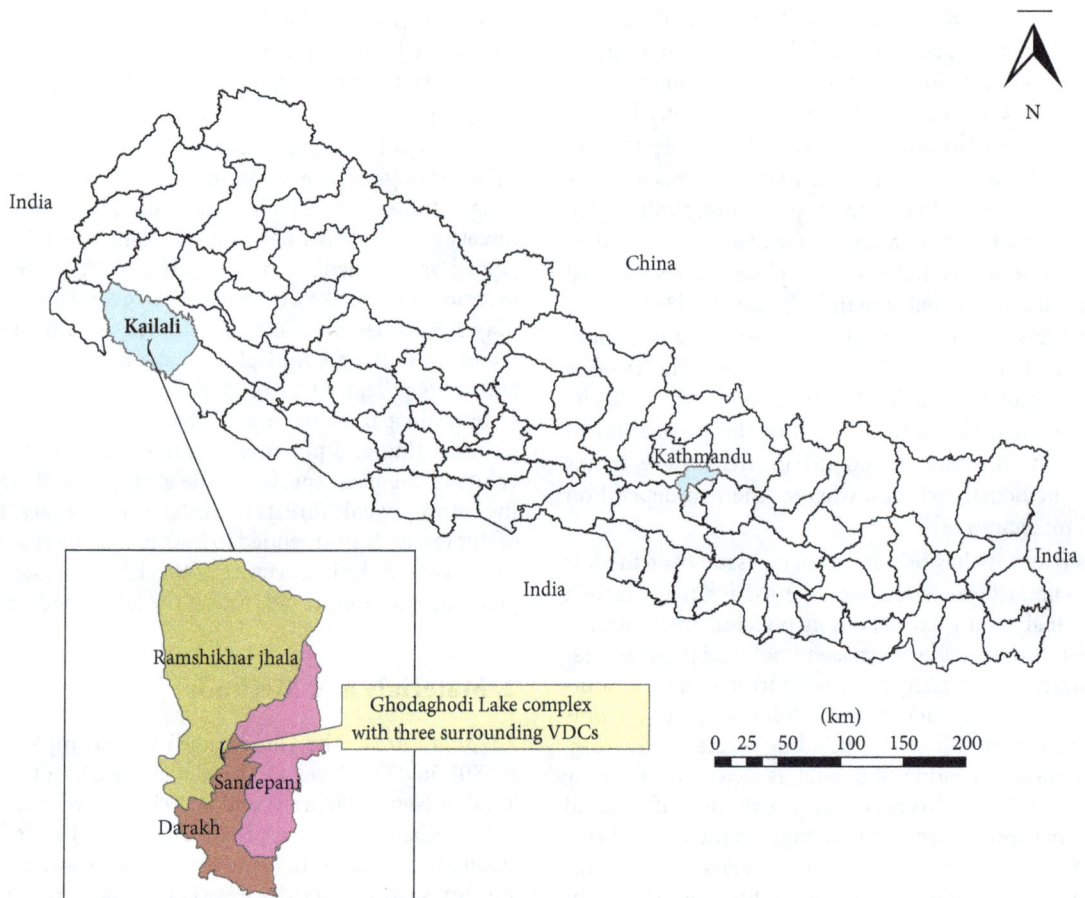

FIGURE 1: Map showing Ghodaghodi Lake complex and adjoining villages.

almost of the same type with very little horizontal spatial variation. The predominant soil type was high in plasticity, of low permeability clay of yellowish to light brown colour. Low permeability of the soil was on the basis that loss of water through seepage and deep percolation seemed to be considerably low. No rock outcrop could be seen in the vicinity of the lake area. The lake is fed by direct precipitation during the monsoon season and by surface flows from the watershed area, ground water springs, and small streams. Water depth varies from 1-2 m during the dry period to 3-4 m during the monsoon season [23].

The wetland complex is bordered by three local politico-administrative units called Village Development Committees (VDC): Sandepani in the East, Darakh in the South, and Ramshikharjhala in the North and West. Around 60% of the total area is used as agricultural land, 37% land is under forest covers, lakes, common pastures, and scrubs; 2% percent under settlements and roads while rivers and streams cover 1% percent of the total land. Farming with traditional use of natural resources is the major occupation of the people living in the lake area. The present uses of lake resources include fishing, livestock grazing and collection of fodder, firewood, and nontimber forest products including medicinal plants. Tharus, an indigenous ethnic group of the lake area, comprising more than 50% of the total population, are the most dependent community on wetland resources [20]. The uses of lake resources by local people, mainly fodder, aquatic macrophytes, livestock grazing, and fishing, were also mentioned by [24]. The total households and population of these VDCs have been increased sharply (73% and 38%, resp.) in the last decade as shown in the Table 1.

2.2. Data Collection. Data were collected during March-April 2007. Therefore the species documented in this study can only be validated in the summer months of Nepal. Data on both terrestrial/riparian and aquatic plants were recorded in the floral category. For aquatic plants, submerged, free-floating, floating leaves, and emergent plants were collected. Sampling was undertaken in such a way that it would cover all the representative habitats of aquatic flora. Shorelines as well as open water areas, floating as well as submerged vegetations were sampled across the lake in transect lines in a boat. Five transect lines, three at Ghodaghodi (719 m, 1673 m, and 1572 m), one at Bainshwa (460 m), and one at Nakharodi (960 m), were established; and quadrat method of random sampling as mentioned by Ludwig and Reynolds [27] was followed to collect macrophytes. Local people and secondary literature [28–31] were consulted for field identification of aquatic macrophytes.

TABLE 1: Household and population increment in the lake adjoining villages.

VDC	Total household		Total population	
	2001	2011	2001	2011
Darakh	1,694	3,153	12,171	17,623
Ramsikharjhala	1,824	3156	13,560	18,016
Sandepani	2,592	4,278	17,956	24,892
Total	6,110	10,587	43,687	60,531

Source: CBS [25, 26].

For terrestrial plants, trees, shrubs and herbs were surveyed. Three transect lines, each from eastern, western, and northern parts, passing through the surrounding forest of Ghodaghodi Lake complex, were demarcated. A total of 37 quadrat plots, also used by Dongol [32], of 10 m × 10 m were laid on the ground at intervals of 250 m along those transect lines in an alternate side. Similarly, nested plots of 3 m × 3 m and 1 m × 1 m on the lower right-hand side corner were laid within the 10 m × 10 m plot for shrub and herb inventory, respectively.

Field identification of the plant species was carried out using standard literature [30, 33–35] and with the help of local people. Herbaria were prepared for unidentified plant samples and brought to the Kathmandu University Laboratory for expert identification. Unidentified species were also taken to the Government accredited National Herbarium Centre at Godavari, Lalitpur.

In the faunal category, the survey was undertaken for fishes, birds, mammals, and reptiles. For fish, five sampling areas were purposively identified, three in Ghodaghodi Lake and two in Bainshwa and Nakharodi. Fishing nets were kept overnight and the fish catch was collected next morning and repeated for three consecutive days. We followed Shrestha [36] for fish specimen identification. Local fish markets located at Sukhad Bazar were surveyed twice in three-day intervals to identify additional species harvested from the lake.

For terrestrial birds, the same plots of terrestrial vegetation assessment were considered. Bird surveys were carried out before vegetation surveys in the morning from 0500 hr to 0700 hr through direct observation via binoculars (8 × 30) and a call count method. Birds were surveyed for a total of five consecutive days. For aquatic birds, the same sampling points used for fish were considered. The survey procedure and period were similar to those of terrestrial birds and in both cases, we followed the literature of BCN and DNPWC [37] for field identification. The same stations were used for the survey of aquatic animals.

For terrestrial wildlife, the transect lines setup for terrestrial plant assessments was used. Each transect was walked three times and wildlife species were observed. This survey was carried out before vegetation surveys so that habitats could remain undisturbed and animals could be spotted. Indirect methods such as faecal, hair, and horns were also used to determine the presence of wildlife species.

Anthropogenic threats as well as additional information on exiting biodiversity and local medicinal uses of plant species were observed and recorded through field observations, focus group discussions (FGDs), and key informant interviews (KII). A total of three focus group discussions, one at each VDC and six key informant interviews, two at each VDC, were conducted. Field observations were made using a recording sheet and camera while a checklist containing all aspects of biodiversity, medicinal uses, and existing threats was developed, pretested, and used for the FGD. A total of 8–10 people including ethnic traditional healers were made available for FGD in order to gather information, crosscheck, and validate. Older members of the community and local schoolteachers were selected for key informant interviews.

3. Results and Discussion

3.1. Plant Diversity

3.1.1. Aquatic Plant Diversity. A total of 45 species of aquatic plants including 9 submerged, 6 free-floating, 21 floating leaved, and 9 emergent species (Table 2) were found in the Ghodaghodi, Bainshwa, and Nakharodi Lakes. Lotus (*Nelumbo nucifera*), water chestnut (*Trapa bispinosa*), water lilies (*Nymphaea nouchali*), and pondweed (*Potamogeton natans*) were the major anchored leaf-floating species. Hydrilla (*Hydrilla verticillata*), hornwort (*Ceratophyllum demersum*), and pondweeds (*Potamogeton* spp.) were the major submerged species. Similarly, feathered mosquito fern (*Azolla imbricate*), duck weed (*Lemna minor*), and Asian watermeal (*Wolffia globosa*) were found to be the dominant free-floating species. An inventory of the Ghodaghodi Lake by IUCN [7] resulted in 107 species of aquatic macrophytes, whereas a seasonal study by Kafle [22] reported only 22 aquatic plant species. A total of seven categories of existing uses were found from the lake aquatic macrophytes. Out of the 45 plant species identified in the wetland complex, 22 species have been used for different purposes, such as human medicines (11 species), fodder (7 species), food (6 species), green manure (8 species), food for fish (5 species), food for duck (5 species), and industrial (2 species). There were still 23 macrophytes species untapped for any uses including medicines, though used in other parts of the country as well as South Asia.

3.1.2. Terrestrial/Riparian Plant Diversity. A total of 54 terrestrial/riparian plant species (Table 3) were found, out of which 25 were trees, 7 shrubs, and 22 herbs. *Shorea robusta* and *Terminalia alata* were the major terrestrial tree species. Other species include *Syzygium cumini, Aegle marmelos, Acacia catechu*, and *Dalbergia latifolia*. The forest also contained lianas and undergrowth, and the secondary layer was formed by a variety of trees including *Mallotus philippensis*. A rare spiny shrub *Gardenia campanulata*, with highly restricted distribution within Nepal, was also found. *Syzygium cumini* and *Syzygium jumbos* were the dominant species of swamp forest, particularly around Nakharodi Lake. However, a total of 137 and 35 terrestrial/riparian plant species were identified by IUCN [7] and Kafle [22], respectively, at Ghodaghodi Lake. Local people in Ghodaghodi Lake complex traditionally use

TABLE 2: Aquatic macrophytes recorded in the Ghodaghodi Lake complex.

Species	Potential uses	Existing uses by local people in the Ghodaghodi Lake area
Submerged		
Ceratophyllum demersum	Human medicine [38, 39]; fish food [40, 41]; green manure [41]	Fish food; duck food; green manure
Hydrilla verticillata	Fish food [42]	Fish food; duck food; fodder; green manure
Limnophila sessiliflora	N/A	NONE
Potamogeton natans	N/A	NONE
Potamogeton crispus	N/A	NONE
Potamogeton pectinatus	N/A	NONE
Utricularia flexuosa	N/A	Human medicine; green manure
Utricularia stellaris	N/A	NONE
Vallisneria spiralis	Fish food [42]	NONE
Free-floating		
Azolla imbricata	Green manure [39, 43]; fodder [39]; duck food [39]	Fish food; duck food; green manure
Hygroryza aristata	Fodder [41]	Food; fodder
Lemna minor	N/A	Fish food; duck food; green manure
Pistia stratiotes	Human medicine [38]; duck food [41, 43]; green manure [38, 43]	Human medicine; green manure
Ricciocarpus natans	N/A	NONE
Wolffia globosa	N/A	NONE
Submerged rooted with floating leaves		
Caldesia parnassifolia	N/A	NONE
Jussiaea repens	N/A	NONE
Nymphaea nouchali	Food [41]	NONE
Nelumbo nucifera	Human medicine [38]; food [41–43]	Human medicine; food; fodder
Nymphoides cristatum	N/A	NONE
Ottelia alismoides	Food [39]	NONE
Potamogeton nodosus	Duck food [42]	NONE
Trapa bispinosa	Human medicine [44]	Human medicine; food
Emergent		
Alisma plantago-aquatica	N/A	NONE
Cassia sophera	Human medicine [45]	NONE
Centella asiatica	Human medicine [40, 46–49]; veterinary medicine [32]	Human medicine; food
Cyperus diffuses	Fodder [42]	NONE
Cyperus rotundus	Human medicine [14, 49]	Human medicine; fodder
Eclipta prostrata	Human medicine [14, 42, 49]; food [38]	Human medicine; fodder
Equisetum debile	Human medicine [39]	Human medicine; green manure
Fimbristylis dichotoma	Fodder [13]	NONE
Hygrophila auriculata	Human medicine [43, 50]; food [41]	Human medicine
Ipomea fistulosa	N/A	NONE
Ipomea aquatica	Human medicine [43, 51]; food [38, 41, 42]	Human medicine; fish food; duck food; food
Limnophila indica	Human medicine [39]	NONE
Ludwigia octovalvis	Human medicine [38, 39]	NONE
Monochoria vaginalis	Human medicine [39, 43]	Food
Phragmites karka	Human medicine [38]; fodder [39]; industrial [41]	Fodder; industrial use

TABLE 2: Continued.

Species	Potential uses	Existing uses by local people in the Ghodaghodi Lake area
Polygonum barbatum	Food [41]	NONE
Ranunculus sceleratus	Human medicine [43, 52]	Human medicine
Rumex dentatus	Human medicine [53]; food [43]; green manure [43]	Food; green manure
Saccharum spontaneum	Human medicine [54]; fodder [41, 43]; industrial [39, 41]	Fodder; industrial
Sonchus asper	Human medicine [55]	NONE
Typha angustifolia	Food [42]; fodder [42]; industrial [39, 42]	NONE

Note. Human and veterinary medicine includes the uses of a plant in different ailments by local people in the study area as well as in other parts of South Asia. Food includes vegetables, pickle, and spices consumed by human. Industrial includes use of a plant for handicraft, oils and tannin, and others that can provide direct cash benefit. N/A: information could not be found in the scientific literature by authors. NONE: no use of a plant in the study area so far.

several terrestrial/riparian plant species to sustain their livelihood. A total of eight categories of existing uses were found from the lake terrestrial plant resources. Out of the 54 plant species identified in the wetland complex, 35 species have been used for different purposes, such as human medicines (29 species), veterinary medicine (5 species), fodder (13 species), fruit (8 species), timber (7 species), firewood (6 species), industrial (4 species), and food (3 species). A total of 19 species were still untapped in the study area though they have been found to be used for different purposes including medicine in other parts of the country as well as South Asia.

3.2. Animal Diversity

3.2.1. Fish Diversity. A total of 19 fish species were found in the Ghodaghodi Lake complex (Table 4). IUCN [7] identified 27 species of fish from the same area.

3.2.2. Avifaunal Diversity. Wetland birds comprise a significant portion of the avian fauna recorded in Nepal [71]. This study found a total of 41 species of birds representing 20 families in the Ghodaghodi Lake complex (Table 5). IUCN [7] documented 140 species of avifauna whereas Kafle [22] reported 60 species. The lower number of bird species in our study compared to others may be because of seasonal effects; that is, IUCN inventory was for six months and Kafle has documented avifauna seen in two seasons.

3.2.3. Mammalian Diversity. A total of 17 species of mammals belonging to 12 families were recorded in the Ghodaghodi Lake complex (Table 6). These included endangered species like fishing cat (*Felis viverrina*) and Asian elephant (*Elephas maximus*), while the common otter (*Lutra lutra*), leopard cat (*Felis bengalensis*), leopard (*Panthera pardus*), blue bull (*Boselaphus tragocamelus*), and spotted deer (*Axix axix*) fall under the vulnerable category. IUCN [7] reported 34 mammalian species from the same complex. Some of the mammals such as jungle hare, spotted deer, and wild boar are used by the local people for bush meat.

3.2.4. Reptile Diversity. This study found a total of 6 species of reptiles representing 5 families in the Ghodaghodi Lake complex (Table 7). It also showed the existence of the vulnerable marsh mugger (*Crocodylus palustris*) in different parts of the Ghodaghodi Lake. It was informed by the local people that marsh mugger often travels near Bainshwa Lake by crossing a small ridge that separates these two lakes. Similarly, the critical endangered red crowned roofed turtle (*Kachuga kachuga*) and near-threatened Indian python (*Python molurus*) were also found to exist in the lake. Ghodaghodi Lake supports the habitat for additional species of turtles including flap-shell turtle (*Lissemys punctata*) and Indian roofed turtle (*Kachuga tecta*).

3.3. Potential Anthropogenic Threats to Ghodaghodi Lake Complex Biodiversity. Freshwater ecosystems are affected by overharvesting of the resources and habitat destruction. The major drivers of the destruction are human demography, excessive resource use, increased water consumption, technological development, and social organization [74]. These changes are responsible for some major alteration like physical restructuring of aquatic ecosystems, introduction of exotic species, discharge of toxic substances, and overharvesting of resources [75]. Bhandari [76] observed conversion of wetland because of overexploitation of resources, pollution of water, invasion of alien species, encroachment to the lake area, and sedimentation as the major categories of threats in Nepal. The Ghodaghodi Lake complex supports several aquatic and terrestrial flora and fauna, contributing to wetland biodiversity conservation in the area. However, as other wetlands, Ghodaghodi Lake complex is no exception, receiving threat from unsustainable harvesting and illegal poaching, expanding population, changes in land use, and overexploitation of lake resources. Most of these threats we observed are anthropogenic in nature and can be broadly categorized into (i) unsustainable harvesting and poaching and (ii) habitat destruction.

3.3.1. Unsustainable Harvesting and Poaching. Local people heavily harvested aquatic and terrestrial plants and fishes. Illegal tree felling and smuggling of *Shorea robusta* timber

TABLE 3: Terrestrial/riparian plants recorded within the forest adjacent to the Ghodaghodi Lake complex.

Species	Potential uses	Existing uses by local people in the Ghodaghodi Lake area
Herb		
Aerva lanata	Human medicine [50]; food [32]	NONE
Achyranthes aspera	Human medicine [40, 49, 56, 57]	Human medicine; veterinary medicine
Ageratum conyzoides	Human medicine [57–59]	Human medicine
Alternanthera sessilis	Human medicine [60]	NONE
Asteracantha longifolia	N/A	Human medicine; fodder
Bacopa monnieri	Food [41, 43]	NONE
Barleria cristata	N/A	NONE
Blumea mollis	Human medicine [61]	Human medicine; fodder
Colocasia antiquorum var. *esculenta*	N/A	NONE
Chenopodium ambrosioides	Human medicine [55]	NONE
Cirsium wallichii	N/A	NONE
Dopatrium junceum	N/A	Human medicine; veterinary medicine; food; fodder
Dichanthium annulatum	N/A	NONE
Echinochloa colonum	N/A	Human medicine
Oxalis corniculata	Human medicine [32, 48]; veterinary medicine [32]; food [32]; fodder [32]	Human medicine; veterinary medicine; food; fodder
Polypogon monspeliensis	N/A	NONE
Phyla nodiflora	Human medicine [32]	Human medicine
Scirpus grossus	Industrial [41]	NONE
Scirpus articulates	N/A	Human medicine
Scoparia dulcis	Human medicine [32, 62]	Human medicine
Solanum nigrum	Human medicine [51, 63]	NONE
Vernonia cinerea	Human medicine [61]	Human medicine
Shrub		
Calotropis gigantea	Human medicine [48, 49, 54, 64]	Human medicine
Colebrookea oppositifolia	N/A	Human medicine; veterinary medicine
Gardenia campanulata	N/A	NONE
Murraya koenigii	Human medicine [14, 49]	Fruit; food; fodder; firewood
Phoenix acaulis	N/A	NONE
Woodfordia fruticosa	Human medicine [48, 54]; veterinary medicine [32]; industrial [32, 65]	Human medicine; veterinary medicine; industrial use
Zizyphus mauritiana	Human medicine [48, 49]	Human medicine; fruit; industrial use
Tree		
Acacia catechu	Human medicine [48, 54]; timber [32]; firewood [32]; industrial [32]	Human medicine; industrial
Aegle marmelos	Human medicine [49]; food [54]; fruit [54]	Human medicine; fruit
Artocarpus lakoocha	Human medicine [54, 62]	Human medicine
Bauhinia purpurea	Human medicine [53]; food [64]; fodder [64]; industrial [32]	Human medicine; fodder; food; firewood; industrial use
Careya arborea	Human medicine [66]	NONE
Cassia fistula	Human medicine [48, 53]	Human medicine; firewood
Dalbergia sissoo	Human medicine [48]; fodder [64]; timber [32]; firewood [32]	Human medicine; timber; fodder
Dalbergia latifolia	N/A	Timber
Dillenia pentagyna	Food [32]	NONE
Ficus racemosa	Fodder [54]; fruit [54]	Fodder

TABLE 3: Continued.

Species	Potential uses	Existing uses by local people in the Ghodaghodi Lake area
Lannea coromandelica	Human medicine [67]	NONE
Mangifera indica	Human medicine [50, 54]	Timber; fruit
Mallotus philippensis	Human medicine [48, 49, 64]; firewood [64]	Human medicine; fodder; firewood
Ougeinia dalbergioides	N/A	NONE
Psidium guajava	Human medicine [48, 68]	Fruit
Schleichera oleosa	Human medicine [32, 68]; food [32]; fruit [32]; industrial [32]	Timber; firewood; fruit
Shorea robusta	Fodder [32]; industrial [32]; firewood [32, 64]	Human medicine; timber
Spondias pinnata	Human medicine [54, 68]; fruit [54]	Human medicine; fruit
Syzygium cumini	Food [50, 54, 64]; firewood [32]; fruits [32, 54]; fodder [32]	Human medicine; timber; fruit; fodder
Syzygium jambos	N/A	NONE
Terminalia bellerica	Human medicine [48, 54]; timber [32]; firewood [32]; industrial [20; 39]	Human medicine
Terminalia alata	Fodder [32, 54]; timber [32, 61]; industrial [65]	Human medicine
Toona ciliata	Human medicine [67]	Human medicine; timber
Trewia nudiflora	Human medicine [69]; food [64]; firewood [32, 64]; fodder [32, 64]; fruit [32]	Firewood; fodder
Zizyphus rugosa	Human medicine [70]	NONE
Climber and vine		
Bauhinia vahlii	Fodder [54]; fruit [54]; industrial [54]	Human medicine; fruit; fodder; industrial
Spatholobus parviflorus	N/A	Fodder

Note. Human and veterinary medicine includes the uses of a plant in different ailments by local people in the study area as well as in other parts of South Asia. Food includes vegetables, pickle, and spices consumed by human. Industrial includes use of a plant for handicraft, oils and tannin, and others that can provide direct cash benefit. N/A: information could not be found in the scientific literature by authors. NONE: no use of a plant in the study area so far.

TABLE 4: Fishes recorded in the Ghodaghodi Lake complex.

Scientific name	Family
Colisa fasciatus	Belonidae
Clarias batrachus	Claridae
Channa striatus	Channidae
Channa punctatus	Channidae
Lepidocephalus guntea	Cobitidae
Channa gachua	Channidae
Puntius gelius	Cyprinidae
Puntius conchonius	Cyprinidae
Puntius chola	Cyprinidae
Puntius sophore	Cyprinidae
Rasbora daniconius	Cyprinidae
Heteropneustes sp.	Saccobranchidae
Labeo boga	Cyprinidae
Mastacembelus pancalus	Mastacembelidae
Xenentodon cancila	Belonidae
Mastacembelus armatus	Mastacembelidae
Mystus vittatus	Bagridae
Notopterus notopterus	Notopteridae

and *Acacia catechu* were common practices found during the study period. Human encroachments were severe in the southeastern parts of Nakharodi Lake; and eastern, southeastern, and northwestern parts of Ghodaghodi Lake. Siwakoti and Karki [20] mentioned illegal cutting and encroachment at the lake area. Mammals and birds suffer from poaching by local people as well as illegal hunters from other areas of the country. It has been reported that local poachers have always been active for killing wild animals in the lake area. Illegal hunting of wild boars, deer, and other wildlife as well as bird trapping and egg collection were prevalent in the area, either by the villagers or professional poachers. Neupane et al. [77] reported similar threats of poaching and illegal hunting of small mammals, fish, and birds at wetlands located in the far Western Nepal.

3.3.2. Habitat Destruction. Habitat destruction affects both harvested species such as aquatic plants and fishes and even nonharvested species of weeds, reptiles, and microorganisms. Large mammals like tiger, sloth bear, and most of the deer species were not sighted in the area for a couple of years. The current population pressure in the area and fragmentation of the forest and increased agricultural land have been identified as factors contributing to this degradation. The local people attributed this to the shrinkage and thinning of forest areas for the establishment of settlements around the lake that have heavily degraded the habitat of these mammals. The other drivers of habitat destruction in the

TABLE 5: Avifauna recorded in the Ghodaghodi Lake complex.

Scientific name	Family
Egretta alba modesta	Ardeidae
Metopidius indicus	Jacanidae
Nettapus coromandelianus	Anatidae
Gallinula chloropus	Rallidae
Fulica atra	Rallidae
Ardea intermedia intermedia	Ardeidae
Anhinga melanogaster	Anhingidae
Phalacrocorax niger	Phalacrocoracidae
Spilornis cheela cheela	Accipitridae
Ardeola grayii	Ardeidae
Accipiter nisus	Accipitridae
Dendrocygna javanica	Anatidae
Buceros bicornis	Bucerotidae
Anastomus oscitans	Ciconiidae
Porphyrio porphyrio	Rallidae
Ardea cinerea	Ardeidae
Amaurornis phoenicurus	Rallidae
Oriolus xanthornus	Oriolidae
Threskiornis melanocephalus	Threskiornithidae
Ardea purpurea	Ardeidae
Dinopium shorii	Picidae
Megalaima asiatica	Capitonidae
Megalaima haemacephala	Capitonidae
Coracias benghalensis	Muscicapidae
Merops philippinus	Meropidae
Merops orientalis	Meropidae
Halcyon smyrnensis	Alcedinidae
Cuculus varius	Cuculidae
Centropus sinensis	Cuculidae
Psittacula eupatria	Psittacidae
Pernis ptilorhyncus	Accipitridae
Long-tailed shrike*	Campephagidae
Rufous hooded oriale*	Oriolidae
Pericrocotus flammeus	Campephagidae
Dicrurus paradiseus	Corvidae
Copsychus malabaricus	Muscicapidae
Cinnyris asiaticus	Nectariniidae
Psittacula krameri	Psittacidae
Anastomus oscitans	Ciconiidae
Pandion haliaetus haliaetus	Accipitridae
Dendrocitta vagabunda	Corvidae

*Common name.

TABLE 6: Mammals recorded in the Ghodaghodi Lake complex.

Scientific name	Family	Common name
Macaca mulatta (LC, II)	Cercopithecidae*	Rhesus monkey
Presbytis entellus (I)	Cercopithecidae*	Hanuman langur
Felis chaus (LC)	Felidae**	Jungle cat
Felis viverrina (EN, II)	Felidae**	Fishing cat
Felis bengalensis (VU, II)	Felidae**	Leopard cat
Panthera pardus (VU, I))	Felidae**	Leopard
Canis aureus (LC)	Canidae*	Jackal
Lutra lutra (NT, I)	Mustelidae*	Otter
Pteropus sp. (LC)	Pteropodidae*	Bat
Funambulus sp.	Sciuridae*	Squirrel
Petaurista petaurista (LC)	Sciuridae*	Flying squirrel
Lepus nigricollis (LC)	Leporidae*	Jungle hare
Boselaphus tragocamelus (VU)	Bovidae*	Blue bull
Axix axix (VU)	Cervidae*	Spotted deer
Sus scrofa (LC)	Suidae*	Wild boar
Elephas maximus (EN, I)	Elephantidae**	Asian elephant
Bandicota indica (LC)	Muridae*	Jungle rat

*Seen. **Reported from focus group discussions (FGD) and key informants interviews (KII).
EN: endangered, VU: vulnerable, NT: near threatened, LC: least concern (IUCN Red List Category); I, II: CITES categories (source: Jnawali et al. [72]).

TABLE 7: Reptiles directly observed in the Ghodaghodi Lake complex.

Scientific name	Family	Common name
Crocodylus palustris (VU, I)	Crocodylidae	Marsh mugger
Python molurus (NT, I)	Pythonidae	Python
Kachuga tecta (LC)	Geoemydidae	Indian roofed turtle
Kachuga kachuga (CR)	Geoemydidae	Red crowned roofed turtle
Lissemys punctata (LC)	Trionychidae	Flap shell turtle
Varanus flavescens (LC, I)	Varanidae	Golden monitor lizard

V: vulnerable, CR: critically endangered, NT: near threatened, LC: least concern (IUCN Red List Category) (source: IUCN [73]); I: CITES category.

lake complex include rites and rituals, unmanaged irrigation systems for agricultural fields, runoff from agricultural fields and siltation, and overgrazing.

The wastes generated from the rites and rituals performed by the religious pilgrims, especially indigenous Tharu community, on the shore of the lake, have contributed significantly to water pollution in the Ghodaghodi Lake complex. They celebrate by sacrificing domestic pigs, goats, chickens, and pigeons on the shore, which ultimately becomes a source of pollution. A few past studies [20, 22, 77] also documented cultural and religious activities as a major source of pollution. Recreational activities in the form of picnicking are a common practice inside the complex. The movement of people and the loud music system played for such recreation negatively impact the faunal species including birds.

During the last two decades, land use of the complex has changed due to the encroachment of migrants from the hilly districts of far western region. Forest area has decreased at the expense of agricultural land while forest cover has been converted into open grazing land, displacing the habitat of important birds and wildlife.

Water extraction for irrigation and water use for buffalo wallowing has severely affected the habitat. This study observed both Ghodaghodi and Nakharodi Lakes having a canal used for irrigation purposes, a practice that becomes even more intense during dry periods. Landowners and farmers downstream from the lake have a vested interest to secure water for irrigation. The lake complex does not have a permanent perennial source of water and this extraction system, especially in the dry season, greatly affects the aquatic life and migratory birds.

The southeastern part of Ghodaghodi Lake and eastern parts of Nakharodi Lake have been heavily affected by livestock grazing. As the occupation of most of the villagers used to be farming as well as rearing livestock, grazing the forests and open grasslands of the lake area is common. However, the trampling of soil and browsing of young palatable species from this traditional practice has damaged the regeneration capacity of native vegetation in the area.

Dichlorvos 76% EC and Endosulfan 35% EC were the major pesticides used, while urea and diammonium phosphate (DAP) were the major fertilizers used in crops near the lake vicinity. The seepage of such pesticides and fertilizers through agricultural runoff has caused eutrophication in the lake. As a result, excessive growth of aquatic macrophytes such as *Ceratophyllum demersum, Nelumbo nucifera, Naja minor, Pistia stratiotes,* and *Hydrilla verticillata* from the water surface was evident. Siwakoti and Karki [20] reported the occurrence of both the natural and human caused eutrophication at Ghodaghodi Lake complex with the invasion of *Ipomea fistulosa* and *Salix* spp. Eutrophication from agricultural runoff and invasion of water hyacinth were also reported by Shrestha [42] at Koshi Tappu wetland.

Siltation emerging from the degradation of upper watershed areas is another reason for habitat destruction. Rapid deforestation and encroachment in the watershed area within Betini forest, a major source of water, is causing reduced water supply to the lake. Overgrazing of the grassland at the watershed has also increased soil erosion and siltation. This finding is similar to that of [77]. In Nepal, nearly 69% of wetlands are threatened by sedimentation and siltation while 61% by agricultural runoff [7].

4. Conclusion and Recommendation

This study, a month long summer season survey, documents the existing biodiversity, the use pattern of floral diversity by the local people, and anthropogenic threats to the overall biodiversity of Ghodaghodi Lake complex. The Ghodaghodi Lake complex was found to be a very rich wetland diversity hotspot in the western low-land region of Nepal, though variations in the number of species than those of previous study by IUCN [7] and Kafle [22] were recorded. Within such a small territory, the existence of such a vast number of both the floral and faunal diversity is really an encouraging sign for conservation. The lake resources, especially aquatic and terrestrial plants, were traditionally been used for different purposes including medicinal, food, fodder, fruit, fuel wood, timber, and industrial. Among them, medicinal

uses were a main priority for the local people because they have been traditionally practiced in the locality and are a good alternative to the health care system. However, the data revealed that most of the plant species were still untouched because of a lack of local knowledge on their potential value and utilization. This shows a scope in the exploration of ethnobotanical knowledge of the plant species in the Ghodaghodi Lake complex. If proper conservation and sustainable utilization is ensured, then the lake resources can have a good economic contribution to the lives of local people.

The paradox is however the threats in the area, mostly anthropogenic in nature, which could lead to an imbalance of the natural wetland ecosystem. If immediate conservation action is not put in place, then there is a likelihood that most of the threatened category animals will be in danger due to the anthropogenic threats identified by this study. The results show that Ghodaghodi Lake complex has an enormous economic and conservational potential and care should be taken to maintain its biodiversity value. The economic potential and multipurpose use of plant species, including medicinal values, could supplement the household income of the local people and be a good incentive for them for conservation. We recommend (1) more exploration of the indigenous lake resource uses including medicinal potential of available plant species, (2) prioritizing in situ biodiversity conservation, and (3) implementing awareness programs at local levels on the consequence of threats to lake biodiversity.

Conflict of Interests

The authors declare that there is no conflict of interests regarding the publication of this paper.

References

[1] W. I. Mitsch and I. G. Gosselink, *Wetlands*, Van Nostrand Reinhold, New York, NY, USA, 1986.

[2] MFSC, *Nepal Biodiversity Strategy*, Ministry of Forest and Soil Conservation, Kathmandu, Nepal, 2002.

[3] P. Dugan, *Wetlands in Danger: A World Conservation Atlas*, Oxford University Press, New York, NY, USA, 1993.

[4] A. R. Joshi and D. P. Joshi, "Endemic plants of Nepal Himalaya: conservation status and future direction," *Mountain Environment and Development*, vol. 1, no. 2, pp. 1–35, 1991.

[5] IUCN, *A Review of the Status and Threats to Wetlands in Nepal*, IUCN Nepal, Kathmandu, Nepal, 2004.

[6] BirdLife International, *Lists of Globally Threatened and Near-Threatened Species in Nepal*, 2010.

[7] IUCN, *The Ghodaghodi Tal Conservation Area: A Community Centred Management Plan*, IUCN Nepal, Kathmandu, Nepal, 1998.

[8] U. R. Sharma, K. J. Malla, and R. K. Uprety, "Conservation and management efforts of medicinal and aromatic plants in Nepal," *Banko Janakari*, vol. 14, pp. 3–11, 2004.

[9] S. Bhattarai, R. P. Chaudhary, and R. S. L. Taylor, "Ethnomedicinal plants used by the people of Manang district, central Nepal," *Journal of Ethnobiology and Ethnomedicine*, vol. 2, article 41, 2006.

[10] R. M. Kunwar and R. W. Bussmann, "Ethnobotany in the Nepal Himalaya," *Journal of Ethnobiology and Ethnomedicine*, vol. 4, article 24, 2008.

[11] R. Acharya and K. P. Acharya, "Ethnobotanical study of medicinal plants used by Tharu Community of Parroha VDC, Rupendehi District, Nepal," *Scientific World*, vol. 7, no. 7, pp. 80–84, 2009.

[12] M. Karki and J. T. Williams, *Priority Species of Medicinal Plants in South Asia*, Medicinal and Aromatic Plants Program in Asia (MAPPA), IDRC/SARO, New Delhi, India, 1999.

[13] S. K. Ghimire, "Sustainable harvesting and management of medicinal plants in the Nepal Himalaya: current issues, knowledge gaps and research priorities," in *Medicinal Plants in Nepal: an Anthology of Contemporary Research*, P. K. Jha, S. B. Karmarachraya, M. K. Chhetri, C. B. Thapa, and B. B. Shrestha, Eds., pp. 25–44, Ecological Society of Nepal (ECOS), Kathmandu, Nepal, 2008.

[14] S. R. Baral and P. P. Kurmi, *A Compendium of Medicinal Plants in Nepal*, vol. 281, Mrs. Rachana Sharma, Maujubahal, Chabahil, Kathmandu, Nepal, 2006.

[15] D. Romport, M. Andreas, and B. Vlasakova, "Monitoring of biodiversity changes in the landscape scale," *Journal of Landscape Ecology*, vol. 1, no. 1, pp. 49–68, 2008.

[16] Millennium Ecosystem Assessment, *Ecosystems and Human Well-Being: Wetlands and Water Synthesis*, World Resources Institute, Washington, DC, USA, 2005.

[17] D. Sonal, R. Jagruti, and P. Geeta, "Avifaunal diversity and water quality analysis of an inland wetland," *Journal of Wetland Ecology*, vol. 4, pp. 1–32, 2010.

[18] K. D. Schuyt, "Economic consequences of wetland degradation for local populations in Africa," *Ecological Economics*, vol. 53, no. 2, pp. 177–190, 2005.

[19] D. Dudgeon, A. H. Arthington, M. O. Gessner et al., "Fresh water biodiversity: importance, threats, status and conservation challenges," *Biological Reviews of the Cambridge Philosophical Society*, vol. 81, no. 2, pp. 163–182, 2006.

[20] M. Siwakoti and J. B. Karki, "Conservation status of Ramsar Sites of Nepal Tarai: an overview," *Botanica Orientalis*, vol. 6, pp. 76–84, 2009.

[21] J. A. McNeely, *Economics and Biological Diversity: Developing and Using Economic Incentives to Conserve Biological Resources*, IUCN, Gland, Switzerland, 1988.

[22] G. Kafle, "Avifaunal survey and vegetation analysis focusing on threatened and near-threatened species on Ghodaghodi Lake of Nepal," A Report Submitted to Oriental Bird Club (OBC), Bedford, UK, 2005.

[23] Y. B. Bam, "Conservation and Sustainable Use of Ghodaghodi Lake System," Technical Report submitted to IUCN Nepal, 2002.

[24] J. P. Sah and J. T. Heinen, "Wetland resource use and conservation attitudes among indigenous and migrant peoples in Ghodaghodi Lake area, Nepal," *Environmental Conservation*, vol. 28, no. 4, pp. 345–356, 2001.

[25] CBS, *National Census Report of 2001*, Central Bureau of Statistics, Government of Nepal, 2001.

[26] CBS, *National Census Report of 2011*, Central Bureau of Statistics, Government of Nepal, 2011.

[27] J. A. Ludwig and J. F. Reynolds, *Statistical Ecology: A Primer of Methods and Computing*, Wiley Press, New York, NY, USA, 1988.

[28] H. Hara, A. O. Chater, and L. H. J. Williams, *An Enumeration of Flowering Plants of Nepal*, vol. III, British Meuseum (Nat. Hist.), London, UK, 1982.

[29] C. D. K. Cook, *Aquatic and Wetland Plants of India*, Oxford University Press, 1996.

[30] J. R. Press, K. K. Shrestha, and D. A. Sutton, *Annotated Checklist of the Flowering Plants of Nepal*, The Natural History Museum, London, UK, 2000.

[31] E. M. Bursche, *A Hand Book of Water Plants*, Oriental Enterprise, Dehradun, India, 1991.

[32] D. R. Dongol, "Economic uses of forest plant resources in Western Chitwan, Nepal," *Banko Janakari*, vol. 12, no. 2, pp. 56–64, 2002.

[33] R. P. Choudhary, *Biodiversity in Nepal: Status and Conservation*, 1998.

[34] N. P. Manandhar, *Medicinal Plants of Nepal Himalaya*, Ratna Pustak Bhandar, Kathmandu, Nepal, 1980.

[35] T. K. Rajbhandary, M. S. Bista, and V. L. Gurung, *Enumeration of the Vascular Plants of West Nepal*, Ministry of Forests and Soil Conservation, Department of Plant Resources, Thapathali, Kathmandu, Nepal, 1994.

[36] J. Shrestha, *Enumeration of Fishes of Nepal*, Biodiversity Profile Project, Publication no. 10, Department of National Parks and Wildlife Conservation, Ministry of Forest and Soil Conservation, Government of Nepal, Kathmandu, Nepal, 1995.

[37] BCN and DNPWC, *Birds of Nepal—An Official Checklist for Nepal*, Bird Conservation Nepal and Department of National Park and Wildlife Conservation, 2006.

[38] B. P. Sarmah, D. Baruah, and B. Bakalial, "Wetland medicinal plants in floodplains of Subansiri and Ranga River of Lakhimpur District, Assam, India," *Asian Journal of Plant Science and Research*, vol. 3, no. 3, pp. 54–60, 2013.

[39] B. Niroula and K. L. B. Singh, "Aquatic plant resources of Betana Wetland, Morang, Nepal," *Our Nature*, vol. 9, pp. 146–155, 2011.

[40] A. R. Joshi and K. Joshi, "Ethnomedicinal plants used against skin diseases in some villages of Kali Gandaki, Bagmati, and Tadi Likhu watersheds of Nepal," *Ethnobotanical Leaflets*, vol. 11, pp. 235–246, 2007.

[41] M. K. Misra, A. Panda, and S. Deenabandhu, "Survey of useful wetland plants of South Odisha, India," *Indian Journal of Traditional Knowledge*, vol. 11, no. 4, pp. 658–666, 2012.

[42] P. Shrestha, "Diversity of aquatic macrophytes in the Koshi Tappu Wildlife Reserves and surrounding areas, Eastern Nepal," in *Environmental and Biodiversity: in the Context of South Asia. 2003-2011*, P. K. Jha, G. P. S. Ghimire, S. B. Karmacharya, S. R. Baral, and P. Lacoul, Eds., 1996.

[43] B. Niroula and K. L. B. Singh, "Contribution to aquatic macrophytes of Biratnagar and adjoining areas, Eastern Nepal," *Ecoprint*, vol. 17, pp. 23–34, 2010.

[44] A. R. Mohammad, O. F. Mohammad, and A. H. Mohammad, "Environment friendly antibacterial activity of water chesnut fruits," *Journal of Biodiversity and Environmental Sciences*, vol. 1, pp. 26–34, 2011.

[45] M. S. Hossan, A. Hanif, B. Agarwala et al., "Traditional use of medicinal plants in Bangladesh to treat urinary tract infections and sexually transmitted diseases," *Ethnobotany Research and Applications*, vol. 8, pp. 61–74, 2010.

[46] R. B. Mahato, "Notes on some plants of ethnomedicinal importance from Palpa district," *Tribhuvan University Journal*, vol. 21, no. 1, pp. 71–76, 1998.

[47] N. P. Balami, "Ethnomedicinal uses of plants among the Newar community of Pharping Village of Kathmandu district, Nepal," *Journal of Tribhuvan University*, vol. 24, no. 1, p. 10, 2004.

[48] A. G. Singh, M. P. Panthi, and D. D. Tewari, "Ethno medicinal plants used by the Tharu and Magar communities of Rupendehi District, Western Nepal," *Current Botany*, vol. 2, no. 2, pp. 30–33, 2011.

[49] R. M. Kunwar, Y. Uprety, C. Burlakoti, C. L. Chowdhary, and R. W. Bussmann, "Indigenous use and ethnopharmacology of medicinal plants in far-west Nepal," *Ethnobotany Research and Applications*, vol. 7, pp. 5–28, 2009.

[50] M. Ayyanar and S. Ignacimuthu, "Ethnobotanical survey of medicinal plants commonly used by Kani tribals in Tirunelveli hills of Western Ghats, India," *Journal of Ethnopharmacology*, vol. 134, no. 3, pp. 851–864, 2011.

[51] N. Dhami, "Ethnomedicinal uses of plants is Western Terai of Nepal: a case study of Dekhatbhuli VDC of Kanchanpur district," in *Medicinal Plants in Nepal: An Anthology of Contemporary Research*, P. K. Jha, S. B. Karmacharya, M. K. Chhetry, C. B. Thapa, and B. B. Shrestha, Eds., pp. 165–177, Ecological Society of Nepal (ECOS), Kathmandu, Nepal, 2008.

[52] M. Z. Chopda and R. T. Mahajan, "Wound healing plants of Jalgaon District of Maharashtra State, India," *Ethnobotanical Leaflets*, vol. 13, pp. 1–32, 2009.

[53] S. S. Ahmad, "Medicinal wild plants from Lahore-Islamabad motorway (M-2)," *Pakistan Journal of Botany*, vol. 39, no. 2, pp. 355–375, 2007.

[54] Y. Uprety, R. C. Poudel, H. Asselin, and E. Boon, "Plant biodiversity and ethnobotany inside the projected impact area of the Upper Seti Hydropower Project, Western Nepal," *Environment, Development and Sustainability*, vol. 13, no. 3, pp. 463–492, 2011.

[55] K. Hussain, A. Shahazad, and S. Z. Hunssnain, "An ethnobotanical survey of important wild medicinal plants of Hattar District Haripur, Pakistan," *Ethnobotanical Leaflets*, vol. 12, pp. 29–35, 2008.

[56] N. P. Manandhar, "Native phytotherapy among the Raute tribes of Dadeldhura district, Nepal," *Journal of Ethnopharmacology*, vol. 60, no. 3, pp. 199–206, 1998.

[57] Y. Uprety, E. K. Boon, R. C. Poudel et al., "Non-timber forest products in Bardiya district of Nepal: indigenous use, trade and conservation," *Journal of Human Ecology*, vol. 30, no. 3, pp. 143–158, 2010.

[58] R. B. Mahato and R. P. Chaudhary, "Ethnomedicinal study and antibacterial activities of selected plants of Palpa district, Nepal," *Scientific World*, vol. 2, pp. 38–45, 2003.

[59] A. G. Singh, A. Kumar, and D. D. Tewari, "An ethnobotanical survey of medicinal plants used in Terai forest of western Nepal," *Journal of Ethnobiology and Ethnomedicine*, vol. 8, p. 19, 2012.

[60] P. P. Sapkota, "Ethnoecological observation of Magar of Bakuni, Baglung, Western Nepal, Dhaulagiri," *Journal of Sociology and Anthropology*, vol. 2, pp. 227–252, 2008.

[61] D. M. A. Jayaweera, *Medicinal Plants (Indigenous and Exotic) Used in Ceylon*, Part 3, The National Science Foundation, Colombo, Sri Lanka, 2006.

[62] M. M. Hasan, E. A. Annay, M. Sintaha et al., "A survey of medicinal plant usage by folk medicinal practitioners in seven villages of Ishwardi Upazilla, Pabna District, Bangladesh," *American-Eurasian Journal of Sustainable Agriculture*, vol. 4, no. 3, pp. 326–333, 2010.

[63] K. Joshi, R. Joshi, and A. R. Joshi, "Indigenous knowledge and uses of medicinal plants in Macchegaun, Nepal," *Indian Journal of Traditional Knowledge*, vol. 10, no. 2, pp. 281–286, 2011.

[64] S. Bhattarai, B. Pant, and C. P. Upadhyaya, "Dependency of Tharu communities on wild plants: a case study of Shankarpur, Kanchanpur district," *Banko Janakari*, vol. 21, no. 1, pp. 35–40, 2011.

[65] T. Krishnamurty, *Minor Forest Product of India*, Oxford & IBH Publishing, New Delhi, India, 1993.

[66] M. Rahmatullah, M. N. K. Azam, M. M. Rahman et al., "A survey of medicinal plants used by garo and non-garo Traditional Medicinal Practitioners in Two Villages of Tangail District, Bangladesh," *American-Eurasian Journal of Sustainable Agriculture*, vol. 5, no. 3, pp. 350–357, 2011.

[67] P. K. Rai and H. Lalramnghinglova, "Ethnomedicinal plant resources of Mizoram, India: implication of traditional knowledge in health care system," *Ethnobotanical Leaflets*, vol. 14, pp. 274–305, 2010.

[68] B. P. Gaire and L. Subedi, "Medicinal plant diversity and their pharmacological aspects of Nepal Himalayas," *Pharmacognosy Journal*, vol. 3, no. 25, pp. 6–17, 2011.

[69] S. A. Mukul, M. B. Uddin, and M. R. Tito, "Medicinal plant diversity and local healthcare among the people living in and around a conservation area of Northern Bangladesh," *International Journal of Forest Usufructs Management*, vol. 8, no. 2, pp. 50–63, 2007.

[70] A. G. Singh and J. P. Hamal, "Traditional phytotherapy of some medicinal plants used by Tharu and Magar communities of Western Nepal, against dermatological disorders," *Scientific World*, vol. 11, no. 11, pp. 81–89, 2013.

[71] H. S. Baral, "Updated status of Nepal's wetland birds," *Banko Janakari*, (Wetland Special Issue), pp. 30–35, 2009.

[72] S. R. Jnawali, H. S. Baral, S. Lee et al., *The Status of Nepal Mammals: The National Red List Series*, The National Red List Series, Department of National Parks and Wildlife Conservation, Kathmandu, Nepal, 2011.

[73] IUCN, IUCN Red List of Threatened Species. Version 2013.2, 2013, http://www.iucnredlist.org/.

[74] R. J. Naiman, J. J. Magnuson, D. M. McKnight, and J. A. Stanford, *The Freshwater Imperative: A Research Agenda*, Island Press, Washington, DC, USA, 1995.

[75] D. J. Rapport and W. G. Whitford, "How ecosystems respond to stress: common properties of arid and aquatic systems," *BioScience*, vol. 49, no. 3, pp. 193–203, 1999.

[76] B. B. Bhandari, "Wise use of wetlands in Nepal," *Banko Janakari*, (Wetland Special Issue): pp. 10–17, 2009.

[77] P. K. Neupane, M. Khadka, R. Adhikari, and D. R. Bhuju, "Lake water quality and surrounding vegetation in Dry Churiya Hills, Far-Western Nepal," *Nepal Journal of Science and Technology*, vol. 11, pp. 181–188, 2010.

6

A Comparison of the South African and United States Models of Natural Areas Management

Daniel S. Licht,[1] Brian C. Kenner,[2] and Daniel E. Roddy[3]

[1] National Park Service, 231 East Saint Joseph Street, Rapid City, SD 57701, USA
[2] Badlands National Park, Interior, SD 57750, USA
[3] Wind Cave National Park, Hot Springs, SD 57747, USA

Correspondence should be addressed to Daniel S. Licht; dan_licht@nps.gov

Academic Editors: P. K. S. Shin and P. M. Vergara

In May-June of 2013 we visited several South African parks and reserves to learn about wildlife and natural areas management in that country. We focused our visit on parks and reserves that are of moderate size (5,000–100,00 ha), comprised of grassland/savanna habitats, located within agrarian landscapes, and enclosed with boundary fences, characteristics similar to several parks and reserves in the Northern Great Plains region of the United States. In this paper we compare the South African model of natural areas management to the United States model. We observed that South African parks and reserves with the aforementioned characteristics are more likely to (1) reintroduce and conserve small, nonviable wildlife populations, (2) reintroduce and conserve top-level predators, (3) have more intensive management of wildlife, (4) manage in partnership across multiple landowners, (5) engage local communities, (6) be self-funding, and (7) restrict visitor movement. The South African model is arguably more effective in conserving biodiversity as measured by conservation of apex predators and natural processes. The differences between the countries appear to be driven in large part by socioeconomic factors. Knowledge of natural areas management in other countries may lead to more innovative and creative models that could benefit biodiversity conservation.

1. Introduction

The United States is often perceived and portrayed as a leader in natural areas management [1], a perception that has some merit considering that the country established Yellowstone National Park in 1872, arguably the world's first national park. Over time a convention and mode of natural areas management evolved in the USA [2]. At the same time other countries were establishing parks and reserves and developing their own management models [1]. South Africa is one such country that is now widely recognized for its innovative and progressive park and reserve management.

With that in mind, the authors went to South Africa, met with colleagues there, and reviewed wildlife and natural areas management in that country. Many of the parks and reserves in South Africa are similar to natural areas in the Northern Great Plains region of the United States in that they are moderate in size (5,000–100,00 ha), they conserve grassland and savanna ecosystems, and they are fenced and/or surrounded by agrarian landscapes. In this paper we qualitatively identify and discuss the most notable differences we observed between the two countries in terms of natural areas management and biodiversity conservation. We did not use a quantitative approach for comparing the effectiveness of biodiversity conservation in the two countries as there are confounding factors such as differing baseline levels of species richness [3] that make such an approach problematic. Our goal is to expose the reader to varying forms of natural areas management in hope that it may lead to better conservation of biodiversity.

2. The South African Model versus the United States Model

It is somewhat misleading to say that a country has a "model" for natural areas management and to starkly compare that model against another country's model, as most countries have a myriad of approaches toward natural areas management and conservation. For example, in South Africa the parks and reserves range from the vast and open Kalahari Gemsbok National Park to small and closed parks near urban areas; management practices and policies vary greatly between these sites. Likewise, within the United States national park management practices vary greatly between Yellowstone National Park and smaller parks near population centers. Yet when comparing "apples to apples," specifically moderate-size grassland/savanna parks located within agrarian landscapes and surrounded by boundary fences, there are significant and noteworthy differences between the two countries. We summarize these differences in Table 1. In the following sections we elaborate on these differences.

2.1. Wildlife Management.

Perhaps the most striking difference between the two countries from a biodiversity perspective is that South African parks and reserves actively reintroduce, conserve, and manage very small populations of wildlife, including apex predators. This is generally done for purposes of (1) restoring an ecological process, and, (2) generating wildlife-related tourism revenue. Notable species that are commonly reintroduced into small parks and reserves in South Africa include the African lion (Panthera leo), African elephant (Loxodonta africana), black rhinoceros (Diceros bicornis), white rhinoceros (Ceratotherium simum), African buffalo (Syncerus caffer), wildebeest (Connochaetes sp.), and African wild dog (Lycaon pictus). Of those, only the black rhinoceros and African wild dog are generally considered as critically endangered in the wild.

Funston [4] compiled a list of private and public protected areas in southern Africa that reintroduced lions since 1992; ten sites were 10,000 ha or less and three were 2,500 ha or less. The smallest site was the privately owned 1,500 ha Madjuma Game Reserve. In several cases the reserves are so small that they could conserve only a single pride of lions comprised of only a few individuals. Many similar size reserves also conserve small populations of elephants, rhinoceroses, and other large-bodied mammals. Even wide-ranging species, such as the African wild dog, have been reintroduced into sites as small as 8,500 ha [5]. Lindsey et al. [5] established minimum reserve sizes for reintroduction of wild dogs (6,500–14,700 ha) based on the criteria of being able to support a single family unit.

The reintroduction of small populations of large-bodied wildlife into small fenced reserves in South Africa is done more for restoring ecological processes or promoting ecotourism than it is for the global conservation of the species (Figure 2). Funston [4] noted that most South African sites with lions are managed to meet the goals for the site (primarily ecotourism); however, he also noted that such sites have the potential for meaningful conservation of the species if metapopulation guidelines are followed. A metapopulation approach, relying on a network of large and medium size reserves, has been proposed for wild dogs and may be critical for survival of the species in South Africa [6].

The reintroduction and conservation of micropopulations of wildlife in small reserves are commonplace in South Africa; however, there are challenges and necessities associated with such a model. Kettles and Slotow [7] list the issues as overpopulation, inbreeding depression, decline of prey and other predator species, conflict with neighboring communities, and, in some cases, spreading disease. For example, a micropopulation of lions caused the decline of some prey species at the small and fenced Madjuma Game Reserve, whereas other prey species increased in abundance following the lion reintroduction; these unanticipated events resulted in reserve managers removing the lion population [8]. Similar undesirable effects have been noted for reintroductions of African wild dogs to small closed reserves [5].

Mitigating these problems requires active intervention in what Funston [4] called "micromanagement." Kettles and Slotow [7] list the possible interventions as "relocation, contraception, hunting, and artificial takeovers." Such micromanagement is often costly and contentious, but usually technologically feasible. In discussing lion management Kettles and Slotow [7] noted that "none of the intervention methods resulted in long-term behavioural or social consequences. Constraints on lion management were more from societal values than biological or technological influences. If applied in the correct manner, at the correct time, all of these interventions, or a combination of them, can assist in achieving management objectives."

One factor that may facilitate the South African micromanagement or hands-on approach is that in South Africa wildlife is not "owned" by the state. Therefore, wildlife can readily be bought, sold, transferred, and managed. In contrast, wildlife in the United States is "owned," or held in trust, by the states or, when specifically identified by Congress, by the federal government. In general, private ownership of wildlife is uncommon and illegal in many cases.

Furthermore, in contrast to the South African model, the reintroduction of micropopulations to small parks and reserves, as well as the associated interventionist management, is discouraged in U.S. parks and reserves. In the case of national parks this reluctance is reaffirmed by National Park Service policies that call for reintroductions only when the population can be "self-perpetuating" [9]. We know of only a few examples that challenge this requirement. One example is Wind Cave National Park which recently reintroduced the black-footed ferret (Mustela nigripes) even though the 2,500 acres of black-tailed prairie dogs (Cynomys ludovicianus) within the park may only support about 30 ferrets and be inadequate for long-term viability. This reluctance and discouragement to conserve small and non-viable populations is why none of the parks and reserves in the Northern Great Plains region of the United States support the full native faunal community whereas many comparable size parks and reserves in South Africa do conserve the full wildlife community.

Perhaps the most striking difference between the two countries and their management models is expressed in terms of apex predators (Figure 1). The United States has no precedence or model for reintroducing and conserving small populations of apex predators in closed protected areas. Wolves (Canis lupus) have been restored in the Western United States, but only to the vast wildernesses of

TABLE 1: Notable differences between natural areas management in South Africa and the United States.

Issue	South Africa	United States
Park/reserve purpose	Conservation of wildlife, especially charismatic megafauna, for ecotourism. Many private reserves were established for commercial purposes.	Conservation of scenery and landscapes, evolved to include the conservation of wildlife and ecosystems for public benefits.
Conservation of small populations	More likely to reintroduce and conserve small nonviable populations.	Policies and traditions often discourage reintroduction of small nonviable populations.
Apex Predators	Apex predators are reintroduced into reserves as small as 5,000 ha.	Apex predators are only reintroduced to large landscapes.
Management Intervention	Very hands-on management, necessitated in part by the small populations and presence of apex predators.	A more hands-off approach, sometimes directed by agency policies.
Active metapopulation approach	Routine transfer of animals between sites for demographic and genetic augmentation, revenue generation, and other reasons.	Less frequent transfer of animals between sites, especially into existing populations. Generally done only when species is threatened with extirpation.
Boundary fences	Routinely used for natural areas and required by law for some species.	Generally discouraged. Used primarily for bison.
Water management	Trend toward removal of anthropogenic water for purposes of restoring ecosystem integrity and heterogeneity; however, many units retain anthropogenic water for wildlife viewing.	Water management generally avoided unless critical for species restoration, sometimes done for purposes of uniform range utilization and increased carrying capacity.
Partnerships across land ownerships	Commonly used to create and expand natural areas. Typically enforced with a legal document. Management often conferred to a single entity.	Less frequently used, rarely with a legal document, and partners maintain their own management.
Park/reserve expansion	Protected area boundaries regularly expand due in large part to new partnerships.	Boundary changes and expansion much less frequent.
Community involvement and benefits	Natural areas are often established for community benefits or have evolved to emphasize those benefits, including resource utilization by local communities. Parks/reserves actively involve local communities via regular and routine meetings.	Community involvement and benefits are more passive. Generally, public engagement is limited to more formal meetings intended to get public input on specific, proposed management actions or plans.
Ecotourism	Ecotourism and hunting are used to justify, create, and operate many reserves.	Ecotourism often viewed as passive benefit of natural areas, but is not typically a primary objective.
Park funding	National parks get about 20% of operating funds from appropriations, rest from gate receipts, wildlife sales, and other sources. Private reserves funded from ecotourism, wildlife sales, and hunting.	Almost all park funding comes from government appropriations. In the case of reserves managed by nonprofit organizations, from donations.
Visitor experiences	Visitation is highly restricted. Visitors, only able to enter/leave within daylight hours and must stay in vehicles in reserves with dangerous animals unless on ranger-led activities.	Visitors generally allowed to freely travel on foot, via vehicle or horseback.
Outreach and interpretation	Small visitor centers with rudimentary interpretive displays and information.	Larger and more state-of-the-art visitor centers utilizing modern technology.

the Yellowstone Ecosystem and the Southwest United States. Nowhere have wolves been reintroduced to parks or reserves similar to what has occurred with lions and wild dogs in South Africa, although the idea has recently been broached in the scientific literature [10]. The reluctance to consider such a model in the USA has many reasons, including the hands-off non-interventionist approach endorsed by agency policies. The current situation at Isle Royale National Park in Lake Superior is noteworthy. In the 1940s wolves got established on the 53,500 ha island park by crossing an ice-bridge from the mainland. The population grew to a long-term average of about 25 animals, but has been in decline for the past several years. The conservation and scientific community is currently debating whether wolves should be translocated to the island

for genetic and/or demographic augmentation of the existing population [11]. Based on our observations there would be little if any hesitation to augment or reintroduce wolves to Isle Royale under the South African model.

A prerequisite for reintroducing lions, elephants, and other large mammals into reserves in South Africa is the establishment of a fence. Electrified boundary fences are required by law for lions, wild dog, and elephants [12]. Fences greatly reduce conflicts with surrounding neighbors and protect people and property from the species; however, they also have the benefit of allowing for the conservation of the species [13]. Were it not for fences the animals would inevitably wander out of the small parks and reserves and probably be killed. Eventually that could lead to extirpation

FIGURE 1: South African reserves as small as 5,000 ha restore lions for purposes of ecosystem stewardship and ecotourism. The restoration of small populations and in some cases single family units requires intensive management. The restoration of apex predators to small and medium size reserves is not done in the United States.

FIGURE 2: A perimeter fence around a South African reserve. In agrarian landscapes fences protect people and property and conserve wildlife populations. Wildlife within fenced African reserves is considered free ranging.

of the park population. Packer et al. [14] suggested that many unfenced African lion populations may decline to near extinction in the next 20–40 years. They also noted that lion populations within fenced reserves are nearer their carrying capacities and are less expensive to manage. Fences are generally viewed negatively by the conservation community in the United States as they result in ecological problems, high costs, and aesthetic impairment. However, fences are used in the Great Plains to keep bison (*Bison bison*) from roaming out of parks and reserves. Boundary fencing has been a critical tool in the conservation of the species.

The differing philosophies and approaches towards wildlife management are epitomized by the use of two terms. In the United States the term "*free ranging*" is generally only applied to an animal outside of a fenced park or reserve; however, in South Africa the same term is often applied to an animal within a fenced reserve, provided the reserve has the minimum land size needed for the animal to meet its life needs. Or in the words of one South African ecologist, a free-ranging animal is one "that is not fed by people" (pers.

comm. Sam Ferreira, Kruger National Park). In the United States the word *metapopulation* typically refers to a wildlife species whose global or regional population is comprised of several subpopulations that naturally interchange animals through dispersal. In contrast, the term *metapopulation* in South Africa generally refers to a population that is managed via anthropogenic translocation of animals between sites.

2.2. Partnerships. Conserving biodiversity often requires landscape approaches. In South Africa there are numerous examples of multiple landowners managing sites in partnership. In some cases the partners are a mix of public and private entities, in other cases the partnership may be comprised only of private landowners (these are sometimes referred to as *conservancies*). Government-sponsored partnerships also occur across national boundaries including between South Africa and neighboring Botswana, Namibia, Mozambique, and Zimbabwe. By establishing partnerships and removing fences between properties the landowners realize economies of scale and increase the site's potential for conserving biodiversity [15].

Welgevonden Game Reserve in north-central South Africa is one of many examples of a private partnership reserve. The 37,500 ha reserve is comprised of approximately 30 landowners who voluntarily became members of the reserve. Some landowners operate resorts and provide tourist activities such as safaris, whereas others maintain their properties for personal pleasure. However, in exchange for access to the entire reserve, all manage their land, operate their facilities, and make use of the reserve under a binding charter. The entire reserve, which hosts the *Big 5* (African lion, African elephant, leopard, rhinoceros, and African buffalo) and other wildlife, is overseen by an elected board of governors. Another example of a multientity partnership is the 55,000 ha Madikwe Game Reserve in the North West Province. The site is run as a three-way partnership between the provincial government, local communities, and the private sector [16]. The success of Madikwe has led to the proposal of even more ambitious partnerships that would benefit local economies as well as biodiversity [17].

In contrast, the wealth of the United States has enabled the establishment of a wide range of government-run parks and reserves, perhaps to the exclusion of creative public-private partnerships. We know of no arrangements in the Northern Great Plains region of the United States similar to those of Welgevonden or Madikwe. Although neighboring land management agencies sometimes have memorandums of understanding and other cooperative documents, the agreements do not have the legal foundation, authority, commitment, or management scope of the multilandowner partnerships commonplace in South Africa. One promising scenario in the United States that could approach what we observed in South Africa is a site in southwestern South Dakota where the potential exists for three adjacent bison herds, one public (Badlands National Park), one tribal (a proposed tribal national park), and one private to be managed as one large herd under a partnership relationship similar to what is practiced in many South African reserves.

2.3. Community Emphasis. We were struck by the emphasis of South African natural areas to provide essential community benefits. From the highest policy statements down to park and reserve operations there is an explicit emphasis in South Africa on working with and for local communities. This emphasis ranges from employing local people to (in the case of private reserves) making donations for community projects and infrastructure. However, it should be noted that this community emphasis is a relatively new model within South Africa. Historically, parks and reserves in that country were established and managed primarily for the pleasure and benefit of the wealthy and privileged white minority. But with the end of apartheid a new natural areas model evolved that has strong socio-economic and community development objectives. Even the vast transboundary parks and reserves build on the theme as a goal of those efforts is to achieve peace and prosperity in the region [18]. Fortunately, this community-emphasis model has facilitated conservation of Africa's charismatic wildlife species and biodiversity conservation.

An example of the new model is the Madikwe Game Reserve. The 1997 management plan for the site states "the approach towards conservation that has been adopted at Madikwe puts the needs of people before that of wildlife and conservation" [16]. The 55,000 ha reserve was initiated in 1991 following a feasibility study that determined that a wildlife reserve, and a wildlife-based tourism business, would be economically more profitable and self-sustaining than the apartheid-era cattle ranching and farming. As a result of the socio-economic assessment, the livestock were removed, wildlife-proof perimeter fences were constructed, and what may be the world's largest wildlife translocation, known as Operation Phoenix, was undertaken. As of 2003 the reserve directly supported almost 450 jobs [19]. Although there have been challenges and criticisms in terms of what types of realized community benefits, infrastructure development, and wildlife management [20, 21], the site has had enormous community, socio-economic, and biodiversity benefits, especially when compared to the prior land use.

We are not suggesting that South African parks and reserves generate more income or create more jobs than United States parks. USA parks provide substantial socio-economic benefits to local communities. For example, Badlands National Park supported 317 jobs in 2011 and brought $22 million to the local economy [22]. Many other USA parks have gateway communities that receive similar benefits from the parks. Yet these benefits are rarely the primary purpose, or even a secondary purpose, of the park or preserve. The socio-economic benefits realized by local communities in the USA are typically, from the management perspective of the agencies, indirect and passive. In contrast, in South Africa many parks and reserves now list community benefits as being a primary explicit objective for the site. And such an approach can result in substantial biodiversity benefits.

2.4. Budgets and Park Operations. In South Africa there is a saying that "the wildlife must pay their way," and wildlife has done that to an amazing extent. The ABSA Group

Economic Research [15] estimated that there were about 5,000 private commercial wildlife ranches and other 4,000 mixed wildlife/livestock farms in South Africa, encompassing 13% of the land area of the country. Many wildlife operations have been started in the past few decades. The country's transition to full democracy in 1994 led to the deregulation of the agricultural sector and a reduction in subsidies, thereby making many marginal rangelands unprofitable for cattle, goat, and sheep operations. Many landowners converted to wildlife operations. Some operations emphasize private hunting and/or generate revenue from wildlife sales. Others cater to nonconsumptive ecotourism, with large predators being critical attractions for ecotourism [23]. Lindsey et al. [24] found that wild dog conservation can pay for itself based on contingency valuation model and that ecotourism should be part of a multifaceted approach to wild dog conservation. Madikwe Game Reserve, established on former rangeland, and many other sites are self-funding operations [16]. South Africa has demonstrated that economics, jobs, and biodiversity can, in some cases, be mutually beneficial.

However, an economics-based model for wildlife conservation does have shortcomings even in South Africa. For example, at Madikwe Game Reserve prey populations are managed at carrying capacity so tourists have a high likelihood of seeing lions and leopards, to the detriment of rare and less competitive habitat specialists such as roan (*Hippotragus equinus*), sable (*Hippotragus niger*), and tsessebe (*Damaliscus lunatus*). Nevertheless, the site still conserves more biodiversity than the prior domestic livestock-based model.

There has been no comparable increase in the number or land area of private wildlife operations in the United States. Farm subsidies and cropland retirement programs remain substantial and temporary cropland set-asides are often supported by the conservation community, helping reinforce agrarian land uses. It has been suggested that the short-term set-asides preclude more beneficial and long-lasting conservation opportunities similar to those practiced in South Africa [25], yet agricultural subsidies and temporary set-asides are likely to remain a driving force in land use in the Great Plains. The availability of large amounts of accessible public land in the United States also makes private for-profit ecotourism operations more challenging. One could also argue that North American wildlife is less charismatic and less able to generate ecotourism revenue than South African wildlife. Wildlife-based operations comprise a relatively small portion of land use in the region. One notable exception is the increase in bison ranches; however, unlike in Africa, the purpose of these ranches is not to provide ecotourism or hunting but for meat production as a better option than traditional cattle ranching. In practice these operations are more similar to livestock operations than they are to wildlife reserves [26]. In spite of the economic obstacles to establishing private reserves in the United States, attempts are being made. The American Prairie Foundation (APF) project in Montana is a noteworthy example and has similarities to the South African model. That project hopes to recreate a vast self-funding prairie reserve out of marginal rangeland. The project has had many successes and continues to grow, although funding continues to primarily come through donations. In contrast to many South African ventures, neither the state nor federal government is an active supporter and promoter of the APF project.

2.5. Visitor Experiences. One dramatic difference between South African parks and reserves and those in the USA is the amount of freedom that visitors have to experience the site. For example, in most South African sites visitors must enter via manned gates during daylight hours. And visitors are rarely allowed to exit their vehicles except at designated spots. The restrictive policy is primarily due to the presence of lions, elephants, rhinoceros, leopards (*Panthera pardus*), and other dangerous animals. While the restrictions are primarily for visitor safety they likely also reduce harassment and disturbance of wildlife. In private reserves vehicle drives usually require a guide; a requirement that provides employment for local people and further protects the wildlife. In the U.S. visitors have much more open access and freedom to move about the site, including traveling via foot without a guide. Such unrestricted travel and nature-viewing experiences are encouraged. In rare cases parks and reserves may close certain areas, often temporarily, to protect critical denning areas, nesting sites, or other critical habitats.

Visitors to parks and reserves in South Africa and in the United States may have different expectations and goals. In South Africa wildlife is typically the star attraction, whereas in the United States wildlife often competes with geologic formations, outdoor recreation, and other features. This difference is illustrated by the fact that in South Africa park and reserve visitors go on "*game drives*" whereas in the USA they go "*sightseeing*." Our use of these terms is of course a generalization, but we believe that it fairly and accurately portrays the differing visitor objectives and experiences in the two countries. Consider that wildlife viewing, although a very popular activity in USA National Parks, does not typically rank as the most important reason that people visit parks [27]. In many parks, including parks established to conserve natural resources, sightseeing and viewing scenery more frequently ranked as the top activity for visitors. In South Africa, wildlife viewing is the top reason people visit most parks and reserves, with the ultimate goal of seeing the *Big 5*.

Based on our observations, interpretive and education materials and programs at parks are more prominent, technological, and state-of-the-art in the United States than they are in South Africa. However, we were struck by the power of some of the rudimentary interpretive displays we saw in South Africa and neighboring Swaziland. We observed large wallboard displays that were hand-drawn and hand-written by local school children. In this day and age of ubiquitous video displays, software programs, and computer-designed posters the presence and power of a home-made interpretive display can be profound.

3. Conclusion

In this paper we summarize and describe notable observed differences in natural areas management between South Africa and the United States, with an emphasis on fenced mid-size natural areas in agrarian and human-dominated landscapes. We have made general statements and conclusions based on our observations, discussions, and research. Our perspective is strongly influenced by our extensive experience with moderate-size national parks in the Great Plains; others may come to slightly different conclusions and areas of emphasis based on their experiences.

From our perspective there are dramatic differences in how the two countries perceive and manage mid-size natural areas. In South Africa the reserves are more socio-economically driven, yet ironically, they are better at conserving biodiversity, at least in terms of conserving large-bodied mammals, apex predators, and natural processes. Reserves in the United States rely less on ecotourism for their existence and therefore have less incentive to conserve charismatic megafauna. However, the South African model also comes with many challenges and costs.

We do not intend to convey that one model is better than the other as each has pros and cons and may be best suited for the milieu in which they occur. Yet other countries' models could and should be considered as they may lead to more innovative and creative approaches toward natural areas management and biodiversity conservation. While we are intrigued by many components of the South African model and would like to see such ideas considered and evaluated in the United States, we recognize that it would take a transformational change in how America views and manages natural areas. Nonetheless, effective twenty-first century conservation and leadership requires consideration of creative and successful models from other countries.

Conflict of Interests

The authors declare that there is no conflict of interests regarding the publication of this paper.

Acknowledgments

This project was made possible in part by a grant from the National Park Foundation, the national charitable partner of America's national parks, and the National Park Service Capital Training Center. The authors thank Dave Druce at Hluhuwe-Imfolozi Provincial Park, Tristan Dickerson at Phinda Private Game Reserve, Sam Ferreira at Kruger National Park, and Jonathan Swart and Andre Burger at Welgevonden Private Game Reserve for their time and information. The authors thank the people within the National Park Service who supported the project.

References

[1] A. Phillips, "Turning ideas on their head: the new paradigm for protected areas," *The George Wright Forum*, vol. 20, no. 2, pp. 8–32, 2003.

[2] R. W. Sellers, *Preserving Nature in the National Parks: A History*, Yale University Press, 1997.

[3] D. S. Licht, R. Slotow, and J. Millspaugh, "A comparison of wildlife management in mid-size parks in South Africa and the

United States," in *Proceedings of the 2007 George Wright Society Conference: Protected Areas in a Changing World*, pp. 300–307, George Wright Society, St. Paul, Minn, USA, 2008.

[4] P. J. Funston, "Conservation and management of lions in southern Africa: status, threats, utilization and the restoration option," in *Management and Conservation of Large Carnivores in West and Central Africa*, B. Croes, R. Buij, H.H. de Iongh, and H. Bauer, Eds., pp. 109–131, Institute of Environmental Sciences, Leiden, The Netherlands, 2008.

[5] P. A. Lindsey, J. T. du Toit, and M. G. L. Mills, "Area and prey requirements of African wild dogs under varying habitat conditions: implications for reintroductions," *South African Journal of Wildlife Research*, vol. 34, no. 1, pp. 77–86, 2004.

[6] P. A. Lindsey and H. T. Davies-Mostert, "South African action plan for the conservation of Cheetahs and African wild dogs," in *Proceedings of the National Conservation Action Planning Workshop*, Bela Bela, South Africa, 2009.

[7] R. Kettles and R. Slotow, "Management of free-ranging lions on an enclosed game reserve," *South African Journal of Wildlife Research*, vol. 39, no. 1, pp. 23–33, 2009.

[8] M. W. Hayward, J. O'Brien, and G. I. H. Kerley, "Carrying capacity of large African predators: predictions and tests," *Biological Conservation*, vol. 139, no. 1-2, pp. 219–229, 2007.

[9] National Park Service, *Management Policies: The Guide to Managing the National Park System*, D.o.t.I. National Park Service, Washington, DC, USA, 2006.

[10] D. S. Licht, J. J. Millspaugh, K. E. Kunkel, C. O. Kochanny, and R. O. Peterson, "Using small populations of wolves for ecosystem restoration and stewardship," *BioScience*, vol. 60, no. 2, pp. 147–153, 2010.

[11] J. A. Vucetich, M. P. Nelson, and R. O. Peterson, "Should Isle Royale wolves be reintroduced? A case study on wilderness management in a changing world," *The George Wright Forum*, vol. 29, no. 1, pp. 126–147, 2012.

[12] R. Slotow, "Fencing for a purpose: a case study of elephants in South Africa," in *Fencing for Conservation: Restriction of Evolutinary Potential or a Riposte to Threatening Processes*, M. J. Somers and M. Hayward, Eds., pp. 91–104, Springer, New York, NY, USA, 2012.

[13] M. W. Hayward and G. I. H. Kerley, "Fencing for conservation: restriction of evolutionary potential or a riposte to threatening processes?" *Biological Conservation*, vol. 142, no. 1, pp. 1–13, 2009.

[14] C. Packer, A. Loveridge, S. Canney et al., "Conserving large carnivores: dollars and fence," *Ecology Letters*, vol. 16, no. 5, pp. 635–641, 2013.

[15] ABSA Group Economic Research, *Game Ranch Profitability in South Africa*, The SA Financial Sector Forum, ABSA, Rivonia, South Africa, 2003.

[16] Madikwe Development Task Team, *The Madikwe Game Reserve Management Plan*, N.W.P. Board, Rustenburg, South Africa, 1997.

[17] C. Ndabeni, M. Shroyer, W. Boonzaaier et al., *The Heritage Park Model: A Partnership Approach to Park Expansion in Poor Rural Areas*, U.F. Service, 2007.

[18] Peace Parks Foundation, "The Global Solution," 2013, http://www.peaceparks.org.

[19] R. Davies, *Madikwe Game Reserve: A Decade of Progress*, North West Parks Board, Rustenburg, South Africa, 2003.

[20] ICS, *Madikwe Game Reserve Management Plan*, N.W.P.T. Board, Rustenburg, South Africa, 2006.

[21] S. Mosidi, "Benefits beyond parks borders: the case of Madikwe Game Park, South Africa," in *Communicating Protected Areas*, D. Hamu, E. Auchincloss, and W. Goldstein, Eds., pp. 109–114, Commission on Education and Communication, IUCN, Gland, Switzerland, 2003.

[22] Y. Cui, E. Mahoney, and T. Herbowicz, "Economic benefits to local communities from national park visitation," in *Natural Resource Report 2013*, National Park Service, Fort Collins, Colo, USA, 2011.

[23] P. A. Lindsey, R. Alexander, M. G. L. Mills, S. Romañach, and R. Woodroffe, "Wildlife viewing preferences of visitors to protected areas in South Africa: implications for the role of ecotourism in conservation," *Journal of Ecotourism*, vol. 6, no. 1, pp. 19–33, 2007.

[24] P. A. Lindsey, R. Alexander, J. T. du Toit, and M. G. L. Mills, "The cost efficiency of wild dog conservation in South Africa," *Conservation Biology*, vol. 19, no. 4, pp. 1205–1214, 2005.

[25] D. S. Licht, *Ecology and Economics of the Great Plains*, University of Nebraska Press, Lincoln, Neb, USA, 1997.

[26] C. C. Gates, C. H. Freese, P. J.P. Gogan, and M. Kotzman, *American Bison: Status Survey and Conservation Guidelines 2010*, International Union for the Conservation of Nature, Gland, Switzerland, 2010.

[27] G. W. Vequist and D. S. Licht, *Wildlife Watching in America's National Parks: A Seasonal Guide*, Texas A&M University Press, College Station, Tex, USA, 2013.

7

Rediscovery of Cameroon Dolphin, the Gulf of Guinea Population of *Sousa teuszii* (Kükenthal, 1892)

Isidore Ayissi,[1,2,3] **Gabriel Hoinsoudé Segniagbeto,**[4] **and Koen Van Waerebeek**[5,6,7]

[1] *Association Camerounaise de Biologie Marine (ACBM), BP 52, Ayos, Cameroon*
[2] *CERECOMA, Specialized Research Center for Marine Ecosystems, c/o Institute of Agricultural Research for Development, P.O. Box 219, Kribi, Cameroon*
[3] *Institute of Fisheries and Aquatic Sciences (ISH) at Yabassi, University of Douala, P.O. Box 2701, Douala, Cameroon*
[4] *Département de Zoologie et de Biologie Animale, Faculté des Sciences, Université de Lomé, Lomé, Togo*
[5] *Conservation and Research of West African Aquatic Mammals (COREWAM), c/o Department of Marine and Fisheries Science, University of Ghana, P.O. Box LG99, Legon, Ghana*
[6] *COREWAM-Senegal, Musée de la Mer de Gorée, IFAN-CH.A.D, Université de Dakar, Dakar, Senegal*
[7] *Peruvian Centre for Cetacean Research (CEPEC), Lima 20, Peru*

Correspondence should be addressed to Koen Van Waerebeek; cepec@speedy.com.pe

Academic Editors: M. Cords, B. Crother, and P. V. Lindeman

Since the 1892 discovery of the Atlantic humpback dolphin *Sousa teuszii* (Delphinidae), a species endemic to coastal western Africa, from a skull collected in Cameroon, not a single record has been documented from the country or neighbouring countries. Increasing concern about the continued existence of the Gulf of Guinea population of *S. teuszii* or "Cameroon dolphin" prompted an exploratory survey in May 2011. Shore-based effort, on foot (30.52 km; 784 min), yielded no observations. Small boat-based surveys (259.1 km; 1008 min) resulted in a single documented sighting of ca. 10 (8–12) Cameroon dolphins in shallow water off an open sandy shore near Bouandjo in Cameroon's South Region. The combination of a low encounter rate of 3.86 individuals $(100 \text{ km})^{-1}$ suggesting low abundance and evidence of both fisheries-caused mortality and of habitat encroachment raises concerns about the Cameroon dolphin's long-term conservation prospect. Our results add to indications concerning several other *S. teuszii* populations that the IUCN status designation of the species as "Vulnerable" may understate its threat level.

1. Introduction

Here we describe coastal survey effort implemented in Cameroon in an attempt to relocate the "Cameroon dolphin," that is, the Gulf of Guinea population of the Atlantic humpback dolphin *Sousa teuszii* (Kükenthal, 1892) [1]. This population has been "lost" to science since 1892 when German zoologist Willy Kükenthal described the new dolphin species *Sotalia teuszii*, later reassigned to the genus *Sousa*. The discovery was based on a single skull collected by the then head of plantations Mr. Eduard Tëusz at Man O'War Bay, near Douala, in Cameroon's Southeast Region. Although a second specimen was collected in 1925 in Senegal, it remained unrecognised till 1965 [2]. No other cases were documented for half a century and Cameroon remained the only known range state of *S. teuszii*, earning the species its vernacular name in several languages, for example, Cameroon (river) dolphin [3–6], Kamerun Delphin, Kamerun-Flussdelphin [6–8], and Dauphin du Cameroun [2, 6]. Finally, in 1943, a specimen was captured in a shark net off M'Bour, Senegal [9], while the first live individual was retrieved from a beach seine at Joal in 1956 [10]. Further *S. teuszii* records followed in western Africa, ranging from Dakhla Bay, Western Sahara, south to Angola [2, 11, 12]. As the species, after its discovery, was not reported again from Cameroon between 1892 and 2011, its historical link with the country faded and the species became known as the Atlantic humpback dolphin [6, 8].

Despite a bizarre mix-up of the Man O'War dolphin skull with a shark-scavenged carcass apparently of an African manatee (*Trichechus senegalensis*), considering that its head showed "paired nostrils" and its stomach contents were exclusively vegetarian [1, 13], there is no reason to doubt that the *S. teuszii* holotype skull shipped to Jena University originated from Cameroon. Uncertainty about a possible herbivorous dolphin persisted until Cadenat [10] found exclusive fish remains in the stomach of a freshly captured animal in Senegal.

An exhaustive review of the biology of *S. teuszii* [11, 12] proposed eight preliminary stocks based largely on distributional grounds, amongst which the "Cameroon Estuary Stock" of unknown status because supported only by the holotype skull [1]. The lack of dedicated survey effort was assumed to be the main reason for the absence of later records in Cameroon. However, when *S. teuszii* was not found in Ghana despite significant port monitoring and the documentation of hundreds of small odontocetes comprising at least 12 species, the possibility of local extirpation could no longer be ignored [14–16].

In 2011, increasing concern about *S. teuszii*'s status in Cameroon prompted an exploratory survey in its coastal waters as reported below. At the same time it was considered crucial to document any anthropogenic threats to *S. teuszii*, in particular to verify whether dolphin captures occurred and whether an active market in cetacean bushmeat for human consumption has become established as it has in Ghana and Nigeria [15–17].

2. Materials and Methods

2.1. Study Area and Ecology. Cameroon's coastline stretches over 402 km from the mouth of the Akwayafe River (N04°40′) bordering Nigeria south to the mouth of the Ntem River (N02°20′) bordering Rio Muni (Equatorial Guinea), between longitudes E08°15′ and E09°30′ (Figure 1). The wide continental shelf occupies ca. 10,600 km². At Douala and Kribi-Campo the 200 m isobath lies some 40 km offshore while off the Rio del Rey basin, it lies up to 80 km offshore. Due to significant erosion, especially in Southwest Region, the estuaries and mangroves are characterised by high turbidity extending up to 30 km offshore from Bakassi and the estuaries of "Bouches du Cameroon" [18, 19].

Coastal surface waters (20–30 m layer) are warm (>24°C) throughout the year [20]. Salinity is low due to high rainfall and a dense river network. Peak salinity reaches 20Ğ at 15 km from Douala port in the dry season and less than 12‰ in the rainy season [21]. The effects of the semidiurnal tides with variable amplitudes (0.3–3 m) are felt in the estuarine complexes. Up to 1973, mangroves covered some 2700 km² and despite much destruction still constitute a predominant feature on Cameroon's coast [22]. Mangrove-lined, wide estuarine channels comprise a primary habitat for *S. teuszii* in Senegal, Gambia, and Guinea [2, 11, 23]. Coastal work in Senegal, Gabon, and Angola demonstrated that this obligate nearshore species can be visually surveyed from small boats and from shore [2, 24, 25]. Both methods were implemented in four areas of Cameroon (Table 1), 7 May–5 June, 2011: (i) Limbe-Idenau coast (Southwest Region; sandy coast and mangrove); Ambas Bay and environs of Bota

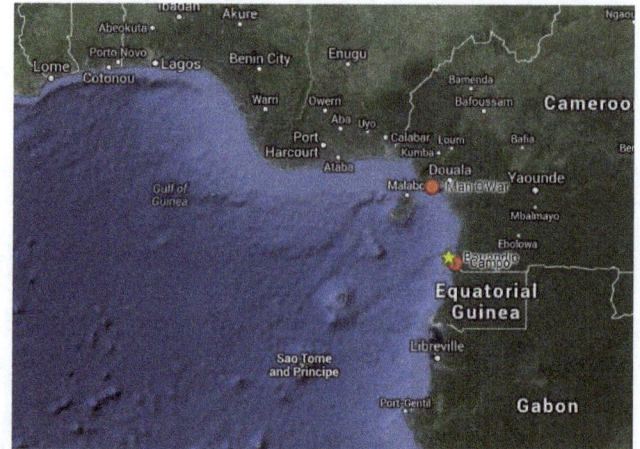

FIGURE 1: Coast of Cameroon as study area in the eastern Gulf of Guinea. Locations of three authenticated records of Cameroon dolphins *Sousa teuszii* are plotted: two specimens (red dots) including the 1892 holotype skull near Douala and a specimen captured at Campo in 2012; single sighting (yellow star) of ca. 10 free-ranging individuals at Bouandjo in 2011.

Islands; (ii) Douala-Edea Wildlife Reserve, southern shores of the Cameroon Estuary (Littoral Region, mangrove channels alternating with open sandy coast); (iii) area north of Ntem River estuary (South Region) bordering Equatorial Guinea; and (iv) Dibamba and Sanaga tidal rivers. The type locality, Man O'War Bay (N03.94839°, E009.22416°), occupied by a military garrison, was not accessible.

2.2. Survey Methods. In five days of small-boat surveys (Table 1), either from a 4 m dug-out canoe with a 8 hp outboard engine or a 4.5 m fibreglass boat with a 40 hp engine, three observers searched 180° ahead mostly by naked eye. Visibility was good to excellent and Beaufort sea state ranged 0–2. Total boat-based survey effort covered 259.1 km over 1,008 min. Shore-based effort (784 min), on foot, covered 30.52 km of beaches, mainly north of the Ntem River. Two observers slowly walked along the high-water line alternately scanning the sea and inspecting flotsam for any skeletal remains washed ashore. At 10 min intervals inshore waters were scanned with 8 × 40 mm binoculars. Distances were calculated with the odometer function of a hand-held GPS (Garmin Oregon 450t).

Several locals at 20 fish landing sites were casually queried regarding any recent cetacean bycatches or sightings [26]. In addition a structured interview, supported by photographs of expected cetaceans, was conducted of the first encountered fisherman at each site who gave coherent and detailed replies ($n = 17$), except at three locations where 3-4 persons were interviewed as a group and their consensus view was summarized. All 25 fishermen interviewed were asked about the presence of coastal cetaceans including Cameroon dolphin, common bottlenose dolphin *Tursiops truncatus,* and humpback whale *Megaptera novaeangliae.*

TABLE 1: Details of visual survey effort for Cameroon dolphin on the coast of Cameroon, from 07 to 28 May, 2011.

Date	Departure point	Destination (arrival time)	Effort	Comments
Boat-based surveys			Total: 259.1 km (1008 min)	
17/05/11	Campo, N02°22.861′, E09°49.465′ (09:25 h)	Ebodjé, N02°34.550′, E09°49.497′ (13:00 h)	22.5 km (215 min)	Coastal strip Campo-Ebodjé. 2 observers, small canoe, 8 HP outboard
21/05/11	Limbé, N04°00.012′, E09°12.551′ (09:40 h)	N04°00.012′, E09°12.551′ (11:05 h)	20.2 km (85 min)	Limbé harbour and environs. 2 observers, canoe, 40 HP outboard
23/05/11	Mouanko, N03°38.371′, E09°48.060′ (14:54)	Mbiako, N03°35.317′, E09°38.934′ (16:28)	33.91 km (94 min)	Downriver Sanaga and short estuary excursion; 3 observers, fibreglass boat; 40 HP outboard
24/05/11	Mbiako, N03°35.317′, E09°38.934′ (10:30 h)	Sitan, N03°28.347′, E09°44.183′ and return to Mbiako (13:14)	42.56 km (164 min)	100–200 m off beach, return 300–500 m off beach; 3 observers, fibreglass boat, 40 HP
24/05/11	Mbiako, N03°35.317′, E09°38.934′ (14:27)	Bolondo, N03°48.731, E09°33.470′ (16:23 h)	35.65 km (116 min)	150–300 m off beach, 3 observers, fibreglass boat, 40 HP
25/05/11	Bolondo, N03°48.731, E09°33.470′ (09:15)	Manoka N03°52.004′, E09°37.761′ (12:00 h), penins.tip, N03°49.273′, E09°32.711′ (14:57)	50.88 km (201 min)	Survey of main creeks and estuary, 3 observers, fibreglass boat, 40 HP
26/05/11	Bolondo, N03°48.731, E09°33.470′ (11:41 h)	Mouanko, N03°38.371′, E09°48.060′ (13:54 h)	53.40 km (133 min)	Coastal strip; 13:03 h entering Sanaga river and upriver survey.
Beach-based surveys			Total: 30.52 km (784 min)	
16/05/11	River Ntem mouth (06:00 h)	River Ntem estuary, right bank (07:30 h)	0 km (90 min)	1 observer, stationary position
16/05/11	N02°21.104′, E09°49.397′ (10:20 h)	N02°24.384′, E09°49.345′ (14:35 h)	6.45 km (255 min)	2 observers, on foot, NE direction
16/05/11	N02°24.384′, E09°49.345′ (14:35 h)	N02°22.861′, E09°49.465′ (15:45 h)	2.95 km (70 min)	2, on foot, SW direction
17/05/11	N02°21.104′, E09°49.397′ (07:30 h)	N02°22.861′, E09°49.465′ (08:35 h)	3.50 km (65 min)	2, on foot, NE direction, low tide
19/05/11	Idenau, N04°13.045′, E08°59.133′ (10:25 h)	N04°13.629′, E08°58.537′ and return to Idenau (11:55 h)	3.75 km (90 min)	3, on foot, NW/NE, rising tide
19/05/11	Seme, N04°03.358′, E09°03.181′ (15:06 h)	N04°03.579′, E09°02.726′ and return	2.22 km (86 min)	2, on foot, high tide
25/05/11	Northern tip peninsula, N03°49.273′, E09°32.711′ (15:12 h)	N03°47.722′ E09°33.724′ and return to Bolondo, N03°48.731, E09°33.470′ (18:20)	11.65 km (128 min)	2, on foot

3. Results and Discussion

No dolphins were seen, nor remains found, during the shore-based effort. The boat-based surveys resulted in a single *S. teuszii* sighting on 17 May, 2011 (11:05 h GMT+01), the first ever sighting record in Cameroon (Figure 2). The encounter lasted for 10 min under optimal conditions, including excellent visibility and 1 Beaufort sea state. With a speed of 9 km·hr^{-1}, the canoe's distance from the open sandy shore fluctuated from 250 to 500 m while avoiding sand banks and submerged rock formations. The somewhat dispersed group of ca. 10 Cameroon dolphins (low, 8; high, 12) was encountered at N02°28.708′, E09°48.661′ near Bouandjo, South Region. The group moved in turbid, shallow water, 250–600 m from the surf zone. The largest three individuals, evidently adults, presented a strongly developed dorsal hump. Smaller, juvenile specimens had a much fainter hump. Behavioural cues suggested foraging as individuals moved independently and rapidly with continuous changes in direction and speed, either subsurface or alternating with up to 4-5 min long dives. Aerial display was limited to a single full-body leap by one adult. Despite our unobtrusive approach, dolphins reacted with avoidance after just a few minutes, scattering in offshore direction, and moved out of sight ca. 600 m from shore. The single sighting translating into a boat-based encounter rate of 0.386 sightings (100 km)$^{-1}$ or 3.86 individuals (100 km)$^{-1}$ suggests that the present abundance of *S. teuszii* in Cameroon may be very low.

Surveys of the Sanaga River on 23 and 26 May 2011 yielded no sightings (Figure 3). During 6 hrs on 3 and 4 June 2011, one of us (I. Ayissi) made small-boat surveys of the tidal Dibamba River but encountered no dolphins. This is in concordance with claims by local fisherfolk who deny the presence of dolphins in the Sanaga and Dibamba Rivers.

The original plan to survey the extensive northern mangrove channels including Rio-del-Rey was frustrated by Idenau port authorities denying permission for security reasons

FIGURE 2: Two adult Cameroon dolphins *Sousa teuszii*, part of a small group of ca. 10 individuals foraging nearshore at Bouandjo, 17 May, 2011. This is the first authenticated sighting and the first evidence since 1892 of the presence of *S. teuszii* in Cameroon. Photo: K. Van Waerebeek.

FIGURE 3: Despite its historical name "Cameroon River dolphin" implying potential riverine habitat, we encountered no dolphins in Cameroon's wide Sanaga River; their absence was confirmed by local fishermen. Photo: K. Van Waerebeek.

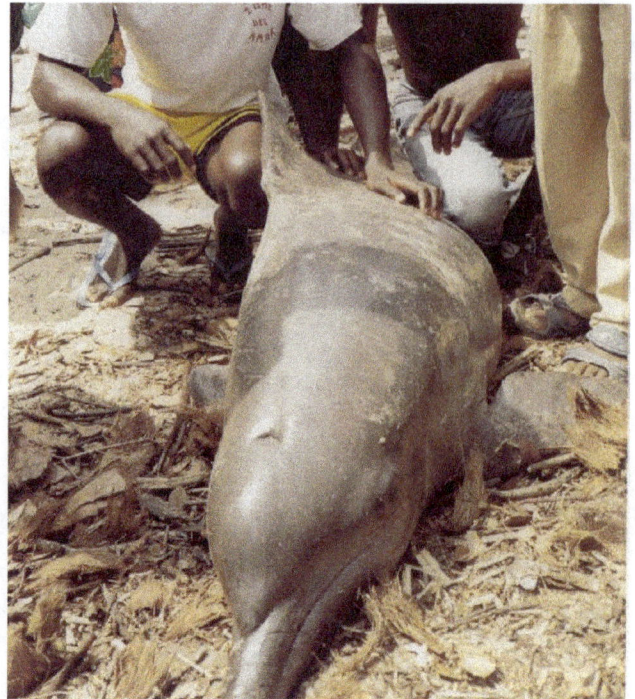

FIGURE 4: A Cameroon dolphin captured by artisanal fishermen near Campo, southern Cameroon, on an indeterminate date in 2012. Photo by unnamed fisherman from Campo.

including piracy. Artisanal fishing effort and nearshore boat traffic were very intense near Idenau. Dozens of large canoes equipped with high-powered outboard engines transported merchandise and passengers to and from Nigeria. These noisy, high-speed crafts would almost certainly disturb and chase away any humpback dolphins, if present.

The only specimen evidence of *S. teuszii* in Cameroon (since 1892) consisted of a dolphin captured and landed by small-scale local fishermen on an indeterminate date in 2012 near Campo (N02°22.861′, E09°49.4654′) in southern Cameroon (Figure 4). Although 81% of interviewed fishermen had seen dolphins from shore, none positively recognised *S. teuszii* from photographs, and 87.5% answered a resolute "never seen." In contrast, almost all fishermen recognised *T. truncatus*, a species not encountered during our surveys. However, a photo of a common bottlenose dolphin landed at Yoyo II (N03°40′, E09°38′) in 2003 and examined

by the authors confirmed the presence of *T. truncatus* in Cameroon.

So-called rapid gillnet bycatch assessments, based primarily on interviews with fishermen, indicated that cetacean bycatches occur but failed to obtain data on species and numbers [27, 28]. The underperformance of interview surveys was consistent with our own findings [26] stressing the need for corroboration through direct monitoring of fisheries backed by specimen and photographic evidence. Besides *S. teuszii* and *T. truncatus*, photo evidence now exists for specimens killed in fisheries interactions of another four cetacean species in Cameroon, that is, humpback whale *Megaptera novaeangliae*, sperm whale *Physeter macrocephalus*, striped dolphin *Stenella coeruleoalba*, and a long-snouted form of common dolphin, *Delphinus* sp. [26].

4. Conclusions

After almost 120 years without humpback dolphin records in Cameroon, documentation of the unique Bouandjo sighting in 2011 and a single specimen captured at Campo in 2012 evidenced the continuing occurrence of Cameroon dolphins in the country. As new cases emerge in the Gulf of Guinea, the "Cameroon Estuary stock" (*sensu* van Waerebeek et al. [2]) may have a wider distribution and population structure will need revisiting. We suggest that a combined lack of surveys and marine mammalogists, low encounter rate, and small, inconspicuous group sizes are the primary reasons for the historical absence of *S. teuszii* records. Low abundance underscores the legitimacy of concern about long-term conservation of Cameroon dolphins. A similar situation was reported in Angola's Namibe province, where only 10 individuals were photo-identified despite extensive field effort [12, 25].

No dolphins were sighted in Cameroon's mangrove channels nor in the Sanaga and Dibamba rivers. While our effort was limited in scope, fishermen living along the river banks reported no dolphins in these rivers. Tidal currents may be very strong: 1–1.5 m/s Influx and up to 2.6 m/s for reflux [29]. Numerous sandbanks falling dry at low tide represent high stranding risk and may discourage humpback dolphins from venturing inside rivers. Nonetheless a final conclusion is premature.

Future surveying and photo-identification efforts should be expanded to cover all seasons and most of the coastline to obtain more robust density estimates. While economical and widely available, dug-out canoes can only offer low eye-height while surveying is restricted to Beaufort sea state 0-1 and low or no swell. The fibreglass boat offered a higher eye-height, faster speed, and superior functionality and stability.

The nearshore or estuarine habitat of S. teuszii magnifies its susceptibility to anthropogenic threats. Bycatch mortality in artisanal fisheries, prey depletion, disturbance, and cetacean bushmeat acquiring market value (fomenting direct captures) are of major concern. With increasing nearshore boat traffic, injurious or lethal collisions between fast craft and slow-swimming Atlantic humpback dolphins are an emerging threat, as documented for Chinese humpback dolphins Sousa chinensis in Hong Kong and Chinese waters [30, 31]. Not only does the boat avoidance behaviour of S. teuszii complicate surveying and photo-identification attempts [2, 8, 25] but dense boat traffic may also lead to Cameroon dolphin communities abandoning home ranges in search of quieter, possibly poorer, feeding areas. Habitat encroachment from urban expansion, port and industrial construction, overfishing [32–34], chemical and acoustic pollution, and ghost nets [35] constitutes additional conservation problems. Scarcity of information on the fine-scale distribution and on stock structure, abundance, and ecology of the Cameroon dolphin prevents an efficient evaluation of the impact of threat factors and hinders efforts to raise public awareness [2, 8, 11, 25].

While the IUCN categorizes S. teuszii as "Vulnerable," the species is listed on CMS Appendix I/II as well as on CITES Appendix I. Most experts consider IUCN's "Vulnerable" classification to be an understatement of its true conservation status. Circumstantial evidence suggests that the species is already endangered in view of its low abundance throughout a fragmented range, unprecedented deterioration of its obligate shallow-water habitat, and significant fisheries-caused mortality coupled with an increased demand for marine bushmeat in western Africa [2, 8, 25, 36]. To date, S. teuszii has not been reported in Ghana or Benin despite significant monitoring of cetacean landings [2, 16, 37] suggesting the species has become very rare and perhaps even locally extinct. Equatorial Guinea and Nigeria are unconfirmed but highly probable range states, considering their extensive mangrove-lined coast. Promotion of the historic significance of S. teuszii as a flagship species for Cameroon and the Gulf of Guinea (the only aquatic mammal taxon first discovered in this region), underscored by its rediscovery, may help foster national awareness for the conservation of aquatic mammals in general. Even so, safeguarding a future for the iconic Cameroon dolphin will prove a formidable challenge and will depend on whether the implementation of proactive conservation measures, in particular installing nearshore marine protected areas, will succeed in the face of accelerating anthropogenic pressures. Inaction could result in the loss of one of Africa's most emblematic marine mammals.

Conflict of Interests

The authors declare that they have no conflict of interests regarding the publication of this paper.

Acknowledgments

Thanks are due to the many persons who facilitated or otherwise helped field work, particularly the D. O. of Mouanko and Campo, the senior staff from Ministry of Environment and Nature Protection including Dr. Ndongo Barthelémy (Inspector General N°2), Dr. Wounissi (Director of Conservation), Mr. Djonou Roland and Mr. Pouth Jean Paul (Cooperation Unit), Mr. Beyiyi Gerard (SDPRN-CMS/Unit), Dr. Ngo Mpeck Marie Laure (National Coordinator GEF-Small Grants Programme), and Dr. Jean Folack (Chief of Station of Specialized Research Center of Marine Ecosystems). The authors thank also field staff from the Ministry of Livestock and Fisheries including Mr. Mouri, Mr. Abessolo, and Ms. Nking Gwendoline, the Chiefs Kema from Mbiako, Edimo from Suellaba, Ewandje from Yoyo I, and Moudjol Jean Bernard from Dibamba, as well as fishers Sebastien from Londji and Victor from Suellaba. Three anonymous reviewers provided useful comments that helped improve the paper. Field work was sponsored by the UNEP/CMS Convention for Migratory Species via a 2011 Small Grant and support received from the Colombus Zoo Conservation Fund, Ohio.

References

[1] W. Kükenthal, "Sotalia tëuszii n. sp., ein pflanzenfressender (?) Delphin aus Kamerun," Zoologische Jahrbücher Abteilung für Systematik, Ökologie und Geographie der Tiere, vol. 6, no. 3, pp. 442–446, 1892.

[2] K. Van Waerebeek, L. Barnett, A. Camara et al., "Distribution, status, and biology of the Atlantic Humpback Dolphin, Sousa teuszii (Kükenthal, 1892)," Aquatic Mammals, vol. 30, no. 1, pp. 56–83, 2004.

[3] W. F. J. Mörzer Bruyns, Field Guide of Whales and Dolphins, Uitgeverij Tor & C.A. Mees, Amsterdam, The Netherlands, 1971.

[4] D. J. Coffey, D. K. Caldwell, and M. C. Caldwell, The Encyclopedia of Sea Mammals, Hart-Davis, MacGibbon, London, UK, 1977.

[5] M. Carwardine, Whales, Dolphins and Porpoises, Eyewitness Handbooks, Dorling Kindersley Limited, London, UK, 1995.

[6] B. M. Culik, Review of Small Cetaceans: Distribution, Behaviour, Migration and Threats, Regional Seas Reports and Studies, no. 177, UNEP/CMS, Bonn, Germany, 2004.

[7] B. Grzimek, *Grzimeks Tierleben, Enzyklopädie des Tierreiches*, Kindler, Zürich, Switzerland, 1970.

[8] K. Van Waerebeek and W. F. Perrin, "Conservation status of the Atlantic humpback dolphin, a compromised future?" in *Proceedings of the 14th Meeting CMS Scientific Council*, Bonn, Germany, March 2007, Document CMS/ScC14/Doc.6.

[9] J. Cadenat, "Observations de cétacés au Sénégal," *Notes Africaines*, vol. 34, pp. 20–23, 1947.

[10] J. Cadenat, "Un delphinidae encore mal connu de la côte occidentale d'Afrique: *Sotalia teuszii* Kükenthal 1892," *Bulletin de l'IFAN*, vol. 18, pp. 555–566, 1956.

[11] K. Van Waerebeek, L. Barnett, A. Camara et al., "Conservation of cetaceans in the Gambia and Senegal 1999–2001, and status of the Atlantic humpback dolphin," WAFCET-2 Report, UNEP/CMS, Bonn, Germany, 2003.

[12] C. R. Weir, K. Van Waerebeek, T. A. Jefferson, and T. Collins, "West Africa's Atlantic humpback dolphin (*Sousa teuszii*): endemic, enigmatic and soon endangered?" *African Zoology*, vol. 46, no. 1, pp. 1–17, 2011.

[13] P.-J. van Beneden, "Un Cétacé fluviatile d'Afrique," *Bulletin de l'Académie Royale des Sciences, des Lettres et des Beaux-Arts de Belgique*, vol. 3, no. 23, pp. 350–355, 1892.

[14] P. K. Ofori-Danson, K. Van Waerebeek, and S. Debrah, "A survey for the conservation of dolphins in Ghanaian coastal waters," *Journal of the Ghana Science Association*, vol. 5, no. 2, pp. 45–54, 2003.

[15] K. Van Waerebeek, P. K. Ofori-Danson, and J. Debrah, "The cetaceans of Ghana, a validated faunal checklist," *West African Journal of Applied Ecology*, vol. 15, pp. 61–90, 2009.

[16] J. S. Debrah, P. K. Ofori-Danson, and K. Van Waerebeek, "An update on the catch composition and other aspects of cetacean exploitation in Ghana," IWC Scientific Committee Document SC/62/SM10, International Whaling Commission, Agadir, Morocco, 2010.

[17] M. Uwagbae and K. Van Waerebeek, "Initial evidence of dolphin takes in the Niger Delta region and a review of Nigerian cetaceans," Scientific Committee Document SC/62/SM1, International Whaling Commision, Agadir, Morocco, 2010.

[18] S. Morin and M. Kuete, "Le Littoral Camerounais: problèmes morphologiques," *Travaux du Laboratoire de Géographie Physique Appliquée. Institut de Géographie, Université de Bordeaux III*, vol. 11, pp. 5–53, 1989.

[19] J. A. Sayer, C. S. Harcourt, and N. M. Collins, Eds., *The Conservation Atlas of Tropical Forest Africa*, Macmillan Publishing Ltd., London, UK, 1992.

[20] A. Crosnier, "Fonds de pêche le long des côtes de la République Fédérale du Cameroun," *Cahiers ORSTOM*, pp. 1–133, 1964.

[21] L. R. Lafond, *Etudes littorales et estuariennes en zone intertropicale humide [Thèse d'Etat en Sciences]*, Université de Paris, Paris, France, 1967.

[22] J. M. Kramkimel and B. Bousquet, *Mangrove d'Afrique et de Madagascar: les mangroves du Cameroun*, CEE, SECA, 1987.

[23] L. L. Bamy, K. Van Waerebeek, S. S. Bah et al., "Species occurrence of cetaceans in Guinea, including humpback whales with Southern Hemisphere seasonality," *Marine Diversity Records*, vol. 3, article e48, pp. 1–10, 2010.

[24] T. Collins, S. Ngouessono, and H. C. Rosenbaum, "A note on recent surveys for Atlantic humpback dolphins, *Sousa teuszii* (Kükenthal, 1892) in the coastal waters of Gabon," Scientific Committee Document SC/56/SM23, International Whaling Commission, 2004.

[25] C. R. Weir, "Distribution, behaviour and photo-identification of Atlantic humpback dolphins *Sousa teuszii* off Flamingos, Angola," *African Journal of Marine Science*, vol. 31, no. 3, pp. 319–331, 2009.

[26] I. Ayissi, K. Van Waerebeek, and G. Segniagbeto, "Report on the exploratory survey of cetaceans and their status in Cameroon," in *Proceedings of the 17th Meeting CMS Scientific Council*, Bergen, Norway, November 2011, Document UNEP/CMS/ScC17/Inf.10.

[27] I. Ayissi, "Rapid gillnet bycatch survey of Cameroon," Tech. Rep., University of Yaoundé, Yaoundé, Cameroon, 2008.

[28] J. E. Moore, T. M. Cox, R. L. Lewison et al., "An interview-based approach to assess marine mammal and sea turtle captures in artisanal fisheries," *Biological Conservation*, vol. 143, no. 3, pp. 795–805, 2010.

[29] J. C. Olivry, "Fleuves et rivières du Cameroun. Collection monographies hydrologiques," *ORSTOM Mémoires*, vol. 9, pp. 1–733, 1986.

[30] T. A. Jefferson, S. K. Hung, and P. K. S. Lam, "Strandings, mortality and morbidity of Indo-Pacific humpback dolphins in Hong Kong, with emphasis on the role of organochlorine contaminants," *Journal of Cetacean Research and Management*, vol. 8, no. 2, pp. 181–193, 2006.

[31] K. Van Waerebeek, A. N. Baker, F. Félix et al., "Vessel collisions with small cetaceans worldwide and with large whales in the Southern Hemisphere, an initial assessment," *Latin American Journal of Aquatic Mammals*, vol. 6, no. 1, pp. 43–69, 2007.

[32] FAO, *Marine Fisheries Resources of Cameroon: A Review of Exploited Fish Stocks*, CECAF/ECAF Series, http://www.fao.org/docrep/003/S4639E/S4639E00.htm.

[33] FAO, "General situation of world fish stocks," http://www.fao.org/newsroom/common/ecg/1000505/en/stocks.pdf.

[34] J. Folack and O. Njifondjou, "Characteristics of marine artisanal fisheries in Cameroon," *IDAF Newsletter*, no. 28, pp. 18–21, 1995.

[35] G. Macfadyen, T. Huntington, and R. Cappell, *Abandoned, Lost or Otherwise Discarded Fishing Gear*, FAO Technical Paper, no. 523, Fisheries and Aquaculture, Rome, Italy, 2009.

[36] P. Clapham and K. Van Waerebeek, "Bushmeat and bycatch: the sum of the parts," *Molecular Ecology*, vol. 16, no. 13, pp. 2607–2609, 2007.

[37] Z. Sohou, J. Dossou-Bodjrenou, S. Tchibozo, F. Chabi-Yaouré, B. Sinsin, and K. Van Waerebeek, "Biodiversity and status of cetaceans in Benin, West Africa: an initial assessment," *West African Journal of Applied Ecology*, vol. 21, no. 1, pp. 121–134, 2013.

8

Diversity of Macrolichens in Bolampatti II Forest Range (Siruvani Hills), Western Ghats, Tamil Nadu, India

P. Balaji[1] and G. N. Hariharan[2]

[1] Department of Botany, Dr. Ambedkar Government Arts College, Vyasarpadi, Chennai, Tamil Nadu 600 039, India
[2] Lichen Ecology and Bioprospecting Laboratory, M. S. Swaminathan Research Foundation III Cross Street, Taramani Institutional Area, Taramani, Chennai, Tamil Nadu 600 113, India

Correspondence should be addressed to P. Balaji; lichenbalaji@gmail.com

Academic Editors: A. R. Atangana and R. Rico-Martinez

An annotated checklist of 103 macrolichen species is provided based on identification of specimens collected from three different vegetation types within the Bolampatti II forest range, Western Ghats, India. Among them, the dominant order is Lecanorales with 47 species, while the dominant family is Parmeliaceae with 40 species. The foremost genus is *Usnea* with 15 species.

1. Introduction

Nearly 80,000 species of fungi are available in nature [1]. Of these, approximately 17% are lichenized, forming symbioses with green algae (Chlorophyta, Viridiplantae) or the so called blue-green algae (Cyanobacteria, Bacteria). These relationships produce symbiotic organisms commonly called lichens [?]. Lichens are an outstandingly successful group of symbiotic organisms exploiting a wide range of habitats throughout the world. About 20,000 species of lichens are so far recorded across the globe, among which the Indian subcontinent harbours 2450 (12.25%) species [3]. Tropical forests, because of their complexity and variety of microhabitats, usually harbour a rich diversity of lichens. Even though they are often small and inconspicuous, especially in the lowland forest, they may play a significant role in the forest ecosystem [4]. It has been estimated that 50% of the Indian lichen biota are currently undescribed [5]. In India, the comprehensive regional treatments (both ecological and systematic) on lichens are still not available, and a few cover areas like portions of Western Ghats (Nilgiri and Palni Hills, Himalayas, and North Eastern regions) [6, 7]. Still, many of the pristine Western Ghats ecosystems remain unexplored to list out the extent and type of distribution of lichens [8]. In view to explore such important organisms, the present study has studied the macrolichens from Bolampatti II forest range, Western Ghats. The primary objective of this paper is to enumerate the macrolichens and their distribution within the forest types of the Bolampatti II forest range.

2. Materials and Methods

2.1. Study Area. Bolampatti II forest range, Coimbatore district, Tamil Nadu (part of Western Ghats—$76°$ $33''$ to $76°$ $46''$ E and $11°$ $2''$ to $10°$ $54''$ N), is a part of the Nilgiri Biosphere Reserve, is commonly known as Siruvani Hills (Figure 1), and is one of the biodiversity hotspots of the world. The study site is located west to Coimbatore city and north of the Walayar valley, is shaped like a horse-shoe opening eastwards, and covers an area of 197.66 km^2. The Bolampatti valley consists of vegetation types progressively from lower to higher altitudes (east to west) dry deciduous forest (DDF) (4a/C2), moist deciduous forest (MDF), and western tropical evergreen forest (WTEF) (1a/C3) [9, 10]. This valley drains eastwards into the Noyilar and its tributaries. The elevation ranges from 458 m (Noyilar base) to 1,986 m on Periyakunjiramalai at the southwest corner and 1,800 m on Vellingiri peak at the northwest. This hill receives a rainfall of 75–200 cm from the northeast monsoon during September to November, with a dry period of six months. The temperature ranges between $10°$C during December and January and $40°$C

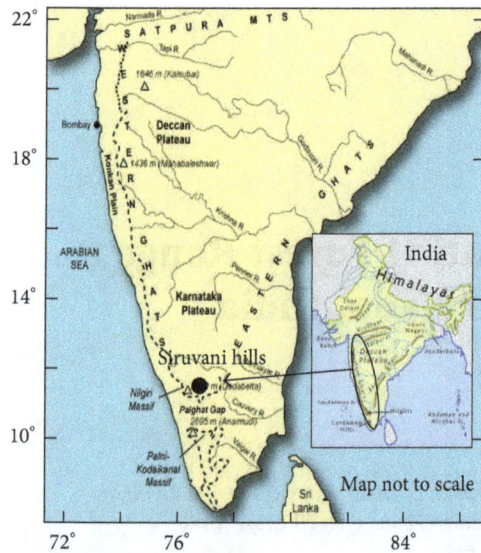

FIGURE 1: Map of Bolampatti II forest range (Siruvani Hills), Western Ghats.

during March to May. The older rock is of Precambrian origin with the formation of mainly Charnockite or Nilgiri gneiss. Soil is mainly of red type in the hills. The foothills generally have sandy loam. The Vellingiri peak is barren and rocky.

2.2. Research Methods. The lichen samples (approximately. 3000 specimens) were collected from various vegetation types, namely, dry deciduous forest (DDF), moist deciduous forest (MDF), and western tropical evergreen forest (WTEF) during 2001–2005 by the authors based on ocular survey on all possible lichen occurring substrates, such as tree bark, rocks, leaves, and soil. Lichens on fallen twigs were collected, since they represent the species that occur on canopy. Each sample was given a field number. The specimens were dried and incorporated into a reference collection that is preserved in the Lichen Ecology and Bioprospecting Laboratory, M. S. Swaminathan Research Foundation (MSSRF), Chennai. The specimens were identified based on the internal and external morphological, reproductive, and chemical features following recent literature [11–18], and lichen taxa are classified based on the systematic arrangement provided by Hawksworth et al. [19]. For each of the lichen species collected, their distribution with respect to growth form, substratum preference, and forest types were provided.

3. Results and Discussion

A total number of 103 species of macrolichens under 27 genera in 9 families in 3 fungal orders were recorded within the Siruvani hills (Table 1; Figures 2(a)–2(f)). Fungal order Lecanorales dominate with 47 species followed by Teloschistales (35) and Peltigerales (21). Out of 9 families of lichens, Parmeliaceae (40 species) is the largest family followed by Physciaceae (20), Collemataceae (16), Caliciaceae

(14), Ramalinaceae (6), and Lobariaceae (4), and three families were with single species each. The largest genus was *Usnea* with 15 species followed by *Heterodermia* and *Parmotrema* with 13 species each, *Leptogium* and *Pyxine* with 11 and 10 species, respectively.

Fourteen (14) genera were with single species each. Among the forest types studied, MDF contain 55 lichens followed by WTEF with 51 and DDF with 48 species, in which, 21 lichen species were specific to DDF while 24 to MDF and 21 to WTEF. Thirty-seven (14) lichen species were common to all the forest types. Seven (7) species were shared between DDF and MDF, while 6 species between DDF and WTEF, and 24 species between MDF and WTEF.

Among the species recorded, 92.2% of lichens were colonized on bark (corticolous), followed by 6% on rock (saxicolous), 0.8% on soil (terricolous), and a single lichen species both on bark and rock substratum. Bark colonizing lichens were maximum in both MDF & WTEF (47% each) followed by DDF (43%). Rock colonizing lichens were maximum in MDF (4) followed by DDF (3) and WTEF (1). Single soil colonizing lichen *Cladonia* sp. was found in MDF. Lichen *Parmotrema praesorediosum* was found colonizing both on bark as well as on rock in DDF type.

The photobiont distribution showed 80% of lichens with *Trebouxia* and 20% of lichens with *Nostoc* sp. *Trebouxia* containing lichens were maximum in WTEF and MDF with 34% each, followed by DDF (32%). *Nostoc*-containing lichens were maximum in MDF (44%) followed by WTEF and DDF (28% each).

Lichens with *Trebouxia* as photobiont dominate tropical regions with an alternating wet and dry period and include macrolichens in Parmeliaceae and macrolichens and crusts in Bacidiaceae (syn. Ramalinaceae), Physciaceae, and Pertusariaceae [20]. The *Trebouxia* containing macrolichens in DDF are also equally present in WTEF and MDF, prevailing in our study sites. It includes lichen families such as Parmeliaceae, Physciaceae, and Caliciaceae. Wolseley and Hawksworth [20] reported that the *Trebouxia* dominated families may occur with increasing dryness and/or disturbance as caused by fire in Thailand.

The *Nostoc*-containing lichens were found dominant in MDF, while there was equal distribution of *Trebouxia* containing lichens in all the forest types studied. Lichen taxa with cyanobacteria are more common in the moister forests; species of *Collema* and *Parmeliella* were found in DDF as well as SEF, and *Leptogium cyanescens* was found in all forest types [21]. The lichen species such as *Collema actinoptychum*, *Collema flaccidum*, and *Collema furfuraceum* are confined only to DDF, while *Collema subflaccidum* is found both in DDF and WTEF. The lichens *Leptogium cyanescens* and *Leptogium denticulatum* are found in all forest types as in the case of lichens of northern Thailand forests [21].

Lichen communities of the seasonal evergreen forest (SEF) and tropical mixed deciduous forest (TMDF) are moisture-dependent and shade-adapted, whereas lichen communities of the DDF are adapted to xerophytic conditions and photophilous [21]. This pattern of distribution can be attributed to the sensitivity of the lichen photobiont to drought or high temperatures, which determines the

TABLE 1: List of lichen species found within the Bolampatti II forest range (Siruvani Hills), Western Ghats, Tamil Nadu, India.

S. no.	Species name	Family	Order	Growth forms	Substratum	DDF	MDF	WTEF
1	*Bulbothrix isidiza* (Nyl.) Hale	Parmeliaceae	Lecanorales	G	1	+	−	−
2	*Bulbothrix tabacina* (Mont. & Bosch) Hale	Parmeliaceae	Lecanorales	G	1	−	+	−
3	*Canoparmelia aptata* (Krempelh.) Elix et Hale	Parmeliaceae	Lecanorales	G	1	−	+	−
4	*Cladonia* sp.	Cladoniaceae	Lecanorales	F	3	−	+	−
5	*Coccocarpia palmicola* (Spreng.) Arv. & D.J. Galloway	Coccocarpiaceae	Peltigerales	A	1	−	+	−
6	*Collema actinoptychum* Nyl.	Collemataceae	Peltigerales	B	1	+	−	−
7	*Collema flaccidum* (Ach.) Ach.	Collemataceae	Peltigerales	B	1	+	−	−
8	*Collema furfuraceum* (Arnold) Du Rietz	Collemataceae	Peltigerales	B	1	+	+	+
9	*Collema rugosum* Kremp.	Collemataceae	Peltigerales	B	1	−	+	−
10	*Collema subflaccidum* Degel.	Collemataceae	Peltigerales	B	1	+	−	+
11	*Dirinaria applanata* (Fée) D.D. Awasthi	Caliciaceae	Teloschistales	C	1	+	−	+
12	*Dirinaria confluens* (Fr.) D.D. Awasthi	Caliciaceae	Teloschistales	C	2	−	+	−
13	*Dirinaria consimilis* (Stirt.) D.D. Awasthi	Caliciaceae	Teloschistales	C	1	+	−	−
14	*Dirinaria picta* (Sw.) Schaer. ex Clem.	Caliciaceae	Teloschistales	C	1	+	+	+
15	*Everniastrum cirrhatum* (Fr.) Hale ex Sipman	Parmeliaceae	Lecanorales	E	1	−	+	−
16	*Heterodermia angustiloba* (Müll. Arg.) D.D. Awasthi	Physciaceae	Teloschistales	D	1	+	+	−
17	*Heterodermia comosa* (Eschw.) Follmann & Redón	Physciaceae	Teloschistales	D	1	−	−	+
18	*Heterodermia diademata* (Taylor) D.D. Awasthi	Physciaceae	Teloschistales	D	1	+	+	+
19	*Heterodermia dissecta* (Kurok.) D.D. Awasthi	Physciaceae	Teloschistales	D	1	+	+	+
20	*Heterodermia hypocaesia* (Yasuda) D.D. Awasthi	Physciaceae	Teloschistales	D	1	+	+	+
21	*Heterodermia isidiophora* (Vain.) D.D. Awasthi	Physciaceae	Teloschistales	D	1	−	+	+
22	*Heterodermia japonica* (K. Satô) Swinscow & Krog	Physciaceae	Teloschistales	D	1	−	+	−
23	*Heterodermia leucomela* (Fée) Swinscow & Krog	Physciaceae	Teloschistales	D	1	−	−	+
24	*Heterodermia microphylla* (Kurok.) Skorepa	Physciaceae	Teloschistales	D	1	−	+	+
25	*Heterodermia obscurata* (Nyl.) Trevis.	Physciaceae	Teloschistales	D	1	−	+	+
26	*Heterodermia pellucida* (D.D. Awasthi) D.D. Awasthi	Physciaceae	Teloschistales	D	1	+	+	+
27	*Heterodermia pseudospeciosa* (Kurok.) W.L. Culb.	Physciaceae	Teloschistales	D	1	+	+	+
28	*Heterodermia speciosa* (Wulfen) Trevis.	Physciaceae	Teloschistales	D	2	+	+	+
29	*Hypotrachyna coorgiana* Patw. & Prabhu	Parmeliaceae	Lecanorales	A	1	−	−	+
30	*Leptogium austroamericanum* (Malme) C.W. Dodge	Collemataceae	Peltigerales	B	1	+	+	−
31	*Leptogium azureum* (Sw. ex Ach.) Mont.	Collemataceae	Peltigerales	B	1	−	+	−
32	*Leptogium cyanescens* (Pers.) Körb.	Collemataceae	Peltigerales	B	1	+	+	+
33	*Leptogium denticulatum* F. Wilson	Collemataceae	Peltigerales	B	1	+	+	+
34	*Leptogium isidiosellum* (Riddle) Sierk	Collemataceae	Peltigerales	B	1	+	+	−
35	*Leptogium marginellum* (Sw.) Gray	Collemataceae	Peltigerales	B	1	−	−	+
36	*Leptogium millegranum* Sierk	Collemataceae	Peltigerales	B	1	−	−	+
37	*Leptogium moluccanum* (Pers.) Vain.	Collemataceae	Peltigerales	B	1	−	+	+
38	*Leptogium phyllocarpum* (Pers.) Mont.	Collemataceae	Peltigerales	B	1	−	+	−
39	*Leptogium pichneum* (Ach.) Nyl.	Collemataceae	Peltigerales	B	1	+	−	−
40	*Leptogium ulvaceum* (Pers.) Vain.	Collemataceae	Peltigerales	B	1	−	−	+
41	*Lobaria japonica* (Zahlbr.) Asahina	Lobariaceae	Peltigerales	A	1	−	+	−
42	*Myelochroa aurulenta* (Tuck.) Elix & Hale	Parmeliaceae	Lecanorales	A	1	+	+	−
43	*Parmelia direagens* Hale	Parmeliaceae	Lecanorales	G	1	+	−	−
44	*Parmelina muelleri* (Vain.) Hale	Parmeliaceae	Lecanorales	G	1	−	−	+
45	*Parmelinella simplicior* (Hale) Elix & Hale	Parmeliaceae	Lecanorales	G	1	−	+	+
46	*Parmelinopsis expallida* (Kurok.) Elix & Hale	Parmeliaceae	Lecanorales	G	1	+	−	−
47	*Parmelinopsis microlobulata* (Awas.) Elix and Hale	Parmeliaceae	Lecanorales	G	1	−	−	+
48	*Parmotrema andinum* (Müll. Arg.) Hale	Parmeliaceae	Lecanorales	G	1	+	−	−
49	*Parmotrema austrosinense* (Zahlbr.) Hale	Parmeliaceae	Lecanorales	G	1	−	+	−

TABLE 1: Continued.

S. no.	Species name	Family	Order	Growth forms	Substratum	Forest types DDF	MDF	WTEF
50	*Parmotrema crinitum* (Ach.) M. Choisy	Parmeliaceae	Lecanorales	G	1	+	−	−
51	*Parmotrema dilatatum* (Vain.) Hale	Parmeliaceae	Lecanorales	G	1	−	+	−
52	*Parmotrema grayanum* (Hue) Hale	Parmeliaceae	Lecanorales	G	2	−	+	−
53	*Parmotrema indicum* Hale	Parmeliaceae	Lecanorales	G	1	−	−	+
54	*Parmotrema melanothrix* (Mont.) Hale	Parmeliaceae	Lecanorales	G	1	+	−	+
55	*Parmotrema mesotropum* (Müll. Arg.) Hale	Parmeliaceae	Lecanorales	G	1	+	−	−
56	*Parmotrema permutatum* (Stirt.) Hale	Parmeliaceae	Lecanorales	G	1	−	+	−
57	*Parmotrema praesorediosum* (Nyl.) Hale	Parmeliaceae	Lecanorales	G	1 & 2	+	−	−
58	*Parmotrema saccatilobum* (Taylor) Hale	Parmeliaceae	Lecanorales	G	1	+	−	+
59	*Parmotrema tinctorum* (Nyl.) Hale	Parmeliaceae	Lecanorales	G	1	+	−	−
60	*Parmotrema xanthinum* (Müll. Arg.) Hale	Parmeliaceae	Lecanorales	G	1	−	−	+
61	*Phaeophyscia hispidula* (Ach.) Moberg	Physciaceae	Teloschistales	C	1	+	−	−
62	*Phyllopsora corallina* (Eschw.) Müll. Arg.	Ramalinaceae	Lecanorales	A	1	+	+	+
63	*Phyllopsora parvifolia* (Pers.) Müll. Arg.	Ramalinaceae	Lecanorales	A	1	+	+	+
64	*Physcia aipolia* (Ehrh. ex Humb.) Fürnr.	Physciaceae	Teloschistales	C	1	+	−	+
65	*Physcia caesia* (Hoffm.) Fürnr.	Physciaceae	Teloschistales	C	1	−	−	+
66	*Physcia dilatata* Nyl.	Physciaceae	Teloschistales	C	2	+	−	−
67	*Physcia dimidiata* (Arnold) Nyl.	Physciaceae	Teloschistales	C	1	+	+	−
68	*Physcia tribacia* (Ach.) Nyl.	Physciaceae	Teloschistales	C	1	+	+	+
69	*Physcia tribacioides* Nyl.	Physciaceae	Teloschistales	C	1	+	−	−
70	*Pseudocyphellaria aurata* (Ach.) Vain.	Lobariaceae	Peltigerales	A	1	−	+	−
71	*Pyxine cocoës* var. *cocoës* (Sw.) Nyl.	Caliciaceae	Teloschistales	C	1	+	+	−
72	*Pyxine cocoës* var. *prominula* (Stirt.) D.D. Awasthi	Caliciaceae	Teloschistales	C	1	+	−	−
73	*Pyxine cognata* Stirt.	Caliciaceae	Teloschistales	C	1	+	−	−
74	*Pyxine consocians* Vain.	Caliciaceae	Teloschistales	C	2	+	−	−
75	*Pyxine himalayensis* D.D. Awasthi	Caliciaceae	Teloschistales	C	1	+	−	−
76	*Pyxine meissneriana* Nyl.	Caliciaceae	Teloschistales	C	1	+	−	−
77	*Pyxine nilgiriensis* D.D. Awasthi	Caliciaceae	Teloschistales	C	2	−	+	−
78	*Pyxine petricola* Nyl.	Caliciaceae	Teloschistales	C	1	+	+	−
79	*Pyxine sorediata* (Ach.) Mont.	Caliciaceae	Teloschistales	C	1	+	+	+
80	*Pyxine subcinerea* Stirt.	Caliciaceae	Teloschistales	C	1	+	−	−
81	*Ramalina baltica* Lettau	Ramalinaceae	Lecanorales	I	1	−	−	+
82	*Ramalina pacifica* Asahina	Ramalinaceae	Lecanorales	I	1	+	+	+
83	*Ramalina roesleri* (Hochst. ex Schaer.) Hue	Ramalinaceae	Lecanorales	I	1	−	+	+
84	*Ramalina* sp.2	Ramalinaceae	Lecanorales	I	1	−	+	−
85	*Rimelia reticulata* (Taylor) Hale and Fletcher	Parmeliaceae	Lecanorales	G	1	+	−	+
86	*Sticta filicina* Ach.	Lobariaceae	Peltigerales	G	1	−	+	−
87	*Sticta weigelii* Isert	Lobariaceae	Peltigerales	G	1	−	+	+
88	*Teloschistes flavicans* (Sw.) Norman	Teloschistaceae	Teloschistales	I	1	−	−	+
89	*Usnea albopunctata* Nyl. apud Crombie	Parmeliaceae	Lecanorales	H	1	−	+	−
90	*Usnea bornmuelleri* J. Steiner	Parmeliaceae	Lecanorales	H	1	−	−	+
91	*Usnea corallina* Motyka	Parmeliaceae	Lecanorales	H	1	−	+	−
92	*Usnea dendritica* Stirt.	Parmeliaceae	Lecanorales	H	1	−	+	+
93	*Usnea galbinifera* Asahina	Parmeliaceae	Lecanorales	H	1	−	−	+
94	*Usnea orientalis* Motyka	Parmeliaceae	Lecanorales	H	1	−	+	+
95	*Usnea pangiana* Stirt.	Parmeliaceae	Lecanorales	H	1	−	−	+
96	*Usnea pectinata* Stirt.	Parmeliaceae	Lecanorales	H	1	−	+	−
97	*Usnea picta* (J. Steiner) Motyka	Parmeliaceae	Lecanorales	H	1	−	−	+
98	*Usnea pictoides* G. Awasthi	Parmeliaceae	Lecanorales	H	1	−	+	−

TABLE 1: Continued.

S. no.	Species name	Family	Order	Growth forms	Substratum	Forest types		
						DDF	MDF	WTEF
99	*Usnea pseudomontis-fuji* Asahina	Parmeliaceae	Lecanorales	H	1	−	+	−
100	*Usnea splendens* Stirt.	Parmeliaceae	Lecanorales	H	1	−	−	+
101	*Usnea stigmatoides* G. Awasthi	Parmeliaceae	Lecanorales	H	1	−	−	+
102	*Usnea undulata* Stirt.	Parmeliaceae	Lecanorales	H	1	−	−	+
103	*Usnea vegae* Motyka	Parmeliaceae	Lecanorales	H	1	−	−	+

Growth form code: A-foliose; B-foliose gelatinous; C-foliose laciniate; D-foliose ribbon like; E-foliose strap shaped; F-foliose two fold; G-foliose typical; H-fruticose cylindrical; I-fruticose strap shaped. Subsratum code: bark-1; rock-2; soil-3. DDF: dry deciduous forest; MDF: moist deciduous forest; WTEF: western tropical evergreen forest.

(a)

(b)

(c)

(d)

(e)

(f)

FIGURE 2: Examples of lichens examined during this study. (a) *Parmotrema praesorediosum,* an epiphytic macrolichen, (b) *Parmotrema mesotropum* (living in an open and dry condition with *Trebouxia* as photobiont), (c) *Ramalina pacifica,* a strap shaped lichen, (d) *Heterodermia isidiophora,* (e) *Pyxine cocoes* var. *cocoes* with black ascomata, and (f) *Leptogium denticulatum,* living in a shade and wet condition with *Nostoc* as photobiont (a cyanolichen).

survival of a lichen thallus [21], and the *Trebouxia* containing lichens are known to survive better in the open and dry [22, 23] condition prevailing in the DDF. The seasonal evergreen and tropical mixed deciduous forests of Thailand [24] were reported to contain more numbers of shade-tolerant *Trentepohlia* containing lichens compared to dry deciduous forests of Thailand. More number of shade-tolerant and moisture-dependent *Nostoc*-containing lichens were observed in MDF type in Bolampatti II forest range also. Lichen families with cyanobacteria and/or chlorococcoid algae include Lobariaceae and Collemataceae, and are more frequent in humid forests. The adaptations to tropical shaded moist conditions are the presence of cyanobacteria and *Trentepohlia* as photobionts [25].

Bergamini et al. [26] state that macrolichens may be good indicators of lichen species richness because of the highly significant relationships. The species turnover within the study sites was very similar for all lichens and macrolichens. The similar trend was also observed in our study sites. In northern Thailand, the lichen species characteristic of disturbance such as *Pyxine consocians* and *Dirinaria consimilis* is also observed in DDF of Bolampatti II forest range. The rate of colonization and growth of these species is an indication of the rate of change [24].

4. Conclusion

The present study revealing the occurrence of 103 macrolichens from 197.66 km^2 forest area indicates the lichen species richness of the study area and their availability for further ecological monitoring. It is desirable to develop location-specific and lichen-centric conservation strategies using this baseline data to protect the valuable and yet poorly studied ecologically-important group called lichens.

Acknowledgments

The authors thank Professor M. S. Swaminathan and the Executive Director, MSSRF, for their encouragement and support, Dr. D. K. Upreti for his critical comments on identification and reconfirmation, Principal Chief Conservator of Forests, Tamil Nadu, and District Forest Officer, Coimbatore, for necessary field permit, field guides Mr. P. Chokkalingam, Mr. P. Radhan, Mr. Senthil and Mr. Selvam for their help, Council of Scientific and Industrial Research, New Delhi, and Government of India for the financial support.

References

[1] J. P. Schmit and G. M. Mueller, "An estimate of the lower limit of global fungal diversity," *Biodiversity and Conservation*, vol. 16, no. 1, pp. 99–111, 2007.

[2] S. T. Bates, A. Barber, E. Gilbert et al., "A revised catalog of Arizona lichens," *Canotia*, vol. 6, no. 1, pp. 26–43, 2010.

[3] S. Nayaka, D. K. Upreti, M. Gadgil et al., "Distribution pattern and heavy metal accumulation in lichens of Bangalore city with special reference to Lalbagh garden," *Current Science*, vol. 84, no. 5, pp. 674–680, 2003.

[4] S. R. Gradstein, "The vanishing tropical rain forest as an environment for bryophytes and lichens," in *Bryophytes and Lichens in a Changing Environment*, J. W. Bates and M. F. Andrew, Eds., pp. 235–258, Clarendon Press, Oxford, UK, 1992..

[5] H. R. Negi, "On the patterns of abundance and diversity of macrolichens of Chopta-Tunganath in the Garhwal Himalaya," *Journal of Biosciences*, vol. 25, no. 4, pp. 367–378, 2000.

[6] K. P. Singh and G. P. Sinha, *Lichen Flora of Nagaland*, Bishen Singh Mahendra Pal Singh, Dehra Dun, India, 1994.

[7] K. P. Singh and G. P. Sinha, "Lichen diversity of the Eastern Himalaya and its conservation," in *Himalayan Microbial Diversity, Part 2, (Recent Researches in Ecology, Environment and Pollution)*, S. C. Sati, J. Saxena, and R. C. Dubey, Eds., vol. 11, pp. 349–359, Today and Tomorrow's Printers and Publishers, New Delhi, India, 1997.

[8] M. Kumar and S. Stephen, "Lichens of Western Ghats—an overview," in *Biology of Lichens*, K. G. Mukerji, B. P. Chamola, D. K. Upreti, and R. K. Upadhyay, Eds., pp. 297–331, 1999.

[9] J. Wilson, "Working plan for the Bolampatti Range of Coimbatore Central Forest Division," Government of Madras, pp.1–155, 1967.

[10] H. G. Champion and S. K. Seth, *A Revised Survey of the Forest Types of India*, Manager of Publications, Delhi, India, 1968.

[11] D. D. Awasthi, "Contributions to the lchen flora of India and Nepal-I. The genus Physcia (Ach.) Vain," *Journal of the Indian Botanical Society*, vol. 39, no. 1, pp. 1–21, 1960.

[12] D. D. Awasthi, *A Monograph of the Lichen Genus Dirinaria*, vol. 2 of *Bibliotheca Lichenologica*, J. Cramer, Lehre, Germany, 1975.

[13] D. D. Awasthi and K. P. Singh, "The lichen flora in the environs of Gangotri and Gomukh, India. I-the macrolichens," *Indian Journal of Forest Research*, vol. 1, pp. 138–146, 1978.

[14] D. D. Awasthi, "Pyxine in India," *Phytomorphology*, vol. 30, pp. 359–379, 1980.

[15] D. D. Awasthi, "Lichen genus Parmelia in India. II. Subgenera Xanthoparmelia (Vain.) Hale and Melanoparmelia (Hue) Essl," *Indian Journal of Forestry*, vol. 4, pp. 198–204, 1981.

[16] G. Awasthi, "Lichen genus usnea in India," *Journal of the Hattori Botanical Laboratory*, vol. 61, pp. 333–421, 1986.

[17] D. D. Awasthi, "A key to the macrolichens of India and Nepal," *Journal of the Hattori Botanical Laboratory*, vol. 65, pp. 207–302, 1988.

[18] S. Huneck and I. Yoshimura, *Identification of Lichen Substances*, Springer, Tokyo, Japan, 1996.

[19] D. L. Hawksworth, P. M. Kirk, B. C. Sutton, and D. N. Pegler, *Dictionary of the Fungi*, CAB International, Wallingford, UK, 8th edition, 1995.

[20] P. A. Wolseley and D. L. Hawksworth, "Adaptations of lichens to conditions in tropical forests of South-East Asia and their taxonomic implications," *Blumea*, vol. 54, no. 1–3, pp. 29–32, 2009.

[21] P. A. Wolseley and B. Aguirre-Hudson, "The ecology and distribution of lichens in tropical deciduous and evergreen forests of Northern Thailand," *Journal of Biogeography*, vol. 24, no. 3, pp. 327–343, 1997.

[22] P. Balaji and G. N. Harihran, "Lichen diversity and its distribution pattern in Tropical Dry Evergreen Forest of Guindy National Park (GNP), Chennai," *The Indian Forester*, vol. 130, no. 10, pp. 1155–1168, 2004.

[23] P. Balaji and G. N. Harihran, "Annonated checklist of lichens of Chennai, Tamil Nadu, India," *Phytotaxonomy*, vol. 5, pp. 1–7, 2005.

[24] P. A. Wolseley and B. Aguirre-Hudson, "Fire in tropical dry forests: corticolous lichens as indicators of recent ecological changes in Thailand," *Journal of Biogeography*, vol. 24, no. 3, pp. 345–362, 1997.

[25] H. J. M. Sipman and R. C. Harris, "Lichens," in *Tropical Rain Forest Ecosystems*, H. Lieth and M. J. A. Werger, Eds., pp. 303–309, Elsevier Science Publishers B.V., Amsterdam, The Netherlands, 1989.

[26] A. Bergamini, C. Scheidegger, S. Stofer et al., "Performance of macrolichens and lichen genera as indicators of lichen species richness and composition," *Conservation Biology*, vol. 19, no. 4, pp. 1051–1062, 2005.

9

Status, Diversity, and Traditional Uses of Homestead Gardens in Northern Bangladesh: A Means of Sustainable Biodiversity Conservation

Bishwajit Roy, Md. Habibur Rahman, and Most. Jannatul Fardusi

Bangladesh Institute of Social Research (BISR), Hasina De Palace, House No. 6/14, Block No. A, Lalmatia, Dhaka 1207, Bangladesh

Correspondence should be addressed to Md. Habibur Rahman; habibmdr@gmail.com

Academic Editors: I. Bisht and H. Ford

A study was conducted to assess the status, ecological diversity, traditional uses, spatial arrangement, and importance of homestead garden for biodiversity conservation of the urban and rural households in Kishoreganj Sadar of northern Bangladesh. Assessment was done by means of multistage random sampling from a total of 80 households using a semistructured questionnaire. A total of 62 plant species belonging to 36 families including 5 threatened species were identified. The majority of the species were used as fruit and food (45%) followed by medicinal plants (38.71%), firewood (32.26%), and timber (29%). Ecological diversity indices indicated that the existing plant species in the homestead gardens in the study area have moderately high biodiversity and species richness. Farmers perceived importance for homestead plant species conservation was for fruit and food (85%) followed by building materials (78.75%), subsistence family income (73.75%), and source of firewood (68.75%). In addition, analysis of existing management regime indicates that growers lack scientific information, almost every household still follows traditional management systems. Finally, a specific homestead forest management plan, conservation of homestead species diversity through scientific management and obtaining training and support from government and NGOs, was found highly demandable by this study.

1. Introduction

Homestead garden is a traditional agroforestry system and an important component in the livelihoods of rural poor, and in the rural economy of the country. During the last 40–50 years, the relative importance has shifted from the traditional forestry to homestead forestry; in such a situation, homestead garden plays a vital role in providing firewood, fodder, medicine, fruit, and timber. It is estimated that about 70% of timber, 90% of firewood, 48% of sawn and veneer logs, and almost 90% of bamboo requirements are met from homestead forests [1].

In Bangladesh, homestead gardens represent a well-established traditional land-use system where natural forest cover is less than 10 percent; homestead gardens, which are maintained by at least 20 million households, represent one possible strategy for biodiversity conservation [2, 3]. The management of the traditional homestead garden has evolved as a response to many factors: cultural, economic and, environmental as well as personal preferences [4]. The conservation of cultivated plants in homestead gardens of Bangladesh not only preserves a vital resource for humankind but plays an important role in household food security, as it is a sustainable source of food, fruits and vegetables [5].

In Bangladesh, there is no specific management plan for the homestead forests [6] which are being traditionally managed by the household owners. Most of the plants grown in homestead garden have multiple uses. These home gardens are some sort of additional income for some families of rural area whereas for most of the families of urban area they act as a medium of nutritional demand fulfill. Millat-E-Mustafa et al. [7] record eight major uses of the homestead forest plants: fruit/food, timber, firewood, spice, fodder, medicine, fencing, and miscellaneous uses. The miscellaneous uses include brooms, handicrafts, shade, ornamental, ceremonial, environmental, and aesthetic. Again, the ecological merits of

Status, Diversity, and Traditional Uses of Homestead Gardens in Northern Bangladesh: A Means of
Sustainable Biodiversity Conservation

115

homestead garden are related to conservation of soil, water, nutrients, and biodiversity [8].

Several studies showed that species diversity in a homestead garden can range from less than five [9–12] to more than 100 [13–15]. In Bangladesh, various studies, for example, [3, 7, 9, 16–23] explore the floristic composition (mainly trees) in the homestead gardens; homestead agroforestry system by [24–26]; homestead plantation and traditional uses by [27, 28]; quantitative structure and silvicultural management by [29–31]; production and services by [32, 33]. Moreover, Motiur et al. [34] studied the role of homestead gardens in rural economy; Alam and Masum [35] and Masum et al. [8] studied the status of homestead garden in an offshore island of Bangladesh, and Akhter et al. [36] studied the role of women in homestead gardens management in the northeastern Bangladesh.

Since the natural forest of Bangladesh is shrinking at an alarming rate due to unprecedented anthropogenic pressure, researchers from across the world have demonstrated homestead gardens' dynamic role in the conservation of biodiversity and provision of necessary daily needs to rural people by turn for urban people. Researchers from across the country and world have explored the quantitative status of homestead garden but not the driving factors which lead people to plant trees in their house premises. Therefore, this study was conducted to evaluate and quantitatively assess the total botanical diversity and the status of homestead garden ecosystem (both rural and urban), and choice of people, spatial arrangement of different species in both urban and rural homestead areas of Kishoreganj Sadar Upazila (subdistrict) of Bangladesh. The study also tried to find out the reasons towards differences in species diversity and biodiversity conservation for both areas and then represented schematically so that we can understand what are the driving factors behind the differences in both areas (rural and urban). In the study areas, homestead gardens are substantial enterprises and play a significant role in household income.

2. Materials and Method

2.1. Study Area Profile. Kishoreganj Sadar Upazila with an area of 193.73 km^2 is bounded by Nandail Upazila on the north, Pakundia and Katiadi upazilas on the south, Karimganj and Tarail upazilas on the east, and Hossainpur and Nandail upazilas on the west. The Main River is Narsunda [37]. Table 1 represents the main features of the study area.

2.2. Research Methods. The study was based on the primary data collected directly from the field during September 2011 to January 2012 through physical measurement. During study, multistage random sampling method was adopted for data collection. From 12 upazilas of Kishoreganj district (administrative unit), Kishoreganj Sadar Upazila was selected purposively. Out of 11 unions in Sadar Upazila, two unions from rural area and only one existing municipality in the study area were firstly selected randomly. Then, two villages from municipality area and one from rest of the unions were randomly selected.

TABLE 1: Description of the study area (Kishoreganj Sadar Upazila).

Items	Description
Location	Kishoreganj Sadar Upazila is an Upazila of Kishoreganj District in the Division of Dhaka with 205 villages. Kishoreganj Municipality was established in 1869 with an area of 19.57 km^2. It has a population of 77165; males 52.51%, females 47.49%. Literacy rate among the municipality people is 59%.
GPS location	It is located between 24°28′ and 24°24′ north latitudes and between 90°46′ and 90°50′ east longitudes.
Population	500208 (density: 1550/km^2); males 51.52%, females 48.48%.
Number of villages	205
Number of households	71178
Average household size	4.21 persons
Literacy	28.30% where the national average was 32.4%.
Land use	Total cultivable land 13766 ha; single crop 17.65%, double crop 60%, and treble crop land 22.35%; land under irrigation 3239 ha. Main crops are paddy, jute, wheat, mustard seed, onion, potato, brinjal, betel leaf, and vegetables. Main exports products are jute and leather.
Main occupations	Agriculture 35.77%, agricultural labourer 17.56%, wage labourer 3.41%, industry 1.22%, business 15.22%, transport 4.97%, construction 1.57%, service 8.72%, and others 11.56%.

2.3. Data Collection from Respondents. Interviews were conducted targeting primarily old-aged or local experienced persons (usually aged between 30 to 70 years). A total of 80 households, that is, 20 households (PSU, primary sampling unit) with a confidence interval of 33% from each village were selected for interviewing (Table 2). A semistructured questionnaire was used for data collection based on the information collected through reconnaissance and pilot survey. All the species found in each household had been accounted for botanical survey. The final survey and eight focused group discussions (two in each PSU for crosschecked) were completed with the participation and informed consent from the members of the households. Responses to open questions were collected on a variety of demographic and socioeconomic indicators: household species composition, choice of species, cultural activities practices in homestead garden, perceived importance for conservation of species, market access of homestead garden products, and so forth. In order to explore spatial arrangement and respondents' choice to plant trees in their homestead garden, we have divided the gardens into five different habitats, namely, (front yard, back yard, homestead boundary, that is, adjacent to the dwelling house, pond bank, and road side) for both rural and urban homesteads. On each topic, the respondents were free to express their views.

TABLE 2: Description of the sample villages.

Items	Urban		Rural	
	PSU* 1 (Solakia)	PSU 2 (Mohinanda)	PSU 3 (Maijkapon)	PSU 4 (Joshodol)
Location	Municipality area	Municipality area	Maijkapon union	Joshodol union
Number of households	160	124	92	110
Household size	7.08	8.03	9.20	6.57
Sampled households	20	20	20	20
Major homestead forest products	Fruits, timber, and ornamental species	Fruits, timber, firewood, and ornamental species	Fruits, timber, bamboo, firewood, and ornamental species	Fruits, timber, bamboo, firewood, and ornamental species
Main occupation	Business, services, and foreign services	Business, services, and foreign services	Agriculture, fisheries, small business, and services	Agriculture, fisheries, small business, and services

*PSU: primary sampling unit.

2.4. Plot Survey. All species present in each sampled homestead garden (average size of 0.05 hectare to 0.25 hectare) were identified and recorded by the botanical name or by local name. All individuals of trees, herbs, and shrubs were counted and recorded except the individuals in hedgerows. No climbers were counted due to the difficulty in differentiating stems. A botanical inventory was conducted only once in each selected home garden. Thus, the seasonal variation in floristic and structure was not assessed. Each species recorded was classified by family, habit (tree, shrub, and herb), and their origin that is exotic or indigenous, and conservation status as followed by [38–45].

2.5. Data Analysis. For the present study, nine ecological indices were used to analyze and to get a clear picture of the species diversity of the study areas, which are listed below.

(1) Species diversity index was calculated according to Odum [46]: SDI = S/N.

(2) Species richness index was measured by Margalef [47]: $R = (S - 1)/\text{Ln}(N)$.

(3) The Shannon-Winner diversity index was calculated following Michael [48]: $H = -\sum P_i \text{Ln} P_i$.

(4) Shannon's maximum diversity index was followed by Kent and Coker [49]: $H_{\text{max}} = \text{Ln}(S)$.

(5) Shannon's equitability index according to Kent and Coker [49]: $E_H = H/H_{\text{max}}$.

(6) Species evenness index was estimated following Pielou [50]: $E = H/\text{Log}(S)$.

(7) Simpsons index was estimated according to Magurran [51]: $D = \sum P_i^2$.

(8) As biodiversity increases, the Simpson index decreases. Therefore, to get a clear picture of species dominance, $D' = 1 - D$ is used.

(9) Family importance value (FIV) index (the FIV index is used to evaluate floristic composition at the species

family level, and it combines richness, density, and dominance) according to the formulae of Mori et al. [52].

(i) Family relative density (%) = (no. of individuals in a family/total no. of individuals) × 100.

(ii) Family relative diversity (%) = (no. of species in a family/total number of species) × 100.

(iii) FIV is the sum of family relative diversity and relative density.

Where S is the total number of species, N is the total number of individuals of all the species, P_i is the number of individuals of one species/total number of individuals in the samples.

3. Results

3.1. Status of Homestead Garden Plants. The study survey recorded 62 plant species belonging to 36 families from the set of 80 surveyed homesteads (Table 3). Among the total species, 53.23%, trees, 22.58% shrub, and 24.19% herbs. Amongst the recorded species, 31 trees, 11 herbs and shrubs species were found common both in urban and rural homesteads. Except this, shrub species was higher in urban homestead, tree species was found higher in rural homestead, and in case of herb species was found the same in both areas. Thereafter, out of the recorded species based on conservation status, two species; namely, *Alocasia indica* and *Terminalia chebula* are vulnerable, *Pteris cretica* is near threatened, *Boehmeria nivea*, *Cinnamomum tamala* are not evaluated, and the rest of the species are of least concern in the context of Bangladesh.

Floristic composition of the homestead flora consists of both native and exotic species. About 18 species were identified as exotic, and some of them have been domesticated. Recently, fruit-bearing species were gradually being replaced by some exotic timber species such as *Bauhinia acuminate*, *Albizia saman*, *Swietenia mahagoni*, *Tectona grandis*, and

Status, Diversity, and Traditional Uses of Homestead Gardens in Northern Bangladesh: A Means of Sustainable Biodiversity Conservation

117

TABLE 3: List of homestead garden plant species with conservation status and uses in Kishoreganj Sadar Upazila.

Family	Scientific name	Local name	English name	Individuals		Life form	Origin	FIV	C.S. (References*)	Uses
				Rural	Urban					
Acanthaceae	Adhatoda vasica Nees	Bashok	Malabar nut	16	12	Sh	I	2.84	LC (6)	6
Amaranthaceae	Alternanthera philoxeroides (Mart.) Grisb.	Helencha	Dwarf copperleaf	21	8	H	I	5.12	LC (6)	1, 6
	Gomphrena globosa L.	Time phol	Globe amaranth	—	13		E		LC (6)	1
Anacardiaceae	Mangifera indica L.	Aam	Mango	38	48	Tr	I	8.53	LC (6)	1, 2, 3, 5
	Spondias pinnata Kurz.	Amra	Wild mango	23	19	Tr	I		LC but gradually disappearing (6)	1, 2, 3
Annonaceae	Annona squamosa L.	Shorifa	Sugar apple	14	11	Tr	I	2.72	LC (6)	1, 3
Apiaceae	Centella asiatica (L.) Urban	Thankuni	Indian pennywort	31	17	H	I	3.63	LC (6)	1, 6
Apocynaceae	Carissa carandas Linn.	Koromcha	Black cherry	11	9	ST	I	2.52	LC (6)	1
Araceae	Colocasia esculenta Schott.	Kachu	Elephant's ear	26	7	Sh	I		LC (11)	1, 6
	Alocasia indica Schott.	Man-kachu	Great-leaved caladium	21	8	Sh	I		V (11)	1, 6
	Cocos nucifera L.	Narikel	Coconut	27	23	Tr	I	13.4	LC (11)	1, 2, 8
	Areca catechu L.	Supari	Betel nut	30	21	Tr	I		LC (11)	1, 2, 7, 8
Asteraceae	Eupatorium odoratum L.	Assampata	Common floss flower	29	19	Sh	E	6.47	LC (6)	6
	Tagetes erecta L.	Gada phol	African marigold	—	28		E		LC (6)	6
Caesalpiniaceae	Bauhinia acuminata L.	Kanchan	Butterfly tree	7	21	Tr	I		LC (7)	2, 3
	Delonix regia (Boj) Raf.	Krishnochura	Flame tree	14	16	Tr	I	8.98	LC (7)	2, 8
	Tamarindus indica L.	Tetul	Tamarind tree	19	19	Tr	E		LC (7)	1, 3
Caricaceae	Carica papaya Linn.	Pepe	Papaya	63	37	Sh	I	5.69	LC (7)	1
Combretaceae	Terminalia bellirica Roxb	Bohera	Belliric myrobalan	8	17	Tr	I	5.27	LC (7)	1, 6
	Terminalia chebula (Gaerth.) Retz.	Haritaki	Chebulic myrobalan	6	15	Tr	I		V (7)	1, 6
Crassulaceae	Kalanchoe pinnata (Lam.) Pers.	Patharkuchi	American life plant	22	28	H	I	3.71	LC (7)	6
Cyperaceae	Maurices microcephalus Presl.	Boro gothori	Perennial sunflower	23	16	H	I	3.27	LC (11)	6
Dilleniaceae	Dillenia indica L.	Chalta	Elephant apple	36	31	Tr	I	4.38	LC (7)	1, 3
Elaeocarpaceae	Elaeocarpus floribundus Blume.	Jalpai	Indian olive	20	37	Tr	I	3.99	LC (7)	1, 2, 3
Euphorbiaceae	Phyllanthus emblica L.	Amalaki	Indian gooseberry	11	28	Tr	I		LC (7)	1, 6
	Phyllanthus acidus (L.) Skeels	Arbori		21	19	Tr	I	8.82	LC (7)	1, 8
	Codiaeum variegatum (Linn.) A. Juss.	Pata bahar	Garden croton	—	13		E		LC (7)	2, 3, 5, 7
Fabaceae	Erythrina variegata Linn.	Mandar	Indian coral tree	17	13	Tr	I	2.92	LC (8)	6
Labiatae	Leucas indica (L.) R. Br. Ex Vatke in Oesterr.	Dondokolos	Dronpushpi	24	11	H	E	3.11	LC (8)	6
Lamiaceae	Ocimum sanctum L.	Tulsi	Sacred basil	19	23	H	I	3.39	LC (8)	4, 8
Lauraceae	Cinnamomum tamala Nees.	Tejpata	Bay leaf	4	7	Tr	I	2.16	NE (8)	8
Lythraceae	Lawsonia inermis L.	Mehndi	Henna	10	27	Sh	E	3.19	LC (8)	8
Malvaceae	Hibiscus rosa-sinensis L.	Joba ful	China rose	23	17	Sh	I	3.31	LC (9)	7, 8

TABLE 3: Continued.

Family	Scientific name	Local name	English name	Individuals Rural	Individuals Urban	Life form	Origin	FIV	C.S. (References*)	Uses
Meliaceae	Swietenia mahagony (L.) Jacq.	Mahagoni	Mahagoni	21	31	Tr	E	6.9	LC (9)	2, 3
	Azadiracta indica A. Juss.	Neem	Indian lilac	14	21	Tr	I		LC (9)	2, 3, 6
	Albizia spp.	Koroi	White siris	23	21	Tr	E		LC (9)	2, 3
Mimosaceae	Albizia saman (Jaq.) Merr.	Raintree	Raintree	26	26	Tr	E	12.6	LC (9)	2, 3
	Mimosa pudica L.	Lazzabati	Bashful mimosa/sensitive plant	13	16	H	I		LC (9)	6
	Streblus asper Lour.	Shewra	Siamese roughbush	9	9	Tr	I		LC (9)	6, 7
Moraceae	Artocarpus heterophyllus Lamk.	Kathal	Jackfruit	31	41	Tr	I	4.58	LC (9)	1, 2, 3, 5
Musaceae	Musa spp.	Kola	Banana	51	13	Sh	E	4.26	LC (11)	1, 5, 7
Myrtaceae	Syzygium cumini (L.) Skeel.	Jam	Black berry	16	26	Tr	I	7.54	LC (9)	1, 2, 3
	Psidium guajava (L.) Bat.	Payera	Common guava	28	33	ST	I		LC (9)	1, 2
Oxalidaceae	Averrhoa carambola L.	Kamranga	Star fruit	23	27	Tr	I	6.04	LC (9)	1, 3
	Oxalis corniculata L.	Amrul	Indian zorale	16	—	H	I		LC (9)	6
Poaceae	Bambusa balcooa Roxb.	Barak bans	Bamboo	13	—	Tr	I		LC (12)	2, 3, 5, 7, 8
	Melocanna baccifera (Roxb.) Kurz.	Muli bans	Bamboo	14	17	Tr	I	9.1	LC (12)	2, 3, 5, 7, 8
	Cynodon dactylon (L.) Pers.	Durbaghass	Grass	36	19	H	I		LC (12)	6
Polypodiaceae	Pteris cretica Wilsonii	Fern	Cretan fern	29	20	H	E	3.67	NT (5)	1, 6
Rhamnaceae	Zizyphus mauritiana Lamk.	Boroi	Indian date	18	23	Tr	I	3.35	LC (10)	1, 5
	Aegle marmelos (L.) Correa.	Bel	Golden apple/stone apple	16	—	Tr	I		LC (10)	1, 3
Rutaceae	Citrus grandis (L.) Osb.	Jambura	Pummelo	19	19	Tr	I	9.82	LC (10)	1, 3
	Citrus limon (L.) Burm. F.	Lebu	Lemon	32	31	Sh	E		LC (10)	1, 6
Solanaceae	Solanum violaceum Ortega, Hort. Mart	Bon-begun	Lemon	29	—	Sh	I	5.31	LC (10)	6
	Cestrum nocturnum L.	Hasnahena	Poison berry	—	18	Sh	E		LC (10)	8
Urticaceae	Boehmeria nivea (L.) Gaudich	Artika	Night jasmine	17	—	H	E	2.4	NE (10)	6
	Nyctanthes arbortristis L.	Shiuli phool	China grass	13	11	Sh	I		LC (10)	8
Verbenaceae	Lantana camara L.	Lantana	Night queen	33	6	Sh	E		LC (10)	6
	Tectona grandis L. f.	Shegun	Lantana	10	13	Tr	E	11.6	LC (10)	2, 3
	Lippia alba (Mill.) Briton et Wilson	Voi ukhra	Bushy mat grass	27	5	Sh	E		LC (10)	6
Zingiberaceae	Zingiber officinale Roscoe	Ada	Ginger	61	8	H	I	8.89	LC (12)	4, 6
	Curcuma longa L.	Holud	Turmeric	58	10	H	I		LC (12)	4, 6

Life form: Tr: tree, H: herb, and Sh: shrub. Origin: E: exotic species, I: indigenous species; FIV: family importance value; C.S.: conservation status, LC: least concern, NT: near threatened, NE: not evaluated, and V: vulnerable. Uses: 1: food/fruit, 2: timber, 3: fuel wood, 4: spice, 5: fodder, 6: medicine, 7: fence, and 8: others. * References: values in the parentheses indicate the volume number of the book entitled *Encyclopaedia of Flora and Fauna of Bangladesh* [38–45].

Status, Diversity, and Traditional Uses of Homestead Gardens in Northern Bangladesh: A Means of
Sustainable Biodiversity Conservation

119

Albizia spp. because of the people's attitude towards earning more money through timber production. However, *Swietenia mahagoni*, *Albizia saman*, *Delonix regia*, and *Tectona grandis* have been domesticated and have a long heritage of introduction.

3.2. Species Family Composition.

Araceae, Mimosaceae, and Verbenaceae families represented the highest numbers of four species followed by Caesalpiniaceae, Euphorbiaceae, Poaceae, and Rutaceae representing three species. Amaranthaceae, Asteraceae, Anacardiaceae, Combretaceae, Meliaceae, Myrtaceae, Oxalidaceae, Solanaceae, and Zingiberaceae denoted two species, and the rest of the families comprised only one species (Table 3). Among them, 20 families (55.56%) represented only one species followed by two species were represented by nine families (25%); three and more than three species are represented by four families (11.11%) and 3 species (8.33%), respectively. The family importance value (FIV) was recorded highest for Araceae (163 individuals, FIV 13.4) followed by Mimosaceae (143 individuals, FIV 12.6), Verbenaceae (118 individuals, FIV 11.6), Rutaceae (117 individuals, FIV 9.82) Poaceae (99 individuals, FIV 9.1), Caesalpiniaceae (96 individuals, FIV 8.98), Zingiberaceae (137 individuals, FIV 8.89) and lowest for Lauraceae (11 individuals, FIV 2.16), followed by Urtiaceae (17 individuals, FIV 2.40), Apocynaceae (20 individuals, FIV 2.52), Annonaceae (25 individuals, FIV 2.72), and Acanthaceae (28 individuals, FIV 2.84).

3.3. Ecological Diversity Indices of Homestead Species.

The result of Shannon-Winner diversity index value was calculated highest for rural homestead garden both of tree (3.39) and of shrub (2.36) species where for herb species, the highest value (2.5) was found for the urban homestead garden. Shannon's maximum diversity index value (3.5) was observed highest for rural homestead tree species. In case of herb (2.56) and shrub (2.48) species, the values were similar for both types of the gardens. Shannon's equitability index (0.97), Simpson's index (0.04), and dominance of Simpson's index (0.96) values were found similar for both the rural and urban homestead tree species. For herb species, Shannon's equitability index value was highest for urban gardens (0.97) where dominance of Simpson's index (0.91) was found similar for both rural and urban gardens. However, in case of shrub species, Shannon's equitability index value was highest for rural gardens (0.95); Simpson's index value (0.93) for urban gardens; and dominance of Simpson's index value (0.89) for rural gardens. However, in case of tree and herb species, species evenness index value was highest (2.24) for urban homestead garden where for shrub the value (2.19) was highest in rural homestead. Species richness index values for tree (4.98) and herb (2.26) species were highest in rural homestead but for shrub species, the value (2.14) was highest in urban homestead (Table 4).

3.4. Spatial Arrangement of Homestead Garden Species.

In the study area, it was observed that homestead gardeners do not follow any specific spatial arrangement pattern and scientific

TABLE 4: Ecological diversity indices of homestead plant species in rural and urban areas of the study area.

Ecological diversity index*	Tree		Herb		Shrub	
	Urban	Rural	Urban	Rural	Urban	Rural
SDI	0.04	0.05	0.07	0.04	0.07	0.04
R	4.59	4.98	2.26	2.03	2.14	1.89
E	2.24	2.23	2.24	2.2	2.13	2.19
H	3.35	3.39	2.5	2.45	2.3	2.36
H_{max}	3.43	3.5	2.56	2.56	2.48	2.48
E_H	0.97	0.97	0.97	0.96	0.93	0.95
D	0.04	0.04	0.09	0.09	0.12	0.11
D'	0.96	0.96	0.91	0.91	0.88	0.89

*H: Shannon-Winner diversity index, H_{max}: Shannon's maximum diversity index, E_H: Shannon's equitability index, SDI: species diversity index, R: species richness index, E: species evenness index, D: Simpsons index, and D': dominance of Simpsons index.

considerations for raising plants. Plants were usually planted in the front, back, and surroundings of the houses. Results revealed that 47.5% of rural gardeners preferred mostly the front yard for species plantation followed by back yard 32.5%, homestead boundary (adjacent to the dwelling house) by 12.5% and 7.5% household owners preferred both pond banks and road side simultaneously. In case of urban area, 35% gardeners preferred balcony and veranda for planting species followed by 30% in front yard, 22.5% in homestead boundary, and 12.5% in rooftop. The reason for this variation was that rural people depend more on homestead forest for their livelihood security as well as certain amount of family income than that of urban households. Except these other factors, those influence the species planting in different sites are beautification of household area, low canopy coverage, land scarcity, and so forth. Ornamental plants, for example, *Gomphrena globosa*, *Tagetes erecta*, *Codiaeum variegatum*, *Ocimum sanctum* and *Nyctanthes arbortristis* and major fruit crop species, for example, *Mangifera indica*, *Artocarpus heterophyllus*, *Citrus limon*, *Psidium guajava*, *Zizyphus mauritiana*, *Citrus grandis*, and *Carica papaya* were usually planted in the front yard of the house (rural and urban) so that the gardeners can keep eye on them. Trees in the homestead-boundary acted as a live fence and windbreak. Tall woody species such as are planted in the back yard for building materials and firewood.

3.5. Choice of Species Grown for Different Usages.

The reasons for growing a variety of fruit species in the homestead gardens are more or less complex. A number of factors determined the farmers' decisions for growing particular species or groups of species. The farmers were keen to grow timber trees for cash income if they already had a successful strategy for deriving income from off-farm labour or from crops. Farmers with large families tend to grow fruit trees. In the surveyed area, homestead plant species generally used for fruit and food, medicines, firewood, timber, and ornamental and beautification purposes were identified. Among them, 45% were fruit and food providing species, 38.71%

medicinal plant species, 32.26% firewood species, 29.03% timber species, 16.13% ornamental, beautification, and spiritual species, 11.29% species are used as both for fodder and fence, and 4.83% species used as spice and vegetables.

3.6. Cultural Practices of Homestead Garden.

3.6. Cultural Practices of Homestead Garden. Farmers generally collect planting materials from homesteads wildings (species that are grown/collected from outside homestead premises), friends and families, relatives, government, and NGO nurseries. No specific spacing is followed in planting of species in homestead garden. Sometimes it was also found to plant herbaceous species like *Zingiber officinale, Curcuma longa* under the layer of shrub like *Carica papaya, Citrus limon,* and so forth in order to make the optimum use of their land. Study figured out the analysis of respondents answers regarding different aspects of the existing management systems of homestead gardens in the study area. During survey, it was found that some households are not engaged in any management/cultural operations in their homestead gardens whereas other households are more or less engaged with the management of homestead gardens. Species were planted usually during the morning and/or afternoon of the day mostly in the monsoon season. Generally, fast growing and species having low crown coverage are selected for the plantation. The results revealed that almost all the households carried out watering (100%) and soil ploughing (94%). Weeding was done out by 85% of respondents as well as fencing (53.75%) and 67.5% respondents did mulching. Consequently about 62% respondents practice thinning or pruning in their homestead garden. Large farmers generally hired labour for doing thinning and pruning operations. But they do very much little care for manuring (46.25%) and applying pesticide (35%) in their homesteads.

3.7. Role of Homestead Gardens in Local Biodiversity Conservation. Homestead gardens have long been the most effective and widespread measure for biodiversity conservation in Bangladesh as due to anthropogenic pressure and land use change the natural forest has been decreasing day by day both in explicit and implicit ways leading to threats to future productivity. Generally, rural communities preferred cultivated and planted multipurpose species that can be served as fruits, vegetables and spices also used as timber. Such kind of choice is the most important factor to homestead gardens conservation in Bangladesh and plays a significant role in forest conservation since all the wood and other non-timber forest products that are harvested in the homestead gardens do not need to be collected from forests. Respondents said that homestead gardens attract a number of bird species like *Streptopelia chinensis, Psittacula krameri, Eudynamys scolopaceus, Micropternus brachyurus, Dinopium benghalense, Oriolus xanthornus, Dicrurus macrocercus, Acridotheres tristis, Corvus splendens, Turdus cafer, Orthotomus sutorius, Copsychus saularis, Nectarinia zeylonica, Anthus campestris, Passer domesticus,* and *Ploceus philippinus* to collect their food and making nest. Moreover, some animal species like squirrel, take shelter

TABLE 5: Perceived importance of homestead garden conservation in the study area.

Items	Very important	Less important	Not important
Food and fruit	85	15	0
Firewood	68.75	23.75	7.5
Building materials	78.75	20	1.25
Subsistence family income	73.75	20	6.25
Medicinal plants	28.75	35	36.25
Ecological balance	28.75	37.5	33.75
Soil erosion control	51.25	37.5	11.25
Others*	36.25	17.5	46.25

*Boundary, ornamental, and spiritual purposes, and so forth.

and collect their food, especially fruit like *Aegle marmelos, Annona squamosa, Areca catechu, Averrhoa carambola, Carica papaya, Carissa carandas, Cocos nucifera, Dillenia indica, Elaeocarpus floribundus, Mangifera indica, Phyllanthus acidus, Phyllanthus emblica, Psidium guajava, Spondias pinnata, Syzygium cumini, Tamarindus indica* and *Zizyphus mauritiana* from the urban and rural homestead gardens. They also mentioned that some birds play a significant role as pollinators or in the control of insect pests. At this time dispersal of seeds, also occurs by the animal, birds and helps in natural regeneration of homestead plants species since natural regeneration is the most important factor for tree diversity conservation. Study also found a number of bamboo, shrub, herb, and climber species which were largely used by the households; also, they give shelter to animal diversity.

3.8. Perceived Importance for Conservation of Homestead Garden Species. To determine the perceived importance of homestead species conservation, farmers were interviewed using a questionnaire; asked to evaluate the importance of mentioned eight functions of trees. The results are presented in Table 5. Likewise, farmers' perceived most importance for homestead plant species conservation was related to fruit and food (85%) followed by building materials (78.75%), subsistence family income (73.75%), and source of firewood (68.75%). The surveyed rural area is affected by monsoon flood every year; as a result soil erosion is a serious problem in this region. Therefore, in order to keep houses above the water level, it is mandatory to raise houses at the highest elevations or fill the land by soil in the dry season, especially throughout the floodplain regions. As a consequence, people are usually concerned about the trees role to protect their homestead land against water-induced soil erosion by binding the soil. However, they were not concerned about ecological importance of forest. Yet the majority of the respondents graded the homestead garden as being "less important" as a means of maintaining ecological balance and soil erosion control (37.5%), followed by a source of medicinal plants (35%). So, it seems that there is still a lack of knowledge in these

Status, Diversity, and Traditional Uses of Homestead Gardens in Northern Bangladesh: A Means of
Sustainable Biodiversity Conservation

121

two categories, and institutional and government and NGOs training and learning programs are necessary to facilitate knowledge.

4. Discussion

Analysis of the existing tree composition structure and richness revealed that homestead forest in the study area has moderately high biodiversity and species richness. However, the number of plant species was higher than those found in other homesteads of Bangladesh by Abedin and Quddus [53] cited from Alam and Masum [35] found in Tangail (52 species), Ishurdi (34 species), Jessore (28 species), Patuakhali (20 species), Rajshahi (28 species), and Rangpur (21 species) districts, respectively. Motiur et al. [31] found 60 species in Sylhet Sadar; Motiur et al. [34] found 58 species in Southwest Bangladesh; Kabir and Webb [3] recorded a total of 419 plant species from southwestern Bangladesh. Alam and Masum [35] recorded a total of 101 species and Masum et al. [8] 142 species in an offshore island (Sandwip Island) of Bangladesh. Millat-E-Mustafa [54] identified 92 perennial plant species in one study conducted in different parts of the country.

The traditional production system of homestead garden in the study area is moderate in terms of level of cultural practices for absence of improved management practices and high-quality variety. Farmers depend usually on naturally growing plants on their homestead boundary. Besides, analysis of existing management regime indicated that the growers lack of scientific information, almost all the household owners still followed traditional homestead forest management systems, whereas a little owner adopted modern practices. Increased tree planting in the homesteads and their appropriate management, including intercropping practices, should be the strategy for enhancing tree cover of the homesteads of study area in order to meet basic needs of its people and maintain environmental balance. Homestead gardens are playing a potential role in biodiversity conservation as well as uplifting the socioeconomic condition by contributing families or household's annual income and providing nutritional diet to families. Variable homestead garden products such as seasonal fruits, firewood, medicinal plants, timber, and vegetables and spices were mostly used by the small and medium household owners for their daily needs but large owners get their products into the market for sale. These findings are also supported by the study of Millat-E-Mustafa [54] for the homestead garden of four regions in Bangladesh. Most of the households were found to prefer mostly food or fruit species (45%) because of the income incentives and family needs, and this was also supported by several researchers [8, 21, 54] across the country.

In the present study, homesteads gardens were largely user oriented, and market access was not fully developed. However, market access for homestead products is essential as, they sell their products easily into the market as well as other forest products. It was shown that most of the producers were selling to their neighbours or local traders. Therefore, they do not get proper price for their products. If they get their products to the market or sell products via retailers, they will get proper prices also, which is very much important for the small household owners, therefore creating a scope for income. Many studies of tropical homestead garden have reported reduced species diversity and stem density in homestead garden with closer proximity to market for example [55]. So, market access condition has great effect on homestead forest management; thus, further study is needed to directly test the influence of market access on the homestead gardens structure of both commercial and subsistence-oriented homestead garden in Bangladesh. However, the homestead gardens of the study area present an excellent example of all embracing multipurpose land-use system and biodiversity conservation.

4.1. Species Diversity and Biodiversity Conservation in Traditional Homestead Garden Farming System. Biodiversity conservation has become a growing concern for all over the world, and it is linked up highly with long-term health and vigour of the biosphere, as an indicator of global environment and also as a regulator of ecosystem functioning [56]. The biological diversity indices revealed that homestead garden could play an important role for carbon sequestration in the future since plant growth is directly proportional to the carbon sequestration capacity of the forest [57, 58]. By studying different literatures of homestead gardens all over Bangladesh, we have developed in our mind that a number of opportunities and drawbacks are influencing the selection of species of homestead garden. Considering all these, we have developed a model of species diversity and biodiversity conservation for both urban and rural surveyed homestead gardens represented in Figure 1. We classified the opportunities and drawbacks of a typical homestead garden for both rural and urban areas separately. Among the opportunities, the most prominent according to our observation were generating income, food security, soil erosion control, timber demand, market access, beautification, cattle fodder, medicinal purposes, and fuel wood species. The drawbacks that are influencing the selection of species were cropland expansion, fast growing species demand, natural calamities, land scarcity, domestic animal, low crown canopy, infrastructure, and so forth. The economical condition of the rural people is not as like as urban people. For this, they usually prefer species that will provide them necessary fuel wood and fodder for their cattle. Plantation of medicinal plant species can help to get remedy from diseases; also regular supply of raw materials to the industry could be an important source of earning money to the farmers. Urban people usually prefer to plant various flowering and ornamental plants such as *Tagetes erecta*, *Gomphrena globosa*, and *Codiaeum variegatum* for ornamental beauty of their houses because they occupy a smaller space in their garden premises, and they do not like such condition that will decrease the beautification view of their house. The trend in gradual replacement of functional plants to ornamentals has also been observed in cases where people became richer [59]. Side-by-side grazing of domestic animals disturbs the diversity of homestead forest species. In this circumstance, introduction of grasses, sedges, and small bushes could be a solution to get remedy to this problem.

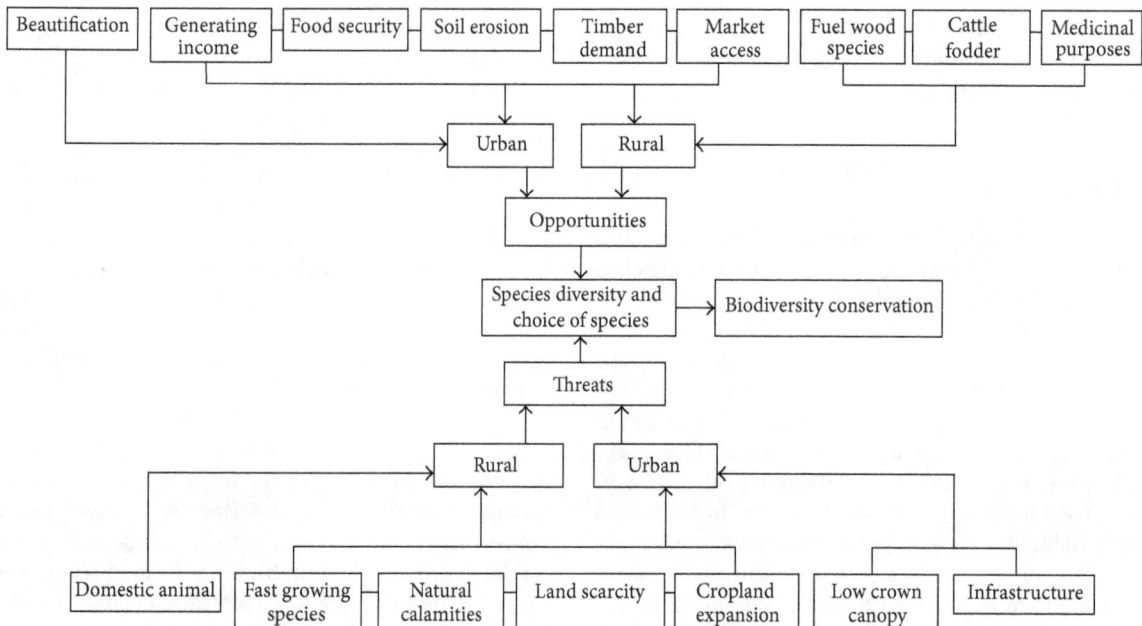

FIGURE 1: Conceptual model of species diversity and biodiversity conservation in homestead garden farming system.

This type of management could help villagers in getting fuel wood without disturbing the main vegetation. Whatsoever, it is now clear that homestead garden is a storehouse of large species diversity and sustainable resource management, and this large species diversity can play an important role in biodiversity conservation.

5. Conclusion

For aesthetic, environmental, and economic perspectives, species planting in homestead garden is desirable. Homestead gardening plays a significant role in both rural and urban landscape planning and management. In this study, we have observed that the homestead plant composition, diversity, and species richness were moderate in Kishoregang Sadar area. Moreover, there was a lack of scientific knowledge of the gardeners, an absence of proper planning, and no specific objectives and goals. Present study did not discuss any economic contribution and market access for homestead; thus, further study is highly recommended and needed to directly test the economic significance and influence of market access on the vegetation structure of both commercial and subsistence-oriented homestead gardens in northern Bangladesh. The moderate domination of fruit species over timber species may be attributed to the gardeners' general perception that fruit species would bring early return as well as the multipurpose nature of fruit species. Homestead garden could provide employment opportunities for both male and female members, resulting in increased family income for better livelihood to a large population in northern Bangladesh. Till now, there is no specific management plan of homestead gardens all over the country although it has tremendous contribution to greening the nature. Considering

the present state of the homestead garden of the study area, this paper suggests that there is a need to establish proper planning and management mechanisms from government for homestead garden. This can be done by providing some incentives and/or training to the owners to be more careful about conserving garden species to improve both rural and urban plant species coverage. It is also recommended that experimentation with new and diversified tree species can play an important role in enhancing the diversity and distribution of homestead garden in the Kishoreganj Sadar area.

Acknowledgments

The authors are highly grateful to each and every respondent who participated in this study for giving their valuable time and information regarding their homestead gardens. They are very much grateful to Mr. Avik Kumar Roy for his consistence support during data collection as well as for giving overall idea about the study area.

References

[1] M. S. Uddin, M. J. Rahman, and M. A. Mannan, "Plant biodiversity in the homesteads of saline area of Southern Bangladesh," in *Proceedings of National Workshop on Agroforestry Research Development of Agroforestry Research in Bangladesh*, M. F. Haq, M. K. Hasan, S. M. Asaduzzaman, and M. Y. Ali, Eds., pp. 45–54, Gazipur, Bangladesh, 2001.

[2] M. Zashimuddin, *Community Forestry for Poverty Reduction in Bangladesh in Forests for Poverty Reduction: Can Community Forestry Make Money?* FAO Regional Office for Asia and the Pacific, Bangkok, Thailand, 2004.

Status, Diversity, and Traditional Uses of Homestead Gardens in Northern Bangladesh: A Means of
Sustainable Biodiversity Conservation

123

[3] M. E. Kabir and E. L. Webb, "Can homegardens conserve biodiversity in Bangladesh?" *Biotropica*, vol. 40, no. 1, pp. 95–103, 2008.

[4] A. J. Southern, *Acquisition of indigenous ecological knowledge about forest gardens in Kandy district, Sri Lanka, [M. Phil. Dissertation]*, University of Wales, Bangor, UK, 1994.

[5] M. B. Uddin and S. A. Mukul, "Improving forest dependent livelihoods through NTFPs and home gardens: a case study from satchari national park," in *Making Conservation Work: Linking Rural Livelihoods & Protected Area Management in Bangladesh*, J. Fox, B. Bushley, S. Dutt, and S. A. Quazi, Eds., pp. 13–35, Nishorgo Program of the Bangladesh Forest Department and East-West Center of University of Hawaii, Dhaka, Bangladesh, 2007.

[6] FAO, "Global forest resource assessment 2010: main report," FAO Forestry Paper 163, Food and Agriculture Organization (FAO) of the United Nations, Rome, Italy, 2010.

[7] M. D. Millat-E-Mustafa, J. B. Hall, and Z. Teklehaimanot, "Structure and floristics of Bangladesh homegardens," *Agroforestry Systems*, vol. 33, no. 3, pp. 263–280, 1996.

[8] K. M. Masum, M. S. Alam, and M. M. Abdullah-Al-Mamun, "Ecological and economical significance of homestead forest to the household of the offshore island in Bangladesh," *Journal of Forestry Research*, vol. 19, no. 4, pp. 307–310, 2008.

[9] M. F. U. Ahmed and S. M. L. Rahman, "Profile and use of multispecies tree crops in the homesteads of Gazipur district, central Bangladesh," *Journal of Sustainable Agriculture*, vol. 24, no. 1, pp. 81–93, 2004.

[10] O. T. Coomes and N. Ban, "Cultivated plant species diversity in home gardens of an amazonian peasant village in northeastern Peru," *Economic Botany*, vol. 58, no. 3, pp. 420–434, 2004.

[11] B. A. Withrow-Robinson and D. E. Hibbs, "Testing an ecologically based classification tool on fruit-based agroforestry in northern Thailand," *Agroforestry Systems*, vol. 65, no. 2, pp. 123–135, 2005.

[12] O. S. Abdoellah, H. Y. Hadikusumah, K. Takeuchi, S. Okubo, and P. Parikesit, "Commercialization of homegardens in an Indonesian village: vegetation composition and functional changes," *Agroforestry Systems*, vol. 68, no. 1, pp. 1–13, 2006.

[13] V. E. Méndez, R. Lok, and E. Somarriba, "Interdisciplinary analysis of homegardens in Nicaragua: micro-zonation, plant use and socioeconomic importance," *Agroforestry Systems*, vol. 51, no. 2, pp. 85–96, 2001.

[14] C. R. Vogl and B. Vogl-Lukasser, "Tradition, dynamics and sustainability of plant species composition and management in homegardens on organic and non-organic small scale farms in Alpine Eastern Tyrol, Austria," *Biological Agriculture and Horticulture*, vol. 21, no. 4, pp. 349–366, 2003.

[15] A. Hemp, "The banana forests of Kilimanjaro: biodiversity and conservation of the Chagga homegardens," *Biodiversity and Conservation*, vol. 15, no. 4, pp. 1193–1217, 2006.

[16] D. K. Das, *List of Bangladesh Village Tree Species*, Forest Research Institute, Chittagong, Bangladesh, 1990.

[17] M. M. Hassan and A. H. Mazumdar, "An exploratory survey of trees on homestead and waste land of Bangladesh," ADAB News, pp. 26–32, 1990.

[18] M. K. Alam and M. Mohiuddin, *Some Potential Multipurpose Trees For Homesteads in Bangladesh*, vol. 2 of *Agroforestry Information Series*, Winrock International, Dhaka, Bangladesh, 1992, Bangladesh Agricultural Research Council (BARC).

[19] M. K. Alam, M. Mohiuddin, and S. R. Basak, "Village trees in Bangladesh: diversity and economic aspects," *Bangladesh Journal of Forest Science*, vol. 25, no. 1–2, pp. 21–36, 1996.

[20] S. A. Khan and M. K. Alam, *Homestead Flora of Bangladesh*, Bangaldesh Agricultural Research Council, International Development Research Cenbtre, Village and Farm Forestry Project (SDC), Dhaka, Bangladesh, 1996.

[21] M. S. Siddiqi and N. A. Khan, "Floristic composition and socioeconomic aspects of rural homestead garden in Chittagong: a case study," *Journal of Forest Science*, vol. 28, no. 2, pp. 94–101, 1999.

[22] A. M. Shajaat Ali, "Homegardens in smallholder farming systems: examples from Bangladesh," *Human Ecology*, vol. 33, no. 2, pp. 245–270, 2005.

[23] M. E. Kabir and E. L. Webb, "Floristics and structure of southwestern Bangladesh homegardens," *The International Journal of Biodiversity Science and Management*, vol. 4, no. 1, pp. 54–64, 2008.

[24] K. U. Ahmad, "Minor fruits in homestead agro forestry," in *Agroforestry Bangladesh Perspective*, M. K. Alam, F. U. Ahmed, and S. M. R. Amin, Eds., pp. 165–169, APAAN; NAWG and BARC, Dhaka, Bangladesh, 1997.

[25] N. M. Islam, *Homestead garden agroforestry in Bangladesh: a case study in Rangpur district [M.S. thesis]*, Agricultural University of Norway, Ås, Norway, 1998.

[26] M. A. Bashar, *Homestead garden Agroforestry: impact on Biodiversity conservation and household food security: a case study of Gajipur district, Bangladesh [M.S. thesis]*, Agricultural University of Norway, Ås, Norway, 1999.

[27] M. S. Alam, M. F. Haque, M. Z. Abedin, and S. Akter, "Homestead trees and household fuel uses in and around the farming systems research site, Jessore," in *Homestead Plantation and Agroforestry in Bangladesh*, M. Z. Abedin, C. K. Lai, and M. O. Ali, Eds., pp. 106–119, BARI, RWEDP and WINROCK, Dhaka, Bangladesh, 1990.

[28] G, Miah, M. Z. Abedin, A. B. M. A. Khair, M. Shahidullah, and A. J. M. A. Baki, "Homestead Plantation and household fuel situation in Ganges floodplain of Bangladesh," in *Homestead Plantation and Agroforestry in Bangladesh*, M. Z. Abedin, C. K. Lai, and M. O. Ali, Eds., pp. 120–135, BARI, Joydebpur, Bangladesh, 1990.

[29] M. A. Momin, M. Z. Abedin, M. R. Amin, Q. M. S. Islam, and M. M. Haque, "Existing homestead plantation and household fuel use pattern in the flood prone tangail region of Bangladesh," in *Homestead Plantation and Agroforestry in Bangladesh*, M. Z. Abedin, C. K. Lai, and M. O. Ali, Eds., pp. 136–145, BARI, Joydebpur, Bangladesh, 1990.

[30] M. Millat-E-Mustafa, Z. Teklehaimanot, and A. K. O. Haruni, "Traditional uses of perennial homestead garden plants in Bangladesh," *Forests Trees and Livelihoods*, vol. 12, no. 4, pp. 235–256, 2002.

[31] R. M. Motiur, J. Tsukamoto, Y. Furukawa, Z. Shibayama, and I. Kawata, "Quantitative stand structure of woody components of homestead forests and its implications on silvicultural management: a case study in Sylhet Sadar, Bangladesh," *Journal of Forest Research*, vol. 10, no. 4, pp. 285–294, 2005.

[32] W. A. Leuschner and K. Khaleque, "Homestead agroforestry in Bangladesh," *Agroforestry Systems*, vol. 5, no. 2, pp. 139–151, 1987.

[33] N. A. Khan, "Social forestry versus social reality: patronage and community-based forestry in Bangladesh," Gatekeeper Series 99, International Institute for Environment and Development (IIED), London, UK, 2001.

[34] R. M. Motiur, Y. Furukawa, I. Kawata, M. M. Rahman, and M. Alam, "Role of homestead forests in household economy and factors affecting forest production: a case study in southwest Bangladesh," *Journal of Forest Research*, vol. 11, no. 2, pp. 89–97, 2006.

[35] M. S. Alam and K. M. Masum, "Status of homestead biodiversity in the offshore Island of Bangladesh," *Research Journal of Agriculture and Biological Sciences*, vol. 1, no. 3, pp. 246–253, 2005.

[36] S. Akhter, M. Alamgir, M. S. I. Sohel, M. P. Rana, S. J. Monjurul Ahmed, and M. S. H. Chowdhury, "The role of women in traditional farming systems as practiced in homegardens: a case study in Sylhet Sadar Upazila, Bangladesh," *Tropical Conservation Science*, vol. 3, no. 1, pp. 17–30, 2010.

[37] H. M. F. Rahman, "Kishoreganj Sadar Upozila," in *Banglapedia: National Encyclopedia of Bangladesh*, http://www.banglapedia.org/.

[38] K. U. Siddique, M. A. Islam, Z. U. Ahmed et al., Eds., *Encyclopaedia of Flora and Fauna of Bangladesh, Vol. 11, Angiosperms: Monocotyledons (Agavaceae-Najadaceae)*, Asiatic Society of Bangladesh, Dhaka, Bangladesh, 2007.

[39] K. U. Siddique, M. A. Islam, Z. U. Ahmed et al., Eds., *Encyclopaedia of Flora and Fauna of Bangladesh, Vol. 5, Bryophytes, Pteridophytes, Gymnosperm*, Asiatic Society of Bangladesh, Dhaka, Bangladesh, 2008.

[40] Z. U. Ahmed, M. A. Hassan, Z. N. T. Begum et al., Eds., *Encyclopaedia of Flora and Fauna of Bangladesh, Vol. 6. Angiosperms: Dicotyledons (Acanthaceae-Asteraceae)*, Asiatic Society of Bangladesh, Dhaka, Bangladesh, 2008.

[41] Z. U. Ahmed, M. A. Hassan, Z. N. T. Begum et al., Eds., *Encyclopaedia of Flora and Fauna of Bangladesh, Vol. 7. Angiosperms: Dicotyledons (Balsaminaceae-Euphorbiaceae)*, Asiatic Society of Bangladesh, Dhaka, Bangladesh, 2008.

[42] Z. U. Ahmed, M. A. Hassan, Z. N. T. Begum et al., Eds., *Encyclopaedia of Flora and Fauna of Bangladesh, Vol. 12. Angiosperms: Monocotyledons (Orchidaceae-Zingiberaceae)*, Asiatic Society of Bangladesh, Dhaka, Bangladesh, 2008.

[43] Z. U. Ahmed, M. A. Hassan, Z. N. T. Begum et al., Eds., *Encyclopaedia of Flora and Fauna of Bangladesh, Vol. 8. Angiosperms: Dicotyledons (Fabaceae-Lythraceae)*, Asiatic Society of Bangladesh, Dhaka, Bangladesh, 2009.

[44] Z. U. Ahmed, M. A. Hassan, Z. N. T. Begum et al., Eds., *Encyclopaedia of Flora and Fauna of Bangladesh, Vol. 9. Angiosperms: Dicotyledons (Magnoliaceae-Ponicaceae)*, Asiatic Society of Bangladesh, Dhaka, Bangladesh, 2009.

[45] Z. U. Ahmed, M. A. Hassan, Z. N. T. Begum et al., Eds., *Encyclopaedia of Flora and Fauna of Bangladesh, Vol. 10. Angiosperms: Dicotyledons (Ranunculaceae-Zygophyllaceae)*, Asiatic Society of Bangladesh, Dhaka, Bangladesh, 2010.

[46] E. P. Odum, *Fundamentals of Ecology*, WB Saunders, Philadelphia, Pa, USA, 1971.

[47] R. Margalef, "Information theory in ecology," *General Systems Yearbook*, vol. 3, pp. 36–71, 1958.

[48] P. Michael, *Ecological Methods For Field and Laboratory Investigation*, McGraw-Hill, New Delhi, India, 1990.

[49] M. Kent and P. Coker, *Vegetation Description and Analysis: A Practical Approach*, WB Saunders, Philadelphia, Pa, USA, 1992.

[50] E. C. Pielou, "Species-diversity and pattern-diversity in the study of ecological succession," *Journal of Theoretical Biology*, vol. 10, no. 2, pp. 370–383, 1966.

[51] A. E. Magurran, *Ecological Diversity and Measurement*, Princeton University Press, Princeton, NJ, USA, 1988.

[52] S. A. Mori, B. M. Boom, A. M. Carvalino, and D. Santos, "The ecological importance of Myrtaceae in eastern Brazilian wet forest," *Biotropica*, vol. 15, pp. 68–70, 1983.

[53] M. Z. Abedin and M. A. Quddus, "Household fuel situation, homestead gardens and agroforestry practice at six agroecologically different locations of Bangladesh," in *Homestead Plantation and Agroforestry in Bangladesh*, M. Z. Abedin, C. K. Lai, and M. O. Ali, Eds., pp. 19–53, Bangladesh Agriculture Research Institute (BARI), Joydebpur, Bangladesh, 1990.

[54] M. Millat-E-Mustafa, "Tropical Homestead gardens: an overview," in *Agroforestry: Bangladesh Perspective*, M. K. Alam, F. U. Ahmed, and S. M. Amin, Eds., pp. 18–133, APAN; NAWG; BAEC, Dhaka, Bangladesh, 1997.

[55] T. Abebe, *Diversity in homegarden agroforestry systems of southern Ethiopia [Ph.D. Dissertation]*, Wageningen University, Wageningen, The Netherlands, 2005.

[56] O. T. Solbrig, "The origin and function of biodiversity," *Environment*, vol. 33, no. 5, pp. 16–38, 1991.

[57] P. Kumar, "Carbon sequestration strategy of Nubra Valley with special reference to agroforestry," DRDO Technology Spectrum, pp. 187–192, 2008.

[58] P. Kumar, S. Gupta, and S. Prakash, "Carbon pool of orchards in siachen sector: socio-economic Carbon sequestration," in *Advances in Agriculture Environment and Health*, S. B. Singh, O. P. Charassia, and S. Yadav, Eds., pp. 225–233, 2008.

[59] L. Christanty, O. S. Abdoellah, G. G. Marten, and J. Iskander, "Traditional agroforestry in West Java: the pekarangan (Homestead garden) and kebun-talun (annual-perennial rotation) cropping systems," in *Traditional Agriculture in Southeast Asia: A Human Ecology Perspective*, G. G. Marten, Ed., pp. 132–158, Westview Press, Boulder, Colo, USA, 1986.

10

Ecological and Economic Importance of Bats (Order Chiroptera)

Mohammed Kasso and Mundanthra Balakrishnan

Department of Zoological Sciences, Addis Ababa University, P.O. Box 1176, Addis Ababa, Ethiopia

Correspondence should be addressed to Mohammed Kasso; muhesofi@yahoo.com

Academic Editors: P. K. S. Shin and P. M. Vergara

Order Chiroptera is the second most diverse and abundant order of mammals with great physiological and ecological diversity. They play important ecological roles as prey and predator, arthropod suppression, seed dispersal, pollination, material and nutrient distribution, and recycle. They have great advantage and disadvantage in economic terms. The economic benefits obtained from bats include biological pest control, plant pollination, seed dispersal, guano mining, bush meat and medicine, aesthetic and bat watching tourism, and education and research. Even though bats are among gentle animals providing many positive ecological and economic benefits, few species have negative effects. They cause damage on human, livestock, agricultural crops, building, and infrastructure. They also cause airplane strike, disease transmission, and contamination, and bite humans during self-defense. Bat populations appear to be declining presumably in response to human induced environmental stresses like habitat destruction and fragmentation, disturbance to caves, depletion of food resources, overhunting for bush meat and persecution, increased use of pesticides, infectious disease, and wind energy turbine. As bats are among the most overlooked in spite of their economical and ecological importance, their conservation is mandatory.

1. Introduction

The order Chiroptera is the second most diverse among mammalian orders, which exhibits great physiological and ecological diversity [1]. They form one of the largest nonhuman aggregations and the most abundant groups of mammals when measured in numbers of individuals [2]. They evolved before 52 million years ago and diversified into more than 1,232 extant species [3]. They are small, with adult masses ranging from 2 g to 1 kg; although most living bats weigh less than 50 g as adults [4]. They have evolved into an incredibly rich diversity of roosting and feeding habits. Many species of bats roost during the day time in foliage, caves, rock crevices, hollows of trees, beneath exfoliating bark, and different man-made structures [2]. During night, they become active and forage on diverse food items like insects, nectar, fruits, seeds, frogs, fish, small mammals, and even blood [3].

The forelimb of a bat is modified into a wing with elongated finger bones joined together by a thin and large (85% of the total body surface area) membrane with rich blood flow [5]. Their wing is an unusual structure in mammals enabling for active unique powered flight. Skin covering the wings of bats not only constitutes a load-bearing area that enables flying but also performs multiple functions like providing a protective barrier against microbes and parasites, gas exchange, thermoregulation, water control, trapping of insects, and food manipulation and for swimming [6]. The powerful flight of bats plays the most important role for their widespread distribution and diversity. This helps in the occurrence of bats in all continents except Antarctica, some Polar Regions, and some isolated oceanic islands. It has also contributed a lot for their extraordinary feeding and roosting habits, reproductive strategies, and social behaviors [2].

Although all bats do not echolocate, in general echolocation is considered as one of the major characteristics of bats. Even if the role of echolocation for plant-visiting bats is not clear, they use wide range (10–200 kHz) of ultrasonic frequencies during foraging. The availability of commercially produced bat detectors contributed a lot in linking data of echolocation with the biology of bats [3].

Bats are an essential natural resource that play great role in providing many ecological and economic services [7]. However, the determination of the ecological and economic values provided by bats is extremely challenging except from the studies on ecosystem services provided directly to the production of goods and services consumed by humans [3, 7].

2. Ecological Importance of Bats

Bats have long been postulated to play important ecological roles in prey and predator, arthropod suppression, seed dispersal, pollination, material and nutrient distribution, and recycle [3].

2.1. As Predators. Bats have diverse patterns of feeding in which some select among available prey while others are generalist predators, feeding on a wide diversity of taxonomic groups. They also opportunistically consume appropriately sized prey depending on availability within a preferred habitat [8]. Their prey size can vary from 1 mm (midges and mosquitoes) to as large as 50 mm long (beetles and large moths) based on the species of bat [8, 9].

Remains of 12 orders or classes of prey belonging to 18 taxonomic families of insects were reported in the diet of bats [10]. The prey items include Acari, Arachnida, Coleoptera, Diptera, Hemiptera, Homoptera, Hymenoptera, Isoptera, Lepidoptera, Neuroptera, Orthoptera, and Trichoptera. They also predate on frogs, fish, small mammals, and even blood of mammals and birds [3]. Some species also eat unusual prey items such as scorpions and spiders [11]. Bats exhibit high species diversity with multiple species forage sympatrically to avoid competition. A resource partition is possible through the use of diverse mechanism like difference in wing shape, body size, and sensory cues [12].

Obtaining accurate estimates of the amount of prey consumed by bats is challenging. However, its amount and type are confirmed as it varies with prey availability, time during night, species, sex, age, and the reproductive status of bats [13, 14]. Variety of approaches like direct observation [15], comparison of pre- and postflight body mass [14], and fecal sample analysis [16] have been used to estimate the amount and type of prey consumed by bats. Results of studies carried out on insectivorous bats indicated that they consume more than 25% of their body mass of insects each night [17]. At the peak night of lactation, a 7.9 g little brown bat (*Myotis lucifugus*) needs to consume 9.9 g of insects which is over 100% of its body mass [18]. At peak lactation, a female Brazilian free-tailed bat (*Tadarida brasiliensis*) consumes insects up to 70% of the body mass each night. It frequently selects nutrient-rich abdomen of moths while discarding the wings, head, and appendages, which greatly increases feeding efficiency and the quantity of insects consumed [14]. This can indicate that maternity colony of one million Brazilian free-tailed bats weighing 12 g each could prey up to 8.4 metric tons of insects in a single night. These studies hint at the immense capability of insect consumption and the potential role of bats in the suppression of arthropod populations [3]. Based on fecal sample analyses, a colony of 300 evening bats (*Nycticeius humeralis*) and 150 big brown bats in Indiana was estimated to consume 6.3 and 1.3 million insects per year, respectively [16, 19].

In this way, an estimated 99% of potential crop pests are limited by natural ecosystems of which some fraction can be attributed to predation by bats [7]. Predation of bats can have direct effects on herbivore communities and indirect effects on plant communities through both density mediated (consumption) and trait-mediated (behavioral) interactions and for nature balance [20].

2.2. Prey for Vertebrates. Although there are relatively few observations of animals feeding on bats, a number of vertebrate predators like fish, amphibians, birds, reptiles, and mammals prey on bats throughout the world [21, 22]. The main bat predators are owls, hawks, falcons, snakes, and mammals such as raccoons, ringtails, and opossums. In some countries like New Zealand, forest-floor dweller bats are frequently predated by the introduced rats, feral cats, and weasel [23]. The larger phyllostomid bats (*Vampyrum spectrum*, *Chrotopterus auritus*, and *Phyllostomus hastatus*) are known to eat smaller bats [24].

Bats generally comprise a relatively small proportion of the diet of most predators. Bats represented only 0.003% of the diet of small falcons and hawks and 0.036% of the diet of owls in Great Britain [21]. Although diurnal raptors feed on bats during twilight hours in some parts of the world [25], nocturnal predation by owls is the most significant predation pressure on bats in temperate regions [21].

Most of the bats are predated on roosting or when they emerge from roosts although sometimes predated during foraging or flying. Large concentrations of bats at roost sites and the relatively predictable patterns of their emergence from roosts, provide significant opportunities for predators to prey on bats [25]. However, strategies like low dependability to roost sites, selection of time, and patterns of emergence from roosts and nocturnal activity are used to minimize the risk of predation [21].

2.3. Hosts for Parasite. Numerous haematophagous ectoparasites live such as bat fleas (Ischnopsyllidae), bat flies (Nycteribiidae), bat mites (Spinturnicidae), and bugs (Cimicidae) on the skin surface and in the fur of bats. These obligate ectoparasites are specialized to their hosts [26]. The skin and hair morphology play important roles in affecting the parasite's life style in terms of adaptation, feeding, movement and egg laying resulted in morphological adaptations with coevolution of both species [5].

The hair density as well as surface structures of bat hairs and the distribution of mast cells are very important for the host defense against parasite infestation. Although the hair density of bats primarily provides protection against unfavorable microclimatic conditions, it also serves as passive antiparasitic defense. The high hair density prohibits infestation by large parasites. However, dense fur in some parts of the host's body may provide a suitable shelter for specialized small parasites [5].

2.4. Pollination. In addition to insect suppression through predation, some bat species primarily the two families of bats (Pteropodidae in the Old World and Phyllostomidae in the New World) play important roles in plant pollination [3]. Although bat pollination is relatively uncommon when compared with bird or insect pollination, it involves an impressive number of economically and ecologically important plants [27]. Particularly, beyond the economic value of plant pollination and seed dispersal services, plant-visiting bats provide important ecological services by facilitating the reproductive success and the recruitment of new seedlings [3]. Many of these plants are among the most important species in terms of biomass in their habitats. For instance, bat-pollinated columnar cacti and agaves are dominant vegetation elements in arid and semiarid habitats of the New World [3].

Bat pollination occurs in more than 528 species of 67 families and 28 orders of angiosperms worldwide [28]. Pteropodid bats are known to pollinate flowers of about 168 species of 100 genera and 41 families and phyllostomid bats pollinate flowers of about 360 species of 159 genera and 44 families [28]. As feeding on nectar and pollen requires relatively specialized morphology (e.g., elongated snout and tongue), relatively few members of these families are obligate pollinators. Unlike predation, which is an antagonistic population interaction, pollination, and seed dispersal are mutualistic population interactions in which plants provide a nutritional reward (nectar, pollen, and fruit pulp) for a beneficial service [3].

2.5. Seed Dispersal.
Seed dispersal is a major way in which animals contribute for ecosystem succession by depositing seeds from one area to another [29]. As 50–90% of tropical trees and shrubs produce fleshy fruits adapted for consumption by vertebrates, the role played by frugivorous bats in dispersing these seeds is tremendous [30].

Countless tropical trees and understory shrubs are adapted for seed dispersal by animals, primarily by bats and birds. Particularly, night-foraging fruit bats are more compliant than birds by covering long distances each night, defecating in flight, and scattering far more seeds across cleared areas [31]. Unlike most seed dispersal by vertebrates that dispersed close to parent plants with only 100–1,000 m away, the seeds dispersed by frugivorous bats were relatively far away (1-2 km) [31]. Furthermore, the flying fox migration for more than 1,000 km across the central belt of the African continent helps to scatter huge numbers of seeds along the way. Unlike birds, bats tend to defecate or spit out seeds during flight and hence facilitate seed dispersal in clear-cut strips [32]. In addition to their tendency to defecate seeds in flight, many bats use one or more feeding roosts each night where they deposit the vast majority of seeds ingested far away from fruiting plants.

Many bat-dispersed seeds are from hardy pioneer plants, the first to grow in the hot, dry conditions of clearings with up to 95% chance of germination. As these plants grow, they provide shelter that helps other, more delicate plants to grow [33]. Fruit-eating bats play an extremely important role in forest regeneration. Tropical frugivorous bats also facilitate tropical forest regeneration and help to maintain species diversity by introducing seeds from outside disturbed areas,

whereas the neotropics frugivorous bats play important role in the early stages of forest succession [32].

The dispersed seeds of palms and figs by bats are also common in many tropical forests. Because they are also eaten by many birds and mammals, figs often act as keystone species in tropical forests [34].

2.6. Soil Fertility and Nutrient Distribution.
Bats play an important ecological role in soil fertility and nutrient distribution due to their relatively high mobility and the use of different habitats for roosting and foraging, which facilitates nutrient transfer within ecosystems [35, 36]. However, the suspected importance of nutrient transfer by bats in overall ecosystem function is probably relatively low when compared with microhabitat conditions [36]. For soil fertility and nutrient distribution, bat guano has a great ecological potential as bats sprinkle it over the landscape throughout the night. Thus, bats contribute a lot in nutrient redistribution, from nutrient-rich sources (e.g., lakes and rivers) to nutrient-poor regions (e.g., arid or upland landscapes) [35]. For instance, a colony of one million Brazilian free-tailed bats (T. brasiliensis) in Texas can contribute to 22 kg of nitrogen in the form of guano.

Bat guano in turn supports a great diversity of organisms including arthropods, fungi, bacteria, and lichens that represent different trophic levels [37]. The diversity of organisms living on guano differs depending on the species and their diet. For example, guano from insectivorous bats is typically inhabited by mites, pseudoscorpions, beetles, thrips, moths, and flies, whereas the guano of frugivorous bats is inhabited by spiders, mites, isopods, millipedes, centipedes, springtails, barklice, true bugs, and beetles [38]. As bats regularly or occasionally roost in caves, bat guano provides the primary organic input to cave ecosystems, which are inherently devoid of primary productivity. They provide essential organic input that supports assemblages of different endemic cave flora and fauna. For example, cave-dwelling salamander and fish populations and invertebrate communities are also highly dependent upon nutrients from bat guano. However, little consideration has been given to the role of bats in supporting entire cave ecosystems [39].

2.7. Bioindicators.
The earth is now subject to climate change and habitat deterioration on a large scale. Monitoring of climate change and habitat loss alone is insufficient to understand the effects of these factors on complex biological communities [40]. Ecosystems are geographically variable and inherently complex whereas responses to anthropogenic changes are in a nonlinear and scale dependent manner. Thus, a broad-scale network of monitoring that captures local, regional, and global components of the earth's biota is critical for understanding and forecasting responses to climate change and habitat conversion [2]. It is therefore important to identify bioindicator taxa that show measurable responses to climate change and habitat loss and that reflect wider-scale impacts on biodiversity [2].

There are three types of bioindicators (biodiversity, ecological, and environmental indicators) [41]. Biodiversity indicators capture responses of a range of taxa and reflect components of biological diversity such as species richness and species diversity. Ecological indicators consist of taxa or assemblages that are sensitive to identified environmental stress factors that demonstrate the effect of those stress factors on biota. Environmental indicators respond in predictable ways to specific environmental disturbances [41].

Biodiversity indicator species have characteristics that can be used as an index of attributes (e.g., presence/absence, population density, and relative abundance) of other species comprising the biota of interest [42]. Thus, these species collectively must have characteristics that make them easily identifiable (stable taxonomy), easy to sample, and show graded responses to habitat degradation that correlate with the responses of other taxa [43]. In addition, as environmental degradation can occur over a variety of scales, monitoring the impacts of such threats through indicator species requires the species that have broad geographic ranges. Bats, as volant taxa, fulfill this criterion better than most other taxa [2].

Bats are excellent ecological indicators of habitat quality. They have enormous potential as bioindicators to both disturbance and the existence of contaminants due to a combination of their size, mobility, longevity, taxonomic stability, observable short and long term effects, trends of populations, and their distribution around the globe [2, 4, 44].

Bat populations are affected by a wide range of stressors that affect many other taxa. In particular, changes in bat numbers or activity can be related to climate change (including extremes of drought, heat, cold, precipitation, cyclone, and sea level rise), deterioration of water quality, agricultural intensification, loss and fragmentation of habitats, fatalities at wind turbines, disease, pesticide use, and overhunting [2]. The magnitude of changes around the globe is quite variable as is the nature of the human activities that alter and fragment landscapes differs from one place to another [45]. As insectivorous bats occupy high trophic levels, they are sensitive to accumulations of pesticides and other toxins, and changes in their abundance may reflect changes in populations of arthropod prey species [1]. High fatalities observed in bats associated with diseases, may provide an early warning of environmental links among contamination, disease prevalence, and mortality. Increased environmental stress can suppress the immune systems of bats and other animals and thus one might predict that the increased prevalence of diseases is a consequence of altered environments [2].

3. Economic Importance of Bats

3.1. Biological Pest Control.
Among the estimated 1,232 extant bat species, over two-thirds are either obligate or facultative insectivorous mammals. They consume nocturnal and crepuscular species of insects from different habitats as such forests, grasslands, agricultural landscapes, aquatic, and wetland habitats [3].

Various species of prominent insect pests have been found in the diet of bats based on identification of insect fragments in fecal samples and stomach contents. They consume enormous quantities of insect pests that cost farmers and foresters billions of dollars annually [46]. These insects include, June beetles (Scarabidae), click beetles (Elateridae), leafhoppers (Cicadelidae), plant hoppers (Delphacidae), the spotted cucumber beetle (Chrysomelidae), the Asiatic oak weevil (Curculionidae), and the green stinkbug (Pentatomidae) [3].

Mexican free-tailed bats (*T. brasiliensis*) feed an estimated one million kilogram of the most costly agricultural pest insects (corn earworm moth) each night [47]. One bat can eat 20 female corn earworm moths in a night and each moth can lay as many as 500 eggs, potentially producing 10,000 crop-damaging caterpillars [46]. About 150 big brown bats also consume enough adult cucumber beetles in one summer to prevent egg-laying that could produce 33 million rootworm larvae and contributing in prevention of agricultural pests damage [16]. Thus, the death of one million bats from the disease called white nose syndrome indicates 660–1,320 metric tons of insects are no longer being consumed each year in affected areas [36]. Millions of Brazilian free-tailed bats each evening consume a wide variety of prey items (12 orders, 35 families) of about 14,000 kg agricultural pests [48, 49]. Based on the dietary composition (minimum number of the total insects per guano pellet), number of specific agricultural pest species in each pellet, and the number of active foraging days per year, a colony of 150 big brown bats (*Eptesicus fuscus*) in the midwestern United States annually consume approximately 600,000 cucumber beetles, 194,000 June beetles, 158,000 leafhoppers, and 335,000 stinkbugs, which are severe crop pests [16].

Bats are just one of several groups of animals that naturally prey upon mosquitoes. A Florida colony of 30,000 southeastern myotis (*Myotis austroriparius*) eats 50 tons of insects annually, including more than 15 tons of mosquitoes [8]. It is also known that northern long-eared bats (*Myotis septentrionalis*) suppress mosquito populations through direct predation [50].

The estimation of the economic importance of bats in agricultural systems is challenging [36]. A common challenge in the study of the use of bats as pest control is the lack of basic ecological information regarding foraging behavior and diet for many species of bats. For example, traditional dietary analyses through fecal or stomach contents have only identified arthropod fragments to the ordinal or familial level, rather than to species [9, 17] and in cases where species identification is possible, it has typically been restricted to hard-bodied insects although recent novel molecular techniques have allowed detection and species identification of both hard and soft bodied insects [51, 52]. However, the value of pest suppression services provided by bats ranges from $12 to $173 per 0.405 ha in Texas [48]. In USA, the estimate value of bats as a result of reduced costs of pesticide applications due to insect pest suppression by bat predation is in the range of $3.7–$53 billion per year excluding the costs of impacts of pesticides on ecosystems [36].

3.2. Pollination.
As pollinators, tropical bats provide invaluable support to many local and national economies [33]. Large-scale cash crops that are originally pollinated or dispersed by bats include wild bananas, mangos, breadfruits,

agave, durians, and petai of which durians and petai currently rely on bats for pollination [7]. Durian, a wildly popular fruit worth more than $230 million per year in southeast Asia, opens its flower at dusk and relies almost exclusively on fruit bats for pollination [7].

Except the "ornamental" bananas with upright flowers that are pollinated by birds, all the rest, including the ancestors of edible bananas, that have horizontal or drooping flowers are pollinated primarily by bats [33]. Their adaptations for bat pollination include nocturnal flowering, a strong and characteristic odor that attracts bats, plus abundant and accessible nectar and pollen. The coevolution of bananas and bats over 50 million years also resulted in adaptations for effective seed dispersal even if other mammals like monkeys feed on fruits and disperse seeds [33]. Although bats are no longer needed to pollinate flowers or disperse the seeds of edible bananas, the ecological services bats provide for their wild relatives are important for conserving its genetic diversity [3].

Agave macroacantha is extremely dependent on nocturnal pollinators for its reproductive success of which bats are especially important for its successful pollination [53]. Some of these pollinators (bats) are migratory, and have been reported to be steadily declining. A continuing decline in the populations of pollinators may hamper the successful sexual reproduction of the plant host and may put its survival under risk [53].

The Mahwa tree or honey tree (*Madhuca indica*) is pollinated by bats. These pollination services highlight one of the highly valued ecosystem services provided by plant-visiting bats both culturally and economically. The timber of this tree is used for making farm cart wheels in India. The flowers are used as food and for preparing a distilled spirit and its sun-dried fruits for human consumption and the oil extracted from flowers and seeds as ingredients for soaps, candles, cosmetics, lubricants, and medicines [54].

Similarly, there are 289 Old World tropical plant species that rely on pollination and seed dispersal services by bats for their propagation [7]. These plants, in turn, contribute to the production of 448 bat-dependent products in a variety of categories such as timber and other wood products (23%); food, drinks and fresh fruit (19%); medicine (15%); dye, fiber, animal fodder, fuel wood, ornamental plants, and others (43%). However, because bat-provided services represent one input within a multi-input production process, only a portion of the total value of the end product can be attributed to bats [7].

The pollination services of bats for 100 food crops by combining the pollination dependence ratios with regional crop production and its prices was determined [55]. Of these, 46 crops depended to some degree on animal pollinators (6 essentially dependent, 13 highly dependent, 13 moderately dependent, and 14 slightly dependent) accounting for 39% of world production value.

Based on the crop production and animal-dependent pollination, the total economic value of bats in global pollination services is estimated to be $200 billion, representing 9.5% of the value of world food crop production in 2005 [55].

3.3. Seed Dispersal. Bats are crucial to the survival of the world's tropical forests. Enormous expanses of rain forest are cleared every year for logging, agriculture, ranching, and other uses. Fruit-eating bats are uniquely suited for dispersing the seeds of "pioneer plants" from which a diverse and healthy forest can reemerge [33]. Thus, the economic value contributed by bats in maintaining forests is tremendous. For instance, the economic value estimate for seed dispersal services provided by bats to the regeneration of giant oak is $212,000 for seeding acorns and $945,000 for planting saplings [56]. The tropical almond tree, *Terminalia catappa*, is one of the bat-dispersed trees with many human uses like shade, fuel-wood, edible nuts, timber, and tannin (extracted from the bark, leaves, roots, and the fruit shell). The large leaves are also used as wrapping material and have also many medicinal uses, including diaphoretic, anti-indigestion, antidysentery and headache [33].

3.4. Guano Mining. Guano from bats has long been mined from caves for use as fertilizer on agricultural crops due to its high concentrations of limiting nutrients like nitrogen and phosphorous [57]. It provides some of the world's finest natural fertilizers [58]. About 950 bat guano products show a market demand for the product. Prices for bat guano organic fertilizer varied from $1.25 to $12.00 per 0.5 kg depending on the size of the package (larger packages have lower unit prices) and the mix of its ingredients [3]. The Mexican free-tailed bat guano has been extracted for fertilizer in thousands of tons from Bracken Cave in Texas alone with the current retail sales ranging from $2.86 to $12.10 per kilogram [58]. In some places, guano harvesting is carried out on a sustainable basis, especially in caves where bats normally migrate elsewhere for a part of their life each year. The bacteria extracted from bat guano have also been used by some companies to improve detergents and other products of great value to humans [58].

3.5. Bush Meat and Medicine. Bats have also long been used for food and medicine [59]. They provide a direct source of human food in many countries [60]. Several anecdotal price information of bat bush meat ranges from $2.50 to $3.50 per bat in Malaysia and $0.43–$10 per bat in Jakarta. Bat bush meat has the highest nutrient (high protein, vitamin and mineral composition) with lowest cost per kg [7, 61, 62]. Several studies have reported on the overhunting of bat for bush meat indicating a need for further conservation [61].

The anticoagulant compound called salivary plasminogen activator (DSPA) found in the saliva of the common vampire bat is used to treat strokes. Unlike alternative medicines, it can be administered even much later after a stroke has occurred and still be effective [63]. Physicians used bats to treat ailments of patients ranging from baldness to paralysis [60, 63].

3.6. Aesthetic and Bat Watching Tourism. Wildlife watching is simply an activity that involves watching wildlife to identify and observe their behavior and appreciate their beauty. It differs from other forms of wildlife-based activities like hunting and fishing [49]. Although perhaps not as widely practiced as bird watching, bat watching is currently growing as a recreational activity [49]. Similar to other wildlife watching tourism, it also generates income in the form of entrance and permit fees, personal payments to the guides, drivers and scouts and payment for accommodation, and other services [49].

The majority of bat watching takes place at cave entrances where bats emergence can be viewed. For this purpose, the charge ranges from $5 to $12 per visitor. For instance, the Congress Avenue Bridge, which is the home to the largest urban bat colony of approximately 1.5 million Mexican free-tailed bats (*T. brasiliensis*) in USA is visited by 200–1,500 visitors per evening with the value of $3 million per year. The spectacular flock emergence of bats from their roost from March to November, to feed and migrate south during the winter months serves as tourist attraction [49, 64].

3.7. Education and Research. Although extremely difficult to quantify, it is important to recognize the extraordinary value of bats to ancient and contemporary traditions and science. The current study of bat echolocation and locomotion has provided inspiration for novel technological advances in biomedical ultrasound, sensors for autonomous systems, and wireless communication and BATMAVs (bat-like motorized aerial vehicles) [65, 66].

Bats contributed a lot to the field of biomimetics, which is the science of modeling cutting-edge technologies based on natural forms [65]. The anticlotting chemicals in the saliva of bats are also currently being investigated as potential anticoagulant for people who are at high risk of blood clots and strokes. In addition, the development of sonar for ships and ultrasound was partly inspired from echolocation that bats use as navigation system to find and follow their prey at night without crashing on trees, buildings, or other obstructions [65, 66].

Particularly, a unique feature of bats that provides potential for future application is their flying ability by their own power. The aerodynamic range of bats includes changing flight direction by turning 180 degrees within just three wing beats while flying at full tilt. They are such quick flyers because of the quickness of their wings that are structured to fold during flight, similar to the way that a human hand folds. Also, their wings are draped by stretchy skin and are powered by special muscles. Ongoing research about the structure of bat wings and the mechanics of bat flight may ultimately lead to the development of technologies that improve the maneuver ability of airplanes [22].

3.8. Bats as Pests. Although bats are grouped among the world's gentlest animals that provide many positive ecological and economic benefits, few of them are considered as pests. They may cause damage on human, livestock, agricultural crops, airplane strike, building, and infrastructure infestation, and rarely become aggressive or bite humans during self-defense [58]. For instance, frugivorous bats that feed on some economically important fruits result in greater loss.

Three species of vampire bats that occur in the New World are major pests feeding mainly on the blood of livestock (cattle, equines, goats, sheep, and pigs), poultry and occasionally humans. They are also responsible in transmitting rabies. Populations of vampire bats have increased sharply in areas of Latin America where European livestock have been introduced [67]. Wounds caused by vampire bats may also be vulnerable to secondary infections [1].

Bat strikes to airplane have been responsible for loss of human lives and damage to materials worldwide resulting in loss of billions of dollars annually [68, 69]. For example, 821 bat strikes were reported in the USA Air Force during 1997–2007 [70] and 327 from 91 airports during 1996–2006 in Australia [71]. From less than 1% of the bat-strike reported in USA, a cumulative damage is more than $825,000 of which more than half is attributable to 5 bat-strike incidents [70]. This high damage is accredited to high body weight (up to 1 kg like flying foxes) and unlike birds, they possess none pneumatized solid and heavy bones. It results in a greater and more concentrated impact force of strike and a greater capacity to perforate an aircraft than bird strikes [69]. Australian flying foxes roost gregariously and emerge from roosts in flocks, which may include thousands of flying foxes, thus the increasing risk of multiple simultaneous strikes. Their major damage to aircraft includes breaking of windscreen, perforation of aircraft skin, and ingestion into engine [71].

Building and house infestation by bat constitutes a serious public health problem [72]. They spoil food and make ceilings, walls, and floors dirty with the accumulation of guano and urine [72]. Besides, they cause discomfort to humans by their distressing noise, offensive odors, and attraction of coprophagous insects. Potential health hazards may result from chitinous remains of finely chewed insects in guano, attack of ectoparasites, drinking water contamination by urine and feces [72].

3.9. Disease Transition and Contamination. Bats are hosts to a range of zoonotic and potentially zoonotic pathogens. They differ from other disease reservoirs because of their unique and diverse lifestyles, including their ability to fly, often highly gregarious social structures, long life spans, and low fecundity rates [73]. They represent a potential epidemiologic of several diseases that can be fatal to humans, including rabies, Ebola, leptospirosis, histoplasmosis, and pseudotuberculosis [72, 74].

Bats are reservoirs of several pathogens, whose spread may be related to physiological stress associated with habitat loss or alteration [75]. The recent die-offs of bats presenting with white nose syndrome may relate to increased levels of environmental stress that render them to be susceptible to fungal infection and viral infections like Henipaviruses, European bat lyssaviruses, rabies, and Ebola virus [74].

Human activities that increase exposure to bats will likely increase the opportunity for infections [73]. Like bird droppings, bat guano can contain a potentially infectious fungus *Histoplasma capsulatum* that causes lung infection known as histoplasmosis [58].

Bat populations appear to be declining presumably in response to human induced environmental stresses like habitat destruction and fragmentation, disturbance to caves, depletion of food resources, overhunting for bush meat and persecution, increased use of pesticides, infectious disease, and wind energy turbine. As bats are among the most overlooked in spite of their economical and ecological importance, their conservation is mandatory.

Acknowledgment

The authors are grateful to Professor Afework Bekele for his valuable comments, suggestions, and corrections on the draft of this paper.

References

[1] A. M. Hutson, S. P. Mickleburgh, and P. A. Racey, *Microchiropteran Bats: Global Status Survey and Conservation Action Plan*, IUCN/SSC chiroptera specialist group, IUCN, Gland, Switzerland, 2001.

[2] G. Jones, D. Jacobs, T. H. Kunz, M. R. Wilig, and P. A. Racey, "Carpe Noctem: the importance of bats as bioindicators," *Endangered Species Research*, vol. 8, pp. 3–115, 2009.

[3] T. H. Kunz, E. B. de Torrez, D. Bauer, T. Lobova, and T. H. Fleming, "Ecosystem services provided by bats," *Annals of the New York Academy of Sciences*, vol. 1223, no. 1, pp. 1–38, 2011.

[4] M. B. Fenton, "Science and the conservation of bats: where to next?" *Wildlife Society Bulletin*, vol. 31, no. 1, pp. 6–15, 2003.

[5] J. P. Madej, L. Mikulova, A. Gorosova et al., "Skin structure and hair morphology of different body parts in the Common Pipistrelle (*Pipistrellus pipistrellus*)," *Acta Zoologica*, vol. 94, no. 4, pp. 478–489, 2012.

[6] A. N. Makanya and J. P. Mortola, "The structural design of the bat wing web and its possible role in gas exchange," *Journal of Anatomy*, vol. 211, no. 6, pp. 687–697, 2007.

[7] M. S. Fujita and M. D. Tuttle, "Flying foxes (Chiroptera: Pteropodidae): threatened animals of key ecological and economic importance," *Conservation Biology*, vol. 5, no. 4, pp. 455–463, 1991.

[8] E. L. P. Anthony and T. H. Kunz, "Feeding strategies of the little brown bat (*Myotis lucifugus*) in southern New Hampshire," *Ecology*, vol. 58, pp. 775–786, 1977.

[9] A. Kurta and J. O. Whitaker Jr., "Diet of the endangered Indiana bat (*Myotis sodalis*) on the northern edge of its range," *The American Midland Naturalist*, vol. 140, no. 2, pp. 280–286, 1998.

[10] M. J. Lacki, J. S. Johnson, L. E. Dodd, and M. D. Baker, "Prey consumption of insectivorous bats in coniferous forests of north-central Idaho," *Northwest Science*, vol. 81, no. 3, pp. 199–205, 2007.

[11] M. Holderied, C. Korine, and T. Moritz, "Hemprich's long-eared bat (*Otonycteris hemprichii*) as a predator of scorpions: whispering echolocation, passive gleaning and prey selection," *Journal of Comparative Physiology A*, vol. 197, no. 5, pp. 425–433, 2011.

[12] D. Fukui, K. Okazaki, and K. Maeda, "Diet of three sympatric insectivorous bat species on Ishigaki Island, Japan," *Endangered Species Research*, vol. 8, no. 1-2, pp. 117–128, 2009.

[13] T. H. Kunz, "Feeding ecology of a temperate insectivorous bat (*Myotis velifer*)," *Ecology*, vol. 55, pp. 693–711, 1974.

[14] T. H. Kunz, J. O. Whitaker, and M. D. Wadanoli, "Dietary energetics of the insectivorous Mexican free-tailed bat (*Tadarida brasiliensis*) during pregnancy and lactation," *Oecologia*, vol. 101, no. 4, pp. 407–415, 1995.

[15] M. B. C. Hickey and M. B. Fenton, "Behavioural and thermoregulatory responses of female hoary bats, *Lasiurus cinereus* (Chiroptera: Vespertilionidae), to variations in prey availability," *Ecoscience*, vol. 3, no. 4, pp. 414–422, 1996.

[16] J. O. Whitaker, "Food of the big brown bat *Eptesicus fuscus* from maternity colonies in Indiana and Illinois," *American Midland Naturalist*, vol. 134, no. 2, pp. 346–360, 1995.

[17] R. A. Coutts, M. B. Fenton, and E. Glen, "Food intake by captive *Myotis lucifugus* and *Eptesicus fuscus* (Chiroptera: Vespertilionidae)," *Journal of Mammalogy*, vol. 54, pp. 985–990, 1973.

[18] A. Kurta, G. Bell, K. Nagy, and T. Kunz, "Energetics of pregnancy and lactation in free-ranging little brown bats (*Myotis lucifugus*)," *Physiological Zoology*, vol. 62, no. 3, pp. 804–818, 1989.

[19] J. O. Whitaker and P. Clem, "Food of the evening bat *Nycticeius humeralis* from Indiana," *The American Midland Naturalist*, vol. 127, pp. 211–217, 1992.

[20] O. J. Schmitz and K. B. Suttle, "Effects of top predator species on direct and indirect interactions in a food web," *Ecology*, vol. 82, no. 7, pp. 2072–2081, 2001.

[21] J. R. Speakman, "The impact of predation by birds on bat populations in the British Isles," *Mammal Review*, vol. 21, no. 3, pp. 123–142, 1991.

[22] M. B. Fenton, "Constraint and flexibility—bats as predators, bats as prey," *Symposia of the Zoological Society of London*, vol. 67, pp. 277–289, 1995.

[23] M. J. Daniel and G. R. Williams, "A survey of the distribution, seasonal activity and roost sites of New Zealand bats," *New Zealand Journal of Ecology*, vol. 7, pp. 9–25, 1984.

[24] G. G. Goodwin and A. M. Greenhall, "A review of the bats of Trinidad and Tobago," *Bulletin of the American Museum of Natural History*, vol. 122, pp. 191–301, 1961.

[25] M. B. Fenton, I. L. Rautenbach, S. E. Smith, C. M. Swanepoel, J. Grosell, and J. van Jaarsveld, "Raptors and bats: threats and opportunities," *Animal Behaviour*, vol. 48, no. 1, pp. 9–18, 1994.

[26] K. Dittmar, M. L. Porter, S. Murray, and M. F. Whiting, "Molecular phylogenetic analysis of nycteribiid and streblid bat flies (Diptera: Brachycera, Calyptratae): implications for host associations and phylogeographic origins," *Molecular Phylogenetics and Evolution*, vol. 38, no. 1, pp. 155–170, 2006.

[27] T. H. Fleming and N. Muchhala, "Nectar-feeding bird and bat niches in two worlds: pantropical comparisons of vertebrate pollination systems," *Journal of Biogeography*, vol. 35, no. 5, pp. 764–780, 2008.

[28] T. H. Fleming, C. Geiselman, and W. J. Kress, "The evolution of bat pollination: a phylogenetic perspective," *Annals of Botany*, vol. 104, no. 6, pp. 1017–1043, 2009.

[29] R. S. Duncan and C. A. Chapman, "Seed dispersal and potential forest succession in abandoned agriculture in tropical Africa," *Ecological Applications*, vol. 9, no. 3, pp. 998–1008, 1999.

[30] H. F. Howe and J. Smallwood, "Ecology of seed dispersal," *Annual Review of Ecology and Systematics*, vol. 13, pp. 201–228, 1982.

[31] M. A. Horner, T. H. Fleming, and C. T. Sahley, "Foraging behaviour and energetics of a nectar-feeding bat, *Leptonycteris curasoae* (Chiroptera: Phyllostomidae)," *Journal of Zoology*, vol. 244, no. 4, pp. 575–586, 1998.

[32] R. Muscarella and T. H. Fleming, "The role of frugivorous bats in tropical forest succession," *Biological Reviews*, vol. 82, no. 4, pp. 573–590, 2007.

[33] I. W. Buddenhagen, "Bats and disappearing wild bananas: can bats keep commercial bananas on supermarket shelves?" *Bats*, vol. 26, pp. 1–6, 2008.

[34] M. Shanahan, S. So, S. G. Compton, and R. Corlett, "Fig-eating by vertebrate frugivores: a global review," *Biological Reviews of the Cambridge Philosophical Society*, vol. 76, no. 4, pp. 529–572, 2001.

[35] E. R. Buchler, "Food transit time in *Myotis lucifugus* (Chiroptera: Vespertilionidae)," *Journal of Mammalogy*, vol. 54, pp. 985–990, 1975.

[36] J. G. Boyles, P. M. Cryan, G. F. McCracken, and T. H. Kunz, "Economic importance of bats in agriculture," *Science*, vol. 332, no. 6025, pp. 41–42, 2011.

[37] G. A. Polis, W. B. Anderson, and R. D. Holt, "Toward an integration of landscape and food web ecology: the dynamics of spatially subsidized food webs," *Annual Review of Ecology and Systematics*, vol. 28, pp. 289–316, 1997.

[38] R. L. Ferreira and R. P. Martins, "Diversity and distribution of spiders associated with bat guano piles in Morrinho cave (Bahia State, Brazil)," *Diversity and Distributions*, vol. 4, no. 5-6, pp. 235–241, 1998.

[39] D. B. Fenolio, G. O. Graening, B. A. Collier, and J. F. Stout, "Coprophagy in a cave-adapted salamander; the importance of bat guano examined through nutritional and stable isotope analyses," *Proceedings of the Royal Society B*, vol. 273, no. 1585, pp. 439–443, 2006.

[40] C. Parmesan, "Ecological and evolutionary responses to recent climate change," *Annual Review of Ecology, Evolution, and Systematics*, vol. 37, pp. 637–669, 2006.

[41] M. A. McGeoch, "The selection, testing and application of terrestrial insects as bioindicators," *Biological Reviews of the Cambridge Philosophical Society*, vol. 73, no. 2, pp. 181–201, 1998.

[42] P. B. Landres, J. Verner, and J. W. Thomas, "Ecological uses of vertebrate indicator species: a critique," *Conservation Biology*, vol. 2, pp. 316–327, 1988.

[43] C. E. Moreno, G. Sánchez-Rojas, E. Pineda, and F. Escobar, "Shortcuts for biodiversity evaluation: a review of terminology and recommendations for the use of target groups, bioindicators and surrogates," *International Journal of Environment and Health*, vol. 1, no. 1, pp. 71–86, 2007.

[44] M. B. Fenton, L. Acharya, D. Audet et al., "Phyllostomid bats (Chiroptera: Phyllostomidae) as indicators of habitat disruption in the neotropics," *Biotropica*, vol. 24, no. 3, pp. 440–446, 1992.

[45] P. M. Vitousek, H. A. Mooney, J. Lubchenco, and J. M. Melillo, "Human domination of Earth's ecosystems," *Science*, vol. 277, no. 5325, pp. 494–499, 1997.

[46] B. W. Keeley and M. D. Tuttle, *Bats in American Bridges*, vol. 4, Bat Conservation International, 1999.

[47] G. F. McCracken, "Bats aloft: a study of high-altitude feeding," *BATS*, vol. 14, pp. 7–101, 1996.

[48] C. J. Cleveland, M. Betke, P. Federico et al., "Economic value of the pest control service provided by Brazilian free-tailed bats in south-central Texas," *Frontiers in Ecology and the Environment*, vol. 4, no. 5, pp. 238–243, 2006.

[49] R. Tapper, *Wildlife Watching and Tourism: A Study on the Benefits and Risks of a Fast Growing Tourism Activity and its Impacts on Species*, UNEP/CMS Secretariat, Bonn, Germany, 2006.

[50] M. H. Reiskind and M. A. Wund, "Experimental assessment of the impacts of northern long-eared bats on ovipositing *Culex* (Diptera: Culicidae) mosquitoes," *Journal of Medical Entomology*, vol. 46, no. 5, pp. 1037–1044, 2009.

[51] G. F. McCracken, V. A. Brown, M. Eldridge, and J. K. Westbrook, "The use of fecal DNA to verify and quantify the consumption of agricultural pests," *Bat Research News*, vol. 46, pp. 195–196, 2005.

[52] E. L. Clare, E. E. Fraser, H. E. Braid, M. H. Fenton, and P. D. Hebert, "Species on the menu of a generalist predator, the eastern red bat (*Lasiurus borealis*): using a molecular approach to detect arthropod prey," *Molecular Ecology*, vol. 18, no. 11, pp. 2532–2542, 2009.

[53] S. Arizaga, E. Ezcurra, E. Peters, F. R. de Arellano, and E. Vega, "Pollination ecology of *Agave macroacantha* (Agavaceae) in a Mexican Tropical Desert: the role of pollinators," *The American Journal of Botany*, vol. 87, no. 7, pp. 1011–1017, 2000.

[54] S. K. Godwa, R. C. Katiyar, and V. R. B. Sasfry, "Feeding value of Mahua (*Madhuca indica*) seed cakes in farm animals," *Indian Journal of Dairy Science*, vol. 49, pp. 143–154, 1996.

[55] N. Gallai, J.-M. Salles, J. Settele, and B. E. Vaissière, "Economic valuation of the vulnerability of world agriculture confronted with pollinator decline," *Ecological Economics*, vol. 68, no. 3, pp. 810–821, 2009.

[56] C. Hougner, J. Colding, and T. Söderqvist, "Economic valuation of a seed dispersal service in the Stockholm National Urban Park, Sweden," *Ecological Economics*, vol. 59, no. 3, pp. 364–374, 2006.

[57] G. E. Hutchinson, "Survey of existing knowledge of biogeochemistry: the biogeochemistry of vertebrate excretion," *Bulletin of the American Museum of Natural History*, vol. 96, pp. 1–554, 1950.

[58] M. D. Tuttle and A. Moreno, *Cave-Dwelling Bats of Northern Mexico: Their Value and Conservation Needs*, Bat Conservation International, Austin, Tex, USA, 2005.

[59] T. P. Eiting and G. F. Gunnell, "Global completeness of the bat fossil record," *Journal of Mammalian Evolution*, vol. 16, no. 3, pp. 151–173, 2009.

[60] S. Mickleburgh, K. Waylen, and P. Racey, "Bats as bushmeat: a global review," *Oryx*, vol. 43, no. 2, pp. 217–234, 2009.

[61] F. O. Abulude, "Determination of the chemical composition of bush meats found in Nigeria," *The American Journal of Food Technology*, vol. 2, no. 3, pp. 153–160, 2007.

[62] R. K. B. Jenkins and P. A. Racey, "Bats as bushmeat in Madagascar," *Madagascar Wildlife Conservation*, vol. 3, pp. 22–30, 2008.

[63] W. D. Schleuning, "Vampire bat plasminogen activator DSPA-alpha-1 (desmoteplase): a thrombolytic drug optimized by natural selection," *Pathophysiology of Haemostasis and Thrombosis*, vol. 31, pp. 118–122, 2000.

[64] L. A. Pennisi, S. M. Holland, and T. V. Stein, "Achieving bat conservation through tourism," *Journal of Ecotourism*, vol. 3, no. 3, pp. 195–207, 2004.

[65] R. Müller and R. Kuc, "Biosonar-inspired technology: goals, challenges and insights," *Bioinspiration and Biomimetics*, vol. 2, pp. 146–161, 2007.

[66] G. Bunget and S. Seelecke, "BATMAV: a 2-DOF bio-inspired flapping flight platform," in *The International Society for Optics and Photonics*, vol. 7643 of *Proceedings of SPIE*, pp. 1–11, 2010.

[67] H. A. Delpietro, N. Marchevsky, and E. Simonetti, "Relative population densities and predation of the common vampire bat (*Desmodus rotundus*) in natural and cattle-raising areas in north-east Argentina," *Preventive Veterinary Medicine*, vol. 14, no. 1-2, pp. 13–20, 1992.

[68] K. M. Brown, R. M. Erwin, M. E. Richmond, P. A. Buckley, J. T. Tanacredi, and D. Avrin, "Managing birds and controlling aircraft in the Kennedy Airport-Jamaica Bay Wildlife Refuge complex: the need for hard data and soft opinions," *Environmental Management*, vol. 28, no. 2, pp. 207–224, 2001.

[69] Transport Canada, *Sharing the Skies: An Aviation Industry Guide to the Management of Wildlife Hazards*, Transport Canada, Ottawa, Canada, 2001.

[70] S. C. Peurach, C. J. Dove, and L. Stepko, "A decade of U.S. Air Force bat strikes," *Humboldt Wild life Care Center*, vol. 3, pp. 199–207, 2009.

[71] J. G. Parsons, D. Blair, J. Luly, and S. K. A. Robson, "Bat strikes in the Australian aviation industry," *Journal of Wildlife Management*, vol. 73, no. 4, pp. 526–529, 2009.

[72] A. M. Greenhall, "Bats: their public health importance and control with special reference to Trinidad," in *Proceedings of the 2nd Vertebrate Pest Control Conference*, vol. 18, pp. 108–116, 1964.

[73] C. H. Calisher, J. E. Childs, H. E. Field, K. V. Holmes, and T. Schountz, "Bats: important reservoir hosts of emerging viruses," *Clinical Microbiology Reviews*, vol. 19, no. 3, pp. 531–545, 2006.

[74] G. Wibbelt, A. Kurth, N. Yasmum et al., "Discovery of herpesviruses in bats," *Journal of General Virology*, vol. 88, no. 10, pp. 2651–2655, 2007.

[75] M. B. Fenton, M. Davison, T. H. Kunz, G. F. McCracken, P. A. Racey, and M. D. Tuttle, "Linking bats to emerging diseases," *Science*, vol. 311, no. 5764, pp. 1098–1099, 2006.

11

Aggressive Waves in the Lemon-Clawed Fiddler Crab (*Uca perplexa*): A Regional "Dialect" in Fiji

Judith S. Weis[1] and Peddrick Weis[1,2]

[1] *Department of Biological Science, Rutgers University, Newark, NJ 07102, USA*
[2] *Department of Radiology, UMDNJ–New Jersey Medical School, Newark, NJ 07101-1709, USA*

Correspondence should be addressed to Peddrick Weis; pw203@umdnj.edu

Academic Editors: P. De los Ríos Escalante, R. Rico-Martinez, and P. K. S. Shin

A population of the lemon-clawed fiddler crab (*U. perplexa*) in Fiji (island of Vanua Levu) was studied for types of communication (i.e., signaling via waving the male's larger claw). Two types of signals were observed. In addition to the expected territorial display of a large and complex vertical wave that conveys its message over a typical distance of 10–40 cm (with large males signaling to other large males over the greatest distance), a short, rapid, and horizontal wave was typically directed over a much shorter distance, rarely exceeding 10 cm. This latter wave type, seemingly of an aggressive nature, differs from the vertically directed aggressive signal observed in an Australian population of this species and thus appears to be a regional "dialect" for this mode of communication.

1. Introduction

Animal communication signals are crucial to many species, as they may convey information to conspecifics about the identity, location, and motivation of the sender. This information is often valuable for reproductive success or survival. In the genus *Uca* (fiddler crabs), the male's enlarged claw is waved in order to attract females for mating, although it may also be used in territorial and agonistic displays and may be exhibited in the absence of a target. Different species have different patterns of claw waving [1, 2], helping females to find conspecific males [3, 4]. Waves have been classified as "vertical" and "lateral" [1]. Vertical waves are described as simple lifts of the claw with little horizontal movement, while lateral waves include complicated lateral components (extending and flexing), in addition to vertical components, generally accompanied by walking leg and body movements. Since waving can have two very different functions, as a signal for mating or for conflict, the purpose of a given waving signal may be uncertain and confusing to both crustacean and human observers.

The courtship waving pattern of the lemon-clawed fiddler crab (*U. perplexa*) in Japan and Australia has been described [1, 5–7]. The typical wave, which has been termed a "lateral" wave, involves extending the claw, then raising it, and then lowering it rapidly to the resting position, taking 1–1.5 seconds. At the same time, the minor claw and front walking legs are also raised and lowered along with the major claw. This is a conspicuous signal that can be detected at a considerable distance, and it has been associated with courtship [1, 5, 8, 9].

How et al. [6] described a second, totally different, type of wave in Australian *U. perplexa*, a vertical wave in which the claw is raised and lowered very rapidly without extending; each wave is about 0.2 seconds and can be repeated in a rapid series. These simple waves had much less elevation, were not accompanied by leg movements, and would be observable only by nearby crabs. These waves were used in territorial interactions as well as courtship, and were elicited by male wanderers during agonistic interactions and female wanderers during close range courtship. Crane [1] stated that rapid vertical motions with a flexed claw are made only during high intensity threat.

In this paper, we describe a different, previously undescribed, type of wave in a population of *U. perplexa* in Fiji and its relation to distance between identity and relative sizes of individuals.

(a) (b)

FIGURE 1: *Uca perplexa* exhibiting (a) a vertical (regular) wave, shown here at top of maneuver and (b) a shoving match that began with rapid horizontal waves (RHW).

2. Methods

Observations were made in December 2009 and January 2010 on a mud flat on the south shore of the Fijian island of Vanua Levu, in the Koro Sun Bay, 13.5 km by road east of the town of Savusavu ($16°47'51''$S $\times$ $179°24'44''$E). At this location, a large colony of *Uca perplexa* was adjacent to a colony of *U. vocans* and a few individuals of *U. tetragonon*. Field observations were done by two experienced individuals who used close-focusing binoculars to scan the colony. Initial observations indicated that two types of waves took place, so both were recorded. Whenever either type of wave was observed, the focal individual was identified as either a small (<10 mm) or a large male (maximum carapace width for this species at this location being ~14 mm). The crab being waved at (target of the wave) was indicated (large or small male *U. perplexa*, female *U. perplexa*, *U. tetragonon*, or *U. vocans*), whenever the recipient was obvious, and the distance between the two crabs was estimated (rulers were placed on the mud flat for reference). When it was not clear who the receiver was, it was so indicated, and no distance was noted. Initial observations examined both types of waves to see which were performed more frequently and by whom. Later observations focused exclusively on the performers and targets of the quick/repeated horizontal wave. Digital imaging involved both still and brief video photography.

Statistical analysis involved Kruskal-Wallis (K-W) one-way analysis of variance. (Nonparametric analysis was required because of the unequal distribution of data.)

3. Results and Discussion

Two very different types of waves were noted. The typical wave, described by Crane [1] and How et al. [6] and termed a "lateral" wave, involves extending the claw horizontally, raising it, and then lowering it down rapidly to the resting position over 1.0–1.5 seconds; simultaneously, the small claw is also raised and the crab rises up on its front pair of walking legs (see Figure 1(a)). For reasons that will be clear, we prefer to call this a "regular" wave rather than a "lateral" wave, since it involves a considerable vertical component. Regular waves

were performed somewhat more by large than small males and were addressed to both males and females. However, they were directed toward other males at a far greater frequency than to females. This may be due, in part, to the greater number of males out on the surface. Large males waving to other large males did so at the greatest distance compared to the other combinations of wavers and receivers, illustrated in Figure 2(a) (K-W statistic = 15.4, $P = 0.009$).

How et al. [6] also described a second wave type, a vertical wave, in which the claw is raised and lowered very rapidly without extending; each wave is about 0.2 seconds. We did not see any waves like this; instead, we saw similarly rapid waves, but they were horizontal instead of vertical. The claw was extended quickly to a 30–40 degree angle (as measured on photographs) and then returned to the resting position. These waves had a duration of ~0.2 sec and were often repeated at intervals of ~0.5 sec, like the "vertical waves" of Australian crabs [6]; however, we did not have available the sophisticated video system of How et al. [6, 7], so the durations and frequencies reported here are estimates. Sequences of over 20 were observed. Like the vertical waves of How et al., these rapid horizontal waves (RHWs) were used primarily in agonistic interactions and were addressed primarily to crabs that were in closer proximity than crabs receiving normal waves (Figure 2(b)). However, unlike the regular waves, there were no significant differences among the groups for the RHW. They were also directed to other males more often than toward females (73% versus 9%), although this was biased by the presence of fewer females, as noted above. Male crabs at a distance might wave back and forth at each other, and then one would continue to approach the other, who would then direct RHWs at the intruder who might either retreat or perform RHWs in response. While retreat was the more common result, RHWs back and forth sometimes led to a "butting match," where the two crabs would push each other until one retreated (Figure 1(b)). In one case, a crab in a butting match was observed to pick up the other and toss it away. RHWs were also addressed toward nearby females, who invariably retreated. No real courtship was observed, and no female was seen to enter a male's burrow. Rather, all females retreated from the waving males.

RHWs were also occasionally directed toward members of other species on the flat, *U. vocans* and *U. tetragonon*, but only when they were in much closer proximity than the targets of intraspecific RHWs (Figure 2(b)). However, these other species ignored the RHWs; they obviously did not "speak the language."

When both kinds of waves were compared for distance (using only male-male data), the differences between wave types are more obvious (Figure 3). K-W one-way ANOVA for Figure 3 = 50.7, $P < 0.001$.

We have examined the two types of waves performed by a population of *U. perplexa* in Fiji. The RHWs appear comparable in speed and function to the "vertical" waves described by Crane [1] and How et al. [6] in this species in Australia. Crane [1] stated that the vertical waves were used for agonistic interactions, which appears to be the message of the RHW described here. These waves tended to be addressed to target crabs that were in closer proximity than targets

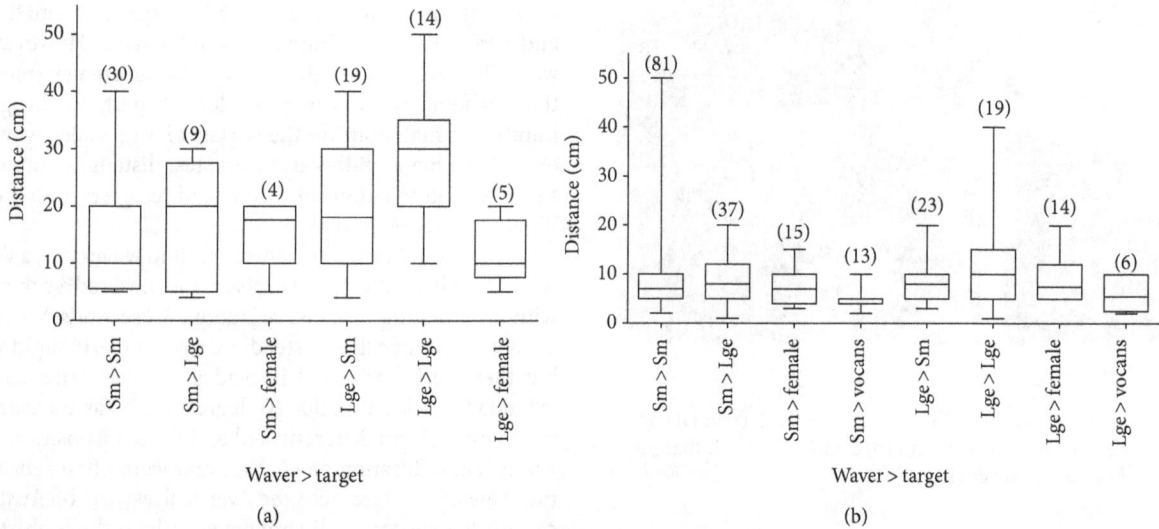

FIGURE 2: Regular waves (a) and RHW (b) in relation to distance, according to target. For (a), Kruskal-Wallis statistic = 15.5, $P = 0.031$; data in (b) are not significantly different. Numbers of observations (n) are in parentheses above graph bars.

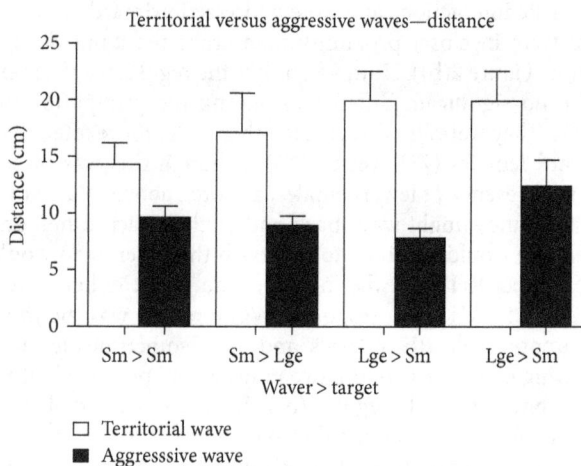

FIGURE 3: Regular versus RHW in relation to distance (males only). Kruskal-Wallis statistic = 50.7, $P < 0.0001$. Kruskal-Wallis one-way ANOVA = 15.4, $P = 0.0084$.

of "regular" waves. As females approached, males increased their intensity of waving but reduced the lateral component of the wave. It is logical that a larger, more conspicuous wave would be directed toward more distant receivers.

The rapid vertical waves of How et al. [6, 7] would be visible only to rather nearby individuals. The same would be true for the RHW of the Fiji crabs, which would probably be visible to a narrow range of observers that were both nearby and directly in front of the displaying crab. While the vertical waves included striking of the chela on the substrate, which might produce surface vibrations audible to other fiddler crabs, such acoustic signals do not appear to be possible with the RHW described. However, the RHW can be (and was) also used to push the other crab away. The forward direction of this Fijian variation is especially amenable to being converted into pushing.

A wave similar to the RHW was recently observed in the related *U. lactea* [10], who called it "lateral-flick waving" and noted that it was directed "mostly to neighbouring resident males," similar to our observations in *U. perplexa*.

Geographic variation in waving (regular or "lateral") by this species has previously been noted by How et al. [11] who described differences between Australian and Japanese populations of *U. perplexa*. However, those geographical differences were subtle, requiring video analysis to differentiate, whereas the difference between a rapid vertical [6] versus horizontal, as described here, is a major distinction in the nature of the wave. The horizontal wave performed in Fiji appears to convey the same aggressive/agonistic message as the vertical wave in Australia. Geographical differences in signaling behaviors could be an initial step toward speciation [12]; however, since our observed difference is not a mating behavior, its relation to speciation is not apparent.

4. Conclusions

A population of the lemon-clawed fiddler crab (*U. perplexa*) in Fiji was found to have an aggressive signal that differs from that of another well-studied population (Australian). This appears to be a regional "dialect" for this mode of communication. The territorial display is, on the other hand, consistent between these two populations.

References

[1] J. Crane, *Fiddler Crabs of the World (Ocypodidae: genus* Uca*)*, Princeton University Press, New Jersey, NJ, USA, 1975.

[2] M. S. Rosenberg, "The systematics and taxonomy of fiddler crabs: a phylogeny of the genus *Uca*," *Journal of Crustacean Biology*, vol. 21, no. 3, pp. 839–859, 2001.

[3] M. Salmon, G. W. Hyatt, K. McCarthy, and J. Costlow, "Display specificity and reproductive isolation in the fiddler crabs, *Uca*

panacea and *U. pugilator*," *Zeitschrift fur Tierpsychologie*, vol. 48, pp. 251–276, 1978.

[4] T. Detto, P. R. Backwell, J. M. Hemmi, and J. Zeil, "Visually mediated species and neighbour recognition in fiddler crabs (*Uca mjoebergi* and *Uca capricornis*)," *Proceedings of the Royal Society B*, vol. 273, no. 1594, pp. 1661–1666, 2006.

[5] Y. Nakasone and M. Murai, "Mating behavior of *Uca lactea perplexa* (Decapoda: Ocypodidae)," *Journal of Crustacean Biology*, vol. 18, no. 1, pp. 70–77, 1998.

[6] M. J. How, J. Zeil, and J. M. Hemmi, "Differences in context and function of two distinct waving displays in the fiddler crab, *Uca perplexa* (Decapoda: Ocypodidae)," *Behavioral Ecology and Sociobiology*, vol. 62, no. 1, pp. 137–148, 2007.

[7] M. J. How, J. M. Hemmi, J. Zeil, and R. Peters, "Claw waving display changes with receiver distance in fiddler crabs, *Uca perplexa*," *Animal Behaviour*, vol. 75, no. 3, pp. 1015–1022, 2008.

[8] M. Murai and P. R. Y. Backwell, "More signalling for earlier mating: conspicuous male claw waving in the fiddler crab, *Uca perplexa*," *Animal Behaviour*, vol. 70, no. 5, pp. 1093–1097, 2005.

[9] M. Murai and P. R. Y. Backwell, "A conspicuous courtship signal in the fiddler crab *Uca perplexa*: female choice based on display structure," *Behavioral Ecology and Sociobiology*, vol. 60, no. 5, pp. 736–741, 2006.

[10] D. Muramatsu, "For whom the male waves: claw-waving display of male fiddler crabs," in *Proceedings of the the Crustacean Society Summer Meeting*, Tokyo, September 2009, abstract [G-E5].

[11] M. J. How, J. Zeil, and J. M. Hemmi, "Variability of a dynamic visual signal: the fiddler crab claw-waving display," *Journal of Comparative Physiology A*, vol. 195, no. 1, pp. 55–67, 2009.

[12] J. Bro-Jørgensen, "Dynamics of multiple signalling systems: animal communication in a world in flux," *Trends in Ecology and Evolution*, vol. 25, no. 5, pp. 292–300, 2010.

12

Population, Ecology, and Threats to Two Endemic and Threatened Terrestrial Chelonians of the Western Ghats, India

Arun Kanagavel,[1] Shiny M. Rehel,[2,3] and Rajeev Raghavan[1]

[1] *Conservation Research Group (CRG), St. Albert's College, Kochi, Kerala 682 018, India*
[2] *Keystone Foundation, Kotagiri, Tamil Nadu 643 217, India*
[3] *Research and Development Centre, Bharathiar University, Coimbatore, Tamil Nadu 641 046, India*

Correspondence should be addressed to Arun Kanagavel; arun.kanagavel@gmail.com

Academic Editors: A. Chistoserdov and L. Luiselli

The Western Ghats part of the Western Ghats-Sri Lanka hotspot harbors two endemic terrestrial chelonians, the Cochin forest cane turtle *Vijayachelys silvatica* and the Travancore tortoise *Indotestudo travancorica*. Population estimates as well as information on the scale and intensity of threats for these chelonians are largely unavailable. This study attempts to address these gaps for two hill ranges of the Western Ghats. Thirty random quadrats at eight forest ranges were surveyed for chelonians and their carapaces recording any found en route and also during opportunistic surveys. Three live *V. silvatica* and 38 *I. travancorica* were subsequently encountered and had overall densities of 0.006 and 0.03 individuals per hectare, respectively. These chelonians were found at quadrats with lower light intensity and soil temperature. Nine carapaces were found during the field surveys: seven the result of human consumption, one trapped in a pit, and another consumed by a wild animal. In addition to field surveys, household surveys in 26 indigenous and nonindigenous human settlements resulted in the observation of one *V. silvatica* and 38 *I. travancorica* including a carapace. Roads were surveyed to assess the threat they posed to chelonians, resulting in the observation of two *I. travancorica* road kills. Increased interactions and discussions between the management authorities and local communities need to be promoted if chelonian conservation is to improve in the landscape.

1. Introduction

The Western Ghats (WG) region in India, part of the Western Ghats-Sri Lanka Biodiversity Hotspot is globally renowned for its diversity of endemic amphibian, reptile, and fish species [1–3]. The two endemic chelonian genera in the WG are represented by the Travancore tortoise (*Indotestudo travancorica*) and the Cochin forest cane turtle (*Vijayachelys silvatica*); both threatened with extinction [4, 5]. The cane turtle is listed as "Endangered" while the Travancore tortoise is "Vulnerable" in the IUCN Red List of Threatened Species [4, 5].

Of these sympatric, cryptic species, Travancore tortoises are known to be more widespread than cane turtles [6, 7]. The Travancore tortoise is found in rocky hills at elevations of 100–1000 m a.s.l. across the southern WG in a multitude of habitats, such as evergreen, semievergreen, bamboo, *Lantana camara* and *Cromolarium glandulosum* bushes, and rubber and teak plantations [8–12]. On the other hand, the Cochin forest cane turtle, known to be a habitat specialist associated with evergreen vegetation, has also been found in semievergreen, deciduous, and bamboo vegetation types at elevations of 180–800 m a.s.l. [12–15]. Cane turtles do not have an affinity to perennial water sources, though they have algal growth on their carapace [12, 15].

Although there are studies which have focused specifically on one or both the species [12, 15], systematic conservation assessments and on-ground conservation action remain lacking. Amidst this scenario of unavailable population estimates and systematic threat assessments [12], we carried out a survey in the forest areas of southern WG, using a combination of ecological and socioeconomic methods, to improve our understanding of the two species.

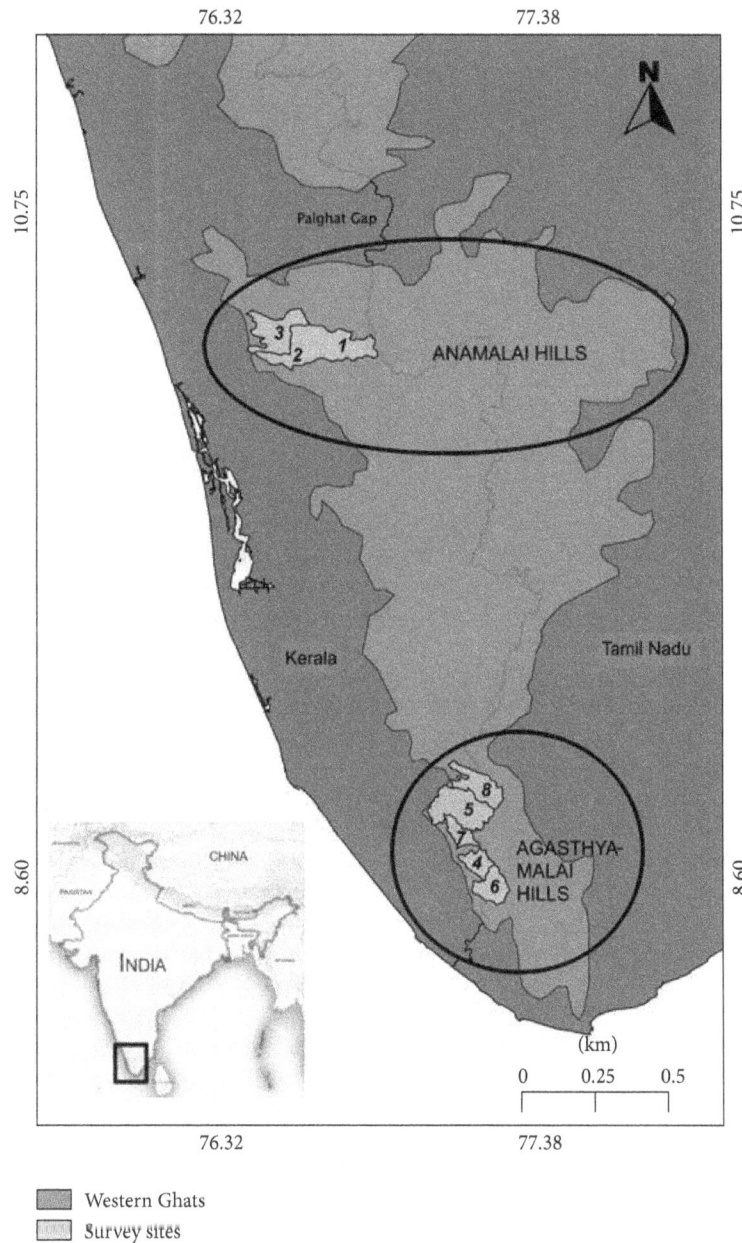

FIGURE 1: Map of the study area showing the locations surveyed for the focal species in the Anamalai and Agasthyamalai Hills of Western Ghats.

2. Material and Methods

2.1. *Study Area.* The study was conducted in two hill ranges of the southern WG, the Anamalai and Agasthyamalai Hills. The Anamalai Hills harbor the largest extent of shola grassland [2]. Tree-felling for cardamom, coffee, and tea plantations during the imperial rule (Congreve 1942 cited in [16]) has led to severe forest fragmentation in the Anamalais, home to numerous indigenous communities [17]. The Agasthyamalai Hills spread over 2112 km^2 are relatively undisturbed and contiguous in comparison to the Anamalais and are known for their plant diversity and unique ecosystems like *Myristica* swamps [2, 18, 19].

2.2. *Quadrat Surveys.* Surveys were conducted in eight forest ranges at four reserve forests (RF) and three wildlife sanctuaries (WLS; Figure 1, Table 1).

Thirty quadrats of two hectares each were visually surveyed for chelonians/shell parts in each of the eight forest ranges anytime between 07:00 h to 21:00 h from December 2010 to December 2011. These quadrats were randomly selected from digitized 1 : 50000, Survey of India toposheets (58 B7, B11, H1, H2, H6) using QGIS 1.7.3. Morphometric measurements (See Supplementary Material available online at http://dx.doi.org/10.1155/2013/341687 Figure S1) were determined using both digital (15 cm, ±0.01 cm accuracy) and dial Vernier calipers (RSK, 30 cm, ±0.02 accuracy).

TABLE 1: Population density and frequency of chelonians/shell parts encountered in the wild and at human settlements.

Number	Protected area	Indotestudo travancorica				Vijayachelys silvatica		
		Density*	Wild	Settlement	Shell	Density*	Wild	Settlement
1	Vazhachal range, Vazhachal RF[a]	0.03	8	5	8	0.02	1	1
2	Athirapilly range, Vazhachal RF[a]	0.03	5	5	1	0	—	—
3	Chalakudy RF[a]	0.03	12	—	—	0.02	1	—
4	Peppara WLS[b]	0	0	9	—	0	—	—
5	Kulathupuzha RF[b]	0.05	7	0	1	0	—	—
6	Neyyar WLS[b]	0.02	1	12	—	0	—	—
7	Palode RF[b]	0.03	2	6	—	0	—	—
8	Shendurney WLS[b]	0.05	3	0	—	0.02	1	—

*Individuals per hectare, [a]Anamalai Hills, and [b]Agasthyamalai Hills.

Measures beyond 25 cm were measured using a tape (150 cm, ±0.1 cm accuracy). The chelonian's weight was determined using a weighing balance (6 kg, ±0.001 kg accuracy).

From each quadrat, time, terrain (whether within leaf litter and/or bark, cavity or in the open), activity of the chelonian when encountered, algal presence, and the number of ticks were recorded. The soil and air temperature as well as the cloacal temperature of the chelonian were determined using a digital thermometer (Eurolab, 50°C–300°C, ±0.1°C accuracy). For cloacal temperature, the thermometer probe was placed at the cloacal aperture. Humidity and light intensity were determined using a digital hygrometer (TempTec 241A, 0–100%, ±1% accuracy) and luxmeter (LX-101, 0–50,000 lux, ±1 lux accuracy), respectively. The geographical coordinates and elevation of the point of location, water sources, and bamboo/cane clusters closest to the encountered chelonians were determined using a Garmin GPS 62S. The distance between the chelonians and the water sources and cane/bamboo clusters was calculated using QGIS 1.7.3. The extent of leaf, herb, shrub, and canopy cover was rated on a qualitative scale of "absent, low, medium, and high." Whether the stratum was wet, moist or dry, rocky terrain, or on a slope at the point of location was also determined. Depth of the leaf litter was determined using a measuring tape (150 cm, ±0.1 cm accuracy). Morphological and environmental data were also collected in case a chelonian/shell part was encountered enroute to a quadrat as well as during opportunistic surveys.

2.3. Household and Road Surveys. To identify households where chelonians were kept, indigenous and nonindigenous settlements were surveyed using a referral sampling strategy (local informants and respondents from parallel interview surveys) [7, 20]. Accompanied by local informants, referred households were visited, and residents were requested to allow both morphometric data (Figure S1) and cloacal temperature to be recorded from the chelonians. A total of 26 settlements (seven nonindigenous and 19 indigenous) were surveyed in the process.

Major highways and smaller roads passing through the forest ranges were surveyed once in each of the eight forest ranges for any road-killed chelonians.

2.4. Analysis. The population density of each species encountered at a site was calculated [21]. The overall population density was the average of the population density of each species across all the sites surveyed. The overall relative abundance of each species was calculated by dividing the total number of individuals of a specific species by the total number of individuals of both the species.

Mann-Whitney U or Kruskal-Wallis test and independent samples t-test or ANOVA were used to test for differences between quadrats that harbored chelonians and those that did not. In case frequencies among any group were less than five, Chi-square and Fisher's exact test were undertaken.

Morphometric differences and differences in habitat preferences between sexes as well as morphometric differences between individuals encountered in the wild and in settlements were also determined. R.2.14.0 was used for the Shapiro-Wilk test and Fisher's exact test, while SPSS 11.5 was used for all the other statistical tests.

3. Results

3.1. Quadrat Surveys. A total of 240 quadrats were surveyed (90 at Anamalai Hills and 150 at Agasthyamalai Hills); 90 each during monsoon and premonsoon and 60 during postmonsoon. The elevation varied from 54 to 1079 m a.s.l., air temperature 20.4 to 36.8°C, soil temperature 18.5 to 34.6°C, humidity 36% to 90% and light intensity 0 to 1026 lux during the survey period. The vegetation types sampled ranged from dry and moist deciduous, semievergreen, evergreen, and riparian forests to grassland and plantations (including cashew, cocoa, coconut, rubber, and teak).

Eighteen out of 240 quadrats surveyed were occupied by the two focal species. Indotestudo travancorica were encountered in 15 quadrats occupying an overall density of 0.03 individuals per hectare, while V. silvatica was encountered at three quadrats occupying an overall density of 0.006 individuals per hectare (Table 1). The overall relative abundance of I. travancorica was 0.83, whereas that of V. silvatica was 0.17.

Light intensity (Mann-Whitney U test, $U = 1682.5$, $P = 0.003$, and $n = 25$) and soil temperature (Mann-Whitney

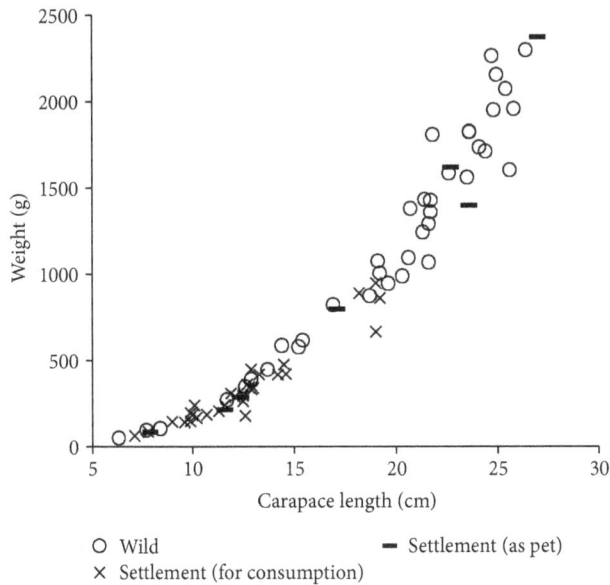

FIGURE 2: Difference in sizes of *Indotestudo travancorica* encountered in the wild and at human settlements.

U test, U = 1450.5, P = 0.04, and n = 38) were found to vary significantly, being lower in quadrats where chelonians occurred (65.4 ± 54.8 lux and 24.2 ± 0.5°C resp.) than where they did not occur (202.1 ± 23.6 lux and 25.4 ± 0.2°C). The two species also differed in the terrains they occupied (Fisher's exact test, χ^2 = 4.3, P = 0.03, and n = 38); *V. silvatica* were encountered on flat terrains (100%), while *I. travancorica* were found more often on sloped terrains (67%).

Twenty-three living and nine carapaces of *I. travancorica* were also found enroute to the quadrats. Six adult charred carapaces were found on a hillock at Vazhachal range, one adult carapace was found beside a forest stream in Kulathupuzha RF, and another adult carapace was found in a pit dug up and later deserted by the Kerala State Electricity Board to lay cable posts in the Athirapilly range. One juvenile *I. travancorica* carapace was found in Vazhachal range with marks made by an unidentified, carnivorous species.

3.2. Household and Road Surveys. One adult female *V. silvatica* and 37 *I. travancorica* (31 juvenile, five adult females, and one adult male) were found in the households that were surveyed (Table 1). Holes were found to be made in the posterior part of the animal's carapace that assisted in tying them to posts. Seven out of the 37 tortoises were pets while the rest, including the single *V. silvatica*, were meant for consumption on attaining larger sizes (Figure 2). Some of the pet tortoises alone weighed above 1000 g and had carapaces longer than 20 cm (Figure 2). An intact shell of *I. travancorica* devoid of any damage, being used as a decorative household item, was also shown.

Two *I. travancorica* road kills were encountered—one juvenile at Athirapilly range close to the ticket counter for tourists and an adult close to the Kanithadam check post at Peppara WLS.

3.3. Vijayachelys silvatica. A total of four adult female *V. silvatica* were found, the details of which are provided in Tables 1 and 2. Due to the small sample size of *V. silvatica* encountered, no further analysis could be undertaken (For indetail information on the associated environmental variables see Supplementary Material Table S1).

3.4. Indotestudo travancorica. A total of 27 adult and 11 juvenile *I. travancorica* were encountered during the field surveys, the details of which are presented in Tables 1 and 3. The sex ratio (male : female) was 1 : 2 in the Anamalai Hills and 1 : 0.7 in the Agasthyamalai Hills (For indetail information on the associated environmental variables see Supplementary Material Table S2).

Morphometric differences were found between male and female *I. travancorica* with respect to shell height (Mann-Whitney U test, U = 34.5, P = 0.009, and n = 35) and bridge length (Mann-Whitney U test, U = 34.5, P = 0.02, and n = 35), being longer in females (8.9 ± 0.3 and 9.3 ± 0.2) than in males (8.0 ± 0.2 and 8.5 ± 0.2).

At the two hill systems, tortoises differed in the habitats they were encountered at (Fisher's exact test, χ^2 = 13.2, P = 0.03, and n = 38), algal growth (Fisher's exact test, χ^2 = 4.4, P = 0.03, and n = 38) and vegetation type (Fisher's exact test, χ^2 = 8.6, P = 0.03, and n = 38). They were found more often within leaflitter and/or bark in the Anamalai Hills (n = 18) than in the Agasthyamalai Hills (n = 4). Only those encountered in the Anamalais (n = 7) had algal growth on their carapace. Thirty-six percent of the tortoises in the Anamalais were found in moist deciduous forests whereas in the Agasthyamalai Hills, 54% were found in an ecotone of moist deciduous and semievergreen forests. Moreover, these chelonians were found in semievergreen forests (32%) only at Anamalai Hills and in evergreen forests (7.7%) and their ecotone with semievergreen forests (15.4%) only at Agasthyamalai Hills.

Significant differences were found between the carapace length (Mann-Whitney U test, U = 270, P < 0.001, and N = 75) and weight (Mann-Whitney U test, U = 249.5, P < 0.001, and N = 75) between tortoises encountered in the wild and those in human settlements. Individuals found at settlements had a relatively smaller carapace length, weighed less, and also did not have any ticks on them in comparison to the wild individuals (Figure 2, Table 3).

3.5. Microhabitat Description. Air (Kruskal-Wallis Test, χ^2 = 17, P < 0.0001, and N = 38), soil (ANOVA, F = 3.8, P = 0.03, and N = 38) and cloacal temperatures (ANOVA, F = 5.5, P = 0.008, and N = 38) varied significantly for the different seasons (Figure 3(a)). Significant differences were found between the time period the wild individuals were encountered and their cloacal temperature (ANOVA, F = 6.4, P = 0.004, and N = 38), air temperature (Kruskal Wallis test, χ^2 = 7.4, P = 0.02, and N = 38), and whether the individuals were present on a slope or not (Fisher's exact test, χ^2 = 6.2, P = 0.04, and N = 38). Cloacal temperatures of wild individuals were lower in the evening followed by morning and higher in the afternoon, (Figure 3(b)). Individuals were

TABLE 2: Morphometrics and cloacal temperatures of *Vijayachelys silvatica* encountered during the study.

Measurements	Wild (*n* = 3)			Settlements (*n* = 1)
	Range	*x*	SD	
Carapace length (CL; cm)	11.5–12.2	11.8	0.4	12.2
Carapace width (CW; cm)	8.3–9.8	9.0	0.8	9.0
Plastron length (PL; cm)	9.3–10.8	10.2	0.8	10.0
Plastron width (PW; cm)	5.5–6.7	6.0	0.6	6.3
Bridge length (BL; cm)	3.7–4.1	3.9	0.2	4.0
Shell height (SH; cm)	4.3–4.5	4.4	0.1	4.0
Weight (g)	192–267	219	41.7	206
Cloacal temperature (°C)	23.3–30.7	26.8	3.7	29.5
Ticks	0–2	1.3	1.2	2
Algal growth	Yes = 3			Yes = 1

TABLE 3: Morphometrics and cloacal temperatures of *Indotestudo travancorica* encountered during the study.

Measurements	Wild (*n* = 38)			Settlements (*n* = 37)			Carapaces (*n* = 10)		
	Range	*x*	SD	Range	*x*	SD	Range	*x*	SD
Carapace length (cm)	6.3–26.4	19.6	5.4	7.1–27.0	13.4	4.5	10.7–28.3	18.0	7.3
Carapace width (cm)	6.2–17.6	12.9	2.8	6.7–17.1	9.8	2.1	8.0–20.6	13.1	4.9
Plastron length (cm)	5.0–20.7	14.9	3.8	5.7–18.0	10.5	2.8	7.2–23.3	17.5	4.8
Plastron width (cm)	2.9–16.8	8.5	2.5	3.3–10.1	6.3	1.6	4.6–14.7	10.5	2.9
Bridge length (cm)	2.8–11.2	7.6	1.9	3.1–10.0	5.7	1.5	3.9–6.6	5.3	1.9
Height (cm)	3.1–10.5	8.1	1.7	3.3–9.9	6.2	1.4	4.6–7.6	6.3	1.5
Weight (g)	53–2298	1207.7	647.6	66–2371	46.5	475.8	NA		
Cloacal temperature (°C)	17.6–33.1	26.7	3.3	23.0–31.7	27.9	2.4	NA		
Ticks	0–7	1.8	2.0	0–2	0.1	0.4	NA		
Algal growth	Yes = 18.4%			Yes = 8.1%			NA		

more likely to be found on slopes in the morning (88.9%) and afternoon (59.3%) than in the evening (0%).

4. Discussion

4.1. Population Estimates. Our estimates suggest low population densities for both species, contrary to previous studies where *V. silvatica* was found at densities between 0.1–0.6 individuals per hectare [12, 15] and *I. travancorica* was known to be more widespread than *V. silvatica* [22]. This could have resulted from a difference in sampling strategy, time of survey, observer bias, or sparse distribution [23]. It could also have been the result of collection of chelonians for consumption by local communities, and the effect of forest fires such as in the case of the Chalakudy Forest Division where three or four major forest fires had broken out (Babu, Chalakudy, pers. comm.) since past fieldwork [15].

Our study suggests that *I. travancorica* is indeed more abundant and widespread than *V. silvatica* in terms of the habitats they occur in. This is also concurrent with parallel interview surveys with local communities [7]. *Indotestudo travancorica* may be more closely associated with water sources than *V. silvatica*, as reported in the literature ([15] but see [12]). *Vijayachelys silvatica* were found at slightly higher humidity, lower light intensity, and mostly where the canopy cover was high in comparison to *I. travancorica*, which was largely associated with medium canopy cover. *Vijayachelys silvatica* were more active on flat terrains while *I. travancorica* may be occupying slopes.

4.2. Vijayachelys silvatica. Similar to past studies [12, 24], most individuals encountered were not found associated with water sources and, being crepuscular, were inactive in leaflitter when encountered in the afternoon. While we encountered them mostly in semievergreen vegetation, *V. silvatica* is known to be a habitat specialist occurring in evergreen and semievergreen vegetation [9, 12]. All the wild individuals were found within or close to bamboo/cane clumps. While few studies suggest that bamboo/cane occurred in *V. silvatica* habitat [9, 14], others suggest there might not be a close association with this vegetation type [12, 24]. The turtles were found to occupy moist areas, which could promote algal growth on their carapace. The wild individuals were also found in areas where the herb and shrub cover were low in contrast to the previous studies [9, 24].

4.3. Indotestudo travancorica. The species is known to be crepuscular [25], which suggests why most of the individuals encountered during our surveys were inactive. Also, the tortoises could indeed be more active in the evenings and during the monsoons as the cloacal temperatures were higher than the air and soil temperatures during these intervals

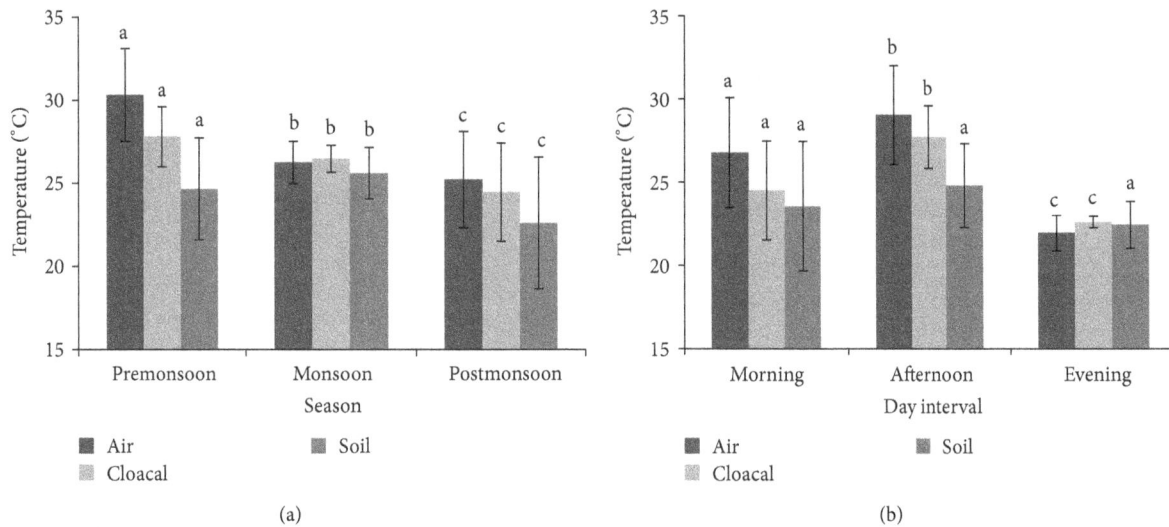

FIGURE 3: Variation in cloacal temperatures of *Indotestudo travancorica* with respect to air and soil temperatures during different (a) seasons (air temperature $P < 0.0001$, cloacal temperature $P = 0.008$ and soil temperature $P = 0.03$) and (b) day intervals (air temperature $P = 0.02$ and cloacal temperature $P = 0.004$). Different superscript letters indicate significant differences.

(Figures 3(a) and 3(b)). Similar to Ramesh [11], we found differences in shell height between gender but not anal fork length, which was not measured by us. We also found differences in bridge length. Individuals were found to be mostly using leaf litter, followed by leaf litter and bark and rock/mud cavities to conceal themselves as observed previously [23]. As observed by Ramesh [11], ticks were found attached either to the carapace or tail area of a large proportion of the tortoises ($n = 24$).

Our results expand the lower altitudinal extent of the species to 85 m. While differences are bound to arise at the two hill ranges due to variations in vegetation, our results suggest that tortoises in the Anamalai Hills utilised leaflitter and bark to conceal themselves more than those in the Agasthyamalai Hills. The tortoises were largely found in deciduous forests and an ecotone of moist deciduous and semievergreen forests, as is the case of previous surveys [9, 10]. However, Ramesh [11] detected more tortoises around wet evergreen forests, which could have resulted from varying sampling intensities in the different forest types, or because species distribution is unaffected by vegetation types but by the overall region as such. No chelonians were found around tree buttresses or thickets of *Lantana camara*/*Chromolaena glandulosum*, unlike previous studies [23, 25]. Our study also agrees with previous investigations suggesting that the species are found in rocky terrains on inclined surfaces [8, 9] and individuals could be moving towards inclined terrains in order to conceal themselves during the day while returning to flat terrains in the evening.

Most individuals were also found within or close to reed bamboo. The tortoises' association with bamboo thickets has been recounted by local communities and to an extent by earlier field surveys, but its high dependence on such vegetation is evident from 91% of their diet being composed of grass and bamboo remains [9, 12].

4.4. Threats and Chelonian Conservation. The study provides evidence that larger individuals of *I. travancorica* are immediately consumed by local communities while younger individuals are collected and reared till they reach sizes suitable for consumption (>1 kg) (Figure 2). Human collection and associated consumption of *V. silvatica* were found to be negligible in comparison to *I. travancorica*, which could be due to the increased difficulty in detection or smaller population sizes [9, 13, 26]. It could also have resulted from local taboos of bad luck and being unable to find any forest produce in the encountered area, similar to perceptions towards the slender loris *Loris lydekkerianus* [27], which inhibits consumption of *V. silvatica*. The consumption of *I. travancorica* from the wild could be affecting some populations as larger sized individuals (>27 cm) [11] were not found during our surveys. Similar to Bhupathy and Choudhury's [10] study undertaken two decades earlier, field surveys at Peppara and Neyyar WLS resulted in either none or a single detection while parallel surveys at human settlements resulted in more detections. Especially since the natural growth and survival rate of *I. travancorica* may not be high [28], efforts need to be undertaken to reduce collection of adult females and supplement existing populations through ex situ initiatives. The frequency of road kills is currently negligible in comparison to the extent of human consumption, which could also have resulted from insufficient surveys. However, vehicular traffic around the Athirapilly waterfalls should be managed beforehand to preempt instances of road kill of chelonians and other species. Chelonian deaths from development activities justify that future initiatives in forest areas need to be thoroughly monitored by the Forest Department, and existing pits need to be filled up immediately to stop them from functioning as death traps for numerous other species too.

The profile of these chelonians also needs to be enhanced through campaigns, in order to improve their appreciation and consequential conservation attention, especially among local communities and individuals with no educational qualification and mid-level incomes [29].

5. Conclusion

In spite of historic and continued use by indigenous communities, chelonians have not been extirpated from the landscape. Existing consumption rates could have probably caused local extirpation around human settlements as most wild chelonians in the area were found far away from human settlements. Utilization by local communities may need to be reduced to prevent further decline of these species [10], and such a policy would need to be developed through in-depth discussions with communities on sustainable harvest and management of this natural resource.

Acknowledgments

The first author would like to thank P. A. Kanagavel and Vijayalakshmi for their assistance in field logistics, S. Rajkumar for assistance during field surveys, Krishna Kumar for information on field sites, K. H. Amita Bachan for sharing shape files, Josin Tharian for toposheets, Susanna Paisley, Peter Bennett and Jonathan Baillie for their help during the project's formulation, Helen Meredith for her comments on the draft of the paper, and the academic editors—A. Chistoserdov and L. Luiselli for their comments which improved the final manuscript. The study was carried out with an official permission of the Kerala State Forest and Wildlife Department (WL 12-7326/2010) and financially supported by the Zoological Society of London (ZSL) Erasmus Darwin Barlow Expedition Grant 2010 and Rufford Small Grant for Nature Conservation (9190-1) to the first author.

References

[1] N. Myers, R. A. Mittermeler, C. G. Mittermeler, G. A. B. da Fonseca, and J. Kent, "Biodiversity hotspots for conservation priorities," *Nature*, vol. 403, no. 6772, pp. 853–858, 2000.

[2] K. S. Bawa, A. Das, J. Krishnaswamy, K. U. Karanth, N. S. Kumar, and M. Rao, *CEPF Western Ghats and Sri Lanka ecosystem profile*, ATREE & Wildlife Conservation Society, Bengaluru, India, 2007.

[3] N. Dahanukar, R. Raghavan, A. Ali, R. Abraham, and C. P. Shaji, "The status and distribution of freshwater fishes of the Western Ghats," in *The Status of Freshwater Biodiversity in the Western Ghats*, S. Molur, K. G. Smith, B. A. Daniel, and W. R. T. Darwall, Eds., pp. 21–48, Compilers, International Union for Conservation of Nature (IUCN) Gland, Switzerland and Zoo Outreach Organization (ZOO), Coimbatore, India, 2011.

[4] Asian Turtle Trade Working Group, *Vijayachelys silvatica* IUCN Red List of Threatened Species Version 2011, IUCN, Gland, Switzerland, 2000.

[5] Asian Turtle Trade Working Group, *Indotestudo travancorica* IUCN Red List of Threatened Species Version 2011, IUCN, Gland, Switzerland, 2000.

[6] V. Deepak and K. Vasudevan, *Endemic Turtles of India*, Wildlife Institute of India, Dehradun, India, 2009.

[7] A. Kanagavel and R. Raghavan, "Local ecological knowledge of the threatened Cochin Forest Cane Turtle *Vijayachelys silvatica* and Travancore Tortoise *Indotestudo travancorica* from the Anamalai Hills of the Western Ghats, India," *Journal of Threatened Taxa*, vol. 4, no. 13, pp. 3173–3182, 2012.

[8] J. Vijaya, "Cane turtle (*Heosemys silvatica*) study project in Kerala," *Hamadryad*, vol. 9, p. 4, 1984.

[9] K. S. Appukuttan, *Cane turtle & Travancore tortoise—A survey report*, KFRI, Peechi, India, 1991.

[10] S. Bhupathy and B. C. Choudhury, "Status, distribution and conservation of the Travancore tortoise, Indotestudo forstenii in Western Ghats," *Journal of the Bombay Natural History Society*, vol. 92, pp. 16–21, 1995.

[11] M. Ramesh, "Relative abundance and morphometries of the Travancore tortoise, *Indotestudo travancorica*, in the Indira Gandhi Wildlife Sanctuary, Southern Western Ghats, India," *Chelonian Conservation and Biology*, vol. 7, no. 1, pp. 108–113, 2008.

[12] K. Vasudevan, B. Pandav, and V. Deepak, *Ecology of Two Endemic Turtles in the Western Ghats*, Wildlife Institute of India, Dehradun, India, 2010.

[13] J. Vijaya, "Status of the forest cane turtle (*Geomyda silvatica*)," *Hamadryad*, vol. 13, p. 10, 1988.

[14] J. Jose, K. K. Ramachandran, and P. V. Nair, "Occurrence of the forest cane turtle Geoemyda silvatica (Reptilia, *Testudines, Bataguridae*) from a myristica swamp of Kulathupuzha forest range, Southern Kerala," *ENVIS Newsletter*, vol. 3, pp. 3–4, 2007.

[15] N. Whitaker and J. Vijaya, "Biology of the forest cane turtle, *Vijayachelys silvatica*, in South India," *Chelonian Conservation and Biology*, vol. 8, no. 2, pp. 109–115, 2009.

[16] M. A. Kumar, *Effect of Habitat Fragmentation on Asian Elephant (Elephas maximus) Ecology and Behaviour Patterns in a Conflict-Prone Plantation Landscape of the Anamalai Hills, Western Ghats, India*, Rufford Maurice Laing Foundation, London, UK, 2006.

[17] M. Chandi, *Tribes of the Anamalais: Livelihood and Resource-Use Patterns of Communities in the Rainforests of the Indira Gandhi Wildlife Sanctuary and Valparai Plateau*, Nature Conservation Foundation, Mysore, India, 2008.

[18] R. Pillay, A. J. T. Johnsingh, R. Raghunath, and M. D. Madhusudan, "Patterns of spatiotemporal change in large mammal distribution and abundance in the Southern Western Ghats, India," *Biological Conservation*, vol. 144, no. 5, pp. 1567–1576, 2011.

[19] T. R. S. Raman and D. Mudappa, "Correlates of hornbill distribution and abundance in rainforest fragments in the Southern Western Ghats, India," *Bird Conservation International*, vol. 13, no. 3, pp. 199–212, 2003.

[20] H. Newing, *Conducting Research in Conservation: Social Science Methods and Practice*, Routledge, London, UK, 2010.

[21] M. K. Soisalo and S. M. C. Cavalcanti, "Estimating the density of a jaguar population in the Brazilian Pantanal using camera-traps and capture-recapture sampling in combination with GPS radio-telemetry," *Biological Conservation*, vol. 129, no. 4, pp. 487–496, 2006.

[22] B. Groombridge, E. O. Moll, and J. Vijaya, "Rediscovery of a rare Indian turtle," *Oryx*, vol. 17, no. 3, pp. 130–134, 1984.

[23] V. Deepak, M. Ramesh, S. Bhupathy, and K. Vasudevan, "*Indotestudo travancorica* (Boulenger 1907)—Travancore tortoise," in *Conservation Biology of Freshwater Turtles and Tortoises: A Compilation Project of the IUCN/SSC Tortoise and*

Freshwater Turtle Specialist Group Chelonian Research Monographs, A. G. J. Rhodin, P. C. H. Pritchard, P. P. van Dijk et al., Eds., pp. 54.51–54.56, IUCN/SSC Tortoise and Freshwater Turtle Specialist Group, Gland, Switzerland, 2011.

[24] E. O. Moll, B. Groombridge, and J. Vijaya, "Rediscription of the cane turtle with notes on its natural history and classification," *Journal of the Bombay Natural History Society*, vol. 83, pp. 112–126, 1986.

[25] M. Ramesh, "Preliminary survey of *Indotestudo travancorica* (*Testudinidae*) at the Indira Gandhi Wildlife Sanctuary, Southern India," *Hamadryad*, vol. 33, pp. 118–120.

[26] A. Kanagavel, "A new record of the Cochin Forest Cane Turtle *Vijayachelys silvatica* (Henderson, 1912) from Shendurney Wildlife Sanctuary, Kerala, India," *Reptile Rap*, vol. 15, pp. 3–6, 2013.

[27] A. Kanagavel, C. Sinclair, R. Sekar, and R. Raghavan, "Moolah, misfortune or spinsterhood? The plight of slender loris *Loris lydekkerianus* in Southern India," *Journal of Threatened Taxa*, vol. 5, no. 1, pp. 3585–3588, 2013.

[28] J. McDougal and C. Castellano, "Husbandry and captive breeding of the Travancore tortoise (Indotestudo travancorica) at the Wildlife Conservation Society," in *Advances in Herpetoculture*, pp. 51–56, 1996.

[29] A. Kanagavel, R. Raghavan, and D. Verissimo, "Beyond the "General Public": implications of audience characteristics for promoting species conservation in the Western Ghats Hotspot, India," *AMBIO*, 2013.

13

A Study on Exploration of Ethnobotanical Knowledge of Rural Community in Bangladesh: Basis for Biodiversity Conservation

Md. Habibur Rahman

Bangladesh Institute of Social Research (BISR), Hasina De Palace, House No. 6/14, Block No. A, Lalmatia, Dhaka 1207, Bangladesh

Correspondence should be addressed to Md. Habibur Rahman; habibmdr@gmail.com

Academic Editors: A. R. Atangana, M. Tigabu, and J. Zhang

Rural home garden is an important traditional source of medicinal plants for daily curative uses throughout Bangladesh. Such knowledge is continuing from generation to generation without documentation. An ethnobotanical investigation was conducted through focus group discussions and households' survey accompanied by field observation to document the indigenous knowledge of herbal medicines being used by the rural communities of Comilla district in Bangladesh. A total of 45 ethnomedicinal plant species belonging to 34 families were found, where trees (37.78%) were the most commonly utilized growth form. Plant resources are used to treat 31 different individual ailments ranging from simple cuts to heart disease. Plants are mainly used to treat dysentery (12 species), cold ailments, cough, and fever (6 species each). For curing ailments, the use of the above ground plant parts was higher (86.44%); particularly fruits (37.29%) and leaves (30.51%) were the most commonly used plant parts. More than half of the medicinal plants are indigenous (71.11%), being edible fruit bearer (48.89%), plants parts suitable for animals and birds (57.78%), and natural regeneration present (64.44%) indicated that medicinal plants play a vital role in biodiversity conservation in the study area.

1. Introduction

Over the past decade, there has been a dramatic increase in the demand for medicinal plants for use in traditional medicine and contemporary and alternative medicine in both developing and developed countries [1]; thus, a large number of people habitually use such medication because herbal treatment is, in some cases, considered relatively cheap [2]. However, its popularity also stems from the efficacy of the treatment in most cases and relative safety, with few or no side effects. Herbal medicines, because of their decentralized nature, are generally easily and quickly available [3]. Now, both developed and developing countries are involved in plants-based herbal medicine system, as modern pharmaceuticals are not accessible for all [4].

Bangladesh is a country that is considered rich in medicinal plants genetic resources by virtue of its favorable agroclimatic condition and seasonal diversity. With productive soils, a tropical climate, and seasonal diversity, Bangladesh contains about 6500 plants species including bryophytes,

pteridophytes, gymnosperms, and angiosperms; among them, 500 plant species have medicinal values [5] and grow in the country's forests, wetlands, homestead forests, and even roadside as indigenous, naturally occurring, or cultivated plants [6, 7]. About 75% (10 million households in over 85,000 villages) of the country's total population lives in rural areas [8], and almost 80% is dependent on natural resources (e.g., medicinal plants) for their primary healthcare [9], with herbal medication remaining a popular and accepted form of treatment [10].

Rural peoples are capable to identify many species of plants that are capable of producing various products, including food, firewood, medicine, forage, and daily needs tools [11], and the customary homestead tree production system also serves as a source of plant products and remedies [12]. Despite such a high demand of herbal medicine, medicinal plants sector is now the most promising business sector in Bangladesh [13] with the presence of more than 500 companies producing herbal medicines [14], and more than 90% of the plants and products needed to meet domestic demand

A Study on Exploration of Ethnobotanical Knowledge of Rural Community in Bangladesh: Basis
for Biodiversity Conservation

147

are imported from other countries, such as India, Nepal, and Pakistan [15].

The indigenous knowledge (IK) (IK develops through sharing experience and is normally passed orally between generations) concerning medicinal plants is lost owing to the change of habitats and culture of rural communities in Bangladesh. The district of Comilla is situated in the eastern part of Bangladesh, a district devoid of natural forests (out of the 64 districts of Bangladesh, 28 districts do not have any natural forests). Rajendrapur Sal forest (*Shorea robusta* C.F. Gaertn) and degraded small Sal patches in *Kotbari* and *Lalmai* hills are most notable because these were the only planted forests in the district, and Sal forest and *Lalmai* hills have historical importance. The district was chosen because it is a transitional zone between the southeastern vast hilly region (i.e., Chittagong Hill Tracts) and other plains regions in the southern and middle of the country. Although there are no significant forest areas, there is a great diversity of medicinal plants in the homestead gardens, roadsides, pond banks, and graveyards in this region. In conjunction with the increasing demand for traditional medicines in the country, research is being conducted outside the forest areas to explore the ethnobotanical documentation of the medicinal plants; for example, [9, 12, 16–27] are common to mention, but no study has been found on the utilization of plants for health-care practices by the rural people in Comilla district of the country. Therefore, the study was undertaken in Burichong Upazila (subdistrict, an administrative entity) of Comilla district of Eastern Bangladesh. Its aim was to assess plant-based ethnomedicinal practice and document IK associated with it, traditional beliefs, and biodiversity conservation related to IK.

2. Materials and Methods

2.1. Study Area. Burichong Upazila with an area of 163.76 sq. km. is located at 23°55′00″N 91°12′64″E (Figure 1). This area is bounded by Brahmanpara Upazila on the north, Comilla Sadar and Barura Upazilas on the south, Tripura state of India on the east, and Debidwar and Chandina Upazilas on the west. It has 37739 units of households. The Upazila supports a population of 259,265; 133,469 male and 125,796 female (population density 1609.98 per sq. km.) with a literacy rate of 49.7%. The Upazila consists of 8 Union Parishads (last stage of administrative entity) and 171 villages. The main rivers are Gomti, Gongur, Titi, and Pagli rivers; the landscape comprises the Tripura valley of India and green agricultural fields. The Upazila is more or less flooded during the rainy season, resulting in deposition of sediments that increase the productivity of land, ultimately contributing to its botanical diversity. The main occupation is agriculture (56.64%), followed by agricultural labour (12.28%), commerce (10.23%), service (8.08%), and so forth [28]. However, Burichong Upazila is considered as one of the most densely homestead forests-covered area with plantations of both indigenous and exotic species and understory covered with rich herbs and shrubs.

2.2. Research Methods. In order to document the utilization of medicinal plants, four field surveys were carried out from early July to late October 2010 in the study area, using a multistage random sampling technique. The surveys were spread across the whole Upazila at random: one in the north (Rajapur), one in the southwest (Mokam), and the other in the middle of the Upazila (Burichong), so as to get maximum information and also to cross check the information provided by the local informants during the earlier visits. From each of the three villages, 30 households (irrespective of socioeconomic condition) were selected randomly for the comprehensive study. Thus, a total of 90 households were selected. Before the household survey, casual field visits were arranged within the villages with local old people, religious leaders, and other key informants to review and document the availability of medicinal plants in the locality. Informal meetings were held in the interviewee's home using the native language (Bengali). The household heads were the key respondents, with help from other family members when necessary. In addition, six focus group discussions (FGDs), two in each village, were arranged in the tea stalls of local market where the rural people usually get together, gossip, and interact in the evening after the daylong business. Information on the local name of plant, plant part used for curing, method of dosage, and administration was recorded. After, the interviews, collected information was cross-checked by the local herbal practitioners locally referred to as *kabiraj*. They had sound knowledge on medicinal plants and were therefore highly rated in the society.

Respondents were interviewed using a semistructured questionnaire and focalized interview to ascertain the plant species and the parts used, for what diseases, the sources they prefer, the reasons for cultivating any plant, and so forth. The plant species used for medicine were firstly identified by local names by the help of *kabiraj* and old-aged persons. The scientific names were obtained by consulting the literature [29, 30]. A final list of the species used for medicinal purposes was cross-checked and prepared based on the study by Dey [29].

3. Results and Discussion

3.1. Medicinal Plants Species Composition. A total of 45 ethnomedicinal plant species including herb, shrub, tree, palm, and vine distributed across 34 families were documented in the study to be used by the rural community for curing different ailments. For the utilization frequency of the plant species, Palmae and Rutaceae appear as the most prominent families (3 species each), followed by Compositae, Cucurbitaceae, Euphorbiaceae, Labiatae, Liliaceae, Myrtaceae, and Zingiberaceae (2 species each) (Table 1). Various researchers across the country, for example, Leguminosae [12, 21], Compositae, Combretaceae, Leguminosae, Liliaceae, and Rutaceae [9], Fabaceae [16, 19], Euphorbiaceae and Lamiaceae [17], Convolvulaceae, Leguminosae, Solanaceae, and Sterculiaceae [22], and Fabaceae and Solanaceae [18] also recorded that the species under these families were frequently used as medicinal plants in rural Bangladesh. Among the recorded

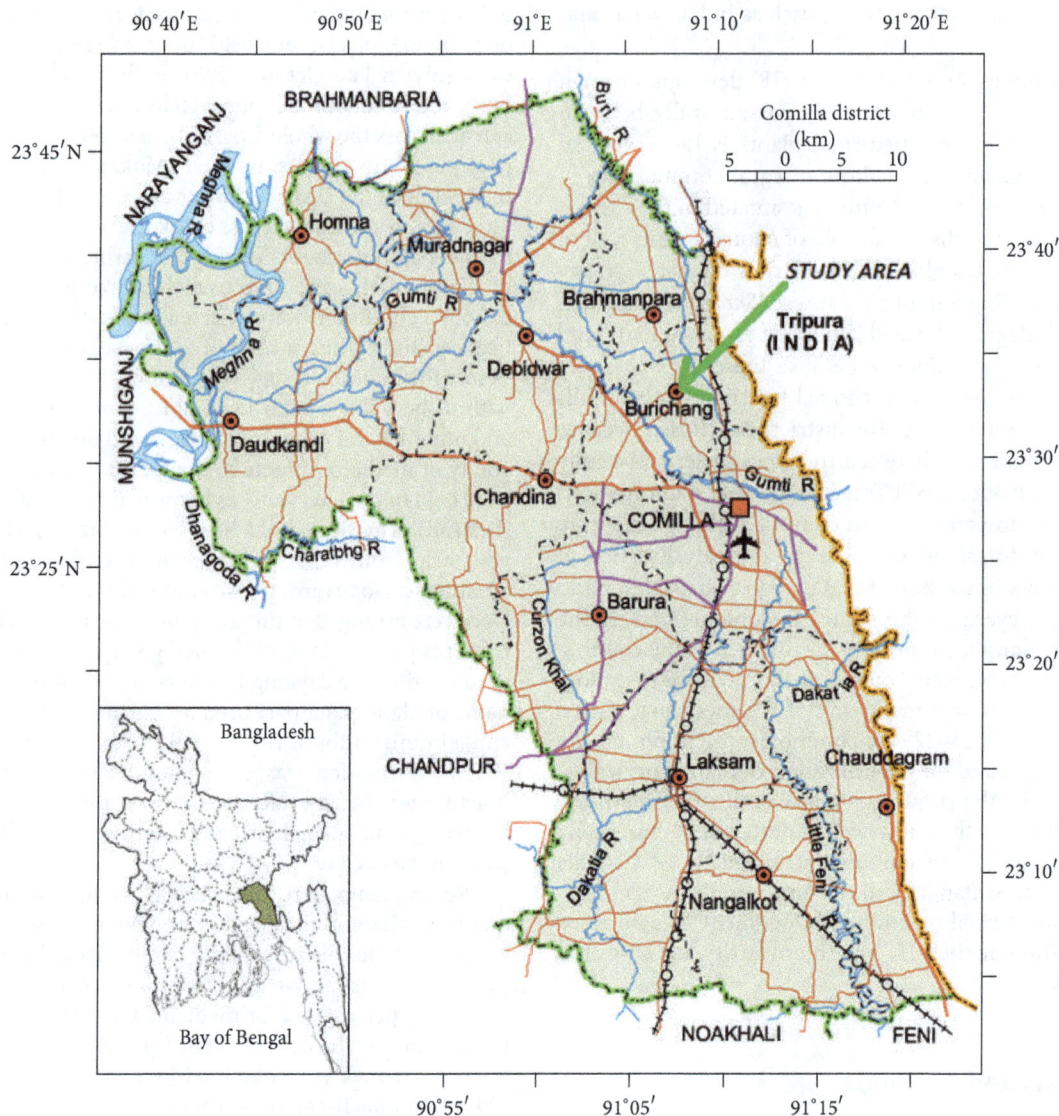

FIGURE 1: Location map of the study area.

plants, trees were the most frequent growth form (37.78%), followed by herbs (24.44%), shrubs (22.22%), vine (8.89%), and palms (6.67%). A similar trend was also observed that trees were the most used growth form of medicinal plants in Bangladesh [2, 9, 31–36], but with a few exceptions [12, 37] where they found that herbs were mainly used as medicinal plants.

3.2. Indigenous Ethnobotanical Knowledge, Pattern, and Ailments. The survey revealed that rural people used various parts of the plants as medicine. The diverse pattern of various parts of medicinal plants (Table 1) reflected greater possession of IK regarding their health care practices by the people. Most of the medicinal plant parts are consumed after macerating, squeezing, grinding, blending, soaking, or boiling, and some are taken raw. Some are applied externally to different body parts for cuts and wounds, scabies, joint pain, piles, skin diseases, and so forth. Fifteen species like *Adhatoda vasica*

Nees., *Artocarpus heterophyllus* Lamk., *Azadirachta indica* A. Juss., *Centella asiatica* (L.) Urban, *Cocos nucifera* L., *Lawsonia inermis* L., *Phoenix sylvestris* Roxb., *Phyllanthus emblica* L., *Piper betle* L., *Plantago ovate*, *Psidium guajava* (L.) Bat., *Ocimum sanctum* L., *Swertia chirata*, *Tamarindus indica* L., and *Terminalia chebula* (Gaerth.) Retz. were used against up to four ailments. One unusual use of fresh mango leaf and piece of branches is as a toothbrush, without any toothpaste, to keep teeth healthy; another is the sun-dried seed of *Syzygium cumini* (L.) Skeel., which, after grinding into powder, is taken with salt as a regular treatment for diabetes. Another frequently used medicinal plant is the vine *Momordica charantia* L., where both the leaf and fruit are used against diabetes. Similarly, used patterns of these plant parts are recorded in rural areas of Bangladesh [9].

Medicinal plants are generally used to treat fever, coughs, cuts and wounds, cold ailments, toothache, hair loss, dandruff, skin diseases, joint pain, stomach problem, dysentery

A Study on Exploration of Ethnobotanical Knowledge of Rural Community in Bangladesh: Basis for Biodiversity Conservation

149

TABLE 1: Composition, parts used, and ailments of recorded medicinal plants in the study area.

Local name	Scientific name	Habit	Family	Frequency (%)[1]	Parts used	Ailments	I/E[2]	F/NF[3]	N/A[4]
Aam	*Mangifera indica* L.	Tree	Anacardiaceae	85	Leaves, tender twig, fruit	Teeth disease, wounds	I	F	N
Ada	*Zingiber officinale* Roxb.	Herb	Zingiberaceae	61	Rhizome	Cough, cold ailments	I	NF	N
Akanda	*Calotropis gigantea* (L.) Ait. f.	Shrub	Asclepiadaceae	42	Leaves, latex	Joint pain, cut and wounds	I	NF	N
Amloki	*Phyllanthus emblica* L.	Tree	Euphorbiaceae	33	Fruit	Dysentery, skin diseases, hair fall, indigestion	I	F	N
Anarosh	*Ananas comosus* (L.) Merr.	Shrub	Bromeliaceae	36	Fruit	Jaundice, worm	E	F	A
Aorboroi	*Phyllanthus acidus* (L.) Skeels	Tree	Euphorbiaceae	29	Fruit	Hair fall, indigestion	E	F	A
Assampata	*Eupatorium odoratum* L.	Shrub	Compositae	47	Green leaves	Antihemorrhoid	E	NF	N
Bashok	*Adhatoda vasica* Nees.	Shrub	Acanthaceae	51	Green leaves	Cough, cold ailments, asthma, dysentery	I	NF	A
Bel	*Aegle marmelos* (L.) Correa.	Tree	Rutaceae	64	Fruit	Weakness, constipation, diarrhea, dysentery	I	F	N
Boroi	*Zizyphus mauritiana* Lamk.	Tree	Rhamnaceae	66	Fruit	Cough, cold, lethargy to food	I	F	N
Chirata	*Swertia chirata* Ham.	Herb	Sapindaceae	13	Whole plant	Gastric pain, diabetes, stomach trouble, fever	E	NF	A
Dalim	*Punica granatum* L.	Tree	Punicaceae	40	Leaves, fruit	Worms in intestine	E	F	N
Durba grass	*Cynodon dactylon* (L) Pers.	Herb	Poaceae	69	Tender leaves	Toothache, cuts and wounds	I	NF	N
Holud	*Curcuma longa* L.	Herb	Zingiberaceae	54	Rhizome	Skin diseases	I	NF	N
Haritoki	*Terminalia chebula* (Gaerth.)Retz.	Tree	Combretaceae	15	Fruit	Constipation, fever, heart disease, cough, urinary problems, loss of appetite	I	F	N
Isabgol	*Plantago ovata*	Herb	Compositae	15	Seed	Heat stock, gastric pain, constipation, male sexual weakness	I	F	N
Jam	*Syzygium cumini* (L.) Skeel.	Tree	Myrtaceae	71	Fruit, seed	Blood purification, diabetes	I	F	N
Jambura	*Citrus grandis* (L.) Osb.	Tree	Rutaceae	59	Fruit	Jaundice, lethargy to food	I	F	A
Jolpai	*Elaeocarpus robustus* Roxb.	Tree	Elaeocarpaceae	52	Fruit	Lethargy to food	E	F	A
Kamranga	*Averrhoa carambola* L.	Tree	Averrhoaceae	66	Fruit	Hair fall, jaundice, weakness	I	F	N
Kanthal	*Artocarpus heterophyllus* Lamk.	Tree	Moraceae	83	Fruit, seed	Wound, diarrhea, constipation, stomach trouble	I	F	N
Khejur	*Phoenix sylvestris* Roxb.	Palm	Palmae	22	Fruit, juice	Constipation, jaundice, weakness	E	F	N
Kachu	*Colocasia esculenta* Schott.	Herb	Araceae	47	Whole plant	Cuts and wounds	I	NF	N
Kola	*Musa sapientum* L.	Shrub	Musaceae	68	Fruit	Dysentery	I	F	N
Korola	*Momordica charantia* L.	Vine	Cucurbitaceae	42	Leaves, fruit	Diabetes, dysentery	I	F	N
Lebu	*Citrus limon* (Linn.) Burm. f.	Tree	Rutaceae	60	Fruit, leaves	Dysentery; indigestion and lethargy to food, weakness	I	F	A
Laijabati	*Mimosa pudica* L.	Herb	Mimosoideae	39	Root	Dysentery, piles	E	NF	N
Maya lata	*Mikania cordata* (Burm. F.) Roxb.	Vine	Compositae	38	Green leaves	Cuts and wounds	E	NF	N

TABLE 1: Continued.

Local name	Scientific name	Habit	Family	Frequency (%)[1]	Parts used	Ailments	I/E[2]	F/NF[3]	N/A[4]
Mehendi	Lawsonia inermis L.	Shrub	Lythraceae	62	Leaves	Dandruff, hair color, skin diseases	I	NF	A
Narikel	Cocos nucifera L.	Palm	Palmae	72	Fruit, juice	Hair fall, burns, dysentery, weakness	I	F	A
Neem	Azadirachta indica A. Juss.	Tree	Meliaceae	69	Green leaves and seed	Skin diseases, chicken pox, fever, dysentery, intestinal worm	I	NF	N
Paan	Piper betle L.	Vine	Piperaceae	17	Green leaves	Dysentery, loss of appetite, indigestion, stomach trouble	I	NF	A
Papaya	Carica papaya L.	Shrub	Caricaceae	72	Fruit	Stomach trouble	I	F	N
Pathor kuchi	Kalanchoe pinnata (Lam.) Pers.	Herb	Crassulaceae	17	Leaves	Cough	E	NF	N
Peyara	Psidium guajava (L.) Bat.	Tree	Myrtaceae	74	Green leaves, fruit	Diarrhea, fever, cuts and wounds	I	F	N
Piaj	Allium cepa L.	Herb	Liliaceae	39	Whole plant	Cold ailments, dandruff	I	NF	A
Roshun	Allium sativum L.	Herb	Liliaceae	26	Whole plant	Heart disease, urinal problem	I	NF	A
Sajna	Moringa oleifera Lamk.	Tree	Moringaceae	31	Fruit, leaves, bark	Joint pain, sexual diseases	E	F	A
Shewra	Streblus asper Lour.	Shrub	Urticaceae	44	Green leaves	Skin disease, joint pain	E	NF	N
Supari	Areca catechu L.	Palm	Palmae	69	Seed	Vomiting	I	F	A
Telakucha	Coccinia cordifolia L.	Vine	Cucurbitaceae	19	Green leaves	Cold ailments, diabetes	E	NF	N
Tetul	Tamarindus indica L.	Tree	Leguminosae	63	Fruit, tender leaves	Loss of appetite, fever, dysentery, lethargy to food	I	F	N
Thankuni	Centella asiatica (L.) Urban	Herb	Umbelliferae	38	Whole plant	Dysentery, diarrhea, gastric pain, piles	I	NF	N
Tokma	Hyptis suaveolens Poit.	Shrub	Labiatae	14	Seed	Gastric pain, fever, burning	I	F	A
Tulsi	Ocimum sanctum L.	Shrub	Labiatae	19	Green leaves, seed	Cough, cold ailments, cuts and wounds	I	NF	A

Note: [1]Frequency: number of household reported medicinal plants found in relation to the total number of species that a particular species was cited.
[2]I/E: I indicates indigenous, and E indicates exotics.
[3]F/NF: F indicates edible fruit/seed bearing plants, and NF indicates non fruit-seed bearing plants.
[4]N/A: N indicates natural regeneration found, and A indicates natural regeneration absent.

A Study on Exploration of Ethnobotanical Knowledge of Rural Community in Bangladesh: Basis
for Biodiversity Conservation

151

TABLE 2: Habit-wise distribution of medicinal plants for different ailments.

Name of ailments	Number of species reported					Total species*
	Tree	Herb	Shrub	Vine	Palm	
Antihemorrhoid	—	—	1	—	—	1 (2.22)
Asthma	—	—	1	—	—	1 (2.22)
Blood purification	1	—	—	—	—	1 (2.22)
Burning	—	—	1	—	1	2 (4.44)
Chicken pox	1	—	—	—	—	1 (2.22)
Cold ailments	1	2	2	1	—	6 (13.33)
Constipation	3	1	—	—	1	5 (11.11)
Cough	2	2	2	—	—	6 (13.33)
Cuts and wounds	1	1	2	1	—	5 (11.11)
Dandruff	—	1	1	—	—	2 (4.44)
Diabetes	1	1	—	2	—	4 (8.89)
Diarrhea	3	1	—	—	—	4 (8.89)
Dysentery	5	2	2	2	1	12 (26.67)
Fever	4	1	1	—	—	6 (13.33)
Gastric pain	—	3	1	—	—	4 (8.89)
Hair fall and color	3	—	1	—	1	5 (11.11)
Heart disease	1	1	—	—	—	2 (4.44)
Indigestion	3	—	—	1	—	4 (8.89)
Jaundice	2	—	1	—	1	4 (8.89)
Joint pain	1	—	2	—	—	3 (6.67)
Lethargy to food	5	—	—	—	—	5 (11.11)
Loss of appetite	2	—	—	1	—	3 (6.67)
Piles	—	2	—	—	—	2 (4.44)
Sexual problem	1	1	—	—	—	2 (4.44)
Skin diseases	2	1	2	—	—	5 (11.11)
Stomach trouble	1	1	1	1	—	4 (8.89)
Toothache	1	1	—	—	1	3 (6.67)
Urinal problem	—	1	—	—	—	1 (2.22)
Vomiting	—	—	—	—	1	1 (2.22)
Worms	1	—	1	—	—	2 (4.44)
Weakness	2	—	—	—	2	4 (8.89)

*Parenthesis shows the percentage value.

and diarrhea (Table 2). Twelve species (26.67%) are used against dysentery (five trees, two herbs, shrubs and vine, and one palm species). Cold ailments, cough, and fever are treated with six species (13.33%); constipation, cuts and wounds, hair fall and color, lethargy to food, and skin diseases are treated with five species (11.11%) each. More than four species (8.89%) are used for treating common conditions of diabetes, diarrhea, gastric pain, indigestion, jaundice, stomach trouble, and weakness. In some cases, a mixture of several species is also used for treating one disease.

For curing ailments, the use of the above ground plant parts was higher (86.44%) than the whole plants (8.47%) and under ground plant parts (5.08%). Out of the above ground plant parts, fruits (37.29%) and leaves (30.51%) were used in the majority of cases, followed by seeds (11.86%) (Table 3). In most cases, the juice from leaves, root, rhizome, and bark is used as medicine, while fruits are eaten raw.

TABLE 3: Utilization of plant parts of the medicinal plant species.

Plants parts used	Individual species	Percentage
Fruits	22	37.29
Leaves	18	30.51
Seed	7	11.86
Whole plant	5	8.47
Root/rhizome	3	5.08
Bark	1	1.69
Juice	2	3.39
Latex	1	1.69

3.3. *Traditional Sources of Medicinal Plants.* Home gardens in the study area are generally maintained for household consumption like supplying fruit, timber, fuel wood, and fodder. Additionally, due to the availability of medicinal plants in their home gardens, they are also dependent on home gardens for their daily herbal medicine. The medicine is generally prepared by elder family members who have good knowledge on the medicinal value of plants, of which those species are usually used to treat common diseases such as cough, cold ailments, and cuts and wounds, all from the plants available in the surrounding home garden, roadside, ponds and canal bank, graveyards, jungle, fallow land, hinterland, and so forth. The villagers prefer graveyards mainly covered by herbs and shrubs, to reduce the problems arising from large tree roots disturbing graves. Hinterlands behind homesteads are usually kept fallow and unproductive; in some cases, they are used for household waste disposal and as space for domestic poultry, with some herbaceous species planted. Stepp and Moerman [38] and Gazzaneo et al. [39] found a similar trend of collecting medicinal plants from anthropogenic habitats by the Maya communities of Mexico and the local herbal specialists of northeastern Brazil, respectively. A study in West Africa showed that, among the harvested species, local community used 90% of the species for medicinal purposes [40]. During the study, the respondents only buy medicinal plant parts, for example, fresh or dried forms from village markets when the species are not prevalent in their home garden. Plants, which have multiple uses, such as vegetables and spices, were cultivated in the homestead and agriculture field. Respondents also shared parts and fruits from plants with each other, so that the demands of neighbors as well as people living further away can easily be met.

3.4. Traditional Beliefs.

Study found that most of the plant parts used for curing ailments are gathered from home gardens, but the people of the study area are destroying the medicinal plants resource due to lack of proper harvesting techniques and lack of awareness about this resource. Conversely, some rural people raise certain species with medicinal properties, particularly *Areca catechu* L., *A. indica*, *Citrus grandis* (L.) Osb., *C. limon* (L.) Burm. f., *C. nucifera*, *L. inermis*, and *O. sanctum* which are usually planted surrounding the homestead especially on the southern side (*A. indica*) that air from the south is purified by its foliage; this is believed and noticed by the local people from the study area.

Aegle marmelos (L.) Correa., *A. indica*, *L. inermis*, *Mangifera indica* L., *O. sanctum*, *P. betle*, *Zizyphus mauritiana* Lamk., and fruit of *C. nucifera* are viewed as sacred and culturally important plants by the Hindu and Muslim religious communities. The leaf of *L. inermis* L. is often used in dyeing the hand palms of bridal couples, women and children on cultural, ceremonial, and religious occasions. While Chowdhury et al. [41] reported the livelihood potential of the commercial farming of *L. inermis* L. in the central part of the country, Cartwright-Jones [42] noticed that *L. inermis* leaf paste is popular as an adornment for weddings and other celebrations in South Asia, the Middle East, and Africa. The orange dye obtained from the leaves is also used for dyeing hair, beards, eyebrows, fingernails, and palms.

The paste made from rhizome of the *Curcuma longa* L. is used in dyeing the full body of bridal couples for both communities. In the Hindu community, leaves of *O. sanctum* are used commonly in their worship. *T. indica* is planted either in the periphery of homesteads or in fallow lands of the backyard with a belief that evil spirits take shelter on its crown. Miah and Rahman [43] also reported on these plants and their cultural, ceremonial, and religious uses having positive effects on the flora of the Muslim and Hindu homestead forests in Bangladesh. In the study area, more than half of the medicinal plants (tree species) are being edible fruit bearers as they are planted purposively by the local people for the seasonal fruit not for the medicinal purposes.

3.5. Ethnobotanical Knowledge Regarding Biodiversity Conservation.

Home gardens have long been the most effective and widespread measure for biodiversity conservation in Bangladesh due to anthropogenic pressure and land-use change affecting the natural forest which has been decreasing day by day both in explicit and implicit ways leading to threats to future productivity. Among the identified species, about 71.11% of plants are indigenous species, and 28.89% of plants are exotic species. To get a quick cash return now, the largest part of home garden owners are interested in the plantation of exotic plant species although it has a negative effect on biodiversity. It was found that 48.89% of plants produce edible fruits and seeds, 28.89% of the plants parts are used as vegetables and spices for daily cooking purposes by the people. Generally, rural communities preferred cultivated and planted multipurpose species that can be served as fruits, vegetables and spices that also can be used as timber. Such kind of choice is the most important factor to home gardens conservation in Bangladesh. This clearly plays a significant role in forest biodiversity conservation since all the wood and other nontimber products that are harvested in the home gardens do not need to be collected from natural forests. Some species (e.g., *P. ovate*, *Kalanchoe pinnata* (Lam.) Pers., *Hyptis suaveolens* Poit., and *O. sanctum* L.) are grown in earthen pots and kept in front of dwelling houses, serving for beautification purposes. About 57.78% of plants are suitable for animals' and birds' conservation because the fruits and seeds of these plants are widely used as food for birds and animals. At this time, dispersal of seeds also occurs and helps in natural regeneration of plants species. In the present study, 64.44% of species are found to be naturally regenerated, and most of these are indigenous (Table 4).

The conservation of plants in home gardens not only preserves a vital resource for humankind but also provides significant economic and nutritional benefits for the rural poor [44]. During the study, it was found that collecting and selling the whole plant of *Allium cepa* L., *Allium sativum* L., *C. asiatica*, *Colacasia esculenta* Schott., and *S. chirata*; the fruit of *M. indica*, *Phyllanthus emblica* L., *Aannas comosus* (L.) Merr., *Phyllanthus acidus* (L.) Skiels, *A. marmelos*, *Z. mauritiana*, *Punica granatum* L., *T. chebula*, *S. cumini*, *C. grandis*, *Elaeocarpus robustus* Roxb., *Averrhoa carambola* L., *A. heterophyllus*, *P. sylvestris*, *Musa sapientum* L., *C. limon*, *M. charantea*, *C. nucifera*, *Carica papaya* L., *P. guajava*, and *T. indica*; the leaves with rhizome of *C. esculenta* and *Zingiber officinale* Roxb.; moreover the seeds of *P. ovate*, *A. heterophyllus*, *A. catechu*, *H. suaveolens* Poit., and *O. sanctum* play a significant role in household's cash income generation as well

A Study on Exploration of Ethnobotanical Knowledge of Rural Community in Bangladesh: Basis
for Biodiversity Conservation

153

TABLE 4: Status of medicinal plants species from biodiversity conservation point of view.

Species category	Individual species	Percentage
Indigenous	32	71.11
Exotics	13	28.89
Plants produce edible fruits/seeds	22	48.89
Plants parts used as spices/vegetables	13	28.89
Plants parts suitable for animals and birds	26	57.78
Natural regeneration present	29	64.44
Natural regeneration absent	16	35.56

as meet the body requirement of vitamins. Such economic contribution of these species also has a significant role in species conservation and poverty reduction in the study area. Several studies [40, 45–48] found that medicinal plants play a key role in sustaining the rural livelihood and contributing to poverty reduction.

3.6. Traditional Management System. In terms of the level of management in cultural practices, traditional production system of home gardens in Bangladesh is very poor. Owners mainly depend on naturally growing trees on the home gardens. However, both men and women play a significant role in decision making in case of choice of species, nursery raising, plantation, silvicultural practices, and management activities of medicinal plants. Mostly women encouraged both their neighbors and family to conserve home gardens by planting diverse plant species especially medicinal plants and by taking proper care of the gardens. It was found that labor-intensive activities like digging holes (78%), pruning (63%), planting species (66%), and fencing (59%) were done by men, while seed selection (74%), watering (79%), fertilizing (52%), and weeding (51%), utilization patterns of medicinal plants (60%), and storage and pest control techniques (67%) were mainly done by women.

Families exchange seeds of medicinal plants among themselves, usually at the time of fruit selection. Cattle browsing, pest and disease attack, low productivity, and poor fertility of seeds with sometimes-human disturbance are common problems expressed by the owner. Women also have a responsibility for pest control and use a simple indigenous technique to pest control in their garden which is the application of ashes to plants infected by pests. For excessive pest and disease attacks, they used pesticide but at a low rate that could not affect human and animals. Recently, majority of the households were interested to know about species suitability, appropriate mixture, and information related to high yielding and more pest-disease resistance varieties. Cultivation of medicinal plants species is an important strategy for conservation and sustainable maintenance of home gardens. However, the home gardens present an outstanding example of all acceptance multipurpose land-use system and biodiversity conservation point of view.

4. Conclusion

This study revealed that there are medicinal plants species that make a significant contribution to the healing of diseases of rural community. Due to the increased dependence on herbal treatment and overexploitation of plants, not only by the *Kabiraj*, but also by most of the local people, anthropogenic pressures on medicinal plants are more increased. From the conservation point of view, the plants diversity is critically depleted due to habitat destruction. Therefore, there is an urgent need for conservation of the genetic diversity of the species with special emphasis on anthropogenic populations. Further work should focus on the thorough phytochemical investigation such as alkaloid extraction and isolation along with few clinical trials. This could help in creating mass awareness regarding the need for conservation of such plants and in promoting ethno-medico-botany knowledge within the region. Besides, the young generation should be motivated to acquire this traditional medicinal knowledge. Both government agencies and nongovernmental organizations have roles to play in this regard. This also contributes to the preservation and enrichment of the gene bank of such economically important species before they are lost forever. While there is an issue on the conservation of biological diversity all over the world, local people are using plants for their health care in a sustainable manner, and this may be the key factor in the conservation of plant diversity. It can be concluded from the study that people inherit a rich traditional knowledge and documentation of this knowledge has provided novel information from the area.

References

[1] S. Lee, C. Xiao, and S. Pei, "Ethnobotanical survey of medicinal plants at periodic markets of Honghe Prefecture in Yunnan Province, SW China," *Journal of Ethnopharmacology*, vol. 117, no. 2, pp. 362–377, 2008.

[2] S. A. Mukul, M. B. Uddin, and M. R. Tito, "Medicinal plant diversity and local healthcare among the people living in and around a conservation area of Northern Bangladesh," *International Journal of Forest Usufructs Management*, vol. 8, no. 2, pp. 50–63, 2007.

[3] S. Elliot and J. Brimacombe, *The Medicinal Plants of Gunung Leuser National Park, Indonesia*, WWF, Gland, Switzerland, 1986.

[4] H. Yineger, E. Kelbessa T Bekele, and E. Lulekal, "Plants used in traditional management of human ailments as Bale Mountains National Park, Southeastern Ethiopia," *Journal of Medicinal Plants Research*, vol. 2, pp. 132–153, 2008.

[5] Z. U. Ahmed, Z. N. T. Begum, M. A. Hassan et al., *Encyclopedia of Flora and Fauna of Bangladesh: Index Volume-Flora*, vol. 13, Asiatic Society of Bangladesh, Dhaka, Bangladesh, 1st edition, 2009.

[6] A. Ghani, *Medicinal Plants of Bangladesh With Chemical Constituents and Uses*, Asiatic Society of Bangladesh, Dhaka, Bangladesh, 2003.

[7] M. M. Haque, "Inventory and documentation of medicinal plants in Bangladesh," in *Medicinal Plants Research in Asia: the Framework and Project Work Plans*, P. A. Batugal, J. Kanniah, L. S. Young, and J. T. Oliver, Eds., vol. 1, pp. 45–47, International

Plant Genetic Resources Institute-Regional Office for Asia, the Pacific and Oceania, Serdang, Selangor DE, Malaysia, 2004.

[8] FAO, *State of the World's forests 2009*, Food and Agriculture Organization of the United Nations, Rome, Italy, 2010.

[9] M. S. H. Chowdhury, M. Koike, N. Muhammed, M. A. Halim, N. Saha, and H. Kobayashi, "Use of plants in healthcare: a traditional ethno-medicinal practice in rural areas of southeastern Bangladesh," *International Journal of Biodiversity Science and Management*, vol. 5, no. 1, pp. 41–51, 2009.

[10] A. Z. M. M. Rashid, *Rare and endangered economic plants of Bangladesh*, 2008, http://www.fao.org/DOCREP/004/Y3660e/y3660e05.htm.

[11] N. T. Vergara, "Indigenous knowledge and biodiversity conservation in forestry," in *Proceedings of The National Workshop on Local Knowledge and Biodiversity Conservation in Forestry Practice and Education*, Visayas State University, Visca, Philippines, October 1998.

[12] M. Abdul Halim, M. S. H. Chowdhury, A. I. Wadud, M. S. Uddin, S. K. Sarker, and M. B. Uddin, "The use of plants in traditional health care practice of the shaiji community in Southwestern Bangladesh," *Journal of Tropical Forest Science*, vol. 19, no. 3, pp. 168–175, 2007.

[13] M. Thomsen, S. Halder, and F. U. Ahmed, *Medicinal and Aromatic Plant Industry Development*, InterCooperation, Dhaka, Bangladesh, 2005.

[14] F. U. Ahmed, *Production, Processing and Marketing of Medicinal Plant in Bangladesh*, Aranyak Foundation, Dhaka, Bangladesh, 2009.

[15] M. Hossain, *Bangladesh Seeks Global Market For Its Medicinal Plants*, Science and Development Network, Dhaka, Bangladesh, 2005.

[16] M. Rahmatullah, A. H. Mollik, M. Ali et al., "An ethnomedicinal survey of Vitbilia village in sujanagar sub-district of pabna district, Bangladesh," *The American-Eurasian Journal of Agriculture and Environmental Science*, vol. 10, no. 1, pp. 106–111, 2011.

[17] M. M. Hasan, E. A. Annay, M. Sintaha et al., "A survey of medicinal plant usage by folk medicinal practitioners in seven villages of Ishwardi Upazilla, Pabna District, Bangladesh," *The American-Eurasian Journal of Sustainable Agriculture*, vol. 4, no. 3, pp. 326–333, 2010.

[18] M. Rahmatullah, M. A. Khatun, N. Morshed et al., "A randomized survey of medicinal plants used by folk medicinal healers of Sylhet division, Bangladesh," *Advances in Natural and Applied Sciences*, vol. 4, no. 1, pp. 52–62, 2010.

[19] M. Rahmatullah, M. A. H. Mollik, M. A. Jilani et al., "Medicinal plants used by folk medicinal practitioners in three villages of natore and Rajshahi districts, Bangladesh," *Advances in Natural and Applied Sciences*, vol. 4, no. 2, pp. 132–138, 2010.

[20] M. Rahmatullah, D. Ferdausi, M. A. H. Mollik, R. Jahan, M. H. Chowdhury, and W. M. Haque, "A survey of medicinal plants used by Kavirajes of Chalna area, Khulna district, Bangladesh," *African Journal of Traditional, Complementary and Alternative Medicines*, vol. 7, no. 2, pp. 91–97, 2010.

[21] A. H. Md. Mahabub Nawaz, M. Hossain, M. Karim, M. Khan, R. Jahan, and M. Rahmatullah, "An ethnobotanical survey of Rajshahi district in Rajshahi division, Bangladesh," *The American-Eurasian Journal of Sustainable Agriculture*, vol. 3, no. 2, pp. 143–150, 2009.

[22] A. H. Md. Mahabub Nawaz, M. Hossain, M. Karim, M. Khan, R. Jahan, and M. Rahmatullah, "An ethnobotanical survey of Jessore district in Khulna division, Bangladesh," *The American-Eurasian Journal of Sustainable Agriculture*, vol. 3, no. 2, pp. 195–201, 2009.

[23] M. Rahmatullah, D. Ferdausi, A. H. Mollik, N. K. Azam, M. Taufiq-Ur-Rahman, and R. Jahan, "Ethnomedicinal survey of bheramara area in Kushtia district, Bangladesh," *The American-Eurasian Journal of Sustainable Agriculture*, vol. 3, no. 3, pp. 534–541, 2009.

[24] M. Rahmatullah, A. Noman, M. S. Hossan et al., "A survey of medicinal plants in two areas of Dinajpur district, Bangladesh including plants which can be used as functional foods," *The American-Eurasian Journal of Sustainable Agriculture*, vol. 3, no. 4, pp. 862–876, 2009.

[25] M. Rahmatullah, A. K. Das, M. A. H. Mollik et al., "An ethnomedicinal survey of Dhamrai sub-district in Dhaka district, Bangladesh," *The American-Eurasian Journal of Sustainable Agriculture*, vol. 3, no. 4, pp. 881–888, 2009.

[26] M. Yusuf, M. A. Wahab, J. U. Choudhury, and J. Begum, "Ethno-medico-botanical knowledge from Kaulkhali proper and Betunia of Rangamati district," *Bangladesh Journal of Plant Taxonomy*, vol. 13, no. 1, pp. 55–61, 2006.

[27] S. N. Uddin, M. Z. Uddin, M. A. Hassan, and M. M. Rahman, "Preliminary ethnomedical plant survey in Khagrachari district, Bangladesh," *Bangladesh Journal of Plant Taxonomy*, vol. 11, no. 2, pp. 39–48, 2004.

[28] "Banglapedia: National Encyclopedia of Bangladesh," *Burichong Upazila*, 2006, http://www.banglapedia.org/HT/B_0659.HTM.

[29] T. K. Dey, *Useful Plants of Bangladesh*, The Ad Communication, Chittagong, Bangladesh, 2nd edition, 2006.

[30] K. D. Das and M. K. Alam, *Trees of Bangladesh*, Bangladesh Forest Research Institute, Chittagong, Bangladesh, 2001.

[31] M. D. Miah and M. S. H. Chowdhury, "Indigenous healthcare practice through medicinal plants from forest by the *Mro* tribe in Bandarban region, Bangladesh," *INDILINGA: African Journal of Indigenous Knowledge System*, vol. 2, pp. 61–73, 2003.

[32] M. A. S. A. Khana, S. Ahmed Mukul, M. Salim Uddin, M. Golam Kibria, and F. Sultana, "The use of medicinal plants in healthcare practices by Rohingya refugees in a degraded forest and conservation area of Bangladesh," *International Journal of Biodiversity Science and Management*, vol. 5, no. 2, pp. 76–82, 2009.

[33] M. S. H. Chowdhury and M. Koike, "Therapeutic use of plants by local communities in and around Rema-Kalenga Wildlife Sanctuary: implications for protected area management in Bangladesh," *Agroforestry Systems*, vol. 80, no. 2, pp. 241–257, 2010.

[34] M. H. Rahman, M. J. Fardusi, and M. S. Reza, "Traditional knowledge and use of medicinal plants by the Patra tribe community in the North-Eastern region of Bangladesh," *Proceedings of the Pakistan Academy of Sciences*, vol. 48, no. 3, pp. 159–167, 2011.

[35] H. Rahman, M. Rahman, M. Islam, and S. Reza, "The importance of forests to protect medicinal plants: a case study of Khadimnagar National Park, Bangladesh," *International Journal of Biodiversity Science, Ecosystems Services and Management*, vol. 7, no. 4, pp. 283–294, 2011.

[36] M. H. Rahman, M. Rahman, B. Roy, and M. J. Fardusi, "Topographical distribution, status and traditional uses of medicinal plants in a tropical forest ecosystem of Northeastern Bangladesh," *International Journal of Forest Usufructs Management*, vol. 12, no. 1, pp. 37–56, 2011.

A Study on Exploration of Ethnobotanical Knowledge of Rural Community in Bangladesh: Basis
for Biodiversity Conservation

155

[37] M. A. S. A. Khan, F. Sultana, M. H. Rahman, B. Roy, and S. I.
Anik, "Status and ethno-medicinal usage of invasive plants in
traditional health care practices: a case study from northeastern
Bangladesh," *Journal of Forestry Research*, vol. 22, no. 4, pp. 649–
658, 2011.

[38] J. R. Stepp and D. E. Moerman, "The importance of weeds in
ethnopharmacology," *Journal of Ethnopharmacology*, vol. 75, no.
1, pp. 19–23, 2001.

[39] L. R. S. Gazzaneo, R. F. Paiva de Lucena, and U. P. de
Albuquerque, "Knowledge and use of medicinal plants by
local specialists in an region of Atlantic Forest in the state
of Pernambuco (Northeastern Brazil)," *Journal of Ethnobiology
and Ethnomedicine*, vol. 1, article 9, 2005.

[40] S. Paré, P. Savadogo, M. Tigabu, J. M. Ouadba, and P. C. Odén,
"Consumptive values and local perception of dry forest decline
in Burkina Faso, West Africa," *Environment, Development and
Sustainability*, vol. 12, no. 2, pp. 277–295, 2010.

[41] M. S. H. Chowdhury, M. M. Rahman, M. Koike et al., "Small-
Scale Mehedi (*Lawsonia inermis* L.) farming in the central
Bangladesh: a promising NTFP-based rural livelihood outside
the forests," *Small-Scale Forestry*, vol. 9, no. 1, pp. 93–105, 2010.

[42] C. Cartwright-Jones, *Developing Guidelines on Henna: A geographical Approach*, TapDancing Lizard Publishing, 2006.

[43] M. D. Miah and M. M. Rahman, "The effect of religious subculture on the stock and diversity of the village forests in the
flood-plain area of Bangladesh," in *Proceedings of The Human
Dimensions of Family, Farm, and Community Forestry International Symposium*, pp. 89–94, Washington State University,
Pullman, Wash, USA, Mar2004.

[44] B. M. Kumar and P. K. R. Nair, *Tropical Homegardens: A Time-
Tested Example of Sustainable Agroforestry*, Springer, Amsterdam, The Netherlands, 2006.

[45] P. Parveen, B. Upadhyay, S. Roy, and A. Kumar, "Traditional uses
of medicinal plants among the rural communities of Churu district in the Thar Desert, India," *Journal of Ethnopharmacology*,
vol. 113, no. 3, pp. 387–399, 2007.

[46] P. M. Shrestha and S. S. Dhillion, "Medicinal plant diversity
and use in the highlands of Dolakha district, Nepal," *Journal of
Ethnopharmacology*, vol. 86, no. 1, pp. 81–96, 2003.

[47] J. Okello and P. Ssegawa, "Medicinal plants used by communities of Ngai Subcounty, Apac District, northern Ugand," *African
Journal of Ecology*, vol. 45, no. 1, pp. 76–83, 2007,

[48] P. Ssegawa and J. M. Kasenene, "Medicinal plant diversity
and uses in the Sango bay area, Southern Uganda," *Journal of
Ethnopharmacology*, vol. 113, no. 3, pp. 521–540, 2007.

14

Landscape Pattern Impacts on the Population Density and Distribution of Black Shama (*Copsychus cebuensis* Steere) in Argao Watershed Reserve, Argao, Cebu, Philippines

Archiebald Baltazar B. Malaki,[1] **Rex Victor O. Cruz,**[2] **Nathaniel C. Bantayan,**[2] **Diomedes A. Racelis,**[2] **Inocencio E. Buot Jr.,**[3] **and Leonardo M. Florece**[4]

[1] *Cebu Technological University, Cebu Campus, Argao 6021, Cebu, Philippines*
[2] *Institute of Renewable and Natural Resources, College of Forestry and Natural Resources, University of the Philippines Los Banos, Laguna 4031, Philippines*
[3] *Institute of Biological Sciences, College of Arts and Sciences and School of Environmental Science & Management (SESAM), University of the Philippines Los Banos, Laguna 4031, Philippines*
[4] *School of Environmental Science and Management, University of the Philippines Los Banos, Laguna 4031, Philippines*

Correspondence should be addressed to Archiebald Baltazar B. Malaki; archlam68@yahoo.com

Academic Editors: I. Bisht, P. De los Ríos Escalante, and R. Rico-Martinez

This study determined the impacts of landscape pattern on population density of *C. cebuensis* within AWR, a conservation priority in Cebu, Philippines. Three land uses were identified, namely, (a) cultivated (3,399 ha/45%); (b) forestlands (3,002 ha/40%); and (c) build-up (1,050 ha/15%). Forest patches at class have irregular/complex shapes; thus the forest areas in AWR are more fragmented and heterogeneous. Estimated population density of *C. cebuensis* was 52 and 53 individuals per hectare in mixed and natural forests. There were only three predictors at the landscape and four at the sampling site level, respectively have able to explain the behavior of the population density of *C. cebuensis*. Relative humidity and canopy cover were having high positive significant correlations while tree basal area has high negative correlation (at landscape). Elevation and canopy cover have positive high significant and significant correlations, while slope and shrub cover have negative significant correlation with *C. cebuensis* population density. The adjusted R^2 values were 0.345 and 0.212 (at landscape and sampling site). These suggest that about 34.5% of the variations of the population density of *C. cebuensis* have been accounted for by the former and only 21.2% by the latter. Preservation and protection of remaining forest fragments within AWR are paramount.

1. Introduction

Landscape ecology explains landscape pattern and interprets its possible ecological impacts particularly on energy and material flow on flora and fauna found at the landscape [1]. Argao watershed is currently experiencing these various changes that threaten biodiversity. It is located at the southeastern part of Cebu, Philippines. Some of the limited natural forest patches of the watershed are sporadically distributed on upland barangays, which are pervasively threatened from sustained illegal cutting of trees, fuel wood gathering, and converting forest to other land uses such as agriculture [2].

In 2001, the International Union for the Conservation of Nature (IUCN) declared Mt. Lantoy, which is located within Argao watershed reserve (AWR), as 71st of the 117 Important Bird Areas (IBA) in the country [3], because of the presence of endangered Cebu Black Shama (*Copsychus cebuensis*) and Cebu Flowerpecker (*Dicaeum quadricolor*).

C. cebuensis is an endemic bird species to the island of Cebu in the Philippines. It has always been considered rare and endangered as it has very small range and population [3]. Conducting detailed ecological research to determine threats and their level of impact, as a case study in Argao watershed (AWR), will facilitate conservation planning and

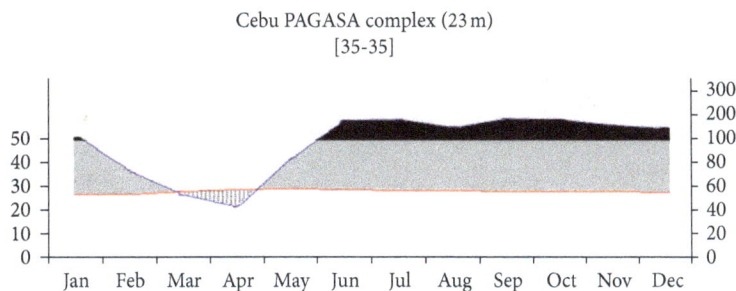

FIGURE 1: Climogram for Mactan, Cebu, showing average monthly temperature and rainfall data from 1973 to 2007. Climatic data were taken
from PAGASA, Mactan [11].

management. The study examined impacts of landscape
pattern in AWR on the population density and distribution
of *C. cebuensis*. Emphasis on: (1) area distribution of different landscape elements constituting AWR; (2) analyzed
landscape pattern or structure in AWR; (3) determined
the relationship between landscape variables and population
density and distribution of *C. cebuensis* in AWR.

2. Materials and Methods

2.1. Description of the Study Site

2.1.1. Topography and Slope. The general topography of the
area is steep to very steep, with rugged terrain ranging from
12% to 60% slope in any direction [4]. The highest elevation
is about 1,000 m above sea level which is located in Barangay
Ablayan, Dalaguete, Cebu, in the southwestern part of the
watershed where headwaters emanate. On the other hand,
one of the highest mountain peaks within the watershed area
is situated in Mt. Lantoy in Barangay Tabayag, Argao, with an
elevation of about 593 m above sea level.

2.1.2. Climatic Type. The climate at AWR is classified under
Climate Type III. Rainy season is from May up to November,
while dry season is from December to April. Figure 1 shows
the climogram of Mactan, Cebu, the nearest weather station
to AWR. The figure shows the average monthly temperature
and rainfall data from 1973 to 2007. The dotted area represents
the dry period. The wet season is represented by the gray-
shaded area, with rainfall less than 100 mm, and black area,
with rainfall more than 100 mm.

2.1.3. Vegetative Cover and Land Use. The area has two
different types of forest cover: the naturally grown trees which
are indigenous or native species and the man-made forest
which is the plantation, previously managed under Southern
Cebu Reforestation Development Project (SCRDP) in the
early 1970s. Some remnants of natural forest fragments are
found at the peak of Mt. Lantoy and at the slopes of Argao
River. Isolated forest patches are still present in barangays
Canbantug, Panadtaran, Usmad, Tabayag, Conalum, and
Cansuje. Mother trees of Ipil (*Intsia bijuga* (Colebr.) Kuntze)
and Molave (*Vitex parviflora* Juss) are native to the area.

Natural vegetation and plantation forests comprise 29%
(1,119.54 ha) of the watershed area [12]. Mahogany (*Swietenia
macrophylla*) and Yemane (*Gmelina arborea*) are the two
exotic species commonly planted in the plantation forest.

*2.1.4. Collection/Procurement of Geographic Information Sys-
tem (GIS) Thematic Layers and Remotely Sensed (RS) Image.*
GIS thematic layers and RS image were obtained to be able
to generate the land use and land cover of the study site as
well as other watershed maps. These data were obtained from
the National Mapping and Resource Information Authority
or NAMRIA. NAMRIA provided three digital images like
AVNIR, SPOT, and LANDSAT. Among these, the image
from SPOT-5 was utilized for this study. SPOT-5 image
was with 2.5 m resolution and taken 2006/2007 available
during conduct of study. Other base maps were acquired
from different agencies including DENR Regional Office-VII,
CENRO-Argao, and LGU-Argao particularly CBRM Office.

*2.1.5. Collecting/Gathering of Other Supplementary Informa-
tion.* Related information on the study topic was collected
from the internet, especially in the directory of open online journal in the UP system to acquire published arti-
cles/materials, books, monograph, and journals, among others which were related to the present work.

2.2. Examination of Areal Distribution of Different Land
Use/Land Cover (LULC) Classes

2.2.1. Land Use/Land Cover (LULC) Generation. This study
utilized the RGB SPOT-5 satellite imagery, true color com-
posites of bands 2, 1, and 3. This is a processed digital
image of 2.5 m, pixel size resolution taken in December
2006/2007, acquired at the National Mapping Resource Information Authority (NAMRIA), Taguig City, Metro Manila. All
GIS simulation modeling was carried out using ARC/INFO
software program version 10 [13], particularly in generating
the land use/land cover classes (LULCC) of the study site.
Supervised using maximum likelihood (ML), the algorithm
was used in image classification. Then ground truthing or
ground validation activities between the two simulation
processes were conducted.

2.2.2. Ground Truthing or Validation. High-resolution digital map like Google Earth was used, and actual study site visitation based on randomly GIS generated geographic coordinates was done to validate the images. This was important to clearly identify the areas covered by heavy clouds or shadows produced on the SPOT-5 satellite image.

2.2.3. Supervised Classification. This is a process used for the extraction of quantitative information from remotely sensed image data with the analyst having prior knowledge of the available classes. This process is performed after the completion of the ground truthing activities, where the information obtained from the latter is used as the basis for the supervised classification technique. Finally, the number of pixels and area for each class and the landscape metrics or indices using ArcGIS/Patch analyst extensions [14] were estimated, and the statistics for each class were shown in graphs and/or tables.

Figure 2 summarizes the foregoing processes.

2.2.4. Area Distribution Analysis of Different Land Use/Land Cover (LULC) Classes. To determine the areal distribution of the different LULC classes, the newly generated LULC map was used in the analysis as input. The area of each LULC class was determined using GIS, specifically the spatial analyst tools. Subsequently, the results were tabulated and qualitatively analyzed.

2.3. Landscape Pattern Analysis (LPA)

2.3.1. Landscape Metrics (LM) Selection. Table 1 summarizes the landscape metrics that were used in quantifying the landscape pattern (LP) in AWR. Specifically, shape metrics intend to measure landscape configuration (LC), where LC refers to the nature of shape of patches in certain class or the entire watershed landscape. It is also an overall measure of how complex or irregular the shape of all the patches is in a class or landscape [5].

2.3.2. Calculation and Analysis of Landscape Metrics (LM) or Indices. The newly generated LULC map of AWR was used as base map to generate and calculate LM or shape metrics using PATCH ANALYST extensions for ArcMap Version 4.2 [14]. The indices or metrics were calculated only at class level since fragmentation is best analyzed at this level rather than at the landscape [5]. Furthermore, the calculation for the percentage occupied by each class with respect to the entire watershed landscape was undertaken with an excel format. Subsequently, qualitative analysis of LP in AWR was undertaken on the basis of calculated metrics or indices value and available literature.

Equations used in the calculation of indices or metrics are as follows.

(1) Mean Shape Index (MSI). MSI is equal to 1 when all patches are circular (for polygons) or square (for rasters or grids), and it increases with increasing patch shape irregularity [5].

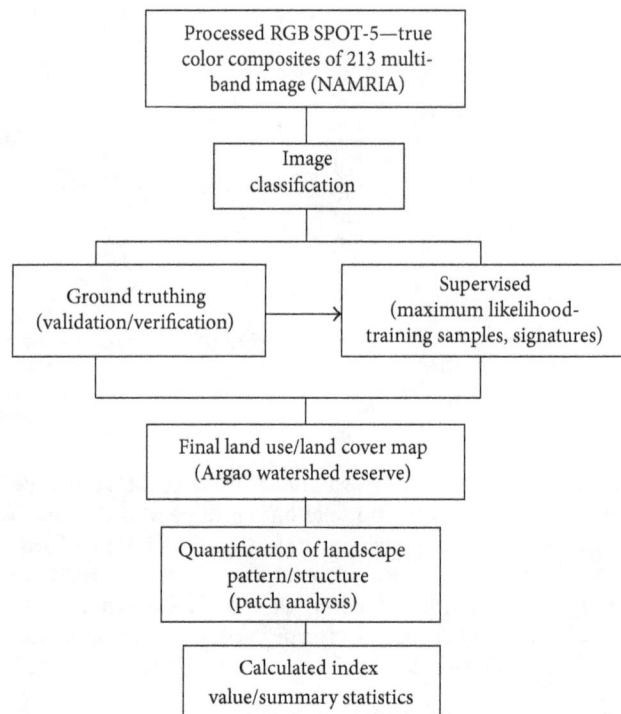

FIGURE 2: GIS simulation processes for land use/land cover classes generation in AWR.

Consider

$$\text{MSI} = \frac{1}{n} \sum_{j=1}^{n} \frac{(0.25 * p_j)}{\sqrt{a_j}}, \tag{1}$$

where p_j is the perimeter of the jth patch and a_j is the area of the jth patch.

(2) Area-Weighted Mean Shape Index (AWMSI). It is equal to 1 when all patches are circular (for polygons) or square (for rasters (grids)), and it increases with increasing patch shape irregularity [5].

Consider

$$\text{AWMSI} = \sum_{j=1}^{n} \frac{0.25 p_j}{\sqrt{a_j}} \frac{a_j}{a_{\text{total}}}, \tag{2}$$

where p_j is thae perimeter of the jth patch and a_j is the area of the jth patch. a_{total} = is the summed area of all patches (Pn $j^{1/4}1 \, a_j$).

(3) Mean Patch Fractal Dimension (MPFD). Index valueapproaches one for shapes with simple perimeters and approaches two when shapes are more complex [5].

Consider

$$\text{FRAC} = \frac{2 \ln (0.25 p_{ij})}{\ln a_{ij}}, \tag{3}$$

where p_{ij} is the perimeter (m) of patch ij and a_{ij} is the area (m^2) of patch ij.

Landscape Pattern Impacts on the Population Density and Distribution of Black Shama (Copsychus cebuensis Steere) in Argao Watershed Reserve, Argao, Cebu, Philippines

159

TABLE 1: Landscape metrics used in quantifying the landscape pattern in Argao watershed [5].

Landscape metrics	Abbreviation	Description
Shape metrics		
(a) Mean shape index	MSI	Patch-level shape index averaged over all patches in the landscape
(b) Area-weighted mean shape index	AWMSI	Mean patch shape index weighted by relative patch size
(c) Mean patch fractal dimension	MPFD	The average fractal dimension of individual patches in the landscape
(d) Area-weighted mean patch fractal dimension	AWMPFD	The patch fractal dimension weighted by relative patch area

TABLE 2: List of landscape variables.

Habitat variables	References	Methods of measurement
(1) Physical features		
(a) Altitude/elevation	Bibby et al. [6]	GPS receiver
(b) Slope gradient	Bibby et al. [6]	Abney hand level
(c) Aspect	Bibby et al. [6]	Compass
(d) Distance/nearest streams/rivers	Bibby et al. [6]	Spatial Analyst (GIS)
(2) Human disturbance		
(a) Dist. from sampling site to forest edge/road		Spatial analyst (GIS)
(b) Dist. from sampling site to a nearest village/community		Spatial analyst (GIS)
(c) Dist. to agriculture/cultivation		Spatial analyst (GIS)
(3) Vegetation structures		
(a) Total height (TH)	Bibby et al. [6]	Haga
(b) Diameter at breast height (DBH)	Bibby et al. [6]/Rosli et al. [7]	Diameter tape
(c) Canopy density	Bibby et al. [6]	Densitometer
(d) BA of trees/ha	Bibby et al. [6]	Computation/equation
(e) Tree density/ha	Rosli et al. [7]/Posa and Sodhi [8]	Computation/equation
(f) % shrub/herb cover	Rosli et al. [7]/Posa and Sodhi [8]	Computation/equation
(g) Number of trees	Rosli et al. [7]/Posa and Sodhi [8]	Ocular observation
(g.1) Diameter classes		Frequency/counts
(g.2) Height classes		Tally/counts
(h) Presence and absence of bamboo thickets	Kennedy et al. [9]	Tally/counts Occular observation
(i) Diversity index (H')	Shannon-Weaver [10]	(1 = presence; 0 = absence) Evenness index value
(1) Microclimate		
(a) Temperature	Rosli et al. [7]/Posa and Sodhi [8]	Digital-min/max thermohygrometer
(b) Relative humidity	Rosli et al. [7]/Posa and Sodhi [8]	EasyViewDigital Light
(c) Light intensity	Rosli et al. [7]/Posa and Sodhi [8]	Meter-Model EA30

(4) Area-Weighted Mean Patch Fractal Dimension (AWMPFD). It is similar to mean patch fractal dimension with the addition of individual patch area weighting applied to each patch [5].

2.4. Relationship between Landscape Variables and Population Density of C. cebuensis

2.4.1. Landscape Variables (LV). To determine the relationship between landscape variables and population density of

C. cebuensis, field survey on the selected forest patches in Argao watershed was conducted. The landscape variables (LV) (Table 2) were broadly categorized into the following: (a) physical, (b) habitat disturbance, (c) vegetation, and (d) microclimate variables. Some of these variables were gathered simultaneously with the bird data, and some LV were gathered indirectly using field data as inputs and equations. Table 2 presents the variables that are considered in this study.

2.4.2. Bird Sampling. Data on the population density of *C. cebuensis* was collected during the breeding season of the target species using distance point count sampling method [6]. A total of 130 circular point count stations, 87 of which were mixed and 43 were natural forest, respectively, with 20 meter radius have been established among the four sampling areas: twenty-six (26) plots in Mt. Lantoy (Brgy. Tabayag), 23 plots in Brgy. Canbantug, 38 in Brgy. Usmad, and 43 plots in Brgy. Cansuje. At each point survey station, the target bird species detected (i.e., either seen or heard) within 20 m radius were recorded for about 10 minutes [15]. The distance of every *C. cebuensis* encountered was estimated when first detected and recorded. Other information also noted such as no. of clusters, time start, time observed, and type of contact, among others. The survey was conducted early in the morning and late in the afternoon (between 0600 H to 1000 H and 1600 H to 1800 H, and no survey was conducted during bad weather especially during rainfall). This period of time is suitable as most of forest birds are active in the early morning and late in the afternoon. All point survey stations were visited just only one time during the whole study period from April 7 to May 28, 2012 (for almost 2-month period).

2.4.3. Regression Analysis. An enter method multiple regression analysis was applied to determine the relationship between the landscape variables (25 predictors were tested in the landscape level analysis and only 10 predictors were finally included in the sampling site) and population density of *C. cebuensis* using

$$Y = Y = b_0 + b_1 X_1 + b_2 X_2 \cdots + b_q X_q + \varepsilon, \quad (4)$$

where Y is the dependent variable (population density of *C. cebuensis*); $X_1, X_2, \ldots, X_q$ are the independent variables or predictors (landscape variables); $b_0, b_1, b_2, \ldots, b_q$ are the partial regression coefficients of independent variables or predictors; and ε is the Random error.

All statistical analyses were undertaken with SPSS 11.5 license to UPLB College, Laguna, Philippines.

3. Results and Discussion

3.1. Land Use/Land Cover (LULC) Analysis

3.1.1. Land Use/Land Cover Map (LULCM). Figure 3 presents the newly generated land use/land cover of Argao watershed. As shown in Table 4 there were three landscape elements or land uses generated, namely, (a) cultivated areas with a total land area of about 3,399 ha or 45%, (b) forest areas with an area of about 3,002 ha or 40%, and (c) built-up areas with 1,050 ha or 15%. It can be observed that AWR is dominated by cultivated areas followed by forest and built-up. This can be attributed to the fact that Argao watershed has higher human population living within and outside the perimeter of the watershed area. According to [16, 17] about 24,096 people (from the municipalities of Argao and Dalaguete) reside within the perimeter of AWR and about 108,646 people reside outside the watershed area [18]. In fact, Argao municipality was considered as one of the biggest municipalities in southern Cebu comprising 45 barangays,

and these barangays are mostly located in the upland areas. In 2010 Argao has a population of 69,503 [16], while the municipality of Dalaguete has 63,239 [18].

The difference between cultivated and forest areas is only about 375 ha or 5%. This figure may be considered as small fraction, yet this is, however, significant due to an increasing human population in the upland or in the watershed area as evident in the present population level. As population grows demand for watershed resources will likely grow especially areas devoted for cultivation or for agriculture, there might be an increasing encroachment into the remaining forest patches, there will be more land use conversion can be expected that would surely tantamount to further forest deforestation and degradation. Ultimately, this condition may seriously affect and would endanger native forest habitat for *C. cebuensis* and other natural resources in the watershed area.

3.2. Landscape Pattern Analysis (Using Shape Metrics)

(1) Area-Weighted Mean Shape Index (AWMSI). Figure 4 shows the calculated index values in AWMSI for three categories or classes. The results show that cultivated areas have the highest computed AWMSI of 65.45, followed by forest class with 15.49, and the lowest were built-up areas with 5.01. All calculated index values are greater than two which serve as benchmark values for patches having more complex and irregular shape [5]. These findings suggest that all three classes or categories have very irregular and/or complex patch shape. These could be attributed to the intense human activities especially in cultivated areas encroaching ultimately into forest areas as observed in this study, thereby exposing the latter to pasture or agriculture, microclimatic alterations that usually occur in the nearby surrounding forest edges through increased penetration of sunlight and wind.

(2) Mean Shape Index (MSI). Figure 5 shows the calculated index values for MSI for all three classes. It can be observed that built-up areas got the highest value of 1.53, while forest and cultivated areas have the same value of 1.39. It can be observed that the values for all the three classes greater than one indicate that the patch shape had increasing irregularity and complexity. This indicates a more fragmented and more heterogeneous watershed landscape [5]. Likewise, this can be attributed to the high level of human activities that usually occur along edges and within the forest interior (i.e., *kaingin* making, infrastructure development, and conversion of forest lands into other uses) which have been observed in the study.

(3) Mean Patch Fractal Dimension (MPFD). Figure 6 presents the computed index value for the three land use classes, of which forest areas had the highest value of 1.85, followed by cultivated lands of 1.84 and built-up areas with 1.73. It was observed that the value for three classes was greater than one and nearly approaches two, suggesting a more complex shape, high patch shape irregularity, and a highly fragmented watershed landscape. These findings further indicate that Argao watershed is more of an edged habitat which can bring impacts on the wildlife species especially forest-interior

Landscape Pattern Impacts on the Population Density and Distribution of Black Shama (Copsychus
cebuensis Steere) in Argao Watershed Reserve, Argao, Cebu, Philippines

161

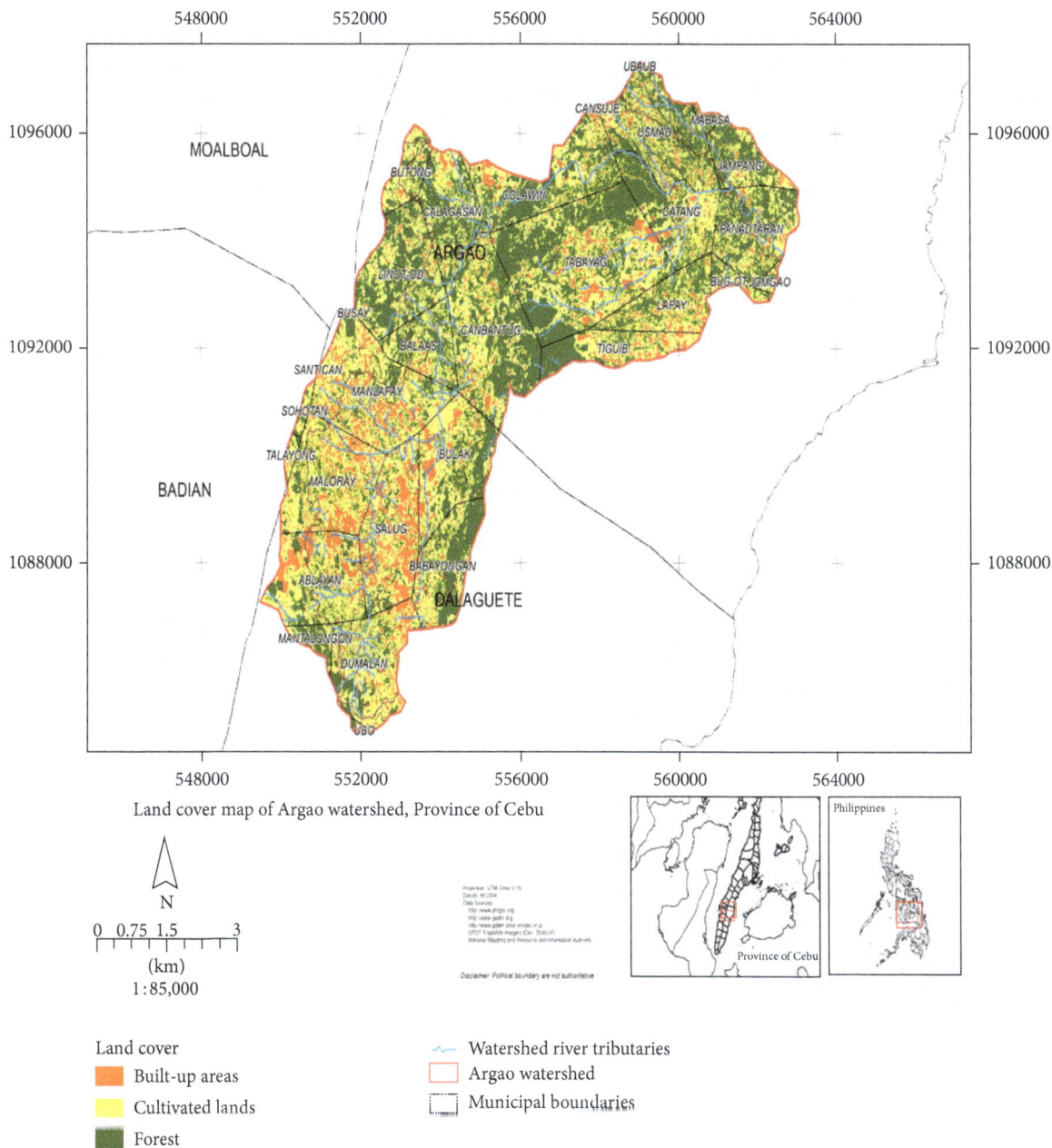

Land cover map of Argao watershed, Province of Cebu

N

0   0.75 1.5   3

(km)

1 : 85,000

Land cover

Built-up areas

Cultivated lands

Forest

Watershed river tributaries

Argao watershed

Municipal boundaries

Figure 3: Newly generated land use/land cover map of AWR from SPOT-5. (Data sources: CENRO, Argao, 2003; CBRMP-LGU, Argao, Cebu, 1999; NAMRIA.)

species like most of insectivorous birds' especially *C. cebuensis* [19].

*(4) Area-Weighted Mean Patch Fractal Dimension
(AWMPFD).* Figure 7 presents the results of the calculated
index values for AWMPFD. It showed that cultivated areas
got the highest index value of 1.65, while forest areas are on
the second place of 1.57, and the lowest are built-up areas with
1.53. All the three values were greater than one and nearly
close to two. These results suggest that the patch shapes of
three classes were complex, further indicating that AWR had

exposed edges, fragmented habitat, and patch shape that tend
to be more irregular and complex and more heterogeneous.

*3.3. Relationship between Landscape Variables and Population
Density of Copsychus cebuensis*

*(1) Population Density Estimates for Habitat or Landscape
Level (Mixed and Natural Forest).* Table 3, however, presents
the estimated population density of *C. cebuensis* (using
DISTANCE 6 release 2 software program) for two habitat

TABLE 3: Summary table for estimating population density on *C. cebuensis*.

Habitat/stratum	Estimates (*D*)	Coefficient of variation (% CV)	Degrees of freedom	95% confidence interval	
Mixed forest	52	10.70	163.71	42.161	64.250
Natural forest	53	10.71	164.12	43.276	65.967

TABLE 4: Pooled density estimates of population density of *C. cebuensis* in four sampling locations.

Sampling locations (Barangay)	Pooled density estimates (*D*)	Coefficient of variation (CV-%)	Degrees of freedom (df)	95% confidence interval	
Tabayag	81.053	38.98	24.24	37.312	176.07
Canbantug	118.55	40.52	22.47	52.860	265.87
Usmad	114.15	20.42	42.18	75.921	171.64
Cansuje	118.61	18.21	50.18	82.514	170.49

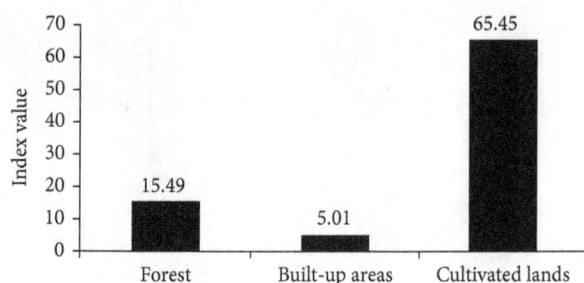

FIGURE 4: Calculated index value for AWMSI at class level.

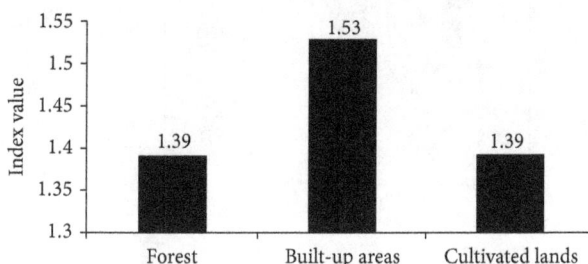

FIGURE 5: Calculated index value for MSI at class level.

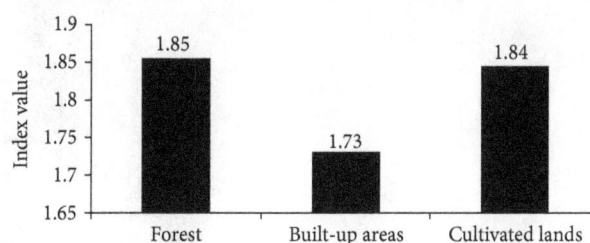

FIGURE 6: Calculated index value for MPFD at class level.

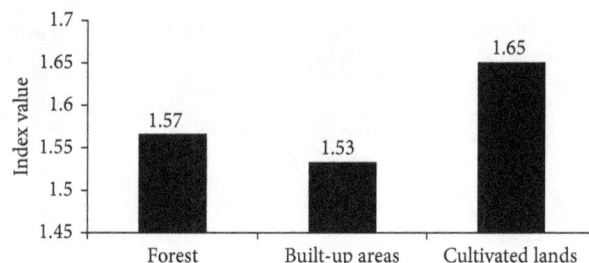

FIGURE 7: Calculated index value for AWMPFD at class level.

types where mixed forest had 52 and 53 individuals—ha^{-1} for natural forest having the same coefficient of variation (CV) of 10% at 95% confidence interval. It is noted that these two habitat types or strata have more or less equal in the density of population per unit area. This indicates that *C. cebuensis* may have been adapted to this type of habitat conditions in the watershed area where they inhabit. These findings further indicate that *C. cebuensis* are still widely distributed among the remaining forest habitat patches within Argao watershed despite of being highly fragmented and despite the fact that they highly disturb watershed landscape.

(2) Population Density Estimates at Sampling Sites. Among the sampling locations, Barangay Cabantug and Barangay Cansuje had the highest pooled population density estimate of *C. cebuensis* with 118 individuals ha^{-1} followed by Barangay Usmad with 114 individuals—ha^{-1}, and the lowest is

Barangay Tabayag with 81 individuals—ha^{-1} (Table 4). It can be observed that, though Barangay Canbantug and Barangay Cansuje have the same population density estimates, they had different %CV of 40% and 18%, respectively. It was observed further that the higher the number of contacts, the lower the %CV, whereas the lower the no. of contacts the higher %CV. According to [6], for a distance sampling method to be adapted especially in point count survey, it needs to have at least 80 to 100 individual contacts in order to achieve more reliable estimates of the density of population for certain organism under consideration. Based on these findings it can be inferred that the *C. cebuensis* are still widely distributed among the fragments of forest habitat patches remaining in Argao watershed.

(3) Regression Analysis between Landscape Variables and Population Density of C. cebuensis for Habitat or Landscape Level (Mixed and Natural Forest). Table 5 presents the results

Landscape Pattern Impacts on the Population Density and Distribution of Black Shama (Copsychus cebuensis Steere) in Argao Watershed Reserve, Argao, Cebu, Philippines

163

TABLE 5: Results of regression analysis between population density estimates of *C. cebuensis* at habitat level (with 25 predictors or landscape variables included).

Model	Predictors (constant)	Code	Unstandardized coefficients		Standardized coefficients	Sig.
			B	STD. error	Beta	
		(Constant)	0.083	44.827		0.999
	(1) Distance from nearest built-up area	DFNBA	−0.001	0.002	−0.052	0.516
	(2) Altitude/elevation	AL	0.000	0.003	0.014	0.882
	(3) Slope	SLP	0.028	0.029	0.075	0.331
	(4) Distance from nearest stream	DFNS	−0.001	0.001	−0.059	0.468
	(5) Temperature	TEM	0.437	0.293	0.146	0.139
	(6) Relative humidity	RH	0.245	0.053	0.400	0.000
	(7) Light intensity	ILUX	$-8.959E-05$	0.000	−0.021	0.805
	(8) Tree basal area	TBA	−9.269	1.566	−0.660	0.000
	(9) Canopy cover	TCPY	0.021	0.006	0.261	0.001
1	(10) Shrub layer	SHBL	0.073	0.043	0.156	0.090
	(11) No. of trees (dbh ≥ 10–19)	NTB	−0.304	1.345	−0.019	0.822
	(12) No. of trees (dbh ≥ 20–29)	NTC	−1.031	4.094	−0.380	0.802
	(13) No. of trees (dbh ≥ 30–39)	NTD	−0.525	4.141	−0.130	0.899
	(14) No. of trees (dbh ≥ 40–49)	NTE	−0.808	4.106	−0.154	0.844
	(15) No. of trees (dbh ≥ 50–59)	NTF	−0.057	4.126	−0.007	0.989
	(16) No. of trees (dbh ≥ 60 cm)	NTG	−0.312	4.421	−0.015	0.944
	(17) No. of trees (<10)	NTS	6.503	4.532	0.407	0.154
	(18) No. of tree (≥10–19)	NTT	0.028	0.239	0.011	0.906
	(19) No. of trees (≥30–39)	NTV	0.000	0.362	0.000	0.999
	(20) No. of trees (≥40–49)	NTW	−2.781	1.998	−0.126	0.167
	(21) No. of trees (≥50)	NTX	−3.495	5.669	−0.047	0.539
	(22) Evenness index value	EIV	18.658	23.786	0.067	0.435

$F(22, 107) = 4.093$.
Sig. of $F = 0.000$.
$R^2 = 0.457$.
Adjusted $R^2 = 0.345$.

of regression analysis between the population density of *C. cebuensis* at habitat or landscape level. The final model shows that adjusted $R^2 = 0.345$, $F(22, 107) = 4.093$, $P < 0.000$ (using the enter method). By looking at the final model (particularly the adjusted R^2 value) this shows that only about 34.5% of the variances of the population density of *C. cebuensis* were being accounted for by the model.

Only three out 25 predictors have been able to explain the behavior of the population density of *C. cebuensis*, namely, (1) relative humidity (RH) and canopy cover (TCPY) having positive high significant correlation (beta) with population density, 0.400 and 0.261, respectively, with *P* values at 0.000 and 0.001 with, respectively, which is <0.01 significant level. These findings indicate that with high RH and TCPY the population density of *C. cebuensis* may increase. While tree basal area (TBA) it has a negative high significant correlation with *C. cebuensis* population density with coefficient (beta) value of −0.660 and a *P* value of 0.000 significant at 0.01 level. This finding indicates that if TBA increases, the population of density of *C. cebuensis* decreases. The study on insectivorous birds and environmental factors in Malaysia's tropical rainforest showed that terrestrial insectivores were sensitive to forest edge and could indicate the quality of forest interior habitats associated with high humidity, dense canopy cover, and deep litter depth [20].

(4) Regression Analysis between Pooled Population Density Estimates of C. cebuensis and Landscape Variables for Sampling Sites. Based on the multiple linear regression analysis (using enter method) the results have shown a final model of adjusted $R^2 = 0.212$, $F(10, 119) = 4.474$, $P < 0.000$ with only 10 predictors included in the final analysis. This indicates that the model has accounted only for 21.2% of the variation of population density of *C. cebuensis* which is lower by 13.3% compared to what is accounted for in the regression done at the habitat level. Table 6 presents the results of the regression done at this level.

It has been observed further that there are four predictors that have explained the behavior of the population density of *C. cebuensis*, namely, (a) elevation, (b) slope, (c) canopy cover,

TABLE 6: Results of the regression analysis between pooled population density estimates in four sampling sites (with 10 predictors included in the analysis).

Model	Predictors	Code	Unstandardized coefficients		Standardized coefficients	Sig.
			B	STD. error	Beta	
	(Constant)		85.333	27.538		0.002
	(1) Elevation	ELEV	0.030	0.007	0.398	0.000
	(2) Slope	SLP	−0.148	0.068	−0.178	0.032
	(3) Temperature	TEM	0.651	0.680	0.097	0.341
	(4) Relative humidity	RH	0.029	0.115	0.022	0.799
1	(5) Light intensity	ILUX	$6.542E - 05$	0.001	0.007	0.938
	(6) Tree density	TDEN	0.007	0.015	0.040	0.626
	(7) Canopy cover	TCPY	0.186	0.091	0.177	0.043
	(8) Shrub cover	SHBL	−0.157	0.078	−0.165	0.047
	(9) Diameter-at-breast-height	DBH	−0.106	0.241	−0.046	0.660
	(10) Total height	TH	−0.466	0.491	−0.101	0.345

$F(10, 119) = 4.474$.
Sig. of $F = 0.000$.
$R^2 = 0.273$.
Adjusted $R^2 = 0.212$.

and (d) shrub cover. Elevation has positive high significant correlation (beta) with population density of C. cebuensis with standardized coefficient of 0.398 and with P value of 0.000, significant at 0.01 level (Table 6). This means that with increasing elevation the density of population of C. cebuensis will likely increase. According to [21] variations in the number of bird occurrences along elevation gradient in the natural and urbanized habitats in western Himalaya, Nainital district of Uttarakhand, India, found out that bird species richness (BSR) varied considerably with elevation. This difference in BSR among study sites could be due to elevation and vegetation differences associated with elevation and not caused by the presence of a group of mid-altitude specialists.

For slope and shrub cover it is negatively correlated with population density of C. cebuensis with coefficients of −0.178 and −0.165 and with values of 0.032 and 0.047, significant at 0.05 level, respectively. These findings indicate that with increasing slope and shrub cover the population density of C. cebuensis will likely decrease. Lastly, for canopy cover it had positive significant correlation with C. cebuensis population density with coefficient of 0.177 and a P value of 0.043 which is significant at 0.01 level. This means that with increasing canopy cover the population density of C. cebuensis will likely increase.

3.4. Conclusion and Recommendations.

AWR is dominated by cultivated land, and forest area at AWR is fragmented and heterogeneous. On the basis of the landscape pattern analysis conducted in AWR using four landscape metrics/indices (shape metric) shows that the landscape pattern in AWR could be characterized as more fragmented and more heterogeneous with patches at classes (especially forest areas) that are increasingly complex and irregular. This condition could affect C. cebuensis subpopulations to become more

fragmented or isolated. However, results show that C. cebuensis subpopulations are still widespread or widely distributed within these remaining forest patches especially those being sampled.

In regression analysis, there are only three out of 25 and four out of 10 have been able to predict or explain the behavior of the population density of C. cebuensis at habitat (landscape) and sampling site levels. Predictors which are positively highly significant at the habitat level were the following: (a) relative humidity (0.400, 0.01 < P value = 0.000) and (b) canopy cover (0.261, 0.01 < P value = 0.001) while tree basal area is negatively highly correlated (−0.660, 0.01 < P value = 0.000). At sampling site, elevation has high positive significant correlation (0.398, 0.01 < P value = 0.000) with C. cebuensis population density, and canopy cover has positive significant correlation (0.177, 0.05 < P value = 0.043), whereas slope and shrub cover have negative significant correlation (−0.178, 0.05 < P value = 0.032 and −0.165, 0.05 < P value = 0.047). Based on the final model and given the two levels of analysis, at landscape, results have shown that 34.5% of the variances of the population density of C. cebuensis have been accounted for, while at the sampling site it accounts only for 21.2%.

Study suggested that regression analysis at landscape or habitat level is better than same analysis at sampling site. Study concludes with a series of sound policies and management recommendations to preserve AWR.

3.4.1. Policy Measures to Protect C. cebuensis Native Habitats.

Integrating the use of indigenous tree species in rehabilitation and reforestation programs whether it is undertaken by public agencies or private entities within study site to preserve native habitat for C. cebuensis taking into account the remaining forest within the watershed area. Promoting and adopting

Landscape Pattern Impacts on the Population Density and Distribution of Black Shama (Copsychus cebuensis Steere) in Argao Watershed Reserve, Argao, Cebu, Philippines

165

landscape approach in rehabilitating degraded habitats. The Department of Environment and Natural Resources (DENR) may have been practicing this approach in rehabilitating degraded areas. This approach, however, focused on the relationship at various scales from species up to landscape. Each and every element or patch is taken into account as a basic unit of the entire system.

3.4.2. Management Aspect. Management and rehabilitation of remaining forest habitat patches within Argao watershed should be based the findings of landscape pattern analysis conducted in this study. Results show that the higher landscape metric value indicates that the forest patches in Argao watershed were highly fragmented. Institute connections on isolated patches, mainly through corridor establishment using riparian forest, is an excellent option of patches connection is imperative. There is a need to rehabilitate riparian vegetation especially those that are highly degraded and sparsely stocked with vegetation. Results show that most of the contacts with *C. cebuensis* were in the valley bottom where relative humidity is high due to the presence of vegetation.

3.4.3. Further Research on Copsychus cebuensis. Follow-up study on the foraging behavior of the target species to ascertain more on the type of insects being preyed about. Follow-up study on the effect of patch area on the population density and/or abundance of the target species. This study was able to survey selected forest habitat patches within AWR. However, the effect of the patch area with the population density of *C. cebuensis* has not been explored.

Continue the monitoring populations in all forest patches in Argao watershed to quantify all existing subpopulations. Since this study had covered only limited number of the remaining forest patches within Argao watershed, it is advised to continue monitor all the forest patches within the area to be able to quantify all existing subpopulations.

Acknowledgments

Thanks are due to the generous funders of this research study: the Department of Science and Technology through the Accelerated Science and Technology Human Resource Development Program (ASTHRDP), Philippines Council for Agriculture, Aquatic and Natural Resources Research and Development (PCAARRD) of the Department of Science and Technology, and the National Mapping and Resource Information Administration (NAMRIA) for providing the satellite image (SPOT-5) used in the landscape pattern analysis.

References

[1] A. Soffianian, S. M. Najafabadi, and V. Rahdari, "Quantifying landscape pattern using RS & GIS techniques for mooteh wildlife refuge," *JWSS—Isfahan University of Technology*, vol. 13, no. 49, pp. 141–150, 2009.

[2] A. B. B. Malaki and I. E. Buot Jr., "Conservation status of indigenous trees in Argao river watershed reserve, Cebu Island, Philippines," *Asia Life Sciences*, vol. 20, supplement 6, pp. 45–59, 2011.

[3] BirdLife International, "Species Factsheet: *Copsychus cebuensis*," 2012, http://www.birdlife.org/.

[4] Community Environment and Natural Resources Office, *Five-Year Development Plan of Argao River Watershed Reserve (ARWR)*, Community Environment and Natural Resources Office, Cebu, Philippines, 2003.

[5] K. Mcgarigal and B. J. Marks, *Spatial Pattern Analysis Program for Quantifying Landscape Structure*, Forest Science Department, Oregon State University, Corvallis, Ore, USA, 1995.

[6] C. Bibby, M. Jones, and S. Marsden, *Expedition Field Techniques: Bird Surveys*, Birdlife International, Cambridge, UK, 2000.

[7] Z. Rosli, M. Zakaria, A. Mohd, A. Yusuf, G. James, and A. Khairulmazmi, "Response of upperstory birds to the environmental variables at different distances from the edge of an isolated forest reserve in Malaysia," *Asia Life Sciences Journal*, vol. 21, no. 1, pp. 65–84, 2012.

[8] M. R. C. Posa and N. S. Sodhi, "Effects of anthropogenic land use on forest birds and butterflies in Subic Bay, Philippines," *Biology Conservation*, vol. 129, no. 2, pp. 256–271, 2006.

[9] R. S. Kennedy, P. C. Gonzales, E. C. Dickinson, H. Miranda, and T. H. Fisher, *A Guide to the Birds of the Philippines*, Oxford University Press, New York, NY, USA, 1st edition, 2000.

[10] C. E. Shannon and W. Weaver, *The Mathematical Theory of Communication*, University of Illinois Press, Urbana, Ill, USA, 1949.

[11] G. O. Cadiz and I. E. Buot Jr., "A checklist of the vascular plants of the mount Tabunan, Cebu, Island, Philippines," *Thailand Natural History Museum Journal*. In press.

[12] P. G. Jakosalem and L. M. Paguntalan, "Strengthened Community-Based Biodiversity Conservation in Selected Sites of Cebu, Philippines "Siloy Project 2007"," Project update (RSG Ref:40. 02. 07), The Ruffourd Small Grant Foundation, 2007, http://www.ruffordsmallgrants.org/.

[13] Environmental Systems Research Institute (ESRI), Inc., *ArcInfo Version 10 Desktop*, Environmental Systems Research Institute (ESRI), Inc., Redlands, Calif, USA, 2010.

[14] R. S. Rempel, A. P. Carr, and D. Kaukinen, *Patch Analyst Extension for ArcMap: Version 4.2*, Ontario Ministry of Natural Resources, Ontario, Canada, 2008.

[15] S. J. Marsden, M. Whiffin, and M. Galetti, "Bird diversity and abundance in forest fragments and Eucalyptus plantations around an Atlantic forest reserve, Brazil," *Biodiversity and Conservation*, vol. 10, no. 5, pp. 737–751, 2001.

[16] Municipal Development Planning Office, *Annual Report. Municipality of Argao*, Municipal Development Planning Office, Cebu, Philippines, 2010.

[17] Municipal Development Planning Office, Annual. Municipality of Dalaguete, Cebu, 2010.

[18] National Statistics Office, *The 2010 Census of Population and Housing*, National Statistics Office, Manila, Philippines, 2010.

[19] A. Ibarra-Macias, W. D. Robinson, and M. S. Gaines, "Experimental evaluation of bird movements in a fragmented neotropical landscape," *Biological Conservation*, vol. 144, no. 2, pp. 703–712, 2011.

[20] H. V. Moradi, M. Zakaria, A. B. Mohd, and E. Yusof, "Insectivorous birds and environmental factors across an edge-interior gradient in tropical rainforest of Malaysia," *International Journal of Zoological Research*, vol. 5, no. 1, pp. 27–41, 2009.

[21] D. Bhatt and K. K. Joshi, "Bird assemblages in natural and urbanized habitats along elevational gradient in nainital district (Western Himalaya) of Uttarakhand State, India," *Current Zoology*, vol. 57, no. 3, pp. 318–329, 2011.

15

Seasonal Colonization of Arbuscular Mycorrhiza Fungi in the Roots of *Camellia sinensis* (Tea) in Different Tea Gardens of India

Chitra Sharma,[1] Rajan K. Gupta,[1] Rakesh K. Pathak,[2] and Kaushal K. Choudhary[3]

[1] *Department of Botany, Government Post Graduate Autonomous College, Rishikesh 249201, Uttarakhand, India*
[2] *Department of Botany, D.A.V. (P.G.) College, Dehradun 248001, Uttarakhand, India*
[3] *Department of Botany, Dr. Jagannath Mishra College, Muzaffarpur 842001, Bihar, India*

Correspondence should be addressed to Kaushal K. Choudhary; kkc1970@gmail.com

Academic Editors: I. Bisht and A. Chistoserdov

Study describes Arbuscular Mycorrhiza (AM) fungi colonization within the roots of cultivated tea plants (*Camellia sinensis*) at four sites, that is, Goodrich, Archadia, IIP, and Vasant Vihar of Doon Valley, Dehradun, India, from April, 2008, to March, 2009. Microscopic study of sterilized and stained root segments showed presence of four species namely *Glomus fasciculatum*, *G. mosseae*, *Gigaspora margarita*, and *Acaulospora scrobiculata* belonging to three genera of mycorrhizal fungi. Maximum AM colonization was observed during April–September and minimum was observed for December-January months of the year. Comparative study of AM fungi colonization at four sites during rainy season showed maximum colonization (100%) at Archadia site having soil with high organic matter, less acidity, and low phosphorus (P) whereas minimum (64.59%) at IIP with low organic matter, more acidity, and high P content. However, no variation in nitrogen content was observed at all four sites. Study suggested a positive relation of percentage root colonization with soil organic matter and negative relation with acidity and P content of soil. Study concludes that the percentage AM colonization is the function of seasonal variation in physicochemical properties of soil and presence of AM inoculums in the soil at a particular time.

1. Introduction

Increasing concern over industry based development and related risk of environment, energy, and food security has stimulated scientists to develop and design biosystems as alternate or supplementary sources of biofertilizer for sustainable practices, particularly for agriculture and remediation of degraded lands. Mycorrhizas—a symbiotic association between the fungi and the roots of higher plants [1]—are receiving more attention worldwide for their ability to enhance the uptake and absorption of relatively immobile nutrients and minerals of plants due to comparatively large surface area of mycelium : root ratio [2]. Mycorrhizal associations occur naturally with more than 95% of terrestrial plants, of which 65% belong to Arbuscular Mycorrhiza (AM) fungi [3, 4]. Inoculation of AM fungi with different plants showed increased uptake of nitrogen (N) by plants [5], promotion of

plant growth [6], exchange of water and mineral with soil [7], resistance to stress and drought and in some cases to soil pathogens [8, 9], and metals toxicity resistance to plants [10]. Establishment of mycorrhizal associations supports up to 80% of N and 90% P requirements of plants [11].

Camellia sinensis is widely used for the cheapest aromatic beverages in the world [12]. Consumption of *C. sinensis* has profound effects on health as it contains more than 700 chemicals such as flavanoids, amino acids, vitamins (C, E, and K), caffeine, and polysaccharides. Beneficial role of *C. sinensis* in normalizing blood pressure, lipid depression activity, prevention of coronary heart diseases, and diabetes by reducing the blood glucose activity as well as the presence of catechins (an antioxidant) has elevated this plant at industry level and has profound effect on economic growth of a country. However, application of chemical fertilizers for quantitative production is adversely affecting the quality of tea. Therefore,

Seasonal Colonization of Arbuscular Mycorrhiza Fungi in the Roots of Camellia sinensis (Tea) in Different Tea Gardens of India

167

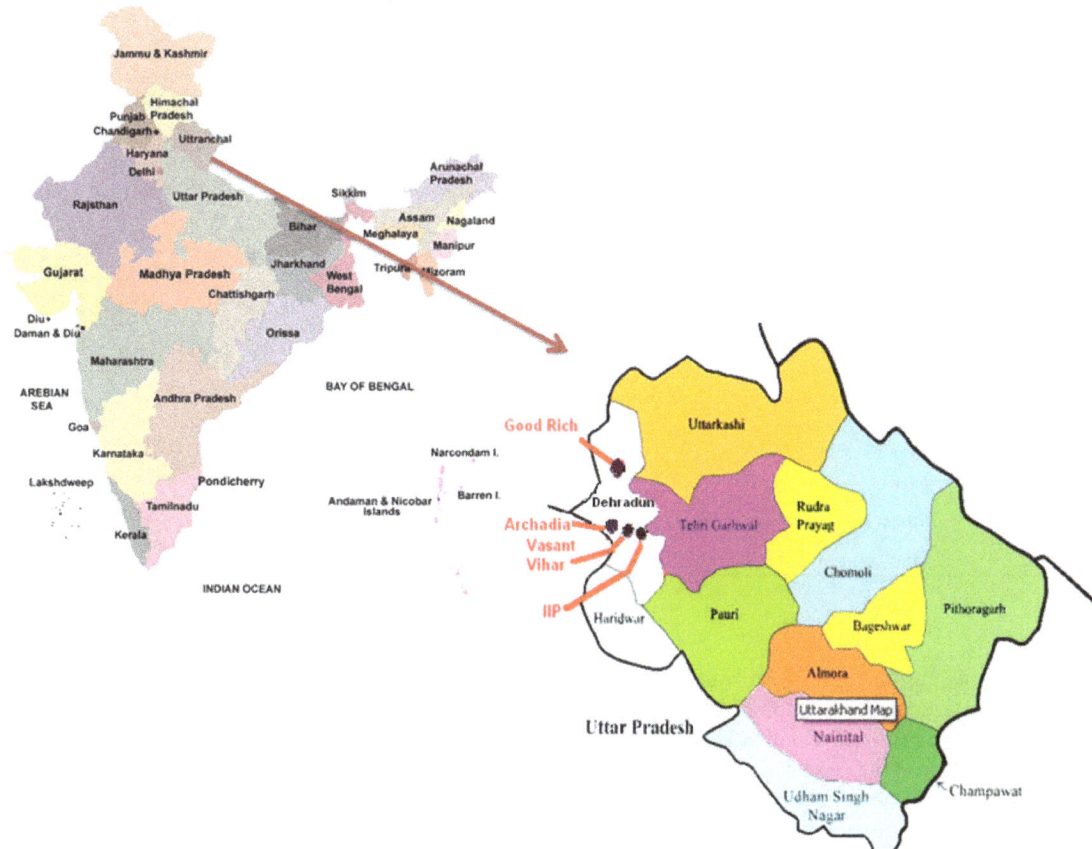

FIGURE 1: Location of four study sites on geographical map.

AM fungi colonizing the tea plants are getting more attention for their ability to support the growth. Although, percentage root colonization of AM fungi with *C. sinensis* roots [13, 14] and with other plants like Pine-Oak forests [15] in the Himalayan region has been variously studied, detailed study on percentage root colonization with respect to seasonal change that might support the annual nutrient management is scarce. Considering the significant nutrient contribution of AM fungi and ultimate requirement of sustainable nutrient management, present study is proposed to enumerate the colonization of AM fungi in the roots of *C. sinensis* during different months for a year.

2. Materials and Methods

2.1. Study Area. The study was conducted on 4th year of tea plantation at four selected sites, namely, Goodrich, Archadia, IIP and Vasant Vihar located at the foot hills of the Himalayas at Doon Valley, Dehradun, India, from April, 2008, to March, 2009. Geographical location of four different sites has been presented in Figure 1. Dehradun is located at the latitude of $30°19'N$ and longitude of $78°04'E$ and is situated between the Himalayas in the north and Shivalik range in the south. Climate of Dehradun is generally temperate and varies from hot in summers to severely cold in winter. Average annual temperature ranges from 5 to 35°C and average annual rainfall is 2073.3 mm. Study was conducted in the area of 100 m × 70 m

for each selected site. The distance between Goodrich (north end of the study site) and IIP (south end of the study site) sites is 73 Km. Archadia and Vasant Vihar are located at a distance of 48 and 66 Km from Goodrich site, respectively.

2.2. Root Sampling. Percentage root colonization of AM fungi was documented with root samples collected from the "root zones" of the *Camellia sinensis* plants at the depth of 15 to 20 cm below the soil surface. Root sampling was done randomly between 8 and 10 am in middle of every month for one year. Eighty root samples were collected in conical flasks for each selected site separately and brought to the laboratory. Study of the root samples for AM colonization was performed on the very next day of the collection. Roots were kept in running tap water for half an hour and washed thrice with sterile double distilled water to avoid the presence of other microbes or AM fungi adsorbed to the root surface.

2.3. Research Methods. AM fungi colonization with the root samples was assessed following the slide method of Giovannetti and Mosse [16]. Root samples were cut into pieces of 1 cm length and stained with 0.05% cotton blue in lactophenol. Excess stain was removed with clear lactophenol. Root segments (1 cm long) were selected randomly from stained samples and mounted on microscopic slides in groups of 10 pieces. Presence and absence of infection were recorded microscopically in each of the 80 root segments for each site

TABLE 1: Percentage root colonization of AM fungi in the roots of tea plants at Archadia, Goodrich, IIP, and Vasant Vihar sites during different months of the year.

	Months	Sites				Level of significance	CD
		Archadia	Goodrich	IIP	Vasant Vihar		
AM colonization	Jan	33.20	31.16	16.63	36.34	***	7.053
	Feb	40.49	31.97	19.46	37.35		
	Mar	59.67	38.45	28.80	40.67		
	Apr	70.97	51.05	35.39	63.75		
	May	70.96	68.75	42.06	85.90		
	Jun	96.86	83.64	64.59	92.69		
	Jul	100.00	83.62	47.49	92.59		
	Aug	78.34	73.49	39.48	86.56		
	Sep	76.31	67.40	29.35	82.42		
	Oct	49.12	52.82	30.85	58.57		
	Nov	39.07	43.96	21.99	52.24		
	Dec	32.50	41.82	21.10	31.20		

* * *: significant at 0.001.

separately for different months of the year and percentage root colonization was calculated. Mycorrhizal fungi were studied for their cellular organization, color, shape and size, wall structure, and position of the vesicles and arbuscules and identified with the help of the literature [17–19]. Micrograph of AM fungi was assessed under a stereomicroscope (Olympus SZ H10 research microscope).

2.4. Analysis. pH of the soil sample was determined in 1:5 suspension of soil: deionised water electrometrically using glass electrode pH meter 335 (Systronics India Limited). Organic carbon was determined by using Walkley and Black's rapid titration method [20] and total nitrogen (N) and phosphorus (P) were determined using the method of Jackson [21].

2.5. Statistical Analysis. Data presented in the tables are the mean values of the analysis. Data produced for different parameters were subjected to multifactor analysis using SPSS software version 10.0.

3. Results

The study describes the differences in monthly colonization of AM fungi in the *C. sinensis* roots round the year at four different sites. Different stages of AM fungi inside root of *C. sinensis* have also been studied and presented in Figure 2.

Extensive study carried out to enumerate the AM fungi colonization in the roots of *C. sinensis* showed differences in percentage root colonization (PRC) during different months of the year at four selected sites (Table 1). Study showed maximum AM fungi colonization at Archadia (100%) followed by Vasant Vihar (92.69%), Goodrich (83.64%), and IIP (64.59%) sites (Table 1). Detailed analysis of percentage root colonization during different months of the year showed that the PRC for all four sites was above/around the annual mean from April to September (hot and humid summer) and

TABLE 2: Statistical analysis of percentage root colonization of AM fungi in the roots of *C. sinensis* at four sites.

Statistical parameters	Sites			
	Archadia	Goodrich	IIP	Vasant Vihar
Mean	62.29	55.68	33.10	63.36
Range	67.50	52.48	47.96	61.49
Coefficient of range	0.51	0.46	0.59	0.50
SD	23.65	19.09	13.78	23.79

below the annual mean from October to March (cold winter). Annual mean was 62.29, 55.68, 33.10, and 63.36 for Archadia, Goodrich, IIP, and Vasant Vihar, respectively (Table 2). AM fungi colonization started to decrease with the onset of cold winter, that is, from the month of October, and reached to lowest during the coldest months of December-January and started to increase with onset of rainy summer. Differential colonization of tea plants with AM fungi during different months of the year indicated the possible role of climatic and edaphic factors.

pH, organic matter, and nitrogen and phosphorus content of soil for four different sites are presented in Table 3. Relation between pH and organic matter of the soil and AM colonization was positive, that is, increase in pH and organic matter was directly related to increase in PRC. However, soil phosphorus has negative relation with percentage root colonization. IIP site with higher P content has low PRC whereas Archadia site with low soil P has high PRC. The P content of different sites might be presented as Archadia < Vasant Vihar < Goodrich < IIP whereas PRC as Archadia > Vasant Vihar > Goodrich > IIP. There were no significant differences in nitrogen status of the soil at different sites.

Statistical analysis of the AM colonization for selected sites around the year showed that AM colonization has great affinity with season and significantly different for all sites

Seasonal Colonization of Arbuscular Mycorrhiza Fungi in the Roots of Camellia sinensis (Tea) in Different Tea Gardens of India

169

(a) Hyphae and vesicle in *C. sinensis* root cortex

(b) AM hyphae penetrating *C. sinensis* root

(c) Vesicle in the root of *C. sinensis*

(d) Hyphae swell apically to form vesicles

(e) Liberating spores and vesicles outside the root cortex

(f) Spore and vesicle in *C. sinensis* root

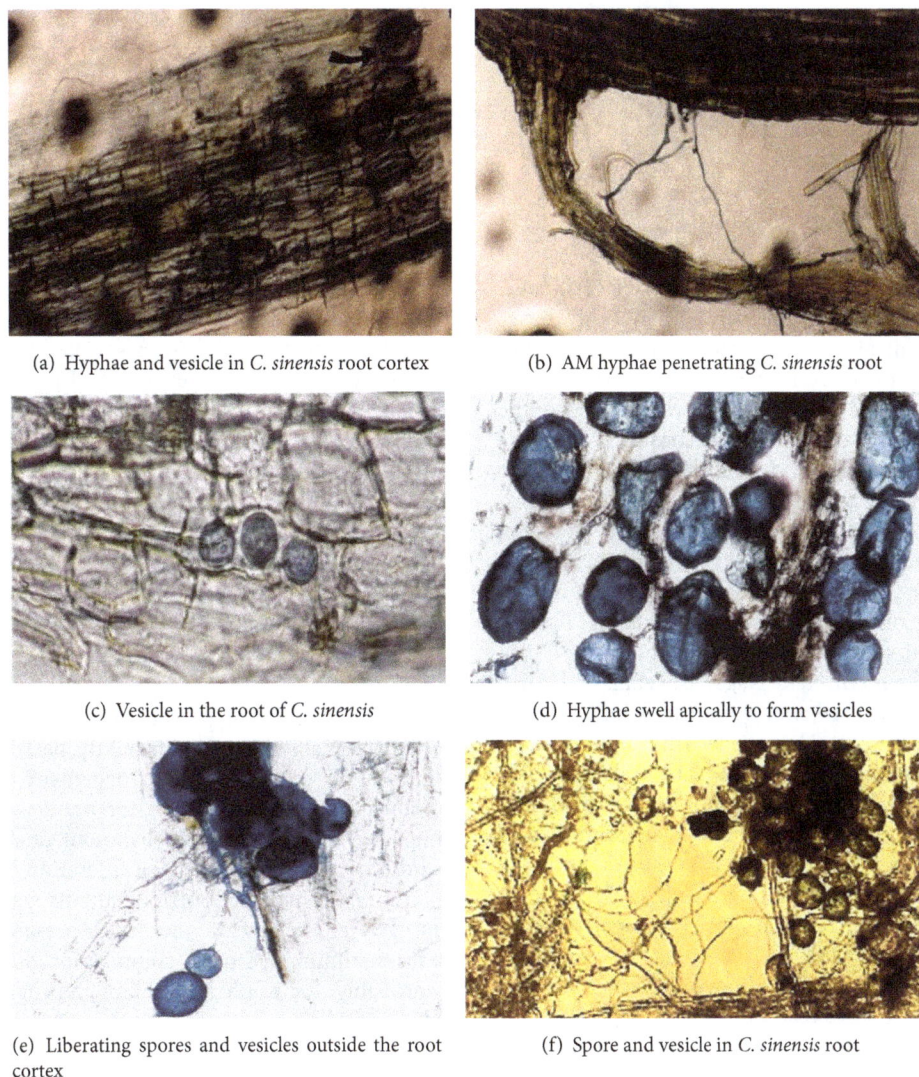

FIGURE 2: Different stages of AM colonization in the roots of *C. sinensis*.

(Table 1). Study gets support from the coefficient of range predicted in Table 2. In spite of significant differences in percentage colonization at four sites ranging from 100% (maximum) at Archadia site to 16.6% (minimum) at IIP site, annual coefficient of range was between 0.46 and 0.59. The study supports that the AM colonization with roots of *C. sinensis* is correlated to seasonal variation in climatic factors and also to AM inocula present at a given time. It was observed that minimum colonization at IIP site in the stress condition of light and temperature during December-January followed the similar trend of minimum colonization during favourable condition of May-June compared to that of other sites. However, the more is the colonization at Archadia site during December-January, the maximum was the colonization during May-June.

Microscopic study showed the presence of four species of AM fungi belonging to three genera of class Glomeromycota and were represented by *Glomus fasciculatum*, *G. mosseae* (family Glomaceae), *Gigaspora margarita*, and *Acaulospora scribiculata* (family Acaulosporeacae). Hyphae and vesicles were the most common structures observed for root segments under microscope (Figure 2(a)). However, hyphae were predominantly present for all selected sites (Figure 2(b)). The root samples of *C. sinensis* showed typical wider and aseptate hyphae growing inter- and intracellularly through the cortex and penetrating to the inner cortex (Figures 2(b) and 2(c)). Vesicles were mostly globose to subglobose or ellipsoidal in shape inside the roots of *C. sinensis* (Figure 2(d)). The arbuscules were observed in cells of the inner cortex. Vesicles with group of spores were also seen in the roots of *C. sinensis* (Figures 2(e) and 2(f)). Further investigation in terms of species abundance showed predominance of *Glomus* sp. followed by *Acaulospora scribiculata* and *Gigaspora margarita*.

4. Discussion

AM fungi have been variously studied for their contribution in degraded land as well as their potential application in agriculturally cultivated plants for enhanced production [22].

TABLE 3: Annual averages of pH, organic matter, and nitrogen and phosphorus content at four sites.

Sites	pH	Organic matter (%)	Nitrogen (%)	Phosphorus (ppm)
Archadia	5.7	2.20	0.21 ± 0.03	12.60 ± 0.44
Goodrich	5.1	1.35	0.17 ± 0.02	25.64 ± 0.09
IIP	4.8	1.20	0.18 ± 0.01	28.17 ± 0.29
Vasant Vihar	5.4	1.55	0.18 ± 0.02	22.55 ± 0.02

The present study describes the colonization of *C. sinensis* roots with AM fungi round the year on monthly basis. Maximum root colonization during hot and humid summer (rainy season, i.e., during April–September) and decrease in colonization with the start of cold winter during the months of October–March were in accordance with the study reported by Chandra and Jamaluddin [23]. They also observed the similar result of maximum AM fungi colonization during rainy season in *C. sinensis* root. Gould et al. [24] also observed the maximum root colonization and spore population density during June and minimum during October-November on reclaimed sites of Archadia. Similar observation of negative relation between P content and AM colonization has been reported by Nagahashi et al. [25]. They reported that increase in P content significantly inhibited the number of branches and total hyphae length. Negative effect of P on AM colonization suggests that plant itself is capable of absorbing sufficient amount of P required for growth and development. Increase in P brings about anatomical and physiological changes in the roots that inhibit or limit the vigorous intraradical spread of AM fungi [25–27]. Positive effect of pH on AM colonization might be compared to the report by Medeiros et al. [28]. They reported the significant increase in root colonization of Sorghum plant with AM fungi with increasing pH. Positive relation of percentage root colonization of AM fungi with organic matter observed in this study was in accordance with Vaidya et al. [29] who reported increased AM population with increase in organic matter. Study reflects that the site Archadia having low phosphorus has high organic matter and high percentage of AM root colonization. In contrast, site IIP with high P content has low organic matter and low percentage of AM root colonization. It might be explained by the fact that high P content in the soil in the root limits the supply of organic carbon to the AM fungus of the colonized root thereby causing reduced extension of hyphae and development of spores [30].

Further investigation of differences in percentage colonization at different sites and differences in range might be attributed to the initial inocula of the AM fungi in the field under natural condition. It is apparent from the study that IIP site with less AM fungi inocula (lower PRC) has minimum percentage of root colonization during favorable condition of May-June; whereas Archadia site with more AM fungi inocula (higher PRC) has maximum percentage of root colonization during favorable condition of May-June (compared to all four sites). Moreover, dominance of *Glomus* sp. followed by *Acaaulospora* and *Gigaspora* sp. is comparable to Zhao et al. [22]. They reported similar observations of dominance of *Glomus* sp. and *Acaulospora* sp. in the tropical rainforests of Xishuangbanna in southwestern China. The differences in percentage root colonization and species abundance might be attributed to change in climatic conditions round the year. Varied diversity of AM fungi associated with different plants is already in the literature [31, 32]. In addition, seasonality, physiochemical properties of soil, host dependence, age of the host plants, and dormancy might play significant role in AM establishment and diversity [33]. The present study will certainly help in maintaining the nutrient status of the soil during lower rate of AM colonization by introducing comparatively high nutrient into the soil to maintain the growth of tea plants during stress period of cold.

5. Conclusion

The present study of percentage root colonization at different tea planted sites on monthly basis might contribute significantly in the future management of nutrient status of soil under field condition for better application of AM fungi for better yield. Establishment of AM fungi in field condition might support the proliferation of some important microbes that actively contribute in nitrogen and other nutrient cycling. The research output of present study concludes that maximum AM fungi colonization during rainy season is probably due to physico-chemical properties of soils and surroundings.

Acknowledgments

The authors wish to thank the owner of Goodrich Tea Estate (Vikasnagar) Mahant Shri Devendra Dass ji Maharaj (Darbar Shri Guru Ram Rai Ji Maharaj), Dehradun, for giving place to set up tea nursery, Mr. Digvijay Singh Rawat, the Manager of Tea Estate, for the tea plants given by them, and the Department of Botany, Govt. P.G. College, Rishikesh, for laboratory facilities. The authors are thankful to the University Grants Commission, New Delhi, and the Uttarakhand Council for Science and Technology, Dehradun, for financial assistance.

References

[1] A. B. Frank, "Über die auf Wurzelsymbiose beruhende Ernährung gewisser Bäume durch unterirdische Pilze," *Ber Deutsch Bot Gesells*, vol. 3, pp. 128–145, 1985.

[2] E. Smith and D. J. Read, *Mycorrhizal Symbiosis*, Academic Press, London, UK, 3rd edition, 2008.

[3] B. Wang and Y.-L. Qiu, "Phylogenetic distribution and evolution of mycorrhizas in land plants," *Mycorrhiza*, vol. 16, no. 5, pp. 299–363, 2006.

[4] M. C. Brundrett, "Mycorrhizal associations and other means of nutrition of vascular plants: understanding the global diversity

Seasonal Colonization of Arbuscular Mycorrhiza Fungi in the Roots of Camellia sinensis (Tea) in Different Tea Gardens of India

171

of host plants by resolving conflicting information and developing reliable means of diagnosis," *Plant and Soil*, vol. 320, no. 1-2, pp. 37–77, 2009.

[5] F. Ganryl, H. G. Diem, J. Wey, and Y. R. Dommergues, "Inoculation with *Glomus mosseae* improves N$_2$ fixation by field-grown soybeans," *Biology and Fertility of Soils*, vol. 1, no. 1, pp. 15–23, 1985.

[6] T. Muthukumar and K. Udaiyan, "Growth of nursery-grown bamboo inoculated with arbuscular mycorrhizal fungi and plant growth promoting rhizobacteria in two tropical soil types with and without fertilizer application," *New Forests*, vol. 31, no. 3, pp. 469–485, 2006.

[7] R. N. Ames, C. P. P. Reid, K. K. Porter, and C. Cambardella, "Hyphal uptake and transport of nitrogen from two ^{15}N leveled source by *Glomus mosseae* arbuscular mycorrhizal fungus," *New Phytologist*, vol. 95, no. 3, pp. 381–396, 1989.

[8] R. M. Augé, "Water relations, drought and vesicular-arbuscular mycorrhizal symbiosis," *Mycorrhiza*, vol. 11, no. 1, pp. 3–42, 2001.

[9] B. A. Sikes, K. Cottenie, and J. N. Klironomos, "Plant and fungal identity determines pathogen protection of plant roots by arbuscular mycorrhizas," *Journal of Ecology*, vol. 97, no. 6, pp. 1274–1280, 2009.

[10] A. Liu, C. Hamel, R. I. Hamilton, B. L. Ma, and D. L. Smith, "Acquisition of Cu, Zn, Mn and Fe by mycorrhizal maize (*Zea mays* L.) grown in soil at different P and micronutrient levels," *Mycorrhiza*, vol. 9, no. 6, pp. 331–336, 2000.

[11] M. G. A. van der Heijden, R. D. Bardgett, and N. M. van Straalen, "The unseen majority: soil microbes as drivers of plant diversity and productivity in terrestrial ecosystems," *Ecology Letters*, vol. 11, no. 3, pp. 296–310, 2008.

[12] K. K. G. Menon, "The *Camellia sinensis* industry in India, how to redesign a native," in *Camellia sinensis Culture Processing and Marketing*, M. J. Mulky and V. S. Sharma, Eds., pp. 3–10, Oxford and IBH Publishing Co. PVT. Ltd., New Delhi, India, 2002.

[13] S. K. Rajan, B. J. D. Reddy, and D. J. Bagyaraj, "Screening of arbuscular mycorrhizal fungi for their symbiotic efficiency with *Tectona grandis*," *Forest Ecology and Management*, vol. 126, no. 2, pp. 91–95, 2000.

[14] R. K. Gupta and C. Sharma, "Diversity of arbuscular mycorrhizal fungi in *Camellia sinensis* in Uttarakhand state, India," *African Journal of Biotechnology*, vol. 9, no. 33, pp. 5313–5319, 2010.

[15] S. Chaturvedi, V. Tewari, S. Sharma et al., "Diversity of arbuscular mycorrhizal fungi in Oak-Pine forests and agricultural land prevalent in the Kumaon Himalayan Hills, Uttarakhand, India," *British Microbiology Research Journal*, vol. 2, no. 2, pp. 82–96, 2012.

[16] M. Giovannetti and B. Mosse, "An evaluation of technologies for measuring vesicular arbuscular mycorrhizal infection in roots," *New Phytologist*, vol. 84, no. 3, pp. 489–500, 1980.

[17] I. R. Hall, "Taxonomy of VA mycorrhizal fungi," in *VA Mycorrhiza*, C. L. Powell and D. J. Bagyaraj, Eds., pp. 57–94, CRC Press, Boca Raton, Fla, USA, 1984.

[18] N. C. Schenck and Y. Perez, *Manual for the Identification of VA Mycorrhizal Fungi*, Synergistic Publications, Gainesville, Fla, USA, 1990.

[19] J. B. Morton and G. L. Benny, "Revised classification of Arbuscular mycorrhizal fungi (Zygomycetes): a new order, Glomales, two new suborders, Glomineae and Gigasporineae, and two new families, Acaulosporaceae and Gigasporaceae, with an emendation of Glomaceae," *Mycotaxon*, vol. 37, no. 1, pp. 471–491, 1990.

[20] A. Walkley and I. A. Black, "An examination of the Degtiareff method for determining soil organic matter and proposed modification of the chromic acid titration method," *Soil Science*, vol. 37, pp. 29–38, 1934.

[21] M. L. Jackson, *Soil Chemical Analysis*, Prentice Hall, Englewood Cliffs, NJ, USA, 1958.

[22] Z. W. Zhao, G. H. Wang, and L. Yang, "Biodiversity of arbuscular mycorrhizal fungi in a tropical rainforest of Xishuangbanna, southwest China," *Fungal Diversity*, vol. 13, pp. 233–242, 2003.

[23] K. K. Chandra and A. Jamaluddin, "Seasonal variation of VAM fungi in tree species planted in coalmine overbunden of Kusmunda (MP)," *Journal of Tropical Forest*, vol. 14, no. 2, pp. 118–123, 1998.

[24] A. B. Gould, J. W. Hendrix, and R. S. Ferriss, "Relationship of mycorrhizal activity to time following reclamation of surface mine land in western Kentucky. I. Propagule and spore population densities," *Canadian Journal of Botany*, vol. 74, no. 2, pp. 247–261, 1996.

[25] G. Nagahashi, D. D. Douds Jr., and G. D. Abney, "Phosphorus amendment inhibits hyphal branching of the VAM fungus *Gigaspora margarita* directly and indirectly through its effect on root exudation," *Mycorrhiza*, vol. 6, no. 5, pp. 403–408, 1996.

[26] F. Amijee, P. B. Tinker, and D. P. Stribley, "The development of endomycorrhizal root systems. VII. A detailed study of effects of soil phosphorus on colonization," *New Phytologist*, vol. 111, no. 3, pp. 435–446, 1989.

[27] J. A. Menge, D. Sterile, D. J. Bagyaraj, E. L. V. Johnson, and R. T. Leonard, "Phosphorus concentrations in plants responsible for inhibition of mycorrhizal infection," *New Phytologist*, vol. 80, no. 3, pp. 575–578, 1978.

[28] C. A. B. Medeiros, R. B. Clark, and J. R. Ellis, "Growth and nutrient uptake of sorghum cultivated with vesicular-arbuscular mycorrhiza isolates at varying pH," *Mycorrhiza*, vol. 4, no. 5, pp. 185–192, 1994.

[29] G. S. Vaidya, K. Shrestha, B. R. Khadge, N. C. Johnson, and H. Wallander, "Study of biodiversity of arbuscular mycorrhizal Fungi in addition with different organic matter in different seasons of Kavre district (central Nepal)," *Scientific World*, vol. 5, no. 5, pp. 76–80, 2007.

[30] D. D. Douds, "Relationship between hyphal and arbuscular colonization and sporulation in a mycorrhiza of *Paspalum notatum* Flügge," *New Phytologist*, vol. 126, no. 2, pp. 233–237, 1994.

[31] F. Y. Wang and S. Y. Shi, "Biodiversity of Arbuscular mycorrhizal fungi in China: a review," *Advances in Environmental Biology*, vol. 2, no. 1, pp. 31–39, 2008.

[32] S. Gaur and P. Kaushik, "Biodiversity of vesicular arbuscular mycorrhiza associated with *Catharanthus roseus*, *Ocimum* spp. and *Asparagus racemosus* in Uttarakhand state of Indian Central Himalaya," *International Journal of Botany*, vol. 7, no. 1, pp. 31–41, 2011.

[33] J. N. Gemma and R. E. Koske, "Seasonal variation in spore abundance and dormancy of *Gigaspora gigantea* in mycorrhizal inoculum potential of a dune soil," *Mycologia*, vol. 80, no. 2, pp. 211–216, 1988.

16

Numerical Taxonomy of Species in the Genus *Mallomonas* (Chrysophyta) from China

Jia Feng and Shulian Xie

College of Life Science, Shanxi University, Taiyuan 030006, China

Correspondence should be addressed to Shulian Xie; xiesl@sxu.edu.cn

Academic Editors: I. Bisht and P. De los Ríos Escalante

Mallomonas is one of the biggest genera of Chrysophyta. In total, 37 species and 2 varieties have been recorded in China. Because of their narrow ecological optimum, species of this genus are considered as valuable bioindicators. However, taxonomy of *Mallomonas* remains unclear. We studied the numerical taxonomy of all the species and varieties recorded in China using Ward's method and the furthest neighbor method based on 52 morphological characters. Shown in the phylogenetic trees, those species could be divided into two major clusters. One cluster includes 5 small clusters and another includes 2. The results of numerical taxonomy are partially consistent with the traditional ones with some divergences. Furthermore, the diversity of silicified scales including shapes and structures was confirmed as the most important character for identification of *Mallomonas* species.

1. Introduction

The genus *Mallomonas*, which was created by Perty in 1852, is reported as one of the biggest genera in Chrysophyta [1, 2]. It is comprised of 163 species around the world so far, of which 37 species and 2 varieties are from China [3–14]. *Mallomonas* are unicellular, free-swimming, freshwater organisms. Because of their narrow ecological optimum, *Mallomonas* species are considered as valuable bioindicators to both recent and historical environments, as well as for biomonitoring [15–18]. Smol suggested that morphological variability of silica structures within individual species could represent an important piece of information for biomonitoring studies [19]. Therefore, it is important for us to make a clear and reasonable classification system of this genus.

The basic taxonomic system of the genus *Mallomonas* was first established in 1933 and has been modified for several times [20–27]. Nowadays, the genus *Mallomonas* is traditionally divided into sections and series in mainly the morphology of scales, which is considered as the most important taxonomic character for species identification and the taxonomy of the species [5, 28]. However, we shall be aware that this classification based mainly on resemblances and the difference in the structures of scales and bristles may be highly artificial, because it is still unknown which scale

characters are stable enough to be selected as reliable taxonomical markers. Also, since it is hard to collect enough samples of unicellular bodies for molecular analysis, phylogenetic studies that used the molecular data were also difficult for generic taxonomy and phylogeny of the genus *Mallomonas*. Therefore, the phylogenetic reliability of individual morphological characteristics remains unclear.

Numerical taxonomy was developed in the late 1950s and had been broadly successful in taxonomy of microalgae, mainly in identification of the subspecies. In numerical taxonomy, several characters are selected equally and used to evaluate the similarity of organisms calculated using mathematical methods. And such similarity could be used to differentiate species and cluster certain species to new groups. As numerical taxonomy simultaneously deals with many characters, it could avoid the limitation of the information of the holotype and subjective factors during identification. And the analyses results could give us one or several characters, which should be selected as the main characters for species identification and taxonomy. In this work, we have carried out the multicharacter analysis of the 39 *Mallomonas* species recorded in China using the numerical taxonomy. Our work provides a comparatively objective description of the system evolution and the taxonomy research on *Mallomonas*.

TABLE 1: The species of *Mallomonas* in China.

No.	Species	Reference
1	*M. acaroides* Perty em. Iwanoff	Kristiansen, 1989, 1990, 2002 [7, 10, 28]; Kristiansen and Tong, 1989 [9]; Wei and Kristiansen, 1994, 1998 [12, 13]; Wei and Yuan, 2001 [14]
2	*M. akrokomos* Ruttner	Kristiansen, 1989, 1990, 2002 [7, 10, 28]; Kristiansen and Tong, 1989 [9]; Wei and Kristiansen, 1994 [12]; Wei and Yuan, 2001 [14]
3	M. alata Asmund et al.	Kristiansen, 1989, 2002 [9, 28]
4	*M. alpina* Pascher et Ruttner em. Asmund et Kristiansen	Kristiansen, 1989, 1990, 2002 [7, 10, 28]; Kristiansen and Tong, 1989 [9]; Wei and Kristiansen, 1994, 1998 [12, 13]; Wei and Yuan, 2001 [14]
5	*M. annulata* (Bradley) Harris	Kristiansen, 1989, 2002 [7, 28]; Kristiansen and Tong, 1989 [9]; Wei and Kristiansen, 1994, 1998 [12, 13]; Wei and Yuan, 2001 [14]
6	*M. areolata* Nygaard	Kristiansen, 1989, 2002 [7, 28]; Wei and Kristiansen, 1994 [12]; Wei and Yuan, 2001 [14]
7	*M. allorgei* (Deflandre) Conrad	Kristiansen, 2002 [28]
8	*M. calceolus* Bradley	Wei and Kristiansen, 1994 [12]; Kristiansen, 2002 [28]
9	*M. caudata* Iwanoff em. Krieger	Wei and Kristiansen, 1994, 1998 [12, 13]; Wei and Yuan, 2001 [14]; Kristiansen, 2002 [28]
10	*M. corymbosa* Asmund et Hilliard	Kristiansen, 1989, 2002 [7, 28]; Wei and Kristiansen, 1994 [12]; Wei and Yuan, 2001 [14]
11	*M. costata* Dürrschmidt em Asmund et Kristiansen	Wei and Kristiansen, 1998 [13]; Wei and Yuan, 2001 [14]; Kristiansen, 2002 [28]
12	*M. crassisquama* (Asmund) Fott	Wei and Kristiansen, 1994, 1998 [12, 13]; Wei and Yuan, 2001 [14]; Kristiansen, 2002 [28]
13	*M. cratis* Harris et Bradley	Kristiansen, 1990 [10]
14	*M. cyathellata* Wujek et Asmund	Wei and Kristiansen, 1994, 1998 [12, 13]; Wei and Yuan, 2001 [14]; Kristiansen, 2002 [28]
15	*M. elongata* Reverdin	Kristiansen, 1989, 2002 [7, 28]; Kristiansen and Tong, 1989 [9]; Wei and Kristiansen, 1994, 1998 [12, 13]; Wei and Yuan, 2001 [14]
16	*M. flora* Harris et Bradley	Wei and Yuan, 2001 [14]; Kristiansen, 2002 [28]
17	*M. grata* Takahashi	Kristiansen, 1989, 2002 [7, 28]; Kristiansen and Tong, 1989 [9]; Wei and Kristiansen, 1994, 1998 [12, 13]; Wei and Yuan, 2001 [14]
18	*M. guttata* Wujek	Kristiansen and Tong, 1989 [9]; Wei and Kristiansen, 1998 [13]; Wei and Yuan, 2001 [14]; Kristiansen, 2002 [28]
19	*M. guttata* var. *simplex* Nicholls	Wei and Yuan, 2001 [14]; Kristiansen, 2002 [28]
20	*M. heterospina* Lund	Kristiansen, 1989, 2002 [7, 28]; Kristiansen and Tong, 1989 [9]; Wei and Kristiansen, 1994 [12]; Wei and Yuan, 2001 [14]
21	*M. insignis* Pénard	Kristiansen, 1989, 2002 [7, 28]
22	*M. mangofera* Harris et Bradley	Kristiansen, 1989, 2002 [7, 28]; Kristiansen and Tong, 1989 [9]; Wei and Kristiansen, 1994, 1998 [12, 13]; Wei and Yuan, 2001 [14]
23	*M. eoa*	Kristiansen and Tong, 1988, 1991 [6, 11]; Kristiansen, 2002 [28]
24	*M. splendens*	Kristiansen and Tong, 1991 [11]; Kristiansen, 2002 [28]
25	*M. matvienkoae* (Matvienko) Asmund et Kristiansen	Kristiansen, 1989, 2002 [7, 28]; Kristiansen and Tong, 1989 [9]; Wei and Kristiansen, 1994, 1998 [12, 13]; Wei and Yuan, 2001 [14]
26	*M. multiunca* Asmund	Wei and Yuan, 2001 [14]; Kristiansen, 2002 [28]
27	*M. multisetigera* Dürrschmidt	Kristiansen, 1989, 2002 [7, 28]; Kristiansen and Tong, 1989 [9]
28	*M. oviformis* Nygaard	Kristiansen, 1989, 2002 [7, 28]; Wei and Kristiansen, 1994 [12]; Wei and Yuan, 2001 [14]
29	*M. papillosa* Harris et Bradley	Kristiansen, 1989, 2002 [7, 28]; Wei and Kristiansen, 1994 [12]; Wei and Yuan, 2001 [14]
30	*M. parvula* Dürrschmidt	Wei and Kristiansen, 1994 [12]; Wei and Yuan, 2001 [14]; Kristiansen, 2002 [28]
31	*M. peronoides* (Harris) Momeu et Péterfi	Kristiansen, 1989, 2002 [7, 28]; Wei and Kristiansen, 1994, 1998 [12, 13]; Kristiansen and Tong, 1989 [9]; Wei and Yuan, 2001 [14]

TABLE 1: Continued.

No.	Species	Reference
32	*M. peronoides* var. *bangladeshica* Takahashi et Hayakawa	Wei and Yuan, 2001 [14]; Kristiansen, 2002 [28]
33	*M. portae-ferreae* Péterfi et Asmund	Wei and Kristiansen, 1994 [12], Wei and Yuan, 2001 [14]; Kristiansen, 2002 [28]
34	*M. pseudocratis* Dürrschmidt	Wei and Yuan, 2001 [14]; Kristiansen, 2002 [28]
35	*M. punctifera* Korshikov	Kristiansen, 1989, 2002 [7, 28]; Kristiansen and Tong, 1989 [9]; Wei and Kristiansen, 1994 [12]; Wei and Yuan, 2001 [14]
36	*M. rasilis* Dürrschmidt	Wei and Kristiansen, 1998 [13]; Kristiansen, 2002 [28]
37	*M. striata* Asmund	Kristiansen, 1989, 1990, 2002 [7, 10, 28]; Kristiansen and Tong, 1989 [9]; Wei and Kristiansen, 1994 [12]
38	*M. tolerans* (Asmund & Hillard) Asmund et Kristiansen	Kristiansen, 1990 [10]
39	*M. tonsurata* Teiling	Kristiansen, 1989, 1990, 2002 [7, 10, 28]; Wei and Kristiansen, 1994, 1998 [12, 13]; Wei and Yuan, 2001 [14]

Furthermore, it has been showed that the shape of the rib and submarginal rib together should be selected as the most important characters, which would help us identify new species and deal with new records.

2. Materials and Methods

The species of *Mallomonas* recorded in China, which included 37 species and 2 varieties as taxonomic operated units, were selected for numerical taxonomy studies (Table 1). Fifty-two morphological characters had been selected and examined according to the following criteria (Table 2). First, all of them were stable and had been commonly used for taxonomic identification in this species. Second, the variability among different taxa could be observed in a preliminary review of herbarium material. Third, such characters should be denoted as dual dates.

Hierarchical cluster analysis is a general approach for cluster analysis, in which the objective is to group together objects or records closed to one another. In this work, hierarchical cluster analysis of all the samples was clustered using the furthest neighbor method and Ward's method as described by Lu [29]. These analyses were performed with the statistical software package "SPSS 12.0" [29].

3. Results

Identification of a species and taxonomy depends on the different shape, size, physiological characters, DNA characters, and living environment, and so forth, of the samples. These characters could be divided into two types, qualitative characters and quantitative characters, respectively. In our work, all the 52 dual characters selected belonged to qualitative characters since they were more directly visual and easier for examination. It could be found that most of these characters with remarkable variation selected according to the principle were shape and structure of the silicified scales, further confirming that the variations of the silicified scales were the most important characters for identification and taxonomic studies in *Mallomonas*, consistent with the related studies before.

TABLE 2: Characters and status used in the study.

No.	Characters and status
1	Scale small (1), no (0)
2	Scale larger (1), no (0)
3	Scale oval or elliptical (1), no (0)
4	Scale round (1), no (0)
5	Scale symmetrical (1), no (0)
6	Large pores on the scale observed with LM (1), absent (0)
7	Body scale with subcircular peculiar appendix (1), absent (0)
8	Dark twisting "vermiform" on the scale (1), absent (0)
9	Scale with appendages "cyathi" (1), absent (0)
10	Scale with scattered cavities of the thickened ridges (1), no (0)
11	Scale consist two or more longitudinal rows of circular unpatterned areas (1), absent (0)
12	Scale consist a single row of large pits along the central longitudinal axis (1), absent (0)
13	Continuous submarginal ribs (1), absent (0)
14	Submarginal parallel ribs on the anterior scale (1), absent (0)
15	Anterior submarginal rib with a wing-like structure (1), absent (0)
16	Distal edges of scales convex (1), no (0)
17	Distal edges of scales concave (1), no (0)
18	Small pores on the distal scale (1), absent (0)
19	Densely spaced papillae on the distal scale (1), absent (0)
20	Two-lateral ribs and not connected on the scale (1), absent (0)
21	V-rib on the scale (1), absent (0)
22	A flower like pattern in the angle of V-rib (1), no (0)
23	Base of the V-rib broadly U-shaped (1), no (0)
24	V-rib with large proximal hood (1), absent (0)
25	Arms of V-rib bend (1), no (0)
26	Arms of the V-rib completely encircle (1), no (0)
27	A secondary layer on scale (1), absent (0)

TABLE 2: Continued.

No.	Characters and status
28	Body dome (1), absent (0)
29	Dome of body scale very broad (1), no (0)
30	Scale dome with parallel striation (1), absent (0)
31	Dome with strongly U-shaped ribs (1), absent (0)
32	A curved serrated bristle attached to dome (1), absent (0)
33	Shield with labyrinthic ornamentation (1), absent (0)
34	A row of minute holes on the shield (1), absent (0)
35	A row of circular pits on the shield (1), absent (0)
36	A dense reticulum of ribs on the shield (1), no (0)
37	Regularly papillae on the shield (1), absent (0)
38	A developed reticulum on distal 2/3 of the shield (1), absent (0)
39	Secondary structure on the shield and flange (1), absent (0)
40	Small pores on the secondary layer of shield (1), absent (0)
41	Large pores on the secondary layer of shield (1), absent (0)
42	A network of circular holes on secondary layer of shield (1), absent (0)
43	An irregular reticulation of ribs on secondary layer of shield (1), no (0)
44	Parallel transverse ribs on secondary layer of shield (1), absent (0)
45	Transverse rib connected run perpendicular (1), no (0)
46	Cyst ovoid (1), no (0)
47	Cyst rhomboidal (1), no (0)
48	Cyst with pores (1), absent (0)
49	Cyst with rims (1), absent (0)
50	Prominent ribs on the anterior flange (1), absent (0)
51	Posterior flange ornamented (1), no (0)
52	Posterior flange covered with reticulation of pores (1), absent (0)

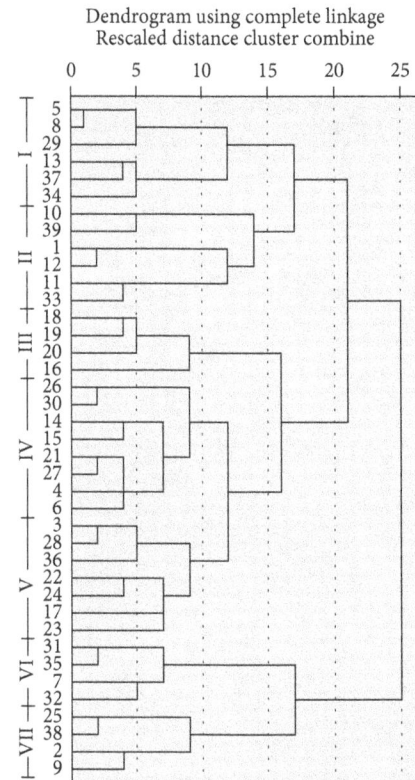

FIGURE 1: Dendrogram using furthest neighbor classification of 39 species of *Mallomonas*.

To examine the relationship and the evolution of *Mallomonas*, two polygenetic trees of 39 species of *Mallomonas* recorded in China were gained from the hierarchical cluster analyses, which were conducted by SPSS 12.0, using Ward's method and furthest neighbor method, respectively. It was clear that the results of the two trees were largely consistent with each other (Figures 1 and 2). According to the two polygenetic trees, firstly, those 39 *Mallomonas* species could be divided into two major groups, with one cluster contains 31 species and the other one contains 8 species. Furthermore, to examine whether there were some characters that could be used to differentiate the upper cluster and the below cluster, all the 52 dual characters used were checked. And the shapes of the rip (V-rip or two lateral rip) and the submarginal rip were found to be the most two important characters to separate the two major clusters.

The upper cluster was composed of five groups: group I and group II together compose one subcluster, and the other three groups compose another one. Based on the result of Ward's method, group I included 6 species, *M. annulata*, *M. calceolus*, *M. cratis*, *M. papillose*, *M. pseudocratis*, and

M. striata; group II included 6 species, *M. acaroids*, *M. crassisquama*, *M. corymbosa*, *M. tonsurata*, *M. costata*, and *M. portae-ferreae*. The characters of group I were that their scale had the dome, and the body of the scale presented ovoid shape. Furthermore, except *M. annulata*, the scales of the other 5 species were tripartite. Based on the taxonomic system set up by Kristiansen, *M. calceolus* and *M. papillose* belong to sect. *Papillos*, while *M. cratis*, *M. pseudocratis*, and *M. striata* belong to sect. *Striatae*. However, according to the result of the furthest neighbor method, *M. costata* and *M. portae-ferreae* also belonged to group I instead of group II. The characters of Group II were that their scales were arranged spirally and were typically tripartite, with densely arranged pores on the base plate. In addition, all of them had V-rib with acute angle. Moreover, the secondary layer of their shield may have ribs and reticulation structures. In the taxonomic system setup by Kristiansen [28], all of them belonged to sect. *Mallomonas*.

The other subcluster of the upper cluster was composed of groups III, IV, and V. Group III included 4 species and they were *M. flora*, *M. guttata*, *M. guttata* var. *simple*, and *M. heterospina*. The characters of this group are that the body scales of these species are spirally arranged, having longitudinal axes that are strongly oblique to the longitudinal axis of the cell. Secondly, their scales are present in an oval shape and the shields contain ribs. Thirdly, the cells are covered with curved bristles. In the taxonomic system setup by Kristiansen [28], *M. flora* belongs to sect. *Striatae*, *M. guttata* and *M. guttata*

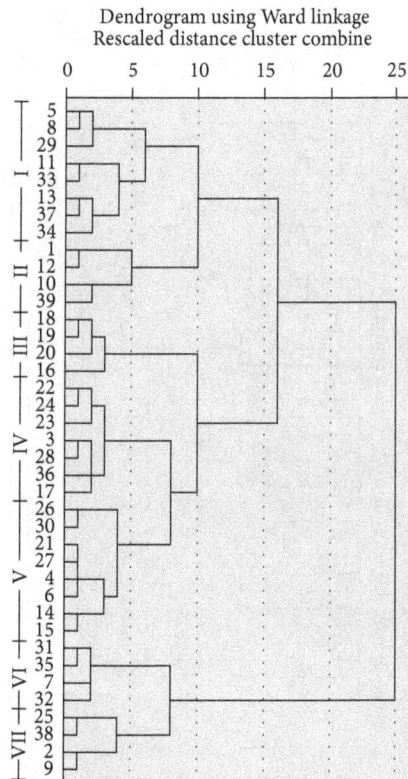

FIGURE 2: Dendroram using Ward's method classification of 39 species of *Mallomonas*.

var. *simple* belong to the species of sect. *Papillose*, and *M. heterospina* belongs to sect. *Heterospinae*, respectively. Group IV includes 8 species and they are *M. alpine*, *M. areolata*, *M. elongate*, *M. multiunca*, *M. multisetigera*, *M. cyathelleta*, *M. insignis*, and *M. parvula*. The characters of this group were that the scales are also spirally arranged. Moreover, their scales are thin, in oval or obovate shape, and contain V-ribs. The base plate of scales is closely spaced by pores. The dome concavity was shallow. The cells are covered with delicate bristles. In the taxonomic system of Kristiansen [28], *M. alpine*, *M. areolata*, *M. elongate*, and *M. cyathelleta* belong to sect. *Mallomonas*, while the others belong to different sections. Group V includes 7 species and they are *M. alata*, *M. oviformis*, *M. rasilis*, *M. mangofera*, *M. splendens*, *M. grata*, and *M. eoa*. The characters of this group are that the bristle of all the species of this group appeared at the apical end of the cell only and was smooth. The scales are rhombic. In the taxonomic system of Kristiansen [28], *M. alata*, *M. grata*, *M. eoa*, and *M. mangofera* belong to sect. *Torquatae*, and *M. oviformis*, *M. rasilis*, and *M. splendens* belong to sect. *Planae*, sect. *Papillosae*, and sect. *Quadrata*, respectively.

When we came to the following cluster, there were only two groups, groups VI and VII. Group VI contains four species, *M. peronoides*, *M. peronoides* var. *bangladeshica*, *M. punctifera*, and *M. allorgei*. These species had mixed characters, such as base plate with pores and scales with pits or papillae. In the taxonomic system of Kristiansen [28], *M. peronoides* and *M. peronoides* var. *bangladeshica* belong to

sect. *Planae*. *M. punctifera* belonged to sect. *Punctiferae*, and *M. allorgei* belonged to sect. *Quadrata*.

The common traits of group VII are that the scales were arranged parallel to their longitudinal axes. Scales were without submarginal rib or V-rib. It included *M. akrokomos*, *M. caudate*, *M. matvienkoae*, and *M. tolerans*. However, based on Kristiansen's system [28], *M. caudate*, *M. matvienkoae*, and *M. tolerans* belonged to sect. *Planae*, while the *M. akrokomos* belonged to sect. *Akrokomae*.

4. Discussion

The taxonomy of the species was not fixed and would change with new methods involved. Numerical taxonomy, which could evaluate all of the characters simultaneously, is one of the popular methods used in the taxonomy of many species since it is more objective than the traditional ones. *Mallomonas* first was established as a genus in 1852. In total, 163 species have been reported, within which 39 species were recorded in China. The taxonomy of this genus had been modified for several times during the last 150 years; however, there were still divergences in the classification on the sections level. In our work, 39 species recorded in China were selected and analyzed by numerical taxonomy methods. When comparing the classification results of the further neighbor methods and Ward's methods, it could be found that they was almost the same to each other. This highly conformity implied that such classification was more appropriate. However, there was also a little difference between the two classification systems. In the classification of the further neighbor, *M. costata* and *M. portae-ferreae* belonged to group I (Figure 1); however, in the classification of Ward's method, they belonged to group II (Figure 2). In addition, when we turned to the traditional taxonomy of the two species, we found that both of them were classified in sect. *Mallomonas*. Therefore, the classification of Ward's method may be more reasonable than that of the further neighbor methods in *Mallomonas*.

When compared with the traditional taxonomy system, it could be concluded that the seven groups of the 39 *Mallomonas* reported in China divided by numerical taxonomy partly supported the traditional taxonomy setup by Kristiansen [28]. For example, all of the species of group II belonged to sect. *Mallomonas*. Similarly, in some other groups, most of the species belonged to one or two sections as described by Kristiansen's system. Three of six species in group I belonged to sect. *Papillos* and the other three belonged to sect. *Striatae*. In addition, four species out of eight in group IV belonged to sect. *Mallomonas*. Nevertheless, the classification suggested by numerical taxonomy showed a difference from the traditional classification. In the classification suggested by numerical taxonomy in our work, several species of one section in the Kristiansen's classification system belonged to different groups. Five species of sect. *Papillos* had been analyzed in the work and three of them belonged to group I and the other two belonged to group III. However, the two, which belonged to group III, were *M. guttata* and its variety, suggesting that this species was remote from the other species of this section. The same difference was seen in sect. *Striatae*, in which the *M. flora* was a little remote from the other species examined. Also, all of the ten species, which were divided into sect. *Mallomonas*, belonged to two different groups, with six of them in group II and the other four in group IV (Figures 1 and 2). Although all of the ten species shared a V-rip with acute angles, the scales of the six species of group II were thinner than that of the four of group IV, which meant that the classification of sect. *Mallomonas* was a little complicated and should be further examined. Taken together, our results of the classification suggested by numerical taxonomy not only partly supported the traditional classification, but also gave some new insights. It could be concluded that the classifications of most of *Mallomonas* species by different methods were matched in a certain degree, but some of them should be seriously considered and redivided by evaluating the numerical taxonomy results and the characters observed.

Several analyses using molecular and morphological data have been performed to begin to understand the phylogenetic relationships of *Mallomonas*. Lavau et al. [30] found *Mallomonas* species that formed a monophyletic clade among 10 *Mallomonas* species, 6 *Synura* species, and 1 *Tessellaria* species based on nuclear SSU rDNA data and scale characteristics. Andersen [31] has performed molecular phylogenetic analysis of the synurophyceae and chrysophyceae using nuclear SSU rDNA and *rbc*L gene. Based on his work, the genus *Mallomonas* was polyphyletic in the *rbc*L phylogenetic tree but was not resolved based on nuclear SSU rDNA sequence data. It had determined the sequences for both the nuclear SSU and LSU rDNA and plastid LSU of RUBISCO (*rbc*L) genes of 19 *Mallomonas* isolates. Bayesian and maximum-likelihood (ML) analyses of the data revealed that *Mallomonas* consists of two strongly supported clades. The results indicated that the sections *Planae* and *Heterospinae* should be combined. The findings also clearly indicated that at least some features of the current infrageneric classification of *Mallomonas* based on scale ultrastructure are not congruent with our molecular phylogeny [28]. It is not so surprising since the relationships between the sections and series lack phylogenetic underpinnings and have been considered as artificial by Kristiansen and Preisig [32]. It is suggested that the current classification of the genus *Mallomonas* at the section level would require some revision [33].

The tools of geometric morphometrics are currently considered the most powerful in biological shape analysis [34–37] and are now widely used in almost all branches of organismal biology [38]. So far in phycology, the landmark-based geometric morphometric methods were used in taxonomic investigations of diatoms [39], macroscopic green algae [40, 41], and the variation in scales of *M. striata* [42]. A wide field for geometric morphometric analyses in the investigation of both synurophycean taxonomy and ecology, especially in species with rich and complex morphology of scales bearing structures, allows the delimitation of landmarks (most *Synura* species, members of the sections *Papillosae*, *Heterospinae*, *Striatae*, and *Pseudocoronatae* of the genus *Mallomonas*, and others) [42].

Although the results acquired by the numerical taxonomy are more objective and explicit, it could not absolutely take the place of the traditional taxonomy. In addition, geometric morphometric studies and molecular information are becoming powerful evidence in taxonology of the algae. Therefore, the numerical taxonomy associated with the morphometric features, geometric morphometry, and sequence analysis would define much clearer groups of *Mallomonas* in advance.

Acknowledgment

This study was financially supported by the National Natural Science Foundation of China (Grant no. 30770162).

References

[1] M. Perty, *Zur Kenntnis kleinster Lebensformen nach Bau, Funktionen, Systemtik mit Spezialverzeichniss der in der Schweiz beobachteten*, Bern, Switzerland, 1852.

[2] R. A. Andersen, "Synurophyceae Classis nov., A new class of algae," *American Journal of Botany*, vol. 74, pp. 337–353, 1987.

[3] B. V. Skvortzov, "Harbin Chrysophyta, Chinae Boreali-Orientalis," *Bullutin Herbin North-Eastern Forestry Academy*, vol. 3, pp. 1–70, 1961.

[4] L. S. Péterfi and B. Asmund, "*Mallomonas portae-ferreae* nova species in the light and electron microscopes," *Studia Universitatis Babes-Bolyai*, vol. 1, pp. 11–18, 1972.

[5] B. Asmund and J. Kristiansen, "The genus *Mallomonas* (Chrysophyceae)," *Opera Botanica*, vol. 85, pp. 1–129, 1986.

[6] J. Kristiansen and D. Tong, "Silica-scaled Chrysophytes of Wuhan, a preliminary note," *Journal of Wuhan Botanical Research*, vol. 6, pp. 97–100, 1988.

[7] J. Kristiansen, "Silica-scaled chrysophytes from China," *Nordic Journal of Botany*, vol. 8, pp. 539–552, 1989.

[8] J. Kristiansen and D. Tong, "*Chrysosphaerella annulata* n. sp., a new scale-bearing Chrysophyte," *Nordic Journal of Botany*, vol. 9, pp. 329–332, 1989.

[9] J. Kristiansen and D. Tong, "Studies on silica-scaled Chrysophytes from Wuhan, Hangzhou, and Beijing, P. R. China," *Nova Hedwigia*, vol. 49, pp. 183–202, 1989.

[10] J. Kristiansen, "Studie on silica-scaled Chrysophytes from Central Asia, from Xinjiang and from Gansu, Qinghai, and Shaanxi Provinces, P. R. China," *Archiv für Protistenk*, vol. 138, pp. 298–303, 1990.

[11] J. Kristiansen and D. Tong, "Investigations on silica-scaled Chrysophytes in China," *Verhandlungen der Internationalen Vereinigung für Theoretische und Angewandte Limnologie*, vol. 24, no. 4, pp. 2630–2633, 1991.

[12] W. Y. X. Wei Yin Xin and J. Kristiansen, "Occurrence and distribution of silica-scaled chrysophytes in Zhejiang, Jiangsu, Hubei, Yunnan and Shandong provinces, China," *Archiv für Protistenkunde*, vol. 144, no. 4, pp. 433–449, 1994.

[13] W. Yin-xin and J. Kristiansen, "Studies on silica-scaled chrysophytes from Fujian Province, China," *Chinese Journal of Oceanology and Limnology*, vol. 16, no. 3, pp. 256–266, 1998.

[14] Y. X. Wei and X. P. Yuan, "Studies on silica-scaled Chrysophytes from the tropics and subtropics of China," *Beihefte zur Nova Hedwigia*, vol. 122, pp. 169–187, 2001.

[15] R. M. M. Roijackers and H. Kessels, "Ecological characteristics of scale-bearing Chrysophyceae from the Netherlands," *Nordic Journal of Botany*, vol. 6, pp. 373–383, 1986.

[16] P. A. Siver, *The Biology of Mallomonas*, Kluwer Academic Publishers, Dordrecht, The Netherlands, 1991.

[17] P. A. Siver, "Inferring the specific conductivity of lake water with scaled chrysophytes," *Limnology and Oceanography*, vol. 38, no. 7, pp. 1480–1492, 1993.

[18] P. A. Siver and A. M. Lott, "Preliminary investigations on the distribution of scaled chrysophytes in Vermont and New Hampshire (USA) lakes and their utility to infer lake water chemistry," *Nordic Journal of Botany*, vol. 20, no. 2, pp. 233–246, 2000.

[19] J. P. Smol, "Application of chrysophytes to problems in paleoecology," in *Chrysophyte Algae. Ecology, Phylogeny and Development*, C. D. Sandgren, J. P. Smol, and J. Kristiansen, Eds., pp. 303–330, Cambridge University Press, Cambridge, UK, 1995.

[20] L. Lwanoff, "Beitrag zur Kenntniss der Morphologie und Systematik der Chrysomonaden," *Bulletin de l'Académie Impériale des Sciences de Saint Pétersbourg*, vol. 11, pp. 247–262, 1899.

[21] W. Conrad, "Le genre Microglena Ehrenberg C G," *Archiv Für Protistenk*, vol. 60, pp. 415–439, 1927.

[22] W. Conrad, "Revision du genre *Mallomonas* Perty et Pseudomallomas Chodat," *Mémoirs du Museum d'histoire Naturelle Belgique*, vol. 56, pp. 1–82, 1933.

[23] P. Bourrelly, "Recherches sur les Chrysophycees, morphologie, phylogenie, systematique," *Revue Algologique Mémoire Serie*, vol. 1, pp. 337–368, 1957.

[24] K. Harris and D. E. Bradley, "An examination of the scales and bristles of *Mallomonas* in the electron microscope using carbon replicas," *Journal of the Royal Microscopical Society*, vol. 76, no. 1-2, pp. 37–46, 1957.

[25] K. Harris and D. E. Bradley, "A taxonomic study of *Mallomonas*," *Journal of General Microbiology*, vol. 22, pp. 750–777, 1960.

[26] L. S. Péterfi and L. Momeu, "Romanian *Mallomonas* species studied in light and electron microscopes," *Nova Hedwigia*, vol. 27, pp. 353–392, 1976.

[27] L. Momeu and L. S. Péterfi, "Taxonomy of *Mallomonas* based on the fine structure of scales and bristles," *Contributions to Botany Cluj-Napoca*, vol. 19, pp. 13–20, 1979.

[28] J. Kristiansen, "The genus *Mallomonas* (Synurophyceae)—a taxonomic survey based on the ultrastructure of silica scales and bristles," *Opera Botanica*, vol. 139, pp. 5–218, 2002.

[29] W. D. Lu, *SPSS for Windows*, Publishing House of Electronics Industry, Beijing, China, 2003.

[30] S. Lavau, G. W. Sounders, and R. Wetherbee, "A phylogenetic analysis of the synurophyceae using molecular data and scale case morphology," *Journal of Phycology*, vol. 33, no. 1, pp. 135–151, 1997.

[31] R. A. Andersen, "Molecular systematic of the Chrysophyceae and Synurophyceae," in *Unravelling the Algae: The Past, Present, and Future of Algal Systematics*, J. Brodie and J. Lewis, Eds., pp. 285–313, CRC Press, Boca Raton, Fla, USA, 2007.

[32] J. Kristiansen and H. R. Preisig, "Süßwasserflora von Mitteleuropa Freshwater Flora of Central Europe," in *Chrysophyte and Haptophyte Algae*, vol. 2 of *Teil/Part2: Synurophyceae*, Springer, Berlin, Germany, 2007.

[33] B. Y. Jo, W. Shin, S. M. Boo, H. S. Kim, and P. A. Siver, "Studies on ultrastructure and three-gene phylogeny of the genus *Mallomonas* (synurophyceae)," *Journal of Phycology*, vol. 47, no. 2, pp. 415–425, 2011.

[34] F. L. Bookstein, *Morphometric Tools for Landmark Data: Geometry and Biology*, Cambridge University Press, Cambridge, UK, 1991.

[35] I. L. Dryden and K. V. Mardia, *Statistical Shape Analysis*, John Wiley & Sons, New York, NY, USA, 1998.

[36] F. J. Rohlf, "Statistical power comparisons among alternative morphometric methods," *American Journal of Physical Anthropology*, vol. 111, pp. 463–478, 2000.

[37] M. L. Zelditch, D. L. Swiderski, D. H. Sheets, and W. L. Fink, *Geometric Morphometrics for Biologists: A Primer*, Elsevier Academic Press, London, UK, 2004.

[38] D. C. Adams, F. J. Rohlf, and D. E. Slice, "Geometric morphometrics: ten years of progress following the 'revolution'," *Italian Journal of Zoology*, vol. 71, no. 1, pp. 5–16, 2004.

[39] B. Beszteri, É. Ács, and L. Medlin, "Conventional and geometric morphometric studies of valve ultrastructural variation in two closely related *Cyclotella* species (Bacillariophyta)," *European Journal of Phycology*, vol. 40, no. 1, pp. 89–103, 2005.

[40] H. Verbruggen, O. De Clerck, E. Cocquyt, W. H. C. F. Kooistra, and E. Coppejans, "Morphometric taxonomy of siphonous green algae: a methodological study within the genus *Halimeda* (Bryopsidales)," *Journal of Phycology*, vol. 41, no. 1, pp. 126–139, 2005.

[41] H. Verbruggen, O. De Clerck, W. H. C. F. Kooistra, and E. Coppejans, "Molecular and morphometric data pinpoint species boundaries in *Halimeda* section *Rhipsalis* (Bryopsidales, Chlorophyta)," *Journal of Phycology*, vol. 41, no. 3, pp. 606–621, 2005.

[42] J. Neustupa and Y. Němcová, "A geometric morphometric study of the variation in scales of *Mallomonas striata* (Synurophyceae, Heterokontophyta)," *Phycologia*, vol. 46, no. 2, pp. 123–130, 2007.

17

Microbial Diversity in Soil under Potato Cultivation from Cold Desert Himalaya, India

Priyanka Sati, Kusum Dhakar, and Anita Pandey

Biotechnological Applications, G. B. Pant Institute of Himalayan Environment and Development, Kosi-Katarmal, Almora, Uttarakhand 263 643, India

Correspondence should be addressed to Anita Pandey; anita@gbpihed.nic.in

Academic Editors: I. Bisht, H. Ford, R. Rico-Martinez, and P. M. Vergara

Mana village (Chamoli district, Uttarakhand, India), situated in high altitudes (3,238 m above mean sea level) of Indian Himalayan region, represents cold desert climatic conditions. At Mana, potato is grown from May to September, while the site remains snow clad for approximately six months (from October to April). Soil samples, collected from Mana potato fields, were analyzed for cultivable microbial diversity along with the chemical and enzymatic properties. The analysis revealed colonization of soil by microflora in moderate numbers (up to 10^7 CFU/g soil) with limited species level. 25 morphologically distinct microbial isolates belonging to Gram +ve and Gram −ve bacteria, actinomycetes, and fungi including yeast were isolated. The bacteria were tentatively identified as species of *Bacillus* and *Pseudomonas,* while the majority of the fungal isolates belonged to the species of *Penicillium*. These microbial isolates possessed plant growth promotion and biocontrol properties assessed mainly in terms of production of indole acetic acid and hydrolytic enzymes and phosphate solubilization. The soil, when used as "inoculum" in plant based bioassays, exhibited positive influence on plant growth related parameters. The limited diversity of cold tolerant microbial species also extends opportunity to understand the resilience possessed by these organisms under low temperature environment.

1. Introduction

Microorganisms are ubiquitous in nature; their distribution is governed by environmental specificities. Extreme environmental conditions are not uncommon, and the microbial diversity of such areas is of particular interest because of the superb adaptability of the native microbes. Due to slow growth rate and difficulty of handling, relatively little attention has been given to cold adapted psychrophiles or psychrotolerant microbes. Decrease in microbial population with a concomitant increase in the altitude has been reported [1]. Under low temperature environments, the importance and distinction between psychrophiles and psychrotrophs or psychrotolerants have also been recognized [2]. Psychrotolerant microbes are important in high-altitude agroecosystems since they survive and retain their functionality at low temperature conditions, while growing optimally at warmer temperatures [3].

The Indian Himalayan region (IHR) occupies special place in the mountain ecosystems of the world. The mountain agroecosystems are characterized by difficult terrain, inadequate infrastructure, inaccessibility and marginal societies, lack of irrigation, severe top soil erosion, and overall external inputs to the system. Agricultural production in the mountains is, to a large extent, influenced by low organic matter, soil moisture status, and colder conditions. Therefore, hill agriculture is, by and large, a low input, low production and subsistence but a sustainable system. The cold adapted microbes that possess various plant growth promotion abilities can be utilized for increased plant production especially in the low temperature environments [4–7].

Potato is grown in more than 150 countries in the world, India being at third place among the ten best producers with approximately 7.5 percent of the world's total production (http://agropedia.iitk.ac.in). In India, potato is also grown in mountain states including Uttarakhand. Potato fields in Mana village (Chamoli district, Uttarakhand) where soil remains influenced by snow for approximately six months of the year (October–April) extend unique opportunity to examine the soil microbial communities from diversity, biotechnological applications, and ecological resilience viewpoints. In the present study, soil samples collected from the cold desert area under potato cultivation in IHR have been analyzed for the diversity of microorganisms with particular reference to their plant growth promoting abilities along with the chemical and enzymatic analyses.

2. Materials and Methods

2.1. Study Location. The soil samples were collected from potato fields at Mana village representing cold desert climatic conditions (30° 46′ 24.8″ N; 79° 29′ 33.4″ E; 3,238 m above mean sea level). The samples were collected from 3 different terraces; 5 samples from each terrace were mixed for obtaining composite samples. The pH of soil was 5.9. The site remains snow clad from October to April, maintaining subzero temperature. Mana that lies 3 Km north of Badrinath (Chamoli district, Uttarakhand, India) is recognized as the last Indian village toward the Indo-Tibet border. The village represents a microecosystem consisting of the local community and the livestock along with physical and organic resources. The local community migrates to lower altitudes (Gopeshwar, Chamoli district) during winters every year and returns when the snow thaws (http://myyatradiary.blogspot.in).

2.2. Soil Analysis (Chemical and Enzymatic). The soil pH and organic carbon, total nitrogen, total phosphorus, and total potassium (percent dry weight basis) contents were determined following standard procedures. Analysis of microbial enzymes, namely, amylase, invertase, and cellulase, was performed following the methods described in Zafar et al. [8]. Urease activity was determined by phenol-hypochlorite method [9], and phosphatase assay was based on pNPP [10].

2.3. Microbial Analysis (Enumeration, Isolation, and Characterization). For enumeration of cultivable microbial communities, isolations were carried out on a range of prescribed media following serial dilution technique. These media included tryptone yeast extract agar (TYA) and *Pseudomonas* isolation agar (for bacteria), actinomycetes isolation agar (for actinomycetes), and potato dextrose agar (for fungi) (all from Himedia, Bombay, India). The plates were incubated in 3 sets at 24°, 14°, and 4°C, and observations were recorded up to 3 weeks. Based on colony morphology, bacteria, actinomycetes, and fungi were carefully picked up from the agar plates; following subculture, the purified isolates were transferred onto slants and glycerol stocks for further use. Morphologically distinct isolates, each given a code number, were subjected to further investigations.

Characterization of bacterial isolates and actinomycetes was carried out following morphological (colony morphology), microscopic (Gram staining), biochemical (utilization of carbon sources and enzyme activity), growth (temperature, pH, and salt tolerance), and cultural (oxygen requirement) characteristics on prescribed media. In case of fungi, the microscopic observations were recorded following staining the cultures with lactophenol cotton blue and observing under microscope (Nikon-Eclipse 50i, Japan). The microbial

cultures were tentatively identified up to genus or species level. All the experiments were performed in triplicates.

2.4. Plant Growth Promotion and Biocontrol Activities of Soil Microorganisms. Qualitative and quantitative estimations of all the microbial isolates for phosphate solubilization were done at 24°C in Pikovskaya's broth medium containing tricalcium phosphate. The solubilized phosphorus in the culture filtrate was determined by using chlorostannous reduced molybdophosphoric acid blue method [11] on the seventh day. Antagonistic activity of the microbial isolates was determined by using the methods described in Chaurasia et al. [12]. The percent inhibition by the production of diffusible and volatile compounds was determined, separately. Antagonistic activities of the microbial isolates were tested against test pathogens, namely, *Fusarium oxysporum* and *F. solani*, on potato carrot agar (PCA). The production of other plant growth regulating activities, namely, ammonia, indole acetic acid (IAA), chitinase, siderophore, and hydrogen cyanide (HCN), was determined following standard procedures as described in Malviya et al. [13].

2.5. Plant Based Bioassays Using Soil Inoculum. The soil collected from the potato fields was used as consortium of "microbial inoculum" following plant based bioassays using the test crops, wheat (*Triticum aestivum*) and lentil (*Lens esculenta*), under net house of the Institute. Seeds were grown in polyethylene bags (20.5 × 8.0 cm; 50 bags for each treatment). The soil was sandy loam with pH H$_2$O 6.7 and 40% (w/w) moisture content. The treatments under consideration were (1) control (seeds without soil inoculum) and (2) seeds inoculated with soil inoculum that was taken from potato field (5 g/bag/seed) at the time of sowing. At harvest (45 days of growth), 10 plants from each treatment were selected randomly, and fresh weight of roots and shoots was taken. Dry weight was taken after drying the roots and shoots in oven at 70°C for 72 h, separately, for each plant. Rhizosphere soil samples, collected from each treatment, were analyzed for colonization of microorganisms including mycorrhizae and endophytes.

2.6. Statistics. Microsoft windows 2003 professional excel program was used to calculate means and standard deviations. One-way ANOVA was performed to determine significant difference between control and inoculated plants (in plant based bioassay).

3. Results and Discussion

Soil under potato cultivation was determined as sandy loam that contained organic carbon (1.75%), phosphorus (0.04%), potassium (1.69%), and nitrogen (0.14%) contents. The enzymatic activity among carbohydrases was measured as amylase 13.3 ± 1.3, invertase 12.4 ± 3.8, and cellulase 15.8 ± 1.9 μg/g soil/h. Phosphatase and urease activities were measured to be 11.9 ± 1.1 and 2.8 ± 0.38 μg/g soil/h, respectively. The value, for soil enzymatic activity, are relatively in lower range as compared to the earlier reports [14]; this can be attributed

TABLE 1: Colony forming units (CFU $\times 10^5$/g soil) of three groups of microbes in soil under potato cultivation.

Medium	Temperature (°C)	Incubation period (weeks)	Bacteria	Actinomycetes	Fungi
TYA	24	01	41.00 ± 3.60	1.66 ± 0.57	1.40 ± 0.55
	14	02	27.33 ± 3.05	1.66 ± 1.15	1.33 ± 0.57
	4	03	32.66 ± 3.05	1.33 ± 1.15	1.00 ± 1.00
PDA	24	01	39.66 ± 2.51	1.33 ± 0.57	1.00 ± 1.00
	14	02	25.33 ± 0.57	2.66 ± 2.08	1.43 ± 1.15
	4	03	32.33 ± 2.51	2.00 ± 1.00	1.33 ± 0.50
AIA	24	01	37.00 ± 2.64	2.66 ± 0.57	1.36 ± 0.58
	14	02	27.66 ± 2.51	3.00 ± 1.00	1.42 ± 0.50
	4	03	32.33 ± 2.88	2.33 ± 1.52	1.33 ± 0.57

to the low activity of soil microbes under extremely low temperature environment.

The microbial colonies on agar plates were observed till 10^{-7} dilution, while 10^{-5} was found to be appropriate for enumeration of colonies. Among three sets of the plates that were incubated at 24, 14, and 4°C, well developed colonies were obtained after one, two, and three weeks of incubation, respectively (Table 1). A total of 25 morphologically distinct isolates, bacteria (14), actinomycetes (3), and fungi (8 (including 1yeast)) were obtained as pure cultures on prescribed media. Amongst bacteria, 11 whitish to cream colonies with smooth and slimy consistency were obtained from TYA plates. These were observed as Gram positive and rod shaped in varied cell arrangement (single, diplobacilli, short to long chains, or clusters in palisade arrangement). Colonies with production of mucoid substances and yellowish to greenish pigment obtained on TYA and Pseudomonas isolation agar plates were observed as Gram −ve oval rods arranged as single cells. Based on colony morphology, microscopy, and biochemical tests including utilization of carbon sources (data not presented), the Gram +ve and Gram −ve bacteria are referred as species of Bacillus and Pseudomonas, respectively. The hard pustules-like colonies with white to gray aerial mycelium developed on TYA, PDA, and AIA plates, along with branched filaments under microscope, were considered under broad category of actinomycetes. These are tentatively referred as species of Streptomyces based on morphology and comparative assessment with the available stock cultures in the laboratory. Based on colony morphology and microscopic features, the eight distinct colonies obtained from PDA were assigned to Penicillium (6), Trichocladium, and yeast (1) (Table 2). The microbial cultures exhibited wide range of tolerance for temperature, pH, and salt concentration (Table 3).

The results on enumeration of microbes indicated the extensive colonization of soil by the major groups of microorganisms, namely, bacteria, actinomycetes, fungi, and yeast. However, these microbial communities were represented by limited number of morphotypes that can be attributed to the selection pressure caused by the stress under extremely low temperature, remaining subzero for almost six months. Potato cultivation under these conditions is also an indicative of the resilience possessed by the crop. Colonization of extreme temperature (low or high) environments by a variety

of microorganisms in Himalayan region has been reported in previous studies [15–17]. Dominance of species of Bacillus in extreme conditions is attributed to the ability to resist the environmental stresses due to their spore forming nature [18]. Similarly, other bacteria, fungi mainly species of Penicillium, actinomycetes, and yeasts have been reported from extreme environments including high altitudes of Himalaya [2, 13, 19–21].

All the microbial isolates exhibited activities related to plant growth promotion and biocontrol as well. Out of 25 isolates, 23 produced IAA, while 15 possessed the ability to solubilize phosphates (1.13 ± 0.02 to 10.46 ± 0.25 μg/mL). Among biocontrol activities, 24 isolates produced ammonia, and 17 produced chitinase. In plate assays, 18 isolates inhibited the growth of test pathogens, Fusarium oxysporum, and F. solani, due to the production of diffusible and volatile antifungal compounds. Morphological abnormalities, as a result of antagonistic microbial activities, were observed under microscope in both the test fungi. None of the isolates produced HCN and siderophore (Table 4, Figure 1(a)).

The beneficial soil microbes influence plant growth through direct or indirect mechanisms. The examples of direct mechanism(s) are growth promotion by providing fixed nitrogen to the host plant, production of phytohormones, and phosphate solubilization. The indirect mechanisms mainly involve biological control of plant pathogens that may be assisted through antibiosis and production of antimicrobial substances, including siderophores, lytic enzymes, and biocidal volatiles [22–26]. Several microorganisms isolated from colder regions in IHR have been characterized for their beneficial plant growth related activities. Species of Bacillus, B. subtilis and B. megaterium in particular, have been investigated for their growth promotion abilities in agricultural [27–29] as well as forest species [30]. Similarly, cold tolerant species of Pseudomonas have been characterized for their growth promotion with particular reference to phosphate solubilization [31–33] and biocontrol abilities [34]. Selected species have also been developed in bioformulations, suitable for field application [35, 36]. The microbial isolates, obtained in the present study, were also found to be positive for the production of a range of hydrolytic enzymes. 24 isolates were positive for lipase, 19 each for protease and xylanase, 14 for amylase, 11 for pectinase, and 9 for cellulase (Figure 1(b)). Production of extracellular cell wall degrading

TABLE 2: Colony morphology and microscopic characters of the microbial isolates.

S. no.	Isolate code	Morphological and microscopic characters	Identification
		Bacteria	
1	Cdpb1	Off white, entire, and slimy colony with 5 mm dia; Gram +ve elongated bacilli in palisade or cluster arrangement; facultative anaerobic	*Bacillus* sp.
2	Cdpb2	Off white, convex, slimy, and round colony with 10 mm dia; Gram +ve bacilli, arranged in long chains and clusters; facultative anaerobic	*Bacillus megaterium*
3	Cdpb4	Off white, entire, smooth, and round colony with 4 mm dia; Gram +ve diplobacilli, arranged in clusters or short chains; facultative anaerobic	*Bacillus* sp.
4	Cdpb5	Off white, entire, and slimy colony with 6 mm dia; Gram +ve bacilli arranged in short chains; facultative anaerobic	*Bacillus* sp.
5	Cdpb7	Off white, entire, and smooth colony with 2-3 mm dia; Gram +ve elongated bacilli, arranged in clusters or short chains; facultative anaerobic	*Bacillus* sp.
6	Cdpb8	Off white, convex, smooth, and round colony with 1-2 mm dia; Gram +ve bacilli, arranged in palisade or short chains; facultative anaerobic	*Bacillus* sp.
7	Cdpb10	Yellow, entire, and smooth colony with 2 mm dia; Gram +ve elongated bacilli, arranged as single or in clusters; facultative anaerobic	*Bacillus* sp.
8	Cdpb13	Off white, convex, slimy, and round colony with 10 mm dia; Gram +ve bacilli, arranged as single and long chains; facultative anaerobic	*Bacillus megaterium*
9	Cdpb16	White, entire, slimy, and round colony with 3-4 mm dia; Gram +ve bacilli, arranged in clusters and long chains; facultative anaerobic	*Bacillus megaterium*
10	Cdpb19	Light yellow, entire, smooth, and round colony with 5 mm dia; Gram −ve arranged in oval or slightly curved rods; facultative anaerobic	*Pseudomonas* sp.
11	Cdpb20	Light yellow, irregular rhizoid, and translucent colony with 4-5 mm dia; Gram +ve diplobacilli, clusters or short chains; facultative anaerobic	*Bacillus* sp.
12	Cdpb22	Greenish yellow, entire, smooth, and round colony with 6-7 mm dia; Gram −ve, oval or slightly curved rods; facultative anaerobic	*Pseudomonas* sp.
13	Cdpb23	Light yellow mucoid colony with 6 mm dia; Gram −ve, small curved rods; facultative anaerobic	*Pseudomonas* sp.
14	Cdpb27	Off white, entire, and slimy colony with 4 mm dia; Gram +ve elongated bacilli, in palisade or cluster arrangement; facultative anaerobic	*Bacillus* sp.
		Actinomycetes	
15	Cdpact28	Aerial mycelium gray, powdery circular colony with 6 mm dia; filaments branched	*Streptomyces* sp.
16	Cdpact29	Aerial mycelium white, rough, and circular colony with 6 mm dia; filaments branched	*Streptomyces* sp.
17	Cdpact30	Aerial mycelium gray, smooth, and circular colony with 6 mm dia; filaments branched	*Streptomyces* sp.
		Fungi	
18	Cdpf2	Pink colony with 15 mm dia; mycelia septate with conidiophores	*Penicillium purpurogenum* (NFCCI2772)
19	Cdpf4	Greenish colony with 17 mm dia; mycelia septate with conidiophores	*Penicillium* sp. (NFCCI2774)
20	Cdpf5	White colony with 10 mm dia; myelia septate with conidiophores	*Penicillium* sp. (NFCCI2775)
21	Cdpf6	White colony with 20 mm dia; myelia septate with conidiophores	*Penicillium* sp. (NFCCI2776)
22	Cdpf7	Yellow colony with 15 mm di; mycelia septate with branched conidiophores	*Penicillium* sp.
23	Cdpf8	Off white colony with 20 mm dia; mycelia septate with branched conidiophores	*Penicillium* sp.
24	Cdpf10	Yellowish green colony with 17 mm dia; septate mycelia	*Trichocladium asperum* (NFCCI2777)
25	Cdpb26	White, entire, smooth, and round colony with 1-2 mm dia; unicellular with budding	Yeast

Dia: diameter; NFCCI: National Fungal Culture Collection of India, Agharkar Research Institute, Pune, India.

TABLE 3: Physiological (temperature, pH, and salt tolerance) characters of the microbial isolates.

S. no.	Isolate code	Temperature (°C)	pH	Salt (%)
		Bacteria		
1	Cdpb1	9–55 (opt. 25)	5–11 (opt. 7)	9
2	Cdpb2	4–55 (opt. 25)	5–11 (opt. 7)	5
3	Cdpb4	4–45 (opt. 25)	5–11 (opt. 7)	7
4	Cdpb5	9–55 (opt. 25)	5–11 (opt. 7)	9
5	Cdpb7	9–45 (opt. 25)	5–11 (opt. 7)	9
6	Cdpb8	4–45 (opt. 25)	5–11 (opt. 7)	9
7	Cdpb10	4–45 (opt. 25)	5–11 (opt. 7)	7
8	Cdpb13	4–55 (opt. 25)	5–11 (opt. 7)	5
9	Cdpb16	4–55 (opt. 25)	5–11 (opt. 7)	5
10	Cdpb19	9–45 (opt. 25)	5–11 (opt. 7)	9
11	Cdpb20	4–55 (opt. 25)	5–11 (opt. 7)	9
12	Cdpb22	9–55 (opt. 25)	5–11 (opt. 7)	5
13	Cdpb23	9–45 (opt. 25)	5–11 (opt. 7)	9
14	Cdpb27	9–45 (opt. 25)	5–11 (opt. 7)	9
		Actinomycetes		
15	Cdpact28	9–45 (opt. 25)	5–11 (opt. 7)	9
16	Cdpact29	9–45 (opt. 25)	5–11 (opt. 7)	9
17	Cdpact30	9–45 (opt. 25)	5–11 (opt. 7)	7
		Fungi		
18	Cdpf2	9–55 (opt. 25)	1.5–14 (opt. 6-7)	7
19	Cdpf4	9–45 (opt. 25)	1.5–14 (opt. 6-7)	7
20	Cdpf5	9–55 (opt. 25)	1.5–14 (opt. 6-7)	7
21	Cdpf6	9–55 (opt. 25)	1.5–14 (opt. 6-7)	7
22	Cdpf7	9–45 (opt. 25)	2–14 (opt. 6-7)	7
23	Cdpf8	9–55 (opt. 25)	2–14 (opt. 6-7)	7
24	Cdpf10	9–45 (opt. 25)	1.5–14 (opt. 6-7)	7
25	Cdpf26	4–55 (opt. 25)	5–11 (opt. 6-7)	7

opt.: optimum.

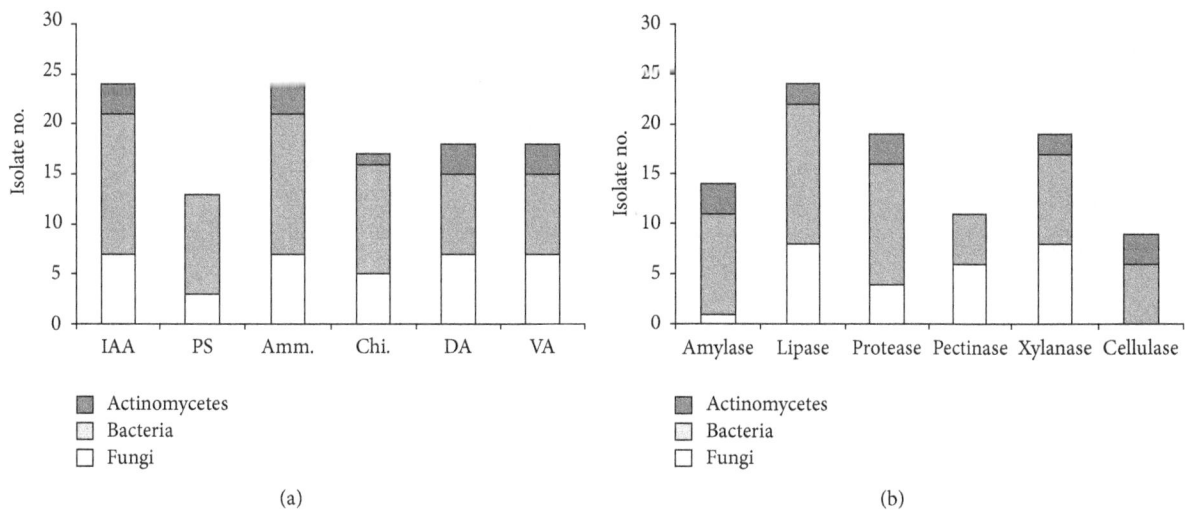

FIGURE 1: (a) Plant growth promotion and biocontrol related characters (IAA: indole acetic acid, PS: phosphate solubilization, Amm.: ammonia, Chi.: chitinase, DA: diffusible antimicrobials, VA: volatile antimicrobials). (b) Enzyme activity of the microbial isolates.

TABLE 4: Plant growth promotion and enzymatic activities of the microbial isolates.

S. no.	Isolate code		Properties
			Bacteria
1	Cdpb1	PGP	+ve for production of IAA, chitinase, ammonia, and antimicrobials; +ve for phosphate solubilization
		Enzymes	+ve for amylase, lipase, protease, pectinase, cellulase, and xylanase
2	Cdpb2	PGP	+ve for production of IAA, chitinase, and ammonia; +ve for phosphate solubilization; −ve for production of antimicrobials
		Enzymes	+ve for amylase, lipase, protease, pectinase, cellulase, and xylanase
3	Cdpb4	PGP	+ve for production of IAA, chitinase, and ammonia; +ve for phosphate solubilization; −ve for production of antimicrobials
		Enzymes	+ve for amylase, lipase, protease, pectinase, and xylanase; −ve for cellulose
4	Cdpb5	PGP	+ve for production of IAA, chitinase, ammonia, and antimicrobials; +ve for phosphate solubilization
		Enzymes	+ve for amylase, lipase, protease, pectinase, and xylanase; −ve for cellulose
5	Cdpb7	PGP	+ve for production of IAA, ammonia, and antimicrobials; −ve for chitinase and phosphate solubilization
		Enzymes	+ve for amylase, lipase, and protease; −ve for pectinase, cellulase, and xylanase
6	Cdpb8	PGP	+ve for production of IAA, ammonia, and antimicrobials; −ve for chitinase and phosphate solubilization
		Enzymes	+ve for amylase, lipase, and protease; −ve for pectinase, cellulase, and xylanase
7	Cdpb10	PGP	+ve for production of IAA and ammonia; −ve for chitinase, antimicrobials, and phosphate solubilization
		Enzymes	+ve for lipase, cellulase, and pectinase; −ve for amylase, protease, xylanase
8	Cdpb13	PGP	+ve for production of IAA, chitinase, and ammonia; −ve for phosphate solubilization and antimicrobials
		Enzymes	+ve for amylase, lipase, and xylanase; −ve for protease, pectinase and cellulose
9	Cdpb16	PGP	+ve for production of IAA, chitinase, antimicrobials, and ammonia; +ve for phosphate solubilization
		Enzymes	+ve for amylase, lipase, protease, and xylanase; −ve for pectinase and cellulose
10	Cdpb19	PGP	+ve for production of IAA, chitinase, and ammonia; +ve for phosphate solubilization; −ve for antimicrobials
		Enzymes	+ve for lipase, protease, and cellulase; −ve for amylase, pectinase, and xylanase
11	Cdpb20	PGP	+ve for production of IAA, chitinase, ammonia, and antimicrobials; +ve for phosphate solubilization
		Enzymes	+ve for lipase, protease, and cellulase, −ve for amylase, pectinase, and xylanase
12	Cdpb22	PGP	+ve for production of IAA, chitinase, and ammonia; +ve for phosphate solubilization; −ve for antimicrobials
		Enzymes	+ve for amylase, lipase, and protease, xylanase; −ve for pectinase and cellulose
13	Cdpb23	PGP	+ve for production of IAA, chitinase, ammonia, and antimicrobials; +ve for phosphate solubilization
		Enzymes	+ve for lipase, protease, and xylanase; −ve for amylase, cellulose, and pectinase
14	Cdpb27	PGP	+ve for production of IAA, chitinase, ammonia, and antimicrobials; +ve for phosphate solubilization
		Enzymes	+ve for amylase, lipase, protease, cellulase, and xylanase; −ve for pectinase
			Actinomycetes
15	Cdpact28	PGP	+ve for production of IAA, ammonia, and antimicrobials; −ve for chitinase and phosphate solubilization
		Enzymes	+ve for amylase, protease, and cellulase, −ve for lipase, xylanase, and pectinase
16	Cdpact29	PGP	+ve production of IAA, chitinase, ammonia, and antimicrobials; −ve for phosphate solubilization
		Enzymes	+ve for amylase, lipase, protease, cellulase, and xylanase; −ve for pectinase
17	Cdpact30	PGP	+ve for production of IAA, ammonia, and antimicrobials; −ve for chitinase and phosphate solubilization
		Enzymes	+ve for amylase, lipase, protease, and cellulase, xylanase; −ve for pectinase

TABLE 4: Continued.

S. no.	Isolate code		Properties
			Fungi
18	Cdpf2	PGP	+ve for production of IAA, chitinase, ammonia, and antimicrobials; +ve for phosphate solubilization
		Enzymes	+ve for amylase, lipase, protease, xylanase, and pectinase; −ve for cellulose
19	Cdpf4	PGP	+ve for production of IAA, ammonia, and antimicrobials; −ve for chitinase and phosphate solubilization
		Enzymes	+ve for lipase, xylanase, and pectinase; −ve for amylase, protease, and cellulose
20	Cdpf5	PGP	+ve for production of IAA, chitinase, ammonia, and antimicrobials; −ve for phosphate solubilization
		Enzymes	+ve for lipase, xylanase, and pectinase; −ve for amylase, protease, and cellulase
21	Cdpf6	PGP	+ve for production of IAA, ammonia and antimicrobials; −ve for chitinase and phosphate solubilization
		Enzymes	+ve for lipase, xylanase, pectinase; −ve for amylase, protease, and cellulose
22	Cdpf7	PGP	+ve for production of chitinase, antimicrobials, and phosphate solubilization; −ve for IAA and ammonia
		Enzymes	+ve for lipase, protease, and xylanase, −ve for amylase, cellulose, and pectinase
23	Cdpf8	PGP	+ve for production of ammonia and antimicrobials; −ve for IAA, chitinase, and phosphate solubilization
		Enzymes	+ve for lipase, protease, xylanase, and pectinase, −ve for amylase, cellulose
24	Cdpf10	PGP	+ve for production of IAA, chitinase, ammonia, and antimicrobials; −ve for phosphate solubilization
		Enzymes	+ve lipase, xylanase, and pectinase; −ve for amylase, protease, and cellulose
25	Cdpf26	PGP	+ve for production of IAA, chitinase, ammonia and phosphate solubilization; −ve for antimicrobials
		Enzymes	+ve for lipase, protease, and xylanase; −ve for amylase, cellulose, and pectinase

PGP: plant growth promotion.

TABLE 5: Effect of inoculation on growth of test crops in net house conditions.

Treatment	Shoot length (cm)	Shoot weight (g)	Root length (cm)	Root weight (g)
		Wheat		
Control	47.17 ± 8.71	0.68 ± 0.22	16.75 ± 3.17	0.20 ± 0.05
Inoculated	62.11 ± 12.31*	1.30 ± 0.80*	21.11 ± 3.87*	0.63 ± 0.39*
		Lentil		
Control	24.29 ± 2.86	0.71 ± 0.20	12.61 ± 2.56	0.10 ± 0.04
Inoculated	28.27 ± 3.63*	0.78 ± 0.30	16.58 ± 3.44*	0.18 ± 0.09*

Values are mean ± SD ($n = 15$); *significant at $P \leq 0.05$.

enzymes has been associated with biocontrol abilities in plant growth promoting microbes [13, 37].

The influence due to the presence of plant growth promoting microbes was demonstrable when the soil was used in form of inoculum representing a "consortium" of beneficial microbes. Inoculation with soil consortium showed positive effects on plant growth related parameters such as length and dry weight of the root and shoot of the test crops, that is, wheat and lentil ($P \leq 0.05$) (Table 5). The inoculation also resulted in stimulation of rhizosphere microorganisms ($P \leq 0.05$), mainly bacteria and actinomycetes. The roots were also observed with moderate colonization by mycorrhizae (up to 30%) along with colonization of endophytes, mainly bacterial and fungal (up to 80%). Use of small proportion of rhizosphere soil in appropriate ratio has been reported for raising healthy seedlings of forest species of Himalayan region [38].

4. Conclusion

The distinct feature of the present study is the geographic location used for potato cultivation where soil remains snow clad for almost six months, before and after crop season. It presented a unique ecological niche, where the crop along with a consortium of beneficial microbes evolves and adapts to the prevailing edaphic and climatic conditions. The analyses of soil for microbial communities indicated toward the importance of selection pressure in the survival and dominance of selected group of microbes under stress conditions. The environment under snow cover is likely to act as a limiting factor for survival of the soil microflora. The best of the survivors then multiply and tend to rapidly increase in number and dominate. As a consequence, although maintaining the low nutrition status in terms of nutrients and enzymes, the soil under potato cultivation was able to colonize higher counts that were enumerated up to 10^{-7} dilution. The dominance of microbes, linked with plant growth promotion and biocontrol activities, allowed the transformation of soil in form of a natural consortium. This consortium consisted of native beneficial microbes mainly belonging to the species of *Bacillus*, *Pseudomonas*, and *Penicillium*, along with actinomycetes and yeast. Potato, being valuable food crop worldwide, has received attention in view of the colonization of plant growth promoting microbes in potato fields [39, 40]. Plant growth promoting microbes are also receiving attention for their associated importance in bioremediation [41, 42].

Conflict of Interests

The authors declare that they have no conflict of interests.

Acknowledgments

Director of GBPIHED is acknowledged for extending the facilities. Ministry of Environment and Forests, Government of India, New Delhi, is acknowledged for financial support.

References

[1] A. Pandey and L. M. S. Palni, "The rhizosphere effect in trees of the Indian Central Himalaya with special reference to altitude," *Applied Ecology and Environmental Research*, vol. 5, no. 1, pp. 93–102, 2007.

[2] R. Margesin, "Effect of temperature on growth parameters of psychrophilic bacteria and yeasts," *Extremophiles*, vol. 13, no. 2, pp. 257–262, 2009.

[3] P. K. Mishra, S. C. Bisht, J. K. Bisht, and J. C. Bhatt, "Cold-tolerant PGPRs as bioinoculants for stress management," in *Bacteria in Agrobiology: Stress Management*, D. K. Maheshwari, Ed., pp. 95–118, Springer, Berlin, Germany, 2012.

[4] A. Pandey, A. Durgapal, M. Joshi, and L. M. S. Palni, "Influence of *Pseudomonas corrugata* inoculation on root colonization and growth promotion of two important hill crops," *Microbiological Research*, vol. 154, no. 3, pp. 259–266, 1999.

[5] P. Trivedi, B. Kumar, A. Pandey, and L. M. S. Palni, "Growth promotion of rice by phosphate solubilizing bioinoculants in a Himalayan location," in *Plant and Soil, Developments in Plant and Soil Sciences, First International Meeting on Microbial Phosphate Solubilization*, E. Velazqez and C. Rodriguez-Barrueco, Eds., vol. 102, pp. 291–299, Springer, 2007.

[6] P. Trivedi, A. Pandey, and L. M. S. Palni, "Bacterial inoculants for field applications under Mountain Ecosystem: present initiatives and future prospects," in *Bacteria in Agrobiology, Plant Probiotics*, D. K. Maheshwari, Ed., pp. 15–44, Springer, 2012.

[7] V. Kumar, P. Singh, M. A. Jorquera et al., "Isolation of phytase-producing bacteria from Himalayan soils and their effect on growth and phosphorus uptake of Indian mustard (*Brassica juncea*)," *World Journal of Microbiology and Biotechnology*, vol. 29, no. 8, pp. 1361–1369, 2013.

[8] Y. Zafar, K. A. Malik, and A. H. Chaudhary, "Activities of some carbohydrases in agricultural soils," *Pakistan Journal of Agricultural Research*, vol. 2, no. 1, pp. 41–45, 1981.

[9] J. K. Fawcett and J. E. Scott, "A rapid and precise method for the determination of urea," *Journal of clinical pathology*, vol. 13, pp. 156–159, 1960.

[10] M. A. Tabatabai and J. M. Bremner, "Use of p-nitrophenyl phosphate for assay of soil phosphatase activity," *Soil Biology and Biochemistry*, vol. 1, no. 4, pp. 301–307, 1969.

[11] M. L. Jackson, *Soil Chemical Analysis*, Prentice-Hall, New Delhi, India, 1967.

[12] B. Chaurasia, A. Pandey, L. M. S. Palni, P. Trivedi, B. Kumar, and N. Colvin, "Diffusible and volatile compounds produced by an antagonistic *Bacillus subtilis* strain cause structural deformations in pathogenic fungi in vitro," *Microbiological Research*, vol. 160, no. 1, pp. 75–81, 2005.

[13] M. K. Malviya, A. Pandey, P. Trivedi, G. Gupta, and B. Kumar, "Chitinolytic activity of cold tolerant antagonistic species of *streptomyces* isolated from glacial sites of Indian Himalaya," *Current Microbiology*, vol. 59, no. 5, pp. 502–508, 2009.

[14] C. Trasar-Cepeda, M. C. Leirós, and F. Gil-Sotres, "Hydrolytic enzyme activities in agricultural and forest soils. Some implications for their use as indicators of soil quality," *Soil Biology and Biochemistry*, vol. 40, no. 9, pp. 2146–2155, 2008.

[15] A. Ghildiyal and A. Pandey, "Isolation of cold tolerant antifungal strains of *Trichoderma* sp. from glacial sites of Indian Himalayan Region," *Research Journal of Microbiology*, vol. 3, no. 8, pp. 559–564, 2008.

[16] K. Rinu and A. Pandey, "Temperature-dependent phosphate solubilization by cold- and pH-tolerant species of *Aspergillus* isolated from Himalayan soil," *Mycoscience*, vol. 51, no. 4, pp. 263–271, 2010.

[17] A. Sharma, A. Pandey, Y. S. Shouche, B. Kumar, and G. Kulkarni, "Characterization and identification of *Geobacillus* spp. isolated from Soldhar hot spring site of Garhwal Himalaya, India," *Journal of Basic Microbiology*, vol. 49, no. 2, pp. 187–194, 2009.

[18] W. Mongkolthanaruk, "Classification of bacillus beneficial substances related to plants, humans and animals," *Journal of Microbiology and Biotechnology*, vol. 22, no. 12, pp. 1597–1604, 2012.

[19] A. Pandey, L. M. S. Palni, and D. Bisht, "Dominant fungi in the rhizosphere of established tea bushes and their interaction with the dominant bacteria under in situ conditions," *Microbiological Research*, vol. 156, no. 4, pp. 377–382, 2001.

[20] A. Pandey, N. Das, B. Kumar, K. Rinu, and P. Trivedi, "Phosphate solubilization by *Penicillium* spp. isolated from soil samples of Indian Himalayan region," *World Journal of Microbiology and Biotechnology*, vol. 24, no. 1, pp. 97–102, 2008.

[21] S. A. Cantrell, J. C. Dianese, J. Fell, N. Gunde-Cimerman, and P. Zalar, "Unusual fungal niches," *Mycologia*, vol. 103, no. 6, pp. 1161–1174, 2011.

[22] B. Lugtenberg and F. Kamilova, "Plant-growth-promoting rhizobacteria," *Annual Review of Microbiology*, vol. 63, pp. 363–383, 2009.

[23] P. N. Bhattacharyya and D. K. Jha, "Plant growth-promoting rhizobacteria (PGPR): emergence in agriculture," *World Journal of Microbiology and Biotechnology*, vol. 28, no. 4, pp. 1327–1350, 2012.

[24] N. Vassilev, B. Eichler-Löbermann, and M. Vassileva, "Stress-tolerant P-solubilizing microorganisms," *Applied Microbiology and Biotechnology*, vol. 95, pp. 851–859, 2012.

[25] B. R. Glick, "Plant growth-promoting bacteria: mechanisms and applications," vol. 2012, Article ID 963401, 15 pages, 2012.

[26] A. Pandey, L. M. S. Palni, and N. Coulomb, "Antifungal activity of bacteria isolated from the rhizosphere of established tea bushes," *Microbiological Research*, vol. 152, no. 1, pp. 105–112, 1997.

[27] K. Rinu and A. Pandey, "*Bacillus subtilis* NRRL B-30408 inoculation enhances the symbiotic efficiency of *Lens esculenta* Moench at a Himalayan location," *Journal of Plant Nutrition and Soil Science*, vol. 172, no. 1, pp. 134–139, 2009.

[28] M. K. Malviya, A. Sharma, A. Pandey, K. Rinu, P. Sati, and L. M. S. Palni, "*Bacillus subtilis* NRRL B-30408: a potential inoculant for crops grown under rainfed conditions in the mountains," *Journal of Soil Science and Plant Nutrition*, vol. 12, pp. 811–824, 2012.

[29] P. Trivedi and A. Pandey, "Plant growth promotion abilities and formulation of *Bacillus megaterium* strain B 388 (MTCC6521) isolated from a temperate Himalayan location," *Indian Journal of Microbiology*, vol. 48, no. 3, pp. 342–347, 2008.

[30] D. Bisht, A. Pandey, and L. M. S. Palni, "Influence of microbial inoculations on *Cedrus deodara* in relation to survival, growth promotion and nutrient uptake of seedling and general soil microflora," *Journal of Sustainable Forestry*, vol. 17, no. 3, pp. 37–54, 2003.

[31] A. Pandey, L. M. S. Palni, P. Mulkalwar, and M. Nadeem, "Effect of temperature on solubilization of tricalcium phosphate by *Pseudomonas corrugata*," *Journal of Scientific and Industrial Research*, vol. 61, no. 6, pp. 457–460, 2002.

[32] P. K. Mishra, S. Mishra, G. Selvakumar et al., "Characterisation of a psychrotolerant plant growth promoting *Pseudomonas* sp. strain PGERs17 (MTCC 9000) isolated from North Western Indian Himalayas," *Annals of Microbiology*, vol. 58, no. 4, pp. 561–568, 2008.

[33] G. Selvakumar, P. Joshi, S. Nazim, P. K. Mishra, J. K. Bisht, and H. S. Gupta, "Phosphate solubilization and growth promotion by *Pseudomonas fragi* CS11RH1 (MTCC 8984), a psychrotolerant bacterium isolated from a high altitude Himalayan rhizosphere," *Biologia*, vol. 64, no. 2, pp. 239–245, 2009.

[34] A. Pandey, L. M. S. Palni, and K. P. Hebbar, "Suppression of damping-off in maize seedlings by *Pseudomonas corrugata*," *Microbiological Research*, vol. 156, no. 2, pp. 191–194, 2001.

[35] P. Trivedi, A. Pandey, and L. M. S. Palni, "Carrier-based preparations of plant growth-promoting bacterial inoculants suitable for use in cooler regions," *World Journal of Microbiology and Biotechnology*, vol. 21, no. 6-7, pp. 941–945, 2005.

[36] P. Trivedi and A. Pandey, "Recovery of plant growth-promoting rhizobacteria from sodium alginate beads after 3 years following storage at 4°C," *Journal of Industrial Microbiology and Biotechnology*, vol. 35, no. 3, pp. 205–209, 2008.

[37] N. Someya, K. Tsuchiya, T. Yoshida, M. T. Noguchi, K. Akutsu, and H. Sawada, "Co-inoculation of an antibiotic-producing bacterium and a lytic enzyme-producing bacterium for the biocontrol of tomato wilt caused by *Fusarium oxysporum* f. sp. lycopersici," *Biocontrol Science*, vol. 12, no. 1, pp. 1–6, 2007.

[38] A. Durgapal, A. Pandey, and L. M. S. Palni, "The use of rhizosphere soil for improved establishment of conifers at nursery stage for application in plantation programmes," *Journal of Sustainable Forestry*, vol. 15, no. 3, pp. 57–73, 2002.

[39] P. Calvo, E. Ormeño-Orrillo, E. Martínez-Romero, and D. Zúñiga, "Characterization of *bacillus* isolates of potato rhizosphere from Andean soils of Peru and their potential PGPR characteristics," *Brazilian Journal of Microbiology*, vol. 41, no. 4, pp. 899–906, 2010.

[40] J. Ghyselinck, S. L. S. Velivelli, K. Heylen et al., "Bioprospecting in potato fields in the central andean highlands: screening of rhizobacteria for plant growth-promoting properties," *Systematic and Applied Microbiology*, vol. 36, no. 2, pp. 116–127, 2013.

[41] X. Zhuang, J. Chen, H. Shim, and Z. Bai, "New advances in plant growth-promoting rhizobacteria for bioremediation," *Environment International*, vol. 33, no. 3, pp. 406–413, 2007.

[42] K. P. Shukla, S. Sharma, N. K. Singh, and V. Singh, "Deciphering rhizosphere soil system for strains having plant growth promoting and bioremediation traits," *Agriculture Research*, vol. 1, no. 3, pp. 251–257, 2012.

18

Using Better Management Thinking to Improve Conservation Effectiveness

Simon A. Black,[1,2] **Jim J. Groombridge,**[1] **and Carl G. Jones**[3,4]

[1] *Durrell Institute of Conservation and Ecology, School of Anthropology and Conservation, University of Kent, Canterbury, Kent CT2 7NZ, UK*
[2] *Department of Human Resources, University of Kent, Canterbury, Kent, UK*
[3] *Durrell Wildlife Conservation Trust, Trinity, Jersey, Channel Islands, UK*
[4] *Mauritian Wildlife Foundation, Grannum Road, Vacoas, Mauritius*

Correspondence should be addressed to Simon A. Black; s.black@kent.ac.uk

Academic Editors: A. Chistoserdov and D. Schmeller

The current paradigm for effective management in biodiversity conservation programmes is dominated by three broad streams of thinking: (i) traditional "command-and-control" approaches which are commonly observed in, but are not exclusive to, bureaucratic government-administered conservation, (ii) more recent notions of "adaptive management," and (iii) emerging "good practice" management frameworks for conservation. Other variations on these themes suggested by the literature tend to endorse additions or enhancement to one or more of these approaches. We argue that instead a more fundamental alternative approach to conservation management is required, based on "systems thinking." The systems thinking approach should encompass (i) an understanding of natural systems, (ii) a sense of how human behaviour is influenced, (iii) an understanding of how knowledge should inform decision-making and problem solving, and (iv) an approach based on an understanding of variation in natural systems. Our argument is that the current paradigms of conservation management fail to address these four fundamentals and therefore do not represent the most effective way to manage conservation programmes. We suggest that the challenge for the conservation community is so great that conservation managers should seriously consider better ways of designing and managing programmes, setting goals, making decisions, and encouraging learning and improvement.

1. Introduction

A number of well-informed commentators have suggested that the dominant management mind-set observed in biodiversity management is one which follows a command-and-control philosophy [1–3]. Command-and-control management is characterised by top-down management hierarchies, functional specialism in teams, decisions made by managers rather than people doing the work, measurement by output (often against targets), and "management by results" (a focus on managing people and budgets) [2, 4–6].

A number of reviews of the conservation literature and the history of conservation successes and failures have identified common shortfalls of a "command-and-control" approach to conservation [1–3]. One reason for the continued widespread use of command-and-control management is that it is the common default mind-set observed in many different industrial, commercial, and public sectors [4–6], largely driven by underlying educational or governmental norms of thinking. This mind-set is therefore readily absorbed by managers, and those working in conservation are no exception [2, 4]. A brief summary of the difficulties with command-and-control thinking includes the points raised by Holling and Meffe [1] who concluded that with this approach it is assumed that the problem is well bounded, clearly defined, relatively simple, and generally linear with respect to cause and effect, yet this set of assumptions does not fit well with the complex, nonlinear and poorly understood natural world. Furthermore, Holling and Meffe [1] issue a range of examples where command-and-control thinking is applied to natural systems and causes negative conservation outcomes, such as reducing genetic health in small populations, the damming of rivers, impacts from monoculture agricultural practices, and fire suppression in national parks, to name a few.

In the past two decades attempts to develop new management approaches in conservation have been offered. Adaptive management [7] is often portrayed as the most advanced approach to improving conservation outcomes [8, 9]; indeed, the Convention on Biological Diversity specifically draws attention to adaptive management [10]. However, at the same time there is a common recognition that this approach is difficult to implement in practice [8]. There is also a suspicion that adaptive management processes and structures (such as decision-making processes and steering groups) are actually command-and-control practices in disguise, with all the inherent problems outlined above. This situation suggests a weakness in adaptive management practice; it appears dependent upon hierarchies where specialists remain in senior positions in committees rather than having a system of decision-making which embraces expertise on the ground [2]. This predicament means that knowledge is not placed in the correct position within governance structures (whether it is knowledge about techniques, sciences, budgets, or legal requirements), and policy and decisions are set that do not relate to the needs of the species, ecosystems, and environmental issues in question.

In this context, since the publication of the original adaptive management methodologies, the conservation community has witnessed several attempts to reframe the approach, including a suggestion of how "soft systems" methodologies could improve the conservation impact of adaptive management [8]. Despite these attempts, we argue that changes fail to question the core ideas around effective management and we suggest that a more fundamental "systems thinking" alternative is required, to set a new direction for effective conservation management.

Over the past decade, a third wave of management approaches has emerged in conservation in the shape of "good management practice" such as the IUCN framework for evaluating protected areas [11] and the Conservation Measures Partnership's open standards for the practice of conservation [12]. These frameworks offer an alternative approach, based on practice encountered in other sectors. However, the "standards" approach has been continually questioned [4, 13–15], for the constraints which this approach imposes on creativity and problem solving and for the bureaucracies that tend to follow the implementation of standards. In short, management standards should not be considered "good practice"; the standards may appear to work well on paper but are devoid of the flexibility that is commonly needed when managing conservation programmes. Conservation would be better served by management practices which can be implemented to the benefit of conservation work (see Table 1).

Conservation is a sector that lacks maturity in its experience of leadership and management thinking [2, 13, 16]; however the prevailing jumbled view of "good management practice" is not a peculiarly conservation-related phenomenon. Within management literature as a whole, the accumulated knowledge of what works and what does not is itself confused by "trends" and "fads" [5, 6, 17]. This confusion presents a problem since the temptation is to copy what appears to work (or at least what is purported to work) in more "management-savvy" sectors. The result is that poor practice from external sectors (e.g., business, government, and public sector) is copied into the conservation community. Instead, we argue that any recommendations of "good management practice" need to (i) be founded on reliable theory, (ii) enable clarity of thinking in many and varied situations, (iii) offer clear ways in how they are applied in practice, and (iv) be applied such that continuous learning and improvement in conservation are achieved.

If a theoretical management base can be established which fits the demands of conservation science, social science, and organisational theory, then a foundation can be developed upon which to base effective conservation management. Deming [18], an outstanding management thinker of recent decades, asserts that "there is nothing as practical as a good theory."

2. Methods for More Effective Management: Theory and Practice

Good management practices need to be based on (i) foundations of reliable theory, (ii) thinking which can be applied in many and varied situations, (iii) approaches which can be clearly applied in practice, and (iv) continuous learning and improvement. We advocate that conservation managers focus on developing their capability across Deming's four areas of competence [5, 18].

2.1. Understanding of Conservation Systems in Terms of the Interplay between Natural Ecosystems and Human-Created Systems. At a fundamental level conservation programs are by their nature open systems [19]. This suggests that conservation managers need to understand their programmes and the interaction of all their elements, that is, the species, ecosystems, landscapes, human communities, work, teams, resources goals, budgets, and plans from a systems perspective [20]. Holling and Meffe [1] assert that "effective natural resource management that promotes long-term system viability must be based on an understanding of the key processes that structure and drive ecosystems and on acceptance of both the natural ranges of ecosystem variation and the constraints of that variation for long-term success and sustainability."

In terms of understanding human organisations as part of the system (i.e., the conservation programme itself and the teams involved) there is a critical need for managers to learn how things work [3, 13]. Several indicators of effective management in relation to people and the organisation of work, decision-making, and problem solving are offered by Black et al., [2]: the setting of clear purpose and vision and having a clear consideration of both project details and the big picture. It is also possible for managers to map out their conservation programme to take account of approaches to organisation management (e.g., leadership, policy, resource management, and people) and natural systems management (e.g., species, ecosystems, and landscapes) and to identify areas for improvement; the Conservation Excellence Model

TABLE 1: Contrasting approaches to management in a conservation context adapted from [2, 13].

Principle	"Command-and-control"	Adaptive management	"Good practice" frameworks	Systems thinking
Design perspective	Top-down hierarchy	Top-down and bottom-up	Top-down	Open system, outside-in
Ethos	Control	Evidence-based	Compliance	Learning
Design of work	Specialist functions in "silos"	Functional specialism with collaboration	Follow elements of the project plan	Understand needs and relevant flows of activity
Approach to change	Reactive projects	Trial and error	Management as a "process" using audit	On-going, integral (part of normal work)
Motivation of people	Extrinsic (reward and reprimand)	Assumed consent	Compliance is required	Intrinsic (self-motivated)
Decision making	Separated from work (carried out in the hierarchy)	Committees take suggestions from practitioners	Made within the boundaries of the framework/project plan	Integrated with work
Measurement	Output, targets, standards: related to budget or plan	Evidence-based practices	Project sustainability	Capability, statistical data: link to purpose
Attitude to stakeholders	Contractual	Participative	Inclusive	Cooperative
Role of managers	Manage people and budgets	Manage activities and decisions	Manage projects and budgets	Act on the system
Attitude to biodiversity	Contractual: only do what is required (e.g. only follow the recovery plan)	Consider impact of actions on biodiversity	Fit needs of biodiversity within project parameters	Always start with "what matters to biodiversity?" (e.g. species, habitats, etc.)

offers a framework to support this type of conceptual mapping [21].

Of course the work of conservation, be it species recovery, habitat restoration, policy implementation, law enforcement, socioeconomic restructuring, education, participatory management, or any other aspect, should become an intrinsic part of the system in which the programme is operating. In an ideal world the legacy of interventions will be left operating seamlessly within the natural systems once the conservation project teams have left. In practice, a hands-on element often remains as is most obviously the case with protected area management as well as numerous species recovery programmes. For example, 84% of US Endangered Species Act species have been identified as currently conservation reliant, whilst other high profile examples from various continents including the Kakapo (*Strigops 171 habroptilus*), California Condor (*Gymnogyps californianus*), Black-footed Ferret (*Mustela 172 nigripes*), Florida Manatee (*Trichechus manatus latirostris*), the African Elephant (*Loxodonta africana*) and Gray Wolf (*Canis lupus*) are well-documented [3, 19, 22]. In addition there are an increasing number of programmes where wider ecosystem recovery is also managed [23]. Consequently, management is ever-present in many conservation programmes and hence should not be regarded as a one-off concern but instead should be considered as a long-term influence with implications which will continue to run alongside the continued viability of the populations, species, and habitats.

Even in the most successful programmes, such as the recoveries of the California Condor and Black-footed Ferret, continued progress with species and ecosystems is reliant upon successful integration of human social systems such as hunting, land use, agricultural practice, or community attitudes towards the species [2, 22]. Although we argue that a conservation managers' job is to understand species and ecosystems and to take action on causes of threat or decline [13], what this means in practice is that they need to learn how to establish measures of species and ecosystem status alongside measures of human impact, to identify priorities. "Systems thinking" enables better use of both natural and social scientific information for problem solving, decision-making, goal-setting, planning, and action [2]. It is a mind-set change from what do we "think" is happening to what do we "know" is happening. However an understanding of systems is only one aspect of importance; it must be linked to the following three other areas of capability [5, 18].

2.2. Using Information Properly, Whether Endowed with Either Scientific Certainty, Imprecision, or Lack of Data. The first area of competence discussed above concerned how managers need to understand how a system operates. This needs to be linked to a second area of capability, an understanding of what "knowledge" really means [24]. Experience (i.e., what I have seen happen before or elsewhere) is subordinate to knowledge (what is happening now) [18]; opinion is different to facts [4], and problem solving, decisions, and work should be informed by knowledge [5, 6, 18]. Conservation has been cited as a "crisis discipline" [25], with a focus on problem solving and mitigation of threats. Knowledge should concern how conservation practitioners develop and regard data (e.g., population size, reproduction, mortality, and threats) and apply analyses (usually temporal and spatial) to guide work in response to circumstances and need.

The sciences of conservation often provide data-rich tapestries of information (genetics, population data, behavioural data, climate data, etc.) Conversely, rare and endangered species can be cryptic or in remote areas so many unknowns may also apply. Similarly, human systems (behaviour, resource use, and organisation) also have an impact on natural systems and vice versa [20], and these human aspects must also be understood. Human and social data may often be examined using social science techniques of which there is also a well-established body of knowledge within conservation.

Many conservation scientists commonly experience the need to handle imprecise data [26]. Hence, it is important that we make assumptions about the knowledge we gain by using a sound theory base [24]. In some areas knowledge will be precise, whilst in others it will be based on much less clear evidence or "belief" [24]. The competence needed by the conservation professional is when to seek precise data and when it is acceptable to rely on reasonable belief. For example, one area commonly accessed when making conservation decisions is social scientific data based on people's perceptions. In conservation projects it may be expedient to obtain perception data (from focus groups, interviews, and surveys) and this may be sensible, but the practitioner should also remain cautious since people's "points of view" (including those of conservation professionals) can be clouded by personal agendas. To put the practicalities of accommodating measurements of human behaviour in stark perspective, the theory of planned behaviour [27] identifies that people's views may not match their actual behaviour. In other words we should not assume knowledge of cause and effect based on perception data—people often say they will do one thing but actually do something else, or say one thing but actually believe something else. We should seek further (or alternative) data to inform our decisions.

Social data in the broadest sense can often be important; yet considerations of perception data specifically should not assume cause and effect linkages, but something more complex [20] which may give clues to how conservation can be improved. Useful analyses will give insights into which actions have an impact on the desired human behaviours. For example, if a community education programme fails to make a difference to people's behaviour towards a threatened species, we cannot assume that the programme is irrelevant nor can we assume that it is ineffective. It may simply have failed due to influences which are not being monitored which impact people's behaviour (e.g., demand for certain agricultural products, availability of alternatives to bush-meat, changes in political circumstances, war, road construction, etc.).

Conversely, there are occasions when attempting to acquire additional knowledge is counterproductive. The

Christmas Island Pipistrelle Bat (*Pipistrellus murrayi*) is an example where data monitoring continued in the face of population decline to the point of extinction [26]. The Po'ouli (*Melamprosops phaeosoma*), a recently extinct species of Hawaiian forest bird, was an example where plenty of data collection effort was applied but was not focused on what was needed to understand how to conserve the species until it was too late [28]. In these instances, reasonable assumptions could have been made (e.g., that the Po'ouli was very rare and in decline or that the Christmas Island Pipistrelle was declining to extinction), to enable early practical decisions about recovery effort to be made. Effort should have been focused in these instances on action to recover the species and subsequent accumulation of new knowledge about effective species recovery [13, 26].

It is not uncommon in conservation to encounter cases where data is deficient, yet the conservation community already has its own premise for these situations, the "Precautionary Principle" [29], which practitioners could benefit from keeping to the forefront of their minds. Spectacular recoveries have been demonstrated in species such as many of the endemic birds of Mauritius, including the Mauritius Kestrel (*Falco punctatus*), Pink Pigeon (*Nesoenas mayeri*), and Echo Parakeet (*Psittacula eques*) through application of reasonable belief of what is possible followed by progressive learning and action based on increased knowledge [22, 30].

2.3. Understanding the Basics of Human Behaviour Relative to Their Work. Understanding systems and having a broad basis for building knowledge overcome two limitations of command-and-control thinking [4, 18]. However the command-and-control paradigm also carries misplaced assumptions about people which also need to be overcome. One is an assumed cause and effect assumption between praise and motivation or sanctions and improvements in poor performance [4–6]. This assumption has been demonstrated not to be the case by numerous behavioural theorists and more recently by combinations of psychology and neuroscience [31]. Similarly, command-and-control assumes that organisations operate as the sum of their parts, so that goals, targets, and objectives can be set for individuals and teams on the assumption that the sum of those goals and targets will result in success, also a frequent misapprehension [5, 6, 18, 31]. Knowing the performance of each element of your programme does not mean one is able to predict the outcomes of the whole [20].

Goal-setting is fraught with difficulty in terms of the psychological mind-set of workers [31]. The command-and-control assumption is that a target is set, which gives people understanding and clarity of purpose [5, 17]. However human responses to targets are far more complex. The most obvious examples of the failure to take account of this are in the goal displacement observed in the Black-footed Ferret programme and the Po'ouli [2, 13]. In the early years of the Black-footed Ferret programme, efforts to achieve goals relating to captive breeding of the animals were highly successful and became increasingly well resources, but parallel building of knowledge for their successful release and survival in the wild fell short, such that many animals were released into the wild, but few survived [3]. For the Po'ouli, goals set to fence protected areas to enable removal of invasive species diverted most programme resources, such that little was available to identify the needs of the bird itself [13]; the result was successful creation of protected, restored habitat, but loss of the endemic bird species for which it was intended. In these cases people were set goals which they followed in competition with peers and other departments involved in the programme [3, 13]. In the end the overall purpose of the programme, to save the species, was fundamentally compromised.

The conservation practitioner therefore needs to develop an understanding of human psychology [5, 18]. For example, human beings are motivated by having a sense of purpose and level of control over their work [31, 32]. This is particularly important in conservation since many programmes involve people on the ground who have either high levels of technical or scientific knowledge or expert understanding of the species and landscapes under consideration (or both), plus high levels of passion and commitment to the work [2, 13]. To deny them input and some level of control over decision-making, and work design is not only demotivating, it is organisationally an inefficient use of their human resource.

Furthermore, when thinking of the involvement of wider communities of people, the same basic understanding of human psychology applies. Community engagement should not simply be for involvement's sake [13] to "do the right thing," without properly understanding why those things should be done in the first place. Social inclusion should be for a particular purpose. For example, the practitioner needs to differentiate the process of social data collection from the usefulness of data. A "good" inclusive social process which engages local communities but gathers poor-quality or unhelpful data is, in the long term, of no value to the conservationist or the community. Similarly, data collected via a "poor" exclusive scientist-driven process which raises suspicion or anxiety in the local community and yet gathers high quality data may also be, in the long term, of little positive benefit. The practitioner should never be tied to method but should consider the psychological needs of the communities and other people who have an interest in the conservation outcome.

Recent [8] descriptions of "social learning" and "communities of practice," concerning a group, or groups, of people who share a concern for something that they do and learn how to do it better through regular interaction, appear helpful as a way to improve conservation effectiveness. From a systems perspective, social learning provides a model for testing and sharing methods and fitting practices into a local socioeconomic or ecosystem context. An example of this combination of social processes and data gathering is the routine collection of skull and tissue samples by hunting communities in the Peruvian Amazon [33], which are analysed by scientists and utilised by communities and conservationists to agree and manage sustainable off-takes.

Clearly, conservationists cannot escape the need to communicate with interested parties to gain relevant socioeconomic and political insights to inform interventions. However, stakeholder interaction may not necessarily assist conservation processes. For example, where extreme and opposing viewpoints occur (e.g., consulting animal rights activists about eradicating invasive mammals), involvement will likely hinder rather than help. Alternatively, decision makers may deliberately consult others to avoid making difficult and unpopular decisions themselves. Training people on how to engage with social groups may itself become a short-term fix.

Although there are calls to train conservation practitioners in methods or skill sets that will enable community involvement [8], we assert that it is far better to train them to understand how human beings function [5], what drives and predicts their behaviour, or at least which strategies to employ if the behaviours of workers or community partners do not fit expected patterns.

2.4. View Management through the Prism of Variation in Natural Systems. The final key area of learning which needs to be examined is how to base problem solving and management of interventions on an understanding of variation in natural systems. This principle was invoked by Deming [18] based on work of practitioners such as Shewhart [34]. Holling and Meffe [1] go as far as suggesting a "Golden Rule" in natural resource management: "*natural resource management should strive to retain critical types and ranges of natural variation in ecosystems. That is, management should facilitate existing processes and variabilities rather than changing or controlling them.*" Making static comparisons of data on species or populations leaves many gaps in knowledge. It is not surprising that the most successful species recovery programmes are those which have been run over the long term, based on a growing knowledge of what works (and what does not) to improve the situation for the species [3, 13, 22, 35].

The majority of conservation practitioners have an appreciation for the need for data. The most progressive amongst them understand how to use that data to inform them on how well their interventions are progressing and what might need to change [2]. This understanding needs to be adapted to take account of longitudinal patterns of distribution, based on an understanding of variation in the natural systems which they study and work on. More challenging, however, is to apply the same thinking to human systems and organisation. Holling and Meffe [1] assert that one of the 6 ways to address the golden rule is to "*examine bureaucracies to identify underlying reasons for their general intransigence and brittleness and promote incentives for alternative behaviours. Develop incentives and rewards for innovation that place streamlining, local solutions, and concern for sustainability above adherence to a command structure.*" This means that conservation managers should identify and manage those "system conditions" [4] which help or hinder improvements in the system. Those conditions might be budgets, rules, targets, role design, management expectations, logistical constraints, policies, methods, information, people with other agendas, or any number of things which can limit or enhance the effectiveness of conservation work. Alternatively, the problem can be reframed and new ways of collaboration be developed. As an alternative management perspective and a better way of working [18], the systems thinking approach can enable conservation managers to see through the bureaucracies of command structures and focus on the needs of species and ecosystems [2].

3. Results and Impacts of Systems Thinking: The Human Dimension

As human beings, conservationists are attempting to influence the natural and social systems which they encounter for the good of threatened species and ecosystems. In Caughley's seminal paper [36] which challenged the two conservation paradigms of his time (small-population versus declining population paradigm), he calls for conservation practitioners to improve the effectiveness of their work by pursuing three things: first, to seek knowledge; second, to identify how to improve the system; and third, to implement action where needed [36]. We endorse his ideas based on our presentation of effective management thinking. We assert that a broad understanding of management based on our suggested four areas of competence delivers the following beneficial results against each of his three factors.

3.1. Conservation Effort and Emphasis Based on Increasing Knowledge. In the 1970s there was a strong belief among many that the California Condor was a relic doomed to extinction and should be allowed to disappear with dignity [2]. With knowledge that the species decline was due to lead poisoning from shot left in abandoned sport-hunting carcasses, N. Snyder and H. Snyder working with colleagues redefined the conservation goals for the species and ensured its recovery [37]. Based on more complete knowledge (rather than assumed beliefs) a new approach for the conservation of the species took shape.

Attempts to upgrade conservation interventions by enhancing one mode of thinking such as adaptive management through addition of other modes of thinking such as soft system methodologies [8] will fail. Adding soft systems onto an incompatible approach risks further entangling professionals in the work of "management" rather than the actual work of conservation. To put it more bluntly, a conservation manager will end up spending time managing plans, coordinating procedures, and running workshops rather than doing the work of conservation. Instead, the conservation community needs to find other ways of incorporating and managing human behaviour and values.

3.2. Clarity of Intervention: How to Improve the System. In the conservation context, when enough knowledge is available, then decisions must be enacted; "decisions must be made whilst there is opportunity to act" [35]. Successful conservation programmes have a good understanding of the species, habitats, landscapes, and the threats that they are experiencing [2, 3, 34, 37]. Conservation practitioners need to examine data objectively and then ask themselves "what is happening?" The simplest "good" solution is not always the best [38]; understand the system first and then act accordingly [18, 36].

Snyder and colleagues made the first inroads into conserving the California Condor after the species had experienced decades of decline by thoroughly examining the impacts on the species and correctly identifying the impact of lead shot left in abandoned carcasses and gut piles that caused fatal lead poisoning [2, 37]. Other examples are apparent elsewhere where improvements have been achieved by a change in management mind-set; the Black-footed Ferret and Florida Manatee are just two examples [3, 13, 26].

3.3. Implementing Conservation: Relevant Methods and Stakeholder Involvement. Evidence-based conservation [39] may offer some insights to assist in knowledge building practices by conservation practitioners. A strong systems thinking approach will help develop evidence-based approaches that can be used to refine conservation practice.

Relationships with stakeholders need to be managed so they have an appropriate role in managing conservation. Social data collection should be both inclusive and useful to communities and to the conservation scientist.

Whilst we recognise the importance of stakeholder interaction, the way in which conservation professionals use people's opinions should be seen as a distinct process. Conservationists often have to take difficult decisions and drive action which is contrary to some interests. Conservation programs need leaders who listen and take advice, but who are prepared to act based on evidence, rather than opinion. Thankfully, conservationists who worked with the California Condor did not concur with influential groups who advocated "extinction with dignity" [40], just as others [41] refused to abandon the Mauritius Kestrel to what some saw as an "all-but-inevitable fate" of extinction [42].

4. Conclusion

The failure to meet the Convention on Biological Diversity 2010 targets [43, 44] illustrates the need to fundamentally change conservation thinking. Conservation managers should, as Caughley [36] suggests, seek knowledge, identify how to improve the system, and implement action where needed. Continuous learning is particularly important for effective conservation, but it demands a sympathetic and proactive management culture. Useful recently suggested concepts of "evidence-based conservation" [39] offer better ways of encouraging conservation learning and a focus on practices which improve results; however methods should always be tested in the context of local knowledge of systems, variation, and people, so managers who use systems thinking will be best placed to use these ideas.

A breakthrough in performance is possible when conservationists manage conservation interventions taking account of natural systems of species and landscapes and impinging human systems (communities, agroforestry, politics, and business). In addition they must also be adept at working within their organisational structures and be able to manage the processes and people within the conservation programme. We suggest that four aspects of thinking should be aligned in the manager's approach; (i) how to understand the overall system being conserved (i.e., an interacting natural and human system); (ii) how to think about available knowledge, information, and data; (iii) how to take consideration of human psychology of people working in and around the programme; and (iv) how to understand variation when managing data in natural systems.

We encourage practitioners to develop a clear long-term vision; be "hands-on" in supporting the conservation team and understanding the work; be mindful of both detail and the big picture; and actively encourage a culture of learning, improvement, and receptiveness to alternative solutions.

References

[1] C. S. Holling and G. K. Meffe, "Command and control and the pathology of natural resource management," *Conservation Biology*, vol. 10, no. 2, pp. 328–337, 1996.

[2] S. A. Black, J. J. Groombridge, and C. G. Jones, "Leadership and conservation effectiveness: finding a better way to lead," *Conservation Letters*, vol. 4, no. 5, pp. 329–339, 2011.

[3] T. W. Clark, R. P. Reading, and A. L. Clarke, *Endangered Species Recovery: Finding the Lessons, Improving the Process*, Island Press, Washington, DC, USA, 1994.

[4] J. Seddon, *Freedom from Command and Control*, Vanguard Press, Buckingham, UK, 2003.

[5] P. R. Scholtes, "The new competencies of leadership," *Total Quality Management*, vol. 10, no. 4-5, pp. S704–S710, 1999.

[6] B. L. Joiner, S. Reynard, and A. Yukihiro, *Fourth Generation Management: The New Business Consciousness*, McGraw Hill, New York, NY, USA, 1994.

[7] C. S. Holling, *Adaptive Environmental Assessment and Management*, John Wiley & Sons, Chichester, UK, 1978.

[8] G. Cundill, G. S. Cumming, D. Biggs, and C. Fabricius, "Soft systems thinking and social learning for adaptive management," *Conservation Biology*, vol. 26, no. 1, pp. 13–20, 2012.

[9] J. M. Dietz, R. Aviram, S. Bickford et al., "Defining leadership in conservation," *Conservation Biology*, vol. 18, no. 1, pp. 274–278, 2004.

[10] R. E. Kenward, M. J. Whittingham, S. Arampatzis et al., "Identifying governance strategies that effectively support ecosystem services, resource sustainability, and biodiversity," *Proceedings of the National Academy of Sciences of the United States of America*, vol. 108, no. 13, pp. 5308–5312, 2011.

[11] M. Hockings, S. Stolton, F. Leverington, N. Dudley, and J. Courrau, *Evaluating Effectiveness: A Framework for Assessing Management Effectiveness of Protected Areas*, International Union for Conservation of Nature, Gland, Switzerland, 2006.

[12] Conservation Measures Partnership, *Open Standards for the Practice of Conservation (Version 2. 0)*, CMP, Washington, DC, USA, 2004.

[13] S. Black and J. Groombridge, "Use of a business excellence model to improve conservation programs," *Conservation Biology*, vol. 24, no. 6, pp. 1448–1458, 2010.

[14] P. Bansal and W. C. Bogner, "Deciding on ISO 14001: economics, institutions, and context," *Long Range Planning*, vol. 35, no. 3, pp. 269–290, 2002.

[15] J. A. Rodríguez-Escobar, J. Gonzalez-Benito, and A. R. Martínez-Lorente, "An analysis of the degree of small companies' dissatisfaction with ISO 9000 certification," *Total Quality Management and Business Excellence*, vol. 17, no. 4, pp. 507–521, 2006.

[16] J. C. Manolis, K. M. Chan, M. E. Finkelstein et al., "Leadership: a new frontier in conservation science," *Conservation Biology*, vol. 23, no. 4, pp. 879–886, 2009.

[17] J. Macdonald, *Calling a Halt to Mindless Change*, American Management Association International, New York, NY, USA, 1998.

[18] W. E. Deming, *Out of the Crisis*, Massachusetts Institute of Technology Center for Advanced Engineering Study, Cambridge, Mass, USA, 1982.

[19] R. L. Wallace, "The Florida Manatee recovery program: uncertain information, uncertain policy," in *Endangered Species Recovery: Finding the Lessons, Improving the Process*, T. W. Clark, R. P. Reading, and A. L. Clarke, Eds., pp. 131–156, Island Press, Washington, DC, USA, 1994.

[20] L. Bertalanffy, *General System Theory: Foundations, Development, Applications*, Revised edition, George Braziller, New York, NY, USA, 1969.

[21] S. A. Black, H. M. R. Meredith, and J. J. Groombridge, "Biodiversity conservation: applying new criteria to assess excellence," *Total Quality Management and Business Excellence*, vol. 22, no. 11, pp. 1165–1178, 2011.

[22] G. Caughley and A. Gunn, *Conservation Biology in Theory and Practice*, Blackwell Science, Cambridge, Mass, USA, 1996.

[23] D. D. Goble, J. A. Wiens, J. M. Scott, T. D. Male, and J. A. Hall, "Conservation-reliant species," *BioScience*, vol. 62, pp. 869–873, 2012.

[24] B. Russell, *Theory of Knowledge*, Encyclopaedia Britannica, Chicago, Ill, USA, 13th edition, 1926.

[25] M. E. Soulé, "What is conservation biology," *BioScience*, vol. 35, no. 11, pp. 727–734, 1985.

[26] T. G. Martin, S. Nally, A. A. Burbidge et al., "Acting fast helps avoid extinction," *Conservation Letters*, vol. 5, no. 4, pp. 274–280, 2012.

[27] I. Ajzen, "The theory of planned behavior," *Organizational Behavior and Human Decision Processes*, vol. 50, no. 2, pp. 179–211, 1991.

[28] J. J. Groombridge, J. G. Massey, J. C. Bruch et al., "An attempt to recover the Po'ouli by translocation and an appraisal of recovery strategy for bird species of extreme rarity," *Biological Conservation*, vol. 118, no. 3, pp. 365–375, 2004.

[29] K. R. Foster, P. Vecchia, and M. H. Repacholi, "Science and the precautionary principle," *Science*, vol. 288, no. 5468, pp. 979–981, 2000.

[30] C. G. Jones, "Conservation management of endangered birds," in *Bird Ecology and Conservation: A Handbook of Techniques*, W. H. Sutherland, I. Newton, and R. E. Green, Eds., pp. 269–302, Oxford University Press, Oxford, UK, 2004.

[31] C. S. Jacobs, *Management Rewired: Why Feedback Doesn't Work and other Surprising Lessons from the Latest Brain Science*, Portfolio Books, New York, NY, USA, 2009.

[32] F. Herzberg, "One more time: how do you motivate employees?" *Harvard Business Review*, vol. 81, no. 1, pp. 87–141, 1976.

[33] J. L. Hurtado-Gonzales and R. E. Bodmer, "Assessing the sustainability of brocket deer hunting in the Tamshiyacu-Tahuayo Communal Reserve, Northeastern Peru," *Biological Conservation*, vol. 116, no. 1, pp. 1–7, 2004.

[34] W. Shewhart, *Economic Control of Quality of Manufactured Product*, Van Nostrand, New York, NY, USA, 1931.

[35] T. Cade and W. Burnham, *Return of the Peregrine: A North American Saga of Tenacity and Teamwork*, The Peregrine Fund, Boise, Idaho, USA, 2003.

[36] G. Caughley, "Directions in conservation biology," *Journal of Animal Ecology*, vol. 63, no. 2, pp. 215–244, 1994.

[37] N. Snyder and H. Snyder, *The California Condor: A Saga of Natural History and Conservation*, Academic Press, London, UK, 2000.

[38] P. Senge, *The Fifth Discipline*, Doubleday, New York, NY, USA, 1990.

[39] W. J. Sutherland, A. S. Pullin, P. M. Dolman, and T. M. Knight, "The need for evidence-based conservation," *Trends in Ecology and Evolution*, vol. 19, no. 6, pp. 305–308, 2004.

[40] D. Phillips and H. Nash, *Condor Question: Captive or Forever Free?* Friends of the Earth, San Francisco, Calif, USA, 1981.

[41] C. G. Jones, W. Heck, R. E. Lewis, Y. Mungroo, G. Slade, and T. Cade, "The restoration of the Mauritius kestrel Falco punctatus population," *Ibis*, vol. 137, supplement 1, pp. S173–S180, 1995.

[42] N. Myers, *The Sinking Ark*, Collins, London, UK, 1979.

[43] S. H. M. Butchart, M. Walpole, B. Collen et al., "Global biodiversity: indicators of recent declines," *Science*, vol. 328, no. 5982, pp. 1164–1168, 2010.

[44] Convention on Biological Diversity, *Addis Ababa Principles and Guidelines for the Sustainable Use of Biodiversity*, Secretariat of the Convention on Biological Diversity, Montreal, Canada, 2004.

19

Exergetic Model of Secondary Successions for Plant Communities in Arid Chaco (Argentina)

Marcos Karlin, Rodrigo Galán, Ana Contreras, Ricardo Zapata, Rubén Coirini, and Eduardo Ruiz Posse

Facultad de Ciencias Agropecuarias, Universidad Nacional de Córdoba, Avenida Valparaíso s/n, Ciudad Universitaria, C.C. 509, 5000 Córdoba, Argentina

Correspondence should be addressed to Marcos Karlin; mkarlin@agro.unc.edu.ar

Academic Editors: A. Chistoserdov, H. Ford, and P. M. Vergara

Ecosystems are open systems where energy fluxes produce modifications over plant communities. According to the state and transition model, plant formations are defined by changes in natural conditions and disturbs. Based on these changes, it is possible to define vectors that show the tendencies of the communities towards other states. Within the subregion of Arid Chaco, mature communities of *Aspidosperma quebracho blanco* represent the quasistable equilibrium communities or "climax," similar to that observed in the Chancaní Natural Reserve (Córdoba, Argentina). Biodiversity values and Lyapunov coefficients were calculated based on plant abundance and cover data. Lyapunov coefficients were calculated as the Euclidean distance of each site with respect to reference condition (community of *Aspidosperma quebracho blanco*), representing for each state the necessary exergy to reach the reference condition. When Lyapunov coefficients decrease in time, it is expected for the system to drive towards a quasistationary state; otherwise, the equilibrium is unstable and becomes less resilient. The diversity of species has a significant effect over the resistance to perturbations but equivocal for the recovery rate. Lyapunov coefficients may be more precise succession indicators than biodiversity indexes, representing the amount of exergy needed for a vegetation state to reach the reference condition.

1. Introduction

Ecosystems are open systems that exchange energy with the surrounding environment, modifying among other components the plant frequencies. Based on these changes it is possible to define vectors that show the tendencies of the succession to reach mature or quasistationary states and their modifications by natural or anthropic disturbances [1]. Concepts of succession are based on the observation and the analysis of the system's states over space and time.

The search for a successional theory that can be empirically verified is still ongoing. From the successional theories of Clements [2], Gleason [3], and Tansley [4], there is no reconciliation between multiple aspects of such theories. The first is based fundamentally on a deterministic mechanism where, starting from an initial vegetation state, succession moves by an autoorganization process through distinguishable phases (communities) reaching a "monoclimax" situation. For Gleason, the main defect of this theory is precisely

the change of phases from community to community. This author claims for a succession given by the substitution species by species [5]. Tansley developed similar concepts as those of Clements; nevertheless, he argues about the possibility of reaching multiple "climax" states through its "policlimax" concept.

Westoby et al. [6] developed a new successional model, originally made for rangelands, though suitable for other ecosystems [7–9]. This is based on the description of a group of discreet vegetation "states" and "transitions" among states. The state changes are triggered by natural events such as fire or by anthropic effects. According to these authors, this model is a practical way to organize the information for the management of natural areas, without following theoretical models of environmental dynamics.

Accounting for the type and number of species defining a subsystem (vegetation states), it is possible to determine the abundance and frequency of the present species in each one, defining the amount of accumulated information. A higher amount of information in each state can be traduced in free energy available for work; that is, information and energy are intimately related [10].

In a determined vegetation state, higher values of abundance, frequency, and richness should mean higher amounts of energy accumulated in such considered state [11].

If an energy gradient is considered, given by the differences in abundance and frequency among each state, higher energy differences (ΔE) mean higher differences among states. Lesser differences in energy mean that the successional states are similar (at least from the thermodynamics point of view) [1].

Based on the second and third laws of thermodynamics, it is possible to define entropy as the amount of energy in the system that cannot be used for work. Entropy, originally defined by Boltzmann from the statistical standpoint for the study of ideal gases, can be used to define "the order of the universe," studying the distribution of the gas particles in a closed recipient

$$S = k \ln \Omega, \qquad (1)$$

see [11], where S is entropy, k the Boltzmann constant, and Ω the possible number of microstates in the system. With a higher number of possibilities, disorder increases.

Within the system, the derivative of entropy (d_iS) can only increase, and therefore

$$d_iS \geq 0, \qquad (2)$$

being entropy a spontaneous process, while the entropy exchange between the system and the environment (d_eS) can be positive, negative, or zero, defining the entropy differential (dS) as

$$dS = d_iS + d_eS. \qquad (3)$$

Therefore, systems' entropy can increase, decrease, or remain constant. This fact can be achieved in dissipative structures [12] depending on the environment conditions.

An analog concept, opposed to entropy, is exergy and information availability [10, 11]. Energy has different qualities depending on its capacity for being used for work, and this quality can be measured as exergy.

A system's dynamic can be defined in relation to the amount of exergy accumulated. Exergy represents Gibbs potential (free energy) accumulated in the system, and available to be used for work [10].

Analogously to entropy, exergy can be defined as

$$Ex_i = T\left(S_i^{\text{eq}} - S\right) \quad [10], \qquad (4)$$

where T is temperature, S is entropy, and S_i^{eq} is its value in the ith thermodynamic equilibrium.

Living systems are dependent on outside energy fluxes to maintain their organization and dissipate energy gradients to carry out these self-organizing processes [13]. A forest is a clear example of dissipative structure, evolving naturally towards a minimal entropy state, reducing the disorder produced by the internal entropy with external energy used for photosynthesis processes [1, 14].

The ecological exergy is a measure of the distance between the ecosystem in its present state and what it would be if it was at equilibrium with the surrounding abiotic environment, moving along an ecological succession. Exergy is used to build up biomass, which in turn stores Eco-Exergy; Eco-Exergy therefore represents a measure of the structural biomass and the information embedded in the biomass [10, 15].

The largest difference between ecosystems and maximum entropic systems is self-organization. Self-organization is the spontaneous emergence of macroscopic order from microscopic disorder, moving towards to a quasistationary equilibrium [16]. The maximum self-organization in an ecosystem is, theoretically, the "climax" state.

Within the subregion of Arid Chaco, mature communities of *Aspidosperma quebracho blanco* would represent the equilibrium or "climax" vegetation states [17], as seen in sectors of the Chancaní Natural Reserve (ChNR) (Córdoba, Argentina).

Plant formations in this subregion are defined by its natural conditions and by anthropic disturbances produced by the historical extraction of natural resources.

Changes in the ecosystem's physiognomy due to natural resources exploitation, such as logging and grazing, or events such as fires, produce a reduction in the quality of plant biomass and, therefore, a reduction of exergy.

Based on the previous concepts, the posed hypothesis is that within the ecosystem degraded states have less exergy than those states nearer to the quasistationary equilibrium ("climax").

In this paper we seek to identify, understand, and quantify vegetation dynamics in the subregion of the Arid Chaco, as well as to explain the secondary succession dynamics of plant communities from a thermodynamic point of view. For this, we will try to reconcile several aspects of the theories of Clements [2], Gleason [3], and Westoby et al. [6] to gain theoretical support for our thermodynamic model.

2. Materials and Methods

This study of ecological successions was conducted at the west of the Sierras de Pocho (Córdoba, Argentina), in the ChNR and surroundings (NW vertex = 31°19′40″S-65°30′07″W; SE vertex = 31°24′36″-65°24′3″), representative area for the different vegetation states in the subregion of Arid Chaco. The ChNR is limited by the Prov. Routes No 28 and No 51 at the north and west, respectively, the Chancaní river at the south, and the Sierras de Pocho at the east (Figure 1).

The area is characterized by continental, mesothermal, semiarid climate, with dry winters, high thermal variations, important rainfall oscillations, and high evapotranspiration rates. Rain oscillates annually between 300 and 600 mm. Mean annual temperature is 21°C, with absolute maximum and minimum of 45 and −5°C, respectively. Annual potential evapotranspiration ranges between 1000 and 1200 mm. Winds run from the north (desiccant winds) and from the south (wet winds), conditioned by the north-south direction of the adjacent mountains [18].

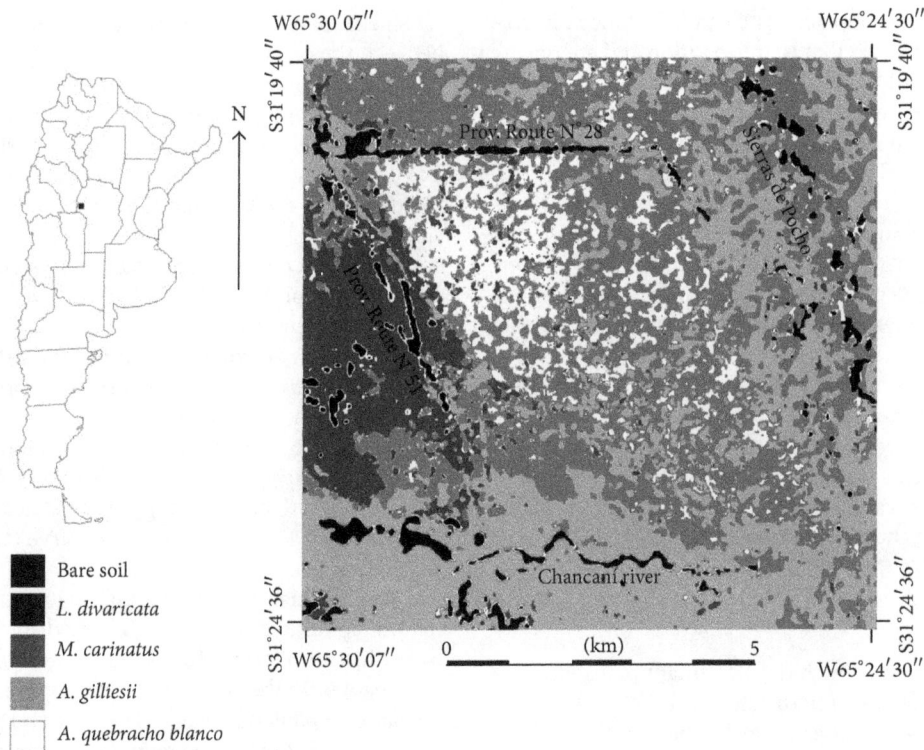

FIGURE 1: Classified image of the study area.

To evaluate the state of secondary successions of this ecosystem, plant dominance and abundance was determined through floristic surveys in sectors of the ChNR and surroundings, previously selected through satellite images, identifying five states and five transitions, according to Westoby et al. [6].

Several homogeneous zones were identified preliminary in the study area through remote sensing techniques, using as cartographic base the chart No 3166-29-1 "Chancaní" of the National Geographic Institute, on a scale 1 : 50.000. Orbital Landsat 7 ETM+ path/row 230 082 images from February 9th 2001 were used. Through visual analysis techniques over false standard color composition (RGB 543) of the image from January 30, 2003, and supported with the Normalized Vegetation Index (NVI), a first approximation of the states and transitions was obtained. With these materials several training sectors were chosen for supervised classification. Each site was geopositioned with GPS and photographed for visual registration. Classification polygons were created and a preclassification was made in order to calculate the statistics and probabilities *a priori* for the sorter of maximal similarity. Then, a classification with the six bands of the satellite was made obtaining an image of five classes. In order to enhance the visual quality of this product, a modal filter was made with a mobile window of 5 per 5 pixels, being the minimum map unit of 150 per 150 m.

Through basic cartography and field surveys, the zones were preliminary identified and classified, defining vegetation states and transitions.

Based on field surveys, 55 species were identified. Preliminary observations suggest that the dominant vegetation is correlated with the physiognomy of each state; for this reason a Random Stratified Sampling method [19] was applied.

Twenty-two floristic surveys were made in an area of 1/4 ha each. Plant surveys were based on the Phytosociological Method of Braun-Blanquet [19, 20]. The number of plant surveys made depended on the variability in the number of previously founded species. In the selected sites, dominance-abundance values were estimated.

The dominance-abundance values were transformed into cover values expressed in percentage based on the middle point of each value in the abundance-dominance scale [21] as follows:

+: very rare or rare individuals with negligible cover,

1: abundant individuals, but with negligible cover,

2: individuals in any number, but covering up to 1/4 of the surface area,

3: individuals in any number, with cover ranging from 1/4 to 1/2 of the surface area,

4: individuals in any number, with cover ranging from 1/2 to 3/4 of the surface area,

5: individuals in any number, with superior cover of 3/4 of the surface area.

The values for each species were smoothed through locally weighted regressions (LOWESS algorithm, with a smoothing probability of 0.5) [22].

Based on this scale, the Shannon-Weaver Biodiversity Index (H) was calculated [21] for each site as

$$H = -\sum_{i=1}^{n} p_i \ln p_i, \tag{5}$$

where $p_i = N_i / \sum_{i=1}^{n} N_i$, n is the number of species in each community, and N_i is the size of the population, defined by its relative frequency according to Wikum and Shanholtzer [21].

With the dominance-abundance values of the four dominant identified communities (*Aspidosperma quebracho blanco*, *Acacia gilliesii*, *Mimozyganthus carinatus*, and *Larrea divaricata*), a discriminant analysis was made to identify classification errors of the surveyed sites. Confidence ellipses were graphed ($P = 0.95$) and finally a definitive classification of the vegetation states was obtained.

In order to quantify the successional series, Lyapunov coefficients (L) values of each state and transition condition were used. Lyapunov coefficients, defined as the property of the behavior of the system in the surrounding of the equilibrium [23], were calculated as the Euclidean distance of each site in relation to the theoretical reference condition ("climax" community of *A. quebracho blanco*), measured through the sum of relative frequencies of each species, analogously to what was referred in (4). The distance between the quasistable equilibrium state (or the relative frequency expected for the ith species corresponding to the reference condition, N_i^*) and the current state (size and current composition of the ith population, N_i) was considered as a measure of the state's disturbance [23] relative to the reference state. Reference values for each species in the community of *A. quebracho blanco*, are referred to as N_i^* in the Lyapunov coefficients [10]

$$L = \left[\sqrt{\sum_{i=1}^{n} (N_i - N_i^*)^2} \right]^2, \quad i = 1, \ldots, n, \tag{6}$$

where N_i is the current situation of the ith species in each site.

H and L values were averaged, standard deviations were determined, and an ANOVA test ($P < 0.05$, LSD Fisher) was made for each site in order to determine significant differences.

Lyapunov coefficients were graphed correlated with the dominance-abundance values of the four dominant species, in order to quantify the succession thermodynamically.

3. Results

In Figure 1 the spatial location of the different vegetation states can be observed.

Bare Soil. Areas with an incipient cover of ruderal plants. It is represented by the adjacent routes, the Chancaní river, areas of rocky outcrop, and a landing strip at the NW of the Reserve. These areas were not considered for the analysis in the present paper.

Communities of L. divaricata (Lar div). Highly degraded areas by overgrazing, clearcutting, and fire, with total elimination

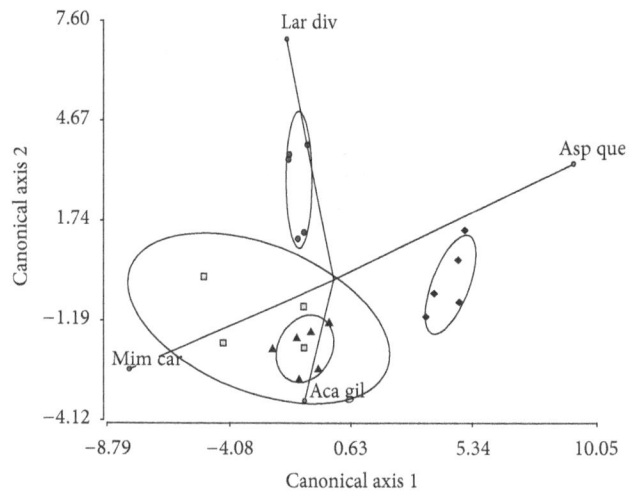

FIGURE 2: Discriminant analysis of the 22 surveyed sites with confidence ellipses ($P = 0.95$). Diamonds: communities of *A. quebracho blanco* ($n = 6$; in the image two sites overlap); triangles: communities of *A. gilliesii* ($n = 6$); squares: communities of *M. carinatus* ($n = 4$); circles: communities of *L. divaricata* ($n = 6$; in the image two sites overlap). Asp que: *A. quebracho blanco*; Aca gil: *A. gilliesii*; Mim car: *M. carinatus*; Lar div: *L. divaricata*.

of tree species. It covers the western part of the study area; it includes the NW area of the ChNR, destined for agriculture in the mid twentieth century.

Communities of M. carinatus (Mim car). Areas where wood was indiscriminately extracted before the Reserve declaration. Besides *M. carinatus*, some herbaceous species such as *Abutilon pauciflorum* are also important. It covers the southwestern and northern areas of the study area.

Communities of A. gilliesii (Aca gil). Slightly degraded areas by wood extraction, though with lesser intensity than the communities of *M. carinatus*. It covers the margins of the Chancaní river, the Sierras de Pocho hillsides at the east, mixed with mountain vegetation. They also form mixed woodlands with *A. quebracho blanco* in the center of the Reserve.

Communities of A. quebracho blanco (Asp que). These areas are considered as the "climax" state, where anthropic disturbances are not important. These communities are codominated by *Cordobia argentea*, a herbaceous species. They cover a northwestern core in the Reserve and form mixed woodlands with *M. carinatus* and *A. gilliesii* in the center of the Reserve.

In Figure 2, vegetation states clusters can be seen limited by confidence ellipses. The closest point (diamond) to the "Asp que" axis corresponds to the reference state. The communities of *A. gilliesii* set inside the confidence ellipse of the communities of *M. carinatus*.

According to the observed in Figures 1 and 2 and to the described disturbs, it is possible to define the transitions.

Overlapping states represent transitions such as *A. quebracho blanco-L. divaricata, A. quebracho blanco-A. gilliesii, A. quebracho blanco-M. carinatus, M. carinatus-L. divaricata, A. gilliesii-M. carinatus.*

There are three main vegetation gradients identified, from the "climax" core to the west, south, and east of the ChNR. The first presents more abrupt changes due to the change in the land use for agriculture, representing the *A. quebracho blanco-L. divaricata* transition. To the south, the gradient *A. quebracho blanco-A. gilliesii-M. carinatus* is more gradual due to the wood selective extraction with higher intensity at the southern limit of the ChNR. To the east, the gradient *A. quebracho blanco-A. gilliesii* is influenced by the Sierras de Pocho acting as an important source of biodiversity in these environments.

According to plant surveys made on representative sites, a modal synthetic phytosociological table was made for the study area, obtaining the mean values for H and L with their standard deviations (Table 1).

No significant differences were observed (ANOVA $P < 0.05$; LSD Fisher) for biodiversity indexes (H) but for Lyapunov coefficients (L). Despite the lack of significant differences for H, the highest values correspond to the communities of *M. carinatus* and *A. gilliesii*, while L values decrease significantly from the communities of *L. divaricata* to communities of *A. quebracho blanco*.

Due to the fact that the dynamics analysis was made in a synchronic way, it is not possible to determine the temporal series between states and transitions, and therefore the analyzed parcels are interpreted as chronosequences [5], quantified through L values. These always decrease as the successional series moves towards a reference state (community of *A. quebracho blanco*). It means an increasing in the accumulated exergy along the series.

Higher values of dominance-abundance of those species which are not present in the reference site promote a higher distance in the succession, and therefore Lyapunov coefficients are equivalent to an entropy measure of the system; a higher L value means a lesser amount of exergy accumulated in the community, and it also means that the subsystem needs a higher amount of exergy to reach the quasistable equilibrium (i.e., "climax").

Assuming the occurrence of a secondary succession defined by the exergy values needed to reach a reference state, in Figure 3 the gradients for each one of the dominant plant populations correlating with L can be seen. The communities of *L. divaricata* would constitute the first step in the successional series; this species appears after an important disturbance increasing its participation until it covers the soil completely. As the other dominant populations begin to increase their frequencies, *L. divaricata* reduces its frequency reaching constant values near $L = 0.25$. *L. divaricata* widely dominates degraded areas, although its frequency reduces to a 25% of the total plant cover with the appearance of *M. carinatus* as dominant species. *M. carinatus* progressively increases its abundance-dominance until it conforms its own community, with maximal frequencies around $L = 0.35$. *A. gilliesii* follows in the series, displacing *M. carinatus*, reaching its maximal participation with $L = 0.28$ and slightly

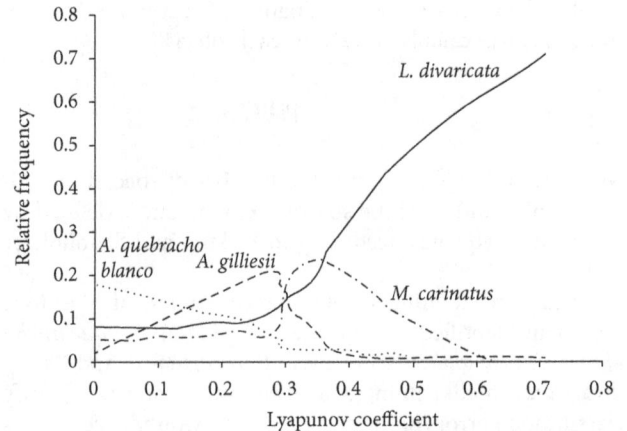

FIGURE 3: Correlation between Lyapunov coefficients and the relative frequencies of the dominant populations. Asp que: *A. quebracho blanco*; Aca gil: *A. gilliesii*; Lar div: *L. divaricata*; Mim car: *M. carinatus*.

decreasing afterwards while denser populations of *A. quebracho blanco* appear. However, *A. gilliesii* do not disappear completely within the communities of *A. quebracho blanco*.

In Figure 3, a transition state between the communities of *M. carinatus* and *A. gilliesii* can be seen near $L = 0.30$, where these species mutually exclude.

4. Discussion

The different vegetation states that can be currently seen in the ChNR, and by extension in most areas of the Arid Chaco subregion, are mainly defined by the intensity of disturbance over the ecosystem. The elimination of the woody biomass has led to the formation of communities of *L. divaricata*, *M. carinatus*, and *A. gilliesii*, competing with herbaceous forage species and affecting natural regeneration of forestry species, forming environments of low productivity.

Figure 2 shows overlapping of communities of *M. carinatus* and *A. gilliesii* because such transitional states are represented by some sites that present similar frequencies of both species.

Analyzing Figure 3, an apparently mature state of the system can be represented by the communities of *A. quebracho blanco*. This state would constitute the "climax" core of the system and is representative of great part of the Arid Chaco [1, 17, 24, 25] defined by the dominated climatic conditions and especially by the characteristics of the azonal soils represented by Typic Torriorthents. The fact that this kind of vegetation state repeats over several points in the subregion suggests an evolution in the successional dynamics towards this "climax" state, in some way deterministic as expressed by Clements in 1916 [2], predicting some aspects of the ecosystem.

Knowing the land use history of the ChNR (see vegetation states in Results section), the study area has suffered two great disturbances of increasing intensity, the first from the climax core towards the south (Chancaní river) by effect of selective logging, and the second by overgrazing, intensive logging, and fire episodes towards the west. The degradation

TABLE 1: Modal synthetic phytosociological table of the study area by state.

State	Community of L. divaricata	Community of M. carinatus	Community of A. gilliesii	Community of A. quebracho blanco
Altitude (masl)	343	402	366	339
Geographic coordinates	S31°22'07.7" W65°28'59.3"	S31°23'31.9" W65°25'59.6"	S31°23'46.4" W65°28'18.4"	S31°22'07.7" W65°28'59.2"
$H(\mu \pm \sigma)^*$	1.70 ± 0.37^a	1.98 ± 0.27^a	1.95 ± 0.17^a	1.68 ± 0.24^a
$L(\mu \pm \sigma)^*$	0.55 ± 0.11^a	0.36 ± 0.03^b	0.29 ± 0.02^b	0.11 ± 0.09^c
Chloris ciliata Sw.	1			
Sporobolus pyramidatus (Lam.) Hitchc.	+			
Digitaria californica (Benth.) Henrard	1	2		
Abutilon pauciflorum A. St.-Hil.		4	+	
Mimozyganthus carinatus (Griseb.) Burkart	1	**3**	1	
Prosopis flexuosa DC.	+	+	1	1
Setaria cordobensis R.A.W. Herrm.	1	1	1	1
Larrea divaricata Cav.	**4**	2	3	2
Aspidosperma quebracho blanco Schltdl.	1	+	+	**2**
Justicia squarrosa Griseb.			2	4
Cordobia argentea (Griseb.) Nied.	1		2	3
Acacia gilliesii Steud.		+	**3**	2
Bromelia urbaniana (Mez) L.B. Sm.			1	2
Celtis ehrenbergiana (Klotzsch) Liebm.	+		+	1
Lycium elongatum Miers			+	
Condalia microphylla Cav.	1		+	
Trichloris crinita (Lag.) Parodi	1	1	+	
Gouinia paraguayensis (Kuntze) Parodi	1	+	+	
Geoffroea decorticans (Gillies ex Hook. & Arn.) Burkart	1		+	
Amphilophium cynanchoides (DC.) L.G. Lohmann	1			
Cercidium praecox (Ruiz & Pav. ex Hook.) Harms	+		+	
Tillandsia duratii var. duratii Vis.		+	1	1

H: shannon-weaver biodiversity index; L: lyapunov coefficient; μ: statistical mean; σ: standard deviation, all based on the total number of surveyed sites. Bold data indicate the representative species of the community. *Different letters indicate significant differences between states ($P < 0.05$; LSD Fisher).

gradient is less intense in the first than in the second. These changes in the vegetation are due to anthropic effects (logging, overgrazing) or natural catastrophic events (fire) and represent secondary successions.

A high magnitude disturbance means high consumption of stored exergy, which is maximal in the mature state and is represented as high density biomass (wood) in the communities of A. quebracho blanco, being L minimal.

The communities of A. gilliesii represent vegetation states into the Arid Chaco but also constitute transitions between this subregion and the Sierra Chaco subregion. This is why they present high values (though without statistical differences with other states) of H. In this case, changes in the dominant vegetation are not due to anthropic effects but to soil changes (Lithic Torriorthents) and climate, generating less favorable conditions for communities of A. quebracho blanco due to lesser soil profile development, water accumulation, and mean temperatures. According to Clements [2] this transition occurs by natural changes, being part of a primary succession.

Holling [26] criticizes the Clementsian model by considering that the species entering in a succession is largely stochastic, that the arrival of the species does not follow a defined order, but these are already in the site (e.g., seed bank), and some disturbing can lead the system to a quasistable state, different than the existing. Then, we will see that some ecological characteristics particular of this subregion contradict these assertions.

According to some observations made in other sites of the subregion [27–30], we can assert that when the succession develops over azonal soils it reaches in most cases to the same vegetation state (communities of A. quebracho blanco), although it may cross through different paths depending of the adjacent ecosystems and the dominant vegetation states. In this study, the natural succession would be the L. divaricata-M. carinatus-A. gilliesii-A. quebracho blanco way, although a regression does not necessarily follow the inverse way, but depends on the level of disturbance (selective logging of firewood).

The exceptions observed in this subregion are the successions facilitated by cattle, which generally end in dense *Prosopis flexuosa* woodlands by seed dispersion effect. However, cattle raising is a relatively recent human activity in this subregion (at least related to the succession scale), so it is possible that finally this woodlands can be replaced by *A. quebracho blanco* woodlands, due to their higher capacity to compete for light in their adult stage [31].

Allelopathic effects of *L. divaricata* restrict the random installation of species, though would act as nurse for species such as *M. carinatus* and *A. gilliesii*, adapted to dense canopies; however, they are more exigent in water and nutriments [32]. *A. quebracho blanco* would also develop under the influence of *L. divaricata*, *M. carinatus* [33], and *A. gilliesii*, according to the observed in Figure 3. However, in Figure 3 it can also be observed that *A. quebracho blanco* frequency increases after a drastic decrease of *M. carinatus* frequency, phenomenon that should be studied in better detail.

Climatic and topographic conditions would restrict the development of invasive allochthonous species. In fact, it is rare to find exotic species outside the influence of water sources.

Lyapunov coefficients adequately define the state of each site. When *L* decreases in time, it is expected that the system moves towards a quasistationary state (communities of *A. quebracho blanco*); otherwise, the equilibrium is unstable and becomes less resilient [1]. Environmental degradation may produce a regression in secondary successions through deforestation, overgrazing, and fire processes, contributing to desertification.

Presumably, vegetation states with higher species abundance and frequency have the capacity to store higher amounts of exergy as biomass and as information [10, 15].

With respect to species diversity, this has a clear and significant effect over disturbance resistance, but equivocal in the recovery rate [34]. The Biodiversity Index can be a good indicator for primary successions but fails modeling secondary successions. Due to this, biodiversity indicators would not be sufficient to define the systems stability, neither would be satisfactory for the evaluation of the system stability. Lyapunov coefficients may be more precise successions indicators than the diversity indexes.

5. Conclusions

In the Arid Chaco subregion it is possible to identify a secondary succession conformed by vegetation states dominated by populations of *L. divaricata*, *M. carinatus*, *A. gilliesii*, and *A. quebracho blanco*.

Biological diversity indexes can be useful for primary successions but fail to predict the behavior of secondary successions. Exergy values could be used for the evaluation of secondary successions produced by environmental degradation in similar areas of the ecoregion and could be applied in similar ecosystems worldwide.

The distance between a representative "climax" state of the subregion, with respect to a reference state, permits estimation of the deterioration degree and the amount of exergy needed to reach a desired state (not necessarily the "climax"). The exergy can be estimated through Lyapunov coefficients as a measure of stability of the system. The exergetic calculus permits the obtaining of comparable values with other successional states in the subregion, though with the values in other ecosystems.

The identification of the vegetation states and transitions conducts to a better knowledge of the ecosystem's dynamic and enables to establish homogeneous areas for the construction of management guidelines for each one.

References

[1] M. S. Karlin, O. A. Bachmeier, A. Dalmasso, J. M. Sayago, and R. Sereno, "Environmental dynamics in salinas grandes, Catamarca, Argentina," *Arid Land Research and Management*, vol. 25, no. 4, pp. 328–350, 2011.

[2] F. E. Clements, *Plant Succession: An Analysis of the Development of Vegetation*, Carnegie Institute of Washington, USA, 1916.

[3] H. A. Gleason, "The Structure and development of the plant association," *Bulletin of Torrey Botanical Club*, vol. 44, no. 10, pp. 463–481, 1917.

[4] A. G. Tansley, "The use and abuse of vegetational concepts and terms," *Ecology*, vol. 16, no. 3, pp. 284–307, 1935.

[5] J. Terradas, *Ecología de la Vegetación*, Omega, Barcelona, Spain, 2001.

[6] M. Westoby, B. Walker, and I. Noy-Meir, "Opportunistic management for rangelands not at equilibrium," *Journal of Range Management*, vol. 42, no. 4, pp. 266–274, 1989.

[7] C. J. Yates and R. J. Hobbs, "Woodland restoration in the Western Australian wheatbelt: a conceptual framework using a state and transition model," *Restoration Ecology*, vol. 5, no. 1, pp. 28–35, 1997.

[8] P. G. Spooner and K. G. Allcock, "Using a state-and-transition approach to manage endangered *Eucalyptus albens* (White Box) woodlands," *Environmental Management*, vol. 38, no. 5, pp. 771–783, 2006.

[9] L. Rumpff, D. H. Duncan, P. A. Vesk, D. A. Keith, and B. A. Wintle, "State-and-transition modelling for Adaptive Management of native woodlands," *Biological Conservation*, vol. 144, no. 4, pp. 1244–1235, 2011.

[10] S. E. Jørgensen and Y. M. Svirezhev, *Towards A Thermodynamic Theory for Ecological Systems*, Elsevier, Amsterdam, The Netherlands, 2004.

[11] B. C. Patten, "An introduction to the cybernetics of the ecosystem: the trophic-dynamic aspect," *Ecology*, vol. 40, no. 2, pp. 221–231, 1959.

[12] I. Prigogine and I. Stengers, *La Nueva Alianza. Metamorfosis de La Ciencia*, Alianza Universidad, Madrid, Spain, 2nd edition, 2004.

[13] E. D. Schneider and J. J. Kay, "Order from disorder: the thermodynamics of complexity in biology," in *What Is Life: The Next Fifty Years. Reflections on the Future of Biology*, M. P. Murphy and L. A. J. O'Neill, Eds., pp. 161–172, Cambridge University Press, Cambridge, UK, 1995.

[14] G. M. Souza, R. V. Ribeiro, M. G. Santos, H. L. Ribeiro, and R. F. Oliveira, "Functional groups of forest succession as dissipative structures: an applied study," *Brazilian Journal of Biology*, vol. 64, no. 3B, pp. 707–718, 2004.

[15] J. Molozzia, F. Salas, M. Callistoa, and J. C. Marques, "Thermodynamic oriented ecological indicators: application of Eco-Exergy and specific Eco-Exergy in capturing environmental changes between disturbed and non-disturbed tropical reservoirs," *Ecological Indicators*, vol. 24, pp. 543–551, 2013.

[16] E. Schrödinger, *¿Qué es la vida?* Tusquets Editores, Barcelona, Spain, 1983.

[17] M. L. Carranza, M. R. Cabido, A. Acosta, and S. A. Páez, "Las comunidades vegetales del Parque Natural Chancaní, Provincia de Córdoba," *Lilloa*, vol. 38, pp. 75–92, 1992.

[18] R. G. Capitanelli, "Clima," in *Geografía Fsica de la Provincia de Córdoba*, J. B. Vázquez, R. A. Miatello, and M. E. Roqué, Eds., pp. 45–138, Editorial Boldt, Córdoba, Argentina, 1979.

[19] D. Mueller-Dombois and H. Ellenberg, in *Aims & Methods of Vegetation Ecology*, John Wiley & Sons, US, 1974.

[20] J. Braun-Blanquet, *Fitosociología. Bases para el estudio de las comunidades vegetales*, Blume, Madrid, Spain, 1979.

[21] D. A. Wikum and G. F. Shanholtzer, "Application of the Braun-Blanquet cover-abundance scale for vegetation analysis in land development studies," *Environmental Management*, vol. 2, no. 4, pp. 323–329, 1978.

[22] J. Di Rienzo, F. Casanoves, L. Gonzalez, M. Tablada, C. Robledo, and M. Balzarini, "Infostat," Facultad de Ciencias Agropecuarias, Universidad Nacional de Córdoba. Statistical software, 2007.

[23] J. Justus, "Ecological and Lyapunov stability," *Philosophy of Science*, vol. 75, no. 4, pp. 421–436, 2008.

[24] E. E. Bonino and P. Araujo, "Structural differences between a primary and a secondary forest in the Argentine Dry Chaco and management implications," *Forest Ecology and Management*, vol. 206, no. 1–3, pp. 407–412, 2005.

[25] M. D. R. Iglesias, A. Barchuk, and M. P. Grilli, "Carbon storage, community structure and canopy cover: a comparison along a precipitation gradient," *Forest Ecology and Management*, vol. 265, pp. 218–229, 2012.

[26] C. S. Holling, "Cross-scale morphology, geometry, and dynamics of ecosystems," *Ecological Monographs*, vol. 62, no. 4, pp. 447–502, 1992.

[27] U. O. Karlin, M. Karlin, and E. J. Ruiz Posse, "Ambientes y vegetación," in *Manejo Sustentable del Ecosistema Salinas Grandes, Chaco Árido*, R. Coirini, M. Karlin, and G. Reati, Eds., pp. 91–118, Encuentro, Córdoba, Argentina, 2010.

[28] H. Calella and R. R. Corzo, *El Chaco Árido de La Rioja. Vegetación y suelos. Pastizales naturales*, Ediciones INTA, Buenos Aires, Argentina, 2006.

[29] D. L. Anderson, J. A. Del Águila, and A. E. Bernardón, "Las formaciones vegetales en la Provincia de San Luis," *Revista de Investigaciones Agropecuarias*, vol. VII, no. 3, pp. 153–183, 1970.

[30] J. Morello, "Provincia fitogeográfica del monte," *Opera Lilloana*, vol. 2, pp. 1–155, 1958.

[31] L. Catalán, *Crecimiento leñoso de Prosopis flexuosa en una sucesión post-agrícola en el Chaco Árido: efectos y relaciones de distintos factores de proximidad [Ph.D. thesis]*, FCEFyN-UNC, Córdoba, Argentina, 2000.

[32] S. A. Páez and D. E. Marco, "Seedling habitat structure in dry Chaco forest (Argentina)," *Journal of Arid Environments*, vol. 46, no. 1, pp. 57–68, 2000.

[33] A. H. Barchuk, M. D. R. Iglesias, and M. N. Boetto, "Spatial association of *Aspidosperma quebracho-blanco* juveniles with shrubs and conspecific adults in the Arid Chaco, Argentina," *Austral Ecology*, vol. 33, no. 6, pp. 775–783, 2008.

[34] D. Tilman, "Biodiversity: population versus ecosystem stability," *Ecology*, vol. 77, no. 2, pp. 350–363, 1996.

20

Ex Situ Conservation of Biodiversity with Particular Emphasis to Ethiopia

Mohammed Kasso and Mundanthra Balakrishnan

Department of Zoological Sciences, Addis Ababa University, P.O. Box 1176, Addis Ababa, Ethiopia

Correspondence should be addressed to Mohammed Kasso; muhesofi@yahoo.com

Academic Editors: I. Bisht and D. Schmeller

Biodiversity encompasses variety and variability of all forms of life on earth that play a great role in human existence. Its conservation embraces maintenance, sustainable utilization, and restoration, of the lost and degraded biodiversity through two basic and complementary strategies called *in situ* and *ex situ*. *Ex situ* conservation is the technique of conservation of all levels of biological diversity outside their natural habitats through different techniques like zoo, captive breeding, aquarium, botanical garden, and gene bank. It plays key roles in communicating the issues, raising awareness, and gaining widespread public and political support for conservation actions and for breeding endangered species in captivity for reintroduction. Limitations of *ex situ* conservation include maintenance of organisms in artificial habitats, deterioration of genetic diversity, inbreeding depression, adaptations to captivity, and accumulation of deleterious alleles. It has many constraints in terms of personnel, costs, and reliance on electric power sources. Ethiopia is considered to be one of the richest centers of genetic resources in the world. Currently, a number of stakeholders/actors are actively working on biodiversity conservation through *ex situ* conservation strategies by establishing gene banks, botanical garden, and zoo.

1. Introduction

According to the Convention on Biological Diversity, biodiversity refers to the variability among living organisms (animals, plants, and microorganisms) including *inter alia*, terrestrial, marine, and other aquatic ecosystems with their ecological complexes. In another expression, biodiversity encompasses the variety and variability of all forms of life on earth that play a great role in human existence [1, 2]. It also includes the ethnical value of biodiversity such as tradition and traditional knowledge of the indigenous and local communities [2] and the diversity within species (genetics), between species and of ecosystems [3].

Genetic diversity refers to the variation within species of any plant, animal or microbes in the functional units of heredity. Species diversity refers to the variety of species within a geographical area, which become central in the evaluation of diversity, and used as a point of reference in biodiversity conservation. Finally, ecosystem diversity refers to the variety of life forms in a given territory or area with all its functional ecological processes, which is often evaluated based on the diversity of all of its components [1].

Biodiversity is important for the maintenance of a healthy environment and used for direct human benefits like food, medicine, and energy. It is also used for recycling of different essential elements, for mitigation of pollution, for protection of watersheds, to mitigate soil erosion and to control excessive variations in climate and catastrophic events. For example, biodiversity provides different services free of charge worth of billions dollar every year for crucial well-being of the society. Some of these services are providing clean water and air, soil formation and protection, pollination, pest control, food, fuel, fibers, medicine, and construction and industry raw materials [4]. Agricultural biodiversity is another important component of biodiversity, which has a more direct link to the well-being and livelihood of mankind than other forms of biodiversity. Food plant and animal species have been collected, used, domesticated, and improved through traditional systems of selection over many generations [5].

However, today much of the lines of evidences are increasingly pointing out a significant global decline in biodiversity by numerous, varied, and interacting drivers [6]. More than half of the habitable surface of the earth has already been significantly altered by human activities. As a consequence, biodiversity of our planet is on the verge of decline and extinction despite our limited and incomplete knowledge on them [7]. Biodiversity loss and extinction processes can occur in two phases. The first phase is known as deterministic and often resulted from human threats such as habitat loss, fragmentation and degradation, direct exploitation of the species, competition from exotic and domestic species, and persecution and killing due to human animal conflicts. The second phase is known as deterministic that resulted from failures in mitigating threats that eventually result in very small, fragmented, and isolated remnant populations. Then these small remnant populations become vulnerable to a number of other, nonhuman caused threats mainly stochastic, genetic (genetic drift and inbreeding), and demographic events [8]. Thus, small, fragmented, isolated populations can find themselves being dragged into an extinction vortex whereby genetic and demographic stochastic events can cause the species to go extinct. During this second phase of the extinction process, very intensive management of populations and individuals is often necessary to prevent extinction [9].

Several human induced impacts are leading to a mass extinction process affecting global biodiversity. The major reasons for rapid diminishing of biodiversity are attributed to conversion of land for agriculture, wild fires, poor management of available land, over-exploitation for food, fuel-wood, medicine, construction, overgrazing by cattle, displacement and loss of landraces, lower yielding varieties, pests and diseases, global climate change, pollution (e.g., acid rain), and gap of scientific knowledge on some of the biological resources [1, 7, 10]. Human beings are destroying biodiversity, particularly during livelihood activities with or without knowledge of the consequences of their actions [6]. Agriculture is one of the most important land-use that results in detrimental environmental consequences from increased use of fertilizers and biocides, land draining, irrigation, and the loss of many biodiversity-rich landscape features [6]. There are many threats to biodiversity as a result of agricultural practice through changes in land-use, replacement of traditional varieties by modern cultivars, agricultural intensification, increased population, poverty, land degradation, and environmental changes (including climate change) [5]. Recent estimates indicate that humans use more than 40% of the terrestrial components and significantly modified global biodiversity [7]. As a consequence, many species of living organisms are classified as threatened today and this has become a central concern for conservation [11].

Conserving biodiversity has economic, social, and cultural values. Conservation of biodiversity is integral to the biological and cultural inheritance of many people and the critical components of healthy ecosystems that are used to support economic and social developments. Moreover, it is used to maintain the earth's genetic library from which society has derived the basis of its agriculture and medicine [5, 12]. The twenty-first century is predicted to be an era of bioeconomy driven by advances of bioscience and biotechnology. Bio-economy may become the fourth economy form after agricultural, industrial, and information technology economies, having far-reaching impacts on sustainable development in agriculture, forestry, environmental protection, industry, food supply, health care and other micro-economy aspects. Thus, a strategic vision for conservation and sustainable use of biodiversity in the 21st century is of far-reaching significance for sustainable development economy and society [13].

Biodiversity conservation refers to the management of human use of biodiversity in order to get the greatest sustainable benefit to present and future generations. Thus, conservation of biodiversity embraces the protection, maintenance, sustainable utilization, restoration, and enhancement of biodiversity [1]. Biodiversity conservation mainly focuses on genetic conservation with its diverse life-support systems (ecosystems) for the connotation of human well-being [3].

Conservation techniques can be grouped into two basic, complementary strategies: *in situ* and *ex situ* [14]. As also outlined in the articles 8 and 9 of the Convention of Biological Diversity (CBD), biodiversity is conserved by two major methods called *in situ* and *ex situ*. The conservation efforts, either *in situ* or *ex situ*, involve the establishment and management of protected areas and relevant research institutes or academic institutions, which establish and manage arboreta, botanical or zoological gardens, tissue culture, and gene banks [1]. The concept of *ex situ* conservation is fundamentally different from that of *in situ* conservation; however, both are important complementary methods for conservation of biodiversity. The principal difference (and hence the reason for the complementarities) between the two lies in the fact that *ex situ* conservation implies the maintenance of genetic materials outside of the "normal" environment where the species has evolved and aims to maintain the genetic integrity of the material at the time of collection, whereas *in situ* conservation (maintenance of viable populations in their natural surroundings) is a dynamic system, which allows the biological resources to evolve and change over time through natural or human-driven selection processes [5].

In situ conservation is defined as conservation of ecosystems and natural habitats, the maintenance of viable populations of the species in their natural surroundings and, in the case of the cultivated species, in the surroundings where they have developed their distinctive properties. *In situ* conservation can be done in farmers' fields, in pasture lands, and in protected areas [15]. For cultivated species, *in situ* conservation concerns the maintenance of the local intra- and inter-population diversity available in various ecological and geographical sites [1, 16]. Thus, it allows ongoing host-parasite coevolution, which is likely to provide material resistance to pests and diseases, and CBD recognized it as a primary approach to conserve biodiversity [4]. However, *in situ* conservation has certain limitations like more difficult access to breeders requiring the application of its complimentary technique. For example, some of the natural habitats or wild habitats are very risky when compared to relatively safe captive environment [9].

The second biodiversity conservation technique receiving the most attention to conserve biodiversity is *ex situ*. *Ex situ* conservation techniques are mostly used to be applied to species with one or some of the following characteristics: endangered species, species with a past, present or future local importance, species of ethno-botanical interest, species of interest for the restoration of local ecosystems, symbolic local species, taxonomically isolated species, and monotypic or oligotypic genera [17]. Intensive conservation and management of populations and individuals can come in many different forms, like translocation, breeding in a fenced wild habitat, supplementary feeding, captive hand rearing of young of wild parents to become pregnant sooner, and captive breeding [9].

2. *Ex Situ* Conservation

Ex situ conservation is a technique of conservation of biological diversity outside its natural habitats, targeting all levels of biodiversity such as genetic, species, and ecosystems [1, 2, 16]. Its concept was developed earlier before its official adoption under the Convention on Biological Diversity signed in 1992 in Rio de Janeiro [2]. In general, *ex situ* conservation is applied as an additional measure to supplement *in situ* conservation, which refers to conservation of biological diversity in its natural habitats [16]. In some cases, *ex situ* management will be central to a conservation strategy and in others it will be of secondary importance [18]. Broadly, *ex situ* conservation includes a variety of activities, from managing captive populations, education and raising awareness, supporting research initiatives and collaborating with *in situ* efforts [19]. It is used as valuable tools in studying and conserving biological resources (plants, animals, and microorganisms) for different purposes [2] through different techniques such as zoos, captive breeding, aquarium, botanical gardens, and gene banks [16].

2.1. Types of Ex Situ Conservation

2.1.1. Zoos. Zoos or zoological gardens or zoological parks in which animals are confined within enclosures or semi-natural and open areas, displayed to the public, and in which they may also breed. They are considered by universal thinkers and environmentalists as important means of conserving biodiversity [19–21]. Zoos attract as many as 450 million visitors each year and so are uniquely placed to have very large educational and economic values [22]. Zoos not only act as places of entertainment and observing animal behavior, but are also as institutions, museums, research laboratories, and information banks of rare animals [23]. Although some people dislike zoos, many people enjoy them. Over the last several decades, zoos have made significant progress in its cooperative management of *ex situ* populations of a variety of biodiversity [9].

Zoos breed many endangered species to increase their numbers. Such captive breeding in zoos has helped to save several species from extinction [19]. Management of animals in zoos includes animal identification, housing, husbandry, health, nutrition as well as addressing and ways of interaction with the public [20]. There are various processes and mechanisms used to determine whether a species or taxon is included within a zoo's collection plan. The frequently used criteria include how the species is valued, according to its uniqueness, contribution to research or education, and conservation status [19]. Zoos help the animal to secure food, shelter, social contact and mates, and to be motivated by desire (appetitive behavior), which is reinforced by pleasure (consummative behavior) [21].

In the past, some zoos paid little attention to the welfare of the animals, and some zoos today have poor environments for animals [24]. They were also once reliant on harvest from the wild to populate their exhibits and reliance on continued wild collection to breeding closed populations [23]. Many zoo animals also became endangered or extinct due to visitor disturbances, unfavorable climate and due to insufficient space [20]. From this aspect, many scholars state on the negative features of keeping animals in zoo as it causes pain, stress, distress, sufferings and evolutionary impacts [21]. Animal welfare, education, conservation, research, and entertainment are major goals of modern zoos, but these can be in conflict. For example, visitors enjoy learning about and observing behavior in captive animals, but visitors often want to observe and interact with the animals in close proximity. Unfortunately, proximity to and interactions with humans induce stress for many species [25]. The same is true for Addis Ababa Lion Zoo Park.

However, progressive zoos are engaged in education, research, and conservation, with the aim of maintaining healthy animals, which behave as if they are in their natural habitats [24]. The current paradigm for managing essential populations is to minimize the rate of genetic decay, slow adaptation to the captive environment, and retain typical behaviors [23, 26]. It is widely accepted that the more generations a population spends in captive breeding, the less suitable it is for attempted restorations in the wild. Hence, population management is designed not to deplete too quickly the resource obtained from the founders [23]. Thus, for true sustainability of the species for the purpose of conservation, display, education, and research, constant refreshing of populations is required [9, 23]. Majority of the current breeding programs base on the genetic management of populations by the analysis of individual pedigrees in order to minimize kinship [9].

2.1.2. Captive Breeding. Captive breeding is an integral part of the overall conservation action plan for a species that helps to prevent extinction of species, subspecies, or population. It is an intensive management practice for threatened individuals, populations, and species by anthropogenic and natural factors [9]. In small and fragmented populations, even if the human caused threats could be magically reversed, the species would still have a high probability of extinction by random demographic and genetic events, environmental variations, and catastrophes. Thus, under sufficient knowledge on the biology and husbandry of the species, captive breeding helps individuals in the relative safety of captivity, under expert care and sound management by providing an insurance against extinction [9]. Stock for reintroduction or reinforcement efforts, opportunities for education, raising of awareness, scientific and husbandry research, and other contributions to conservation are also possible through captive breeding [9, 27].

Environmental enrichment strategies are used to improve both physiological and psychological welfares of captive animals, which can be achieved by increasing the expression of natural behavior and decreasing abnormal behaviors. Successful environmental enrichment includes the improvement of enclosure design and the provision of feeding devices, novel objects, appropriate social groupings, and other sensory stimuli [27]. The minimum requirement for successful *ex situ* management, particularly in the captive populations, is the inclusion of as much of the genetic diversity present in wild populations. Genetic sustainability (retention of 90% of the gentic diversity of the wild population for 100 years) in captive breeding is maintained if consideration is given on number of founders, population growth rate, effective population size, and duration of the captive program [19]. However, even if at least 30 founders in captive breeding are recommended to ensure the representation of large enough proportion of the genetic diversity of the wild population, for critically endangered species, actively removing individuals from the wild population to serve as founders may compromise the survival of the wild population [9]. For example, the Arabian Oryx captive breeding program was based on fewer founders and grew to a couple of thousand individuals through breeding management, which helped to reduce risks.

However, there are several challenges (biological and environmental) that are limiting factors to the attainment of the goal of captive breeding for many species [19]. One of the major challenges is a circular consequence of small-population management that has inherent genetic and demographic problems due to genetic diversity loss and demographic stochasticity [19]. In addition, individuals that are well adapted to the circumstances in captivity may also be less well adapted to the circumstances in the wild and may show lower fitness upon reintroduction [28]. Most notably, within the captive environment, housing and husbandry will also have significant impacts on birth and death rates [19].

2.1.3. Aquarium. An aquarium is an artificial habitat for water-dwelling animals. It can also be used to house amphibians or large marine mammals and plant species for tourist attractions. It is usually found in zoos or marine parks with different size. The 15,750 described species of freshwater fish comprise around 25% of living vertebrate species diversity and a key for global economic and nutritional resources of which more than 11% is threatened (60-extinct, 8-extinct in the wild and 1679-threatened) [22]. Fresh waters (0.3%) of available global surface water support 47–53% of all extant fish species that are threatened by overfishing, pollution, habitat loss, damming, alien invasive species, and climate change. This requires world's zoos and aquariums to identify the potential targets (species or areas) for *in situ* and *ex situ* conservation program [18, 22]. Aquarium is used to admire

at home by hobbyists, to portray as public exhibits, to provide large quantities of human food and animal fodder [18].

Fishes are often overlooked within the development of conservation priorities. This leads to the low focus on meaningful conservation efforts rather than giving more attention for their importance to food supply and livelihoods. For example, it provides job opportunity for over 60 million people, as source of food for over 200 million people in Africa, for US$1.5 billion income from trade of 4000 species global ornamental fish industry, and many are displayed in the world's public zoos and aquaria to a global audience of as many as 450 million people per year [22, 29]. However, despite the clear value of freshwater fish diversity, wetland habitats and their associated freshwater-fish species continue to be lost or degraded at an alarming rate [30]. One recommendation is for aquariums to set up sustainable breeding program that prioritizes threatened species (VU, EN, and CR) and those classified as EW to support species conservation *in situ* and aid the recovery of species via collaborative reintroduction or translocation efforts when appropriate [22].

2.1.4. Botanical Gardens. Botanical gardens consist of living plants, grown out of doors or under glass in greenhouses and conservatories. They are used to grow and display plants primarily for scientific and educational purposes. They also include herbarium, lecture rooms, laboratories, libraries, museum and experimental or research plantings. It can be taxonomic collection of a particular family, genus or group of cultivars, native plants, wild relatives, medicinal, aromatic, or textile plants [4]. There are over 2,000 botanic gardens, holding 80,000 plant species in their living collections and receiving hundreds of millions of visitors per annum [14, 31]. Furthermore, they have valuable and distinctive mix of officials dedicated to plant research, systematics, conservation education, and public awareness [31]. They are now extremely well networked both among themselves and with other professionals, conservation organizations, and nongovernmental organizations (NGOs) [31]. They provide different services for sectors that utilize and conserve plant diversity like agriculture, forestry, pharmaceutical and biofuel industries, protected area management, and ecotourism. They have a unique opportunity as visitor attraction places and scientific institutions for documentation and conservation of plant diversity by shaping and mobilizing citizens to the current environmental challenges [31]. They also play a great role in attaining target of the Global Strategy for Plant Conservation for 2020 to cultivate 75% of the world's threatened plant species in *ex situ* [14]. Botanical gardens give opportunity for arable plants to be grown under relatively modified environmental conditions (intense cultivation, relatively high fertility, and high levels of disturbance) [14].

However, most of the cultivated taxa are held in a small number of collections and mostly only in small populations. Lack of genetic exchange and stochastic processes in small populations make them susceptible to detrimental genetic effects [14]. The low number of *ex situ* populations in most botanical gardens poses a fundamental problem for conservation. The total *ex situ* breeding collection is therefore very small with respect to the stated aim of conserving regional gene pools [14]. The striking lack of information on source populations casts doubt on the value of using such *ex situ* populations for potential reintroductions. They also require testing for fitness and similarity to wild populations before they are brought to the field [14]. Thus, conservation actions of botanic gardens such as training and capacity building, needs to be better understood and better coordinated [31].

2.1.5. Gene Banks. Genome resource banking is another management technique used for biodiversity conservation. Different types of gene banks have been established for the storage of biodiversity, depending on the type of materials conserved. These include seed banks (for seeds), field gene banks (for live plants), *in vitro* gene banks (for plant tissues and cells), pollen, chromosome, and deoxyribonucleic acid (DNA) banks for animals (living sperm, eggs, embryos, tissues, chromosomes, and DNA) that are held in short term or long term laboratory storage; usually cryopreserved or freeze-dried [32]. Currently, there are about 7.4 million PGRFA accessions conserved in over 1750 gene banks [5]. The Genome 10K (G10K) project aims to sequence the genomes of 10,000 vertebrates, of which 4,000 will be fish. To date, 60 fish genomes have been sequenced in laboratories worldwide and added to the database, and a further 100 are targeted. The Frozen Ark database holds details of 28,060 frozen DNA samples. Among these 6,997 are from species listed in the IUCN Red List [18]. The principal aim of gene bank conservation is to maintain genetic diversity alive as long as possible and to reduce the frequency of regeneration that may cause the loss of genetic diversity [5].

With the rapid development in the field of molecular genetics and genomics, DNA is becoming more and more in demand for molecular studies and is one of the most requested materials from gene banks. Establishing DNA storage facility as a complementary "backup" to traditional *ex situ* collections has been suggested [5]. Some efforts have been made to establish DNA banks for endangered animals [33], and a few plant DNA banks in different parts of the world such as the Missouri Botanic Garden, Kew Royal Botanic Garden and Australian Plant DNA Bank. Many research groups are already developing their own archives of extracted genomic DNA. Recently, the Global Biodiversity Information Facility in Germany has established a DNA bank network, which provides DNA samples of complementary collections (microorganisms, protists, plants, algae, fungi, and animals) [5].

Seeds are usually the most convenient and easiest material to collect and to maintain in a viable state for long periods of time and that makes it preferred for conservation in gene banks [14, 34]. Seed banking techniques rely on the storage of dried seeds of threatened or other plants at low temperatures as the most important factors influencing seed longevity are temperature, seed moisture content, and relative humidity [14, 35, 36]. Seeds are typically conserved at moisture content between 3 and 7 percent and stored at 4 degrees Celsius for short-term conservation, and between −18 and −20 degrees Celsius for long-term conservation [5, 34]. Current research

is showing that there exists variability in seed longevity for different species being conserved under similar conditions. In addition, it has been found that the type of seed (endospermic or non-endospermic) and intraspecific variation may also affect accessions longevity [5]. In addition, high initial quality seeds are a major prerequisite for ensuring seed longevity in seed banks [5].

Plants that cannot be conserved as seeds because of their recalcitrant nature (i.e. seeds that are desiccation and/or cold sensitive) or are clonally propagated are traditionally conserved as live plants in *ex situ* field gene banks. But, field gene banks present real logistical challenges; they require large areas and are costly, they are vulnerable to pests and diseases, natural disasters, political unrest, extreme weather, fire, vandalism, and theft, and they often are at risk due to policy changes on land use [5].

In vitro conservation refers to one type of gene bank known as slow-growth conservation method. It involves culturing of different parts of the plant (meristem, tissues, and cells) into pathogen-free sterile culture in a synthetic medium with growth retardants, which has been cited as a good way of complementing and providing backup to field collections [5].

The other genome conservation technique is cryopreservation, in which living tissues are conserved at very low temperatures (−196°C) in liquid nitrogen to arrest mitotic and metabolic activities [5]. It is now realized that cryopreservation method can offer greater security for long-term, cost effective conservation of plant genetic resources, including orthodox seeds [5]. The storage in liquid nitrogen clearly prolonged shelf life of lettuce seeds with half-lives projected as 500 and 3400 years for fresh lettuce seeds stored in the vapor and liquid phases of liquid nitrogen, respectively [5].

2.2. Advantages of Ex Situ Conservation. It is generally preferred to conserve threatened species *in situ*, because evolutionary processes are more likely to remain dynamic in natural habitats [14, 19]. However, considering the rate of habitat loss worldwide, *ex situ* cultivation is becoming increasingly important [14]. Further more, as many of the taxa are located outside natural parks or reserves, *in situ* measures are not enough to assure their conservation. Translocation, introduction, reintroduction, and assisted migrations are species conservation strategies that are attracting increasing attention, especially in the face of climate change [37].

As approximately 450 million people per year visit zoos and aquaria globally, their education and marketing services play a key role in communicating the issues, raising awareness, changing behavior, and gaining widespread public and political support for conservation actions. Zoos support conservation by educating the public, raising money for conservation programs, developing technology that can be used to track wild populations, conducting scientific research, advancing veterinary medicine, and developing animal handling techniques [22]. By studying animals in captivity and applying that knowledge to their husbandry, zoos can provide valuable and practical information that may be difficult or impossible to gather from the wild [24]. Zoos and aquaria have significant roles to play in improving public awareness of the issue facing species and their habitats; for example, through presentation of maps and photographs of species recently extinct as a result of anthropogenic impacts. A similar display of threatened species, even if not currently in the collections of the zoo, would help convey to the public the magnitude of the threat facing the species [18, 22]. It also reaches a wide cross-section of the society, because zoo audiences are not limited to those who are already passionately interested in wildlife and because many zoo visitors are children. Some of these children may become committed conservationists. Some may grow up to be oil company tycoons, politicians, or movie stars, with great potential influence. Some may even live next door to a poacher or wildlife dealer. Thus, instilling an interest in conservation of wildlife in people from all walks of life while they are young is one vital role zoos can play [38]. It is often claimed that zoos perform valuable conservation work by breeding endangered species and returning them to the wild. Zoos can also be used for businesses that make money. This means that animals are often bred for commercial purposes because the public like to see new-born animals. Such breeding leads to a surplus of animals, and in order to keep numbers down sold to private collectors, circuses, or even research laboratories. A zoo with good and attractive entertainments encourages initial visits and subsequent returns to the zoo, which is used to get more revenue for conservation efforts, research, and general animal care and welfare and also to develop more positive perceptions of animals in zoos and become more supportive of conservation efforts [25, 39].

2.3. Disadvantages of Ex Situ Conservation.

Some *ex situ* conserved collections showed lower resistance levels, although still others showed higher resistance levels than their *in situ* conserved counterparts mainly due to the high evolutionary drive and complex nature of evolutionary scenario [40].

The behavior of animals in the zoo may be affected by the frequent arrival of large number of people, who are unfamiliar to the animals [41]. Animals housed in artificial habitats are confronted by a wide range of potentially offensive environmental challenges such as artificial lighting, exposure to loud or aversive sound, arousing odors, and uncomfortable temperatures or substrates. In addition, confinement-specific stressors such as restricted movement, reduced retreat space, forced proximity to humans, reduced feeding opportunities, maintenance in abnormal social groups, and other restrictions of behavioral opportunity are considered [42]. However, over the course of the twentieth century, as knowledge of wildlife biology improved, zoo animals began to be kept in more natural surroundings and social groupings, and diets and veterinary care began to improve. Thus, survival and breeding rates of captive populations improved [38]. Evidence mainly from studies of rodents and primates strongly indicates that prenatal stress can impair stress-coping ability and is able to cause a disruption of behavior in aversive or conflict-inducing situations. Prenatally stressed animals show retarded motor development, reduced exploratory and play behavior, and impairments of learning ability, social

behavior, and sexual and maternal behavior. Prenatal stress may also affect the sex ratio at birth and the reproductive success [43].

Although populations of some species managed in *ex situ* may have the best hope for their long-term survival, they might be challenged if not properly managed during translocation and reintroduction with the effects of climate change [22]. Some species may lose their biological integrity particularly on morphology. For example, an experimental study on black-footed ferrets (*Mustela nigripes*) in *ex situ* indicated a decrease of 5–10% body size than pre-captive, *in situ* animals [44]. In other words, the small cage size and environmental homogeneity inhibit mechanical stimuli necessary for long bone development. Thus, in the absence of such an environment, "unnatural" morphologies can result that may contribute to poor fitness or perhaps even for domestication and reintroduction and relocation [44]. It would be very difficult to reintroduce some zoo-reared animals to their natural habitats because, after generations of captivity, many have lost the necessary skills to survive in their original habitats [22].

For naturally out-breeding species, the high levels of inbreeding in captivity often have negative effects on life history traits related to reproduction and survival [11]. It makes the population in captivity deteriorate due to loss of genetic diversity, inbreeding depression, genetic adaptations to captivity, and accumulation of deleterious alleles [17]. For plants, ecological shifts, small population size, genetic drift, inbreeding, and gardener-induced selection may negatively affect population structure after several generations of *ex situ* cultivation [16, 45, 46]. These factors could seriously put at risk the success of *ex situ* conservation [17].

Captive breeding of threatened species has used increasingly sophisticated technologies and protocols in recent years [47]. Although, this has blurred the dichotomy between *in situ* and *ex situ* species management, the value of captive breeding as a conservation tool remains controversial [48]. It is recognized that *ex situ* conservation has many constraints in terms of personnel, costs, and reliance on electric power sources (especially in many developing countries where electricity power can be unreliable) for gene banks. It requires high facilities and financial investments. It cannot also conserve all of the thousands of plant and animal species that make up complex ecosystems such as tropical rainforests [49]. Capture of individuals from the wild for captive breeding or translocation some times can have detrimental effects on the survival prospects of the species as a whole through disease infection [50].

Even though the management of irreplaceable animal populations in zoos and aquarium has focused primarily on minimizing genetic decay with the use of advanced technologies, recent analyses have shown that as most zoo programs are not projected to meet the stated goals due to lack of achieving "sustainability" of the populations [23]. Thus, managing zoo populations as comprehensive conservation strategies for the species requires research on determinants of various kinds of genetic, physiological, behavioral, and morphological variations, and their roles in population viability, development of an array of management techniques, tools, and training of managers [23].

2.4. Challenges to Ex Situ Conservation. *Ex situ* conservation requires different kinds and levels of intensity of management, and a multistakeholder approach like the input from experts on aquarium and zoo husbandry, *ex situ* breeding, gene-banking, reintroduction, and habitat restoration [51]. Other expert input may include taxonomy, ecology and conservation, ethnography, and sociology. For outreach program, there is a need to liaise with local communities and national government fisheries and wildlife departments; with international (nongovernmental and intergovernmental) conservation bodies [18].

The most important challenges of applying *ex situ* conservation (captive breading) are the difficulty in recognizing the right time, identifying the precise role of the conservation efforts within the overall conservation action plan, and setting realistic targets in terms of required time span, population size, founder numbers, resources, insurance of sound management and cooperation, and the development of much needed new technical methods and tools [9]. In captive breeding to achieve the retention of 90% of the wild gentic diversity, it is necessary to incorporate sufficient number of founders, careful pair combinations and management [9]. Evidence also exists, which demonstrates that manipulation of housing and husbandry variables can also have significant positive influence on animal reproduction in captivity [19].

In many cases, *ex situ* populations are founded from only a few individuals, which cause genetic bottlenecks. Small populations are exposed to threats such as stochastic demographic events as well as genetic effects, including loss of genetic diversity, inbreeding depression or accumulation of new, potentially deleterious mutations [11]. More specific problems in garden populations include poorly documented or even unknown sources of material, accidental hybridization of material from various localities, and or unintended selection for traits more suited to garden conditions [14]. In every region, most of the cooperatively managed breeding programs have too few animals, too few animals in appropriate situations for breeding, too few successful breeders, too few founders, and too many animals with undocumented ancestries and/or too little cooperation with scientifically designated breeding recommendations. These deficiencies are resulting in declining populations or declining gentic diversity or both [23, 52].

Problems associated with small founder populations such as inbreeding depression, removal of natural selection, and rapid adaptation to captivity pose considerable challenges for managers of captive populations of threatened species [48]. Equally, reintroduction of captive-bred stock to the wild may require implementation of rigorous protocols that embrace acclimation, pre- and post-release training, health screening, genetic management, long-term monitoring, and involvement of local stakeholders [53, 54]. Shortfalls in implementing such protocols may jeopardize the likelihood of achieving success [47].

Inbreeding due to the mating between two related individuals is unavoidable in small, fragmented, or isolated populations typical of many threatened species, and it can lead to a significant reduction in fitness. The deleterious effects of inbreeding on individual fitness can be large and may be an important factor contributing to population extinction. Inbreeding depression has potential significance for the management and conservation of endangered species [55]. As populations get smaller, the probability increases for all offspring in a given generation are of the same sex [19].

Evaluating the long-term efficiency of *ex situ* conservation is important, but is complicated because of the difficulty of finding more than one sample of a documented (origin and cultivation) *ex situ* population and its corresponding still-existing *in situ* source population [46].

Animal translocations are usually risky and expensive, and a number of biological and nonbiological factors can influence success. Biological considerations include knowledge of genetics, demography, behavior, disease, and habitat requirements. It also includes legal framework, fiscal and intellectual resources, monitoring capacity, goal of the translocation, logistic challenges, and organizational structure of decision making [56].

The regeneration process is one of the most critical steps and a major challenge in gene bank management, during which there is the highest probability for genetic erosion [57]. It is equally important to understand how different conservation methods (seed, field, and cryopreservation) and their management can affect or change the genetic make up, thereby reducing the effective population size (Ne). This will also contribute to decision-making process for determining which methods to use for conservation of the wide diversity [5].

If people are discouraged or prevented from interacting with the resident animals, fewer visitors attend, decreasing public financial support. The visitors' noise and crowding become a source of stress for many species that affects both their welfare and the enjoyment of the visitor [25].

3. *Ex Situ* Conservation Practice in Ethiopia

Ethiopia is considered to be one of the richest centers of genetic resources in the world. It is believed that indigenous crops such as teff (*Eragrostis tef*), Noug (*Guizotia abyssinica*), and Enset (*Ensete ventricosum*) were first domesticated in Ethiopia. Numerous major crop species including durum wheat (*Triticum durum*), barley (*Hordeum vulgare*), sorghum (*Sorghum bicolor*), sesame (*Sesamum indicum*), castor (*Ricinus communis*), and coffee (*Coffea arabica*) are also known to show significant diversity in the Ethiopian region [58]. Almost 85% of the populations of Ethiopia live in rural areas and most of this population depends directly or indirectly on biodiversity. Biodiversity also plays a crucial role in the different sectors like energy, agriculture, forestry, fisheries, wildlife, industry, health, tourism, commerce, irrigation, and power [15].

The records on biodiversity conservation efforts in Ethiopia date back to the days of Emperor Zera-Yakob (1434–1468 E.C.). The Emperor brought juniper seedlings from Wof Washa of North Shewa and planted in Managesha-Suba area.

Modern conservation intervention began by Emperor Menilik in 1908 E.C. and eventually evolved to the establishment of protected areas in the 1960s [4]. Currently, a number of stakeholders are actively working on biodiversity related issues at the federal government level. These include Institute of Biodiversity Conservation (IBC), Ethiopian Institute of Agricultural Research (EIAR), Ethiopian Wildlife Conservation Authority (EWCA), Ministry of Agriculture and Rural Development (MoARD), Ministry of Science and Technology (MoST), Higher Learning Institutions (HLIs), particularly Addis Ababa University (AAU), and offices in various regional states of Ethiopia [59]. *Ex situ* conservation as complementary to the rehabilitation and restoration of degraded ecosystems and the recovery of threatened species was started 1976 with the establishment of IBC [4].

Ex situ conservation activities mostly focus on high socioeconomic value and internationally important crop types that are considered to be facing immediate danger of genetic erosion [15]. The collections held at IBC are mostly of indigenous landraces some of which are not seen today in farmlands. The collections of root crops, medicinal plants, weedy species, and wild relatives of cultivated species are still relatively inadequate within the existing *ex situ* collections [60]. However, appropriate emphasis is being placed on conservation and sustainable use of all forms of plant biological resources [60]. Since the establishment of IBC, systematic crop germplasm exploration and collection operations have been undertaken in the different administrative regions of the country, covering a wide range of agroecological zones. Collection priorities were set based on factors like economic importance, degree of genetic erosion and diversity, researchers' needs, the rate of diffusion of improved varieties, clearing of natural vegetation, agricultural policies, natural disasters, and resettlement program [4].

Currently about 68,014 seed accessions of 200 plant species, 6,704 accessions of 205 species of forestry, medicinal, forage and pasture plants (in field gene bank), 290 species of microbial genetic resources, and three semen of threatened breed of domestic animals are conserved by IBC. About 90% of the total germplasm holdings in the gene bank consist of field crops. The total collection is composed of cereal seeds, pulses, oil crops, spices, and species of medicinal and industrial value. Aside from the crop collections, the gene bank also holds 650 collections of micro-organisms. Over 9,000 accessions of horticultural crops, medicinal plants, and herbs are maintained in field gene banks. The type and nature of collection missions and number and lists of plant species and landraces collected have been documented in manuals and reports. Regular monitoring activities are performed for seed viability [15].

For plant species with recalcitrant and intermediate storage behavior, there are ten field gene banks under IBC control and small sized fields in the various research stations of the Ethiopian Institute of Agricultural Research (EIAR) and at universities. The plan for the immediate future is to increase the number of field gene banks in different agroecological zones. Community gardens, backyards, and holy places are being considered for inclusion in the future plan. Spices, vegetables and medicinal plants require management on

a large scale and with the full involvement of the local communities [15]. The initiative at national level is still in its infancy and there is currently no well-established national botanical garden in Ethiopia including the Gulele Botanical Garden Center [4, 59]. The Gulele Botanic Garden Center was established through the Proclamation No. 18/2005 E.C. in October 30/2002 E.C. in a 705 hectare land at Gulele and Kolfe-Kernayo subcities. It was established with a vision to see the center to be developed as an exemplary garden in terms of education, ecotourism attraction and center for originality of the Ethiopian plant species, and to be a place of research and nurturing of plant species. The center also has a mission to provide persistent ecotourism services to tourists by taking care of plant species and carrying out educational and research works [59].

Although there is no well-established zoo or zoological garden in Ethiopia, the Addis Ababa Lion Zoo Park can be dominantly cited [59]. The Addis Ababa Lion Zoo Park was established in 1948 with five founder lions presented to Emperor Haile Selassie as gifts. The park accommodates lions with cubs, tortoises, baboons, monkeys, apes, and ducks. A team of international researchers has provided the first comprehensive DNA evidence from 15 (eight males and seven females) samples of Addis Ababa lion indicating the genetically unique samples that requires immediate conservation action. Both microsatellite and mitochondrial DNA data suggest that the zoo lions are genetically distinct from all existing lion populations for which comparative data exist.

Desiccation-intolerant seeds and species that do not readily produce seeds are conserved *ex situ* in field gene banks. For example, accessions of coffee (*Coffea arabica*), root crops such as yam (*Dioscorea bulbifera*) and "Oromo dinich" (*Coleus edulis*), and spices like ginger (*Zingiber officinale*) and Ethiopian cardamom (*Aframomum corrorima*) are conserved in agro-ecological zones in field gene banks [60].

The need for action for global biodiversity conservation is now well understood, and government agencies, nongovernmental organizations, and botanic gardens have all been working in various ways to promote environmental sustainability and reduce species and habitat loss [31].

Seed banking is the major *ex situ* conservation method employed in Ethiopia. There are three major seed banks operating in Ethiopia. The National Tree Seed Project processes seeds from a narrow range of tree species and uses short-term storage facilities. It aims to cater for the annual seed demand from commercial and small-scale forestry enterprises. Of the 70 species regularly collected and processed, 20 are indigenous. The Forage Genetic Resources Centre maintained by the Consultative Group on International Agricultural Research at the International Livestock Research Institute maintains long-term conservation of a wide range of native and exotic forage species. The Institute for Biodiversity Conservation and Research holds active collections of seeds mainly for research and distribution and as a base collection for long-term conservation [60].

For security reasons, the collected and stored germplasm need to be conserved in duplicate gene banks. However, except for the limited samples of the Ethiopian germplasm held by the Consultative Group on International Agricultural

Research, United States Agency for International Development, and the Nordic Gene Bank, majority of the Ethiopian collections are still kept in a single copy at the National Gene Bank. Greater efforts need to be made to store duplicate collections to avoid future genetic erosion [15]. For the continuing power supply, the Ethiopian Gene Bank has independent power supply in the form of a stand by generator to overcome power cuts [15].

Lack of adequate knowledge with respect to collection, handling, and treatment of seeds often impedes the planting of indigenous trees and shrubs. Inadequate work has been done on establishing the seed storage behavior of native species resulting in only limited availability of *ex situ* conservation seed collections especially with respect to native forest species [60] and lack of alternative storage facilities for the existing conventional cold rooms (e.g., *in vitro* and cryo-preservation methods) [4]. The current holdings of the IBC gene bank reach over 60,000 accessions of plant species. Some collections are in the medium-term storage mainly due to insufficient seed samples [15].

4. Conclusion

Biodiversity plays a great role in human existence and in healthy function of natural systems although it is on the way of depletion dominantly due to anthropogenic activities. This requires conservation of biodiversity either in *in situ* or *ex situ* or both methods in combination based on the conservation situation and its objective. Although *in situ* conservation is more encouraged to be used for biodiversity conservation, *ex situ* conservation is recommended as it complements through different techniques like zoo, captive breeding, aquarium, botanical garden, and gene bank. *Ex situ* conservation has its own advantages, disadvantages, and challenges making decision on its application by evaluating advantages, disadvantages and challenges. Although, Ethiopia is rich in biodiversity resources, more people depend on it for their livelihood directly or indirectly causing a great loss. Even if the conservation of biodiversity in Ethiopia has long-time history, its progress, coverage, and enforcement of the rule for conservation seem to be weak. Despite of good progress made in gene bank conservation, it is yet to be developed. In the same way, attention should be given for developing a National Zoological Park and a Botanical Garden.

Acknowledgments

The authors are extremely grateful to Dr. Habte Jebessa for his valuable comments, suggestions and corrections on the draft of this paper. Our appreciation also goes to Professor. Afework Bekele for his remarkable encouragements and critical comments on the draft of this paper.

References

[1] T. I. Borokini, A. U. Okere, A. O. Giwa, B. O. Daramola, and W. T. Odofin, "Biodiversity and conservation of plant genetic resources in Field Gene-bank of the National Centre for Genetic Resources and Biotechnology, Ibadan, Nigeria," *The International Journal of Biodiversity and Conservation*, vol. 2, pp. 37–50, 2010.

[2] M. Antofie, "Current political commitments' challenges for *ex situ* conservation of plant genetic resources for food and agriculture," *Analele Universității din Oradea—Fascicula Biologie*, vol. 18, pp. 157–163, 2011.

[3] C. A. Tisdell, "Core issues in the economics of biodiversity conservation," *Annals of the New York Academy of Sciences*, vol. 1219, no. 1, pp. 99–112, 2011.

[4] Institute of Biodiversity Conservation, *National Biodiversity Strategy and Action Plan*, Institute of Biodiversity Conservation, Federal Democratic Republic of Ethiopia, Addis Ababa, Ethiopia, 2005.

[5] M. E. Dulloo, D. Hunter, and T. Borelli, "Ex situ and in situ conservation of agricultural biodiversity: major advances and research needs," *Notulae Botanicae Horti Agrobotanici Cluj-Napoca*, vol. 38, no. 2, pp. 123–135, 2010.

[6] J. Young, C. Richards, A. Fischer et al., "Conflicts between biodiversity conservation and human activities in the central and eastern European countries," *Ambio*, vol. 36, no. 7, pp. 545–550, 2007.

[7] H. Debela, "Human influence and threat to biodiversity and sustainable living," *Ethiopian Journal of Education and Sciences*, vol. 3, no. 1, pp. 85–95, 2007.

[8] R. C. Lacy, "Considering threats to the viability of small populations using individual-based models," *Ecological Bulletin*, vol. 48, pp. 39–51, 2000.

[9] K. Leus, "Captive breeding and conservation," *Zoology in the Middle East*, vol. 54, supplement 3, pp. 151–158, 2011.

[10] K. A. Wilson, J. Carwardine, and H. P. Possingham, "Setting conservation priorities," *Annals of the New York Academy of Sciences*, vol. 1162, pp. 237–264, 2009.

[11] R. Frankham, J. D. Ballou, and D. A. Briscoe, *Introduction to Conservation Genetics*, Cambridge University Press, Cambridge, UK, 2002.

[12] T. Getachew, "Biodiversity hotspots: pitfalls and prospects," *IBC News Letter*, vol. 1, pp. 14–16, 2012.

[13] H. Huang, "Plant diversity and conservation in China: planning a strategic bioresource for a sustainable future," *Botanical Journal of the Linnean Society*, vol. 166, no. 3, pp. 282–300, 2011.

[14] C. Brutting, I. Hensen, and K. Wesche, "Ex situ cultivation affects genetic structure and diversity in arable plants," *Plant Biology*, vol. 15, pp. 505–513, 2013.

[15] Institute of Biodiversity Conservation, *Ethiopia: Second Country Report on the State of PGRFA to FAO*, FAO, Rome, Italy, 2007.

[16] E. D. Kjaer, L. Graudal, and I. Nathan, *Ex Situ Conservation of Commercial Tropical Trees: Strategies, Options and Constraints*, Danida Forest Seed Centre, Humlebaek, Denmark, 2001.

[17] J. Hakansson, *Genetic Aspects of Ex Situ Conservation: Introductory Paper*, Department of Biology, Linköping University, 2004.

[18] G. M. Reid, T. C. Macbeath, and K. Csatadi, "Global challenges in freshwater-fish conservation related to public aquariums and the aquarium industry," *International Zoo Yearbook*, vol. 47, pp. 6–45, 2013.

[19] V. A. Melfi, "Ex situ gibbon conservation: status, management and birth sex ratios," *International Zoo Yearbook*, vol. 46, no. 1, pp. 241–251, 2012.

[20] C. Ratledge, "Towards conceptual framework for wildlife Tourism," *Tourism Management*, vol. 22, pp. 31–40, 2001.

[21] J. Balcombe, "Animal pleasure and its moral significance," *Applied Animal Behaviour Science*, vol. 118, no. 3-4, pp. 208–216, 2009.

[22] S. F. Carrizo, K. G. Smith, and W. R. T. Darwall, "Progress towards a global assessment of the status of freshwater fishes (Pisces) for the IUCN Red List: application to conservation programmes in zoos and aquariums," *International Zoo Yearbook*, vol. 47, pp. 46–64, 2013.

[23] R. C. Lacy, "Achieving true sustainability of zoo populations," *Zoo Biology*, vol. 32, pp. 19–26, 2013.

[24] R. L. Eaton, "An overview of zoo goals and exhibition principles," *International Journal for the Study of Animal Problems*, vol. 2, pp. 295–299, 1981.

[25] E. J. Fernandez, M. A. Tamborski, S. R. Pickens, and W. Timberlake, "Animal-visitor interactions in the modern zoo: conflicts and interventions," *Applied Animal Behaviour Science*, vol. 120, no. 1-2, pp. 1–8, 2009.

[26] L. A. Dickie, "The sustainable zoo: an introduction," *International Zoo Yearbook*, vol. 43, no. 1, pp. 1–5, 2009.

[27] A. M. Claxton, "The potential of the human-animal relationship as an environmental enrichment for the welfare of zoo-housed animals," *Applied Animal Behaviour Science*, vol. 133, no. 1-2, pp. 1–10, 2011.

[28] S. E. Williams and E. A. Hoffman, "Minimizing genetic adaptation in captive breeding programs: a review," *Biological Conservation*, vol. 142, no. 11, pp. 2388–2400, 2009.

[29] R. J. Whittington and R. Chong, "Global trade in ornamental fish from an Australian perspective: the case for revised import risk analysis and management strategies," *Preventive Veterinary Medicine*, vol. 81, no. 1-3, pp. 92–116, 2007.

[30] D. Dudgeon, A. H. Arthington, M. O. Gessner et al., "Freshwater biodiversity: importance, threats, status and conservation challenges," *Biological Reviews of the Cambridge Philosophical Society*, vol. 81, no. 2, pp. 163–182, 2006.

[31] S. Blackmore, M. Gibby, and D. Rae, "Strengthening the scientific contribution of botanic gardens to the second phase of the Global Strategy for Plant Conservation," *Botanical Journal of the Linnean Society*, vol. 166, no. 3, pp. 267–281, 2011.

[32] A. G. Clarke, "The Frozen Ark Project: the role of zoos and aquariums in preserving the genetic material of threatened animals," *International Zoo Yearbook*, vol. 43, no. 1, pp. 222–230, 2009.

[33] O. A. Ryder, A. McLaren, S. Brenner, Y.-P. Zhang, and K. Benirschke, "DNA banks for endangered animal species," *Science*, vol. 288, no. 5464, pp. 275–277, 2000.

[34] C. W. Vertucci, "Predicting the optimum storage conditions for seeds using thermodynamic principles," *Journal of Seed Technology*, vol. 17, pp. 41–52, 1993.

[35] R. H. Ellis and E. H. Roberts, "Improved equations for the prediction of seed longevity," *Annals of Botany*, vol. 45, no. 1, pp. 13–30, 1980.

[36] J. B. Dickie, R. H. Ellis, H. L. Kraak, K. Ryder, and P. B. Tompsett, "Temperature and seed storage longevity," *Annals of Botany*, vol. 65, no. 2, pp. 197–204, 1990.

[37] M. L. Moir, P. A. Vesk, K. E. C. Brennan et al., "Considering extinction of dependent species during translocation, *ex situ* conservation, and assisted migration of threatened hosts," *Conservation Biology*, vol. 26, no. 2, pp. 199–207, 2012.

[38] S. Christie, "Why keep tigers in zoos?" in *Tigers of the World: The Science, Politics and Conservation of Panthera Tigris*, R. Tilson and P. Nyhus, Eds., pp. 205–214, Elsevier Inc., Amsterdam, The Netherlands, 1998.

[39] U. S. Anderson, A. S. Kelling, R. Pressley-Keough, M. A. Bloomsmith, and T. L. Maple, "Enhancing the zoo visitor's experience by public animal training and oral interpretation at an otter exhibit," *Environment and Behavior*, vol. 35, no. 6, pp. 826–841, 2003.

[40] H. R. Jensen, A. Dreiseitl, M. Sadiki, and D. J. Schoen, "The Red Queen and the seed bank: pathogen resistance of *ex situ* and *in situ* conserved barley," *Evolutionary Applications*, vol. 5, no. 4, pp. 353–367, 2012.

[41] G. R. Hosey, "How does the zoo environment affect the behaviour of captive primates?" *Applied Animal Behaviour Science*, vol. 90, no. 2, pp. 107–129, 2005.

[42] K. N. Morgan and C. T. Tromborg, "Sources of stress in captivity," *Applied Animal Behaviour Science*, vol. 102, no. 3-4, pp. 262–302, 2007.

[43] B. O. Braastad, "Effects of prenatal stress on behaviour of offspring of laboratory and farmed mammals," *Applied Animal Behaviour Science*, vol. 61, no. 2, pp. 159–180, 1998.

[44] S. M. Wisely, R. M. Santymire, T. M. Livieri et al., "Environment influences morphology and development for *in situ* and *ex situ* populations of the black-footed ferret (*Mustela nigripes*)," *Animal Conservation*, vol. 8, no. 3, pp. 321–328, 2005.

[45] K. Helenurm and L. S. Parsons, "Genetic variation and the reintroduction of *Cordylanthus maritimus ssp. maritimus* to Sweetwater Marsh, California," *Restoration Ecology*, vol. 5, no. 3, pp. 236–244, 1997.

[46] D. Lauterbach, M. Burkart, and B. Gemeinholzer, "Rapid genetic differentiation between *ex situ* and their *in situ* source populations: an example of the endangered *Silene otites* (Caryophyllaceae)," *Botanical Journal of the Linnean Society*, vol. 168, no. 1, pp. 64–75, 2012.

[47] A. Balmford, G. M. Mace, and N. Leader-Williams, "Designing the ark: setting priorities for captive breeding," *Conservation Biology*, vol. 10, no. 3, pp. 719–727, 1996.

[48] R. A. Griffiths and L. Pavajeau, "Captive breeding, reintroduction, and the conservation of amphibians," *Conservation Biology*, vol. 22, no. 4, pp. 852–861, 2008.

[49] R. J. Probert, M. I. Daws, and F. R. Hay, "Ecological correlates of *ex situ* seed longevity: a comparative study on 195 species," *Annals of Botany*, vol. 104, no. 1, pp. 57–69, 2009.

[50] L. M. Clayton, E. J. Milner-Gulland, D. W. Sinaga, and A. H. Mustari, "Effects of a proposed *ex situ* conservation program on *in situ* conservation of the babirusa, an endangered suid," *Conservation Biology*, vol. 14, no. 2, pp. 382–385, 2000.

[51] W. G. Conway, "Buying time for wild animals with zoos," *Zoo Biology*, vol. 30, no. 1, pp. 1–8, 2011.

[52] N. C. Ellstrand and D. R. Elam, "Population genetic conseqences of small population size: implications for plant conservation," *Annual Review of Ecology and Systematics*, vol. 24, pp. 217–242, 1993.

[53] A. A. Cunningham, "Disease risks of wildlife translocations," *Conservation Biology*, vol. 10, no. 2, pp. 349–353, 1996.

[54] R. P. Reading, T. W. Clark, and B. Griffith, "The influence of valuational and organizational considerations on the success of rare species translocations," *Biological Conservation*, vol. 79, no. 2-3, pp. 217–225, 1997.

[55] L. I. Wright, T. Tregenza, and D. J. Hosken, "Inbreeding, inbreeding depression and extinction," *Conservation Genetics*, vol. 9, no. 4, pp. 833–843, 2008.

[56] B. Miller, K. Ralls, R. P. Reading, J. M. Scott, and J. Estes, "Biological and technical considerations of carnivore translocation: a review," *Animal Conservation*, vol. 2, no. 1, pp. 59–68, 1999.

[57] L. Laikre, L. C. Larsson, A. Palmé, J. Charlier, M. Josefsson, and N. Ryman, "Potentials for monitoring gene level biodiversity: using Sweden as an example," *Biodiversity and Conservation*, vol. 17, no. 4, pp. 893–910, 2008.

[58] N. I. Vavilov, "The origin, variation, immunity and breeding of cultivated plants," *Chronica Botanica*, vol. 13, pp. 1–366, 1951.

[59] S. Demissew, "How has government policy post-global strategy for plant conservation impacted on science? The Ethiopian perspective," *Botanical Journal of the Linnean Society*, vol. 166, no. 3, pp. 310–325, 2011.

[60] B. Girma, T. Pearce, and D. Abebe, "Biological diversity and current *ex situ* conservation practices in Ethiopia," in *Seed Conservation Turning Science into Practice*, R. D. Smith, J. B. Dickie, S. H. Linington, H. W. Pritchard, and R. J. Probert, Eds., pp. 849–856, Kew Publishing, Kew, UK, 2003.

Permissions

The contributors of this book come from diverse backgrounds, making this book a truly international effort. This book will bring forth new frontiers with its revolutionizing research information and detailed analysis of the nascent developments around the world.

We would like to thank all the contributing authors for lending their expertise to make the book truly unique. They have played a crucial role in the development of this book. Without their invaluable contributions this book wouldn't have been possible. They have made vital efforts to compile up to date information on the varied aspects of this subject to make this book a valuable addition to the collection of many professionals and students.

This book was conceptualized with the vision of imparting up-to-date information and advanced data in this field. To ensure the same, a matchless editorial board was set up. Every individual on the board went through rigorous rounds of assessment to prove their worth. After which they invested a large part of their time researching and compiling the most relevant data for our readers. Conferences and sessions were held from time to time between the editorial board and the contributing authors to present the data in the most comprehensible form. The editorial team has worked tirelessly to provide valuable and valid information to help people across the globe.

Every chapter published in this book has been scrutinized by our experts. Their significance has been extensively debated. The topics covered herein carry significant findings which will fuel the growth of the discipline. They may even be implemented as practical applications or may be referred to as a beginning point for another development. Chapters in this book were first published by Hindawi Publishing Corporation; hereby published with permission under the Creative Commons Attribution License or equivalent.

The editorial board has been involved in producing this book since its inception. They have spent rigorous hours researching and exploring the diverse topics which have resulted in the successful publishing of this book. They have passed on their knowledge of decades through this book. To expedite this challenging task, the publisher supported the team at every step. A small team of assistant editors was also appointed to further simplify the editing procedure and attain best results for the readers.

Our editorial team has been hand-picked from every corner of the world. Their multi-ethnicity adds dynamic inputs to the discussions which result in innovative outcomes. These outcomes are then further discussed with the researchers and contributors who give their valuable feedback and opinion regarding the same. The feedback is then collaborated with the researches and they are edited in a comprehensive manner to aid the understanding of the subject.

Apart from the editorial board, the designing team has also invested a significant amount of their time in understanding the subject and creating the most relevant covers. They scrutinized every image to scout for the most suitable representation of the subject and create an appropriate cover for the book.

The publishing team has been involved in this book since its early stages. They were actively engaged in every process, be it collecting the data, connecting with the contributors or procuring relevant information. The team has been an ardent support to the editorial, designing and production team. Their endless efforts to recruit the best for this project, has resulted in the accomplishment of this book. They are a veteran in the field of academics and their pool of knowledge is as vast as their experience in printing. Their expertise and guidance has proved useful at every step. Their uncompromising quality standards have made this book an exceptional effort. Their encouragement from time to time has been an inspiration for everyone.

The publisher and the editorial board hope that this book will prove to be a valuable piece of knowledge for researchers, students, practitioners and scholars across the globe.

List of Contributors

David Sylvester Kacholi
Department of Biological Sciences, Dar es Salaam University College of Education (DUCE), P.O. Box 2329, Dar es Salaam, Tanzania

Ezekiel Edward Mwakalukwa
Department of Food and Resource Economics, Faculty of Science, University of Copenhagen, Rolighedsvej 23, 1958 Frederiksberg C, Denmark
Department of Forest Biology, Faculty of Forestry and Nature Conservation, Sokoine University of Agriculture, P.O. Box 3010, Chuo Kikuu, Morogoro, Tanzania

Henrik Meilby and Thorsten Treue
Department of Food and Resource Economics, Faculty of Science, University of Copenhagen, Rolighedsvej 23, 1958 Frederiksberg C, Denmark

Bin Mushambanyi Théodore Munyuli
Academic Affairs and Research Program, Cinquantenaire University (UNIC/Lwiro), D.S. Bukavu, South-Kivu Province, Democratic Republic of Congo
Departments of Biology and Environment, National Center for Research in Natural Sciences (CRSN/Lwiro), D.S. Bukavu, South-Kivu Province, Democratic Republic of Congo
Centre of Research for Health Promotion (CRPS), Department of Nutrition and Dietetics, Institute of Higher Education in Medical Techniques (ISTM/Bukavu), P.O. Box 3036, Bukavu, South-Kivu Province, Democratic Republic of Congo
Department of Natural Resources and Environmental Economics, Faculty of Natural Resources and Environmental Sciences, Namasagsali Campus, Busitema University, P.O. Box 236, Tororo, Uganda

Takuo Nagaike
Yamanashi Forest Research Institute, Saisyoji 2290-1, Fujikawa, Yamanashi 400-0502, Japan

Eiji Ohkubo
Yamanashi Gakuin Junior College, Sakaori 2-4-5, Kofu, Yamanashi 400-8575, Japan

Kazuhiro Hirose
Minami-Alps City Office, Ogasawara 376, Minami-Alps, Yamanashi 400-0306, Japan

Pramod Lamsal
Himalayan Geo-En. Pvt. Ltd., 133 Mokshya Marga, Kathmandu 4, Nepal

Krishna Prasad Pant and Kishor Atreya
Kathmandu University, P.O. Box 6250, Dhulikhel, Nepal

Lalit Kumar
University of New England, Armidale, NSW 2351, Australia

Daniel S. Licht
National Park Service, 231 East Saint Joseph Street, Rapid City, SD 57701, USA

Brian C. Kenner
Badlands National Park, Interior, SD 57750, USA

Daniel E. Roddy
Wind Cave National Park, Hot Springs, SD 57747, USA

Isidore Ayissi
Association Camerounaise de Biologie Marine (ACBM), BP 52, Ayos, Cameroon
CERECOMA, Specialized Research Center for Marine Ecosystems, c/o Institute of Agricultural Research for Development, P.O. Box 219, Kribi, Cameroon
Institute of Fisheries and Aquatic Sciences (ISH) at Yabassi, University of Douala, P.O. Box 2701, Douala, Cameroon

Gabriel Hoinsoudé Segniagbeto
Departement de Zoologie et de Biologie Animale, Faculte des Sciences, Universite de Lome, Lome, Togo

Koen Van Waerebeek
Conservation and Research of West African Aquatic Mammals (COREWAM), c/o Department of Marine and Fisheries Science, University of Ghana, P.O. Box LG99, Legon, Ghana
COREWAM-Senegal, Musee de la Mer de Gore, IFAN-CH.A.D, Universite de Dakar, Dakar, Senegal
Peruvian Centre for Cetacean Research (CEPEC), Lima 20, Peru

P. Balaji
Department of Botany, Dr. Ambedkar Government Arts College, Vyasarpadi, Chennai, Tamil Nadu 600 039, India

G. N. Hariharan
Lichen Ecology and Bio prospecting Laboratory, M. S. Swaminathan Research Foundation III Cross Street, Taramani Institutional Area, Taramani, Chennai, Tamil Nadu 600 113, India

Bishwajit Roy, Md. Habibur Rahman and Most. Jannatul Fardusi
Bangladesh Institute of Social Research (BISR), Hasina De Palace, House No. 6/14, Block No. A, Lalmatia, Dhaka 1207, Bangladesh

Mohammed Kasso and Mundanthra Balakrishnan
Department of Zoological Sciences, Addis Ababa University, P.O. Box 1176, Addis Ababa, Ethiopia

Peddrick Weis
Department of Biological Science, Rutgers University, Newark, NJ 07102, USA
Department of Radiology, UMDNJ–New Jersey Medical School, Newark, NJ 07101-1709, USA

Judith S. Weis
Department of Biological Science, Rutgers University, Newark, NJ 07102, USA

Arun Kanagavel and Rajeev Raghavan
Conservation Research Group (CRG), St. Albert's College, Kochi, Kerala 682 018, India

Shiny M. Rehel
Keystone Foundation, Kotagiri, Tamil Nadu 643 217, India
Research and Development Centre, Bharathiar University, Coimbatore, Tamil Nadu 641 046, India

Archiebald Baltazar B. Malaki
Cebu Technological University, Cebu Campus, Argao 6021, Cebu, Philippines

Rex Victor O. Cruz, Nathaniel C. Bantayan and Diomedes A. Racelis
Institute of Renewable and Natural Resources, College of Forestry and Natural Resources, University of the Philippines Los Banos, Laguna 4031, Philippines

Inocencio E. Buot Jr.
Institute of Biological Sciences, College of Arts and Sciences and School of Environmental Science & Management (SESAM), University of the Philippines Los Banos, Laguna 4031, Philippines

Leonardo M. Florece
School of Environmental Science and Management, University of the Philippines Los Banos, Laguna 4031, Philippines

Chitra Sharma and Rajan K. Gupta
Department of Botany, Government Post Graduate Autonomous College, Rishikesh 249201, Uttarakhand, India

Rakesh K. Pathak
Department of Botany, D.A.V. (P.G.) College, Dehradun 248001, Uttarakhand, India

Kaushal K. Choudhary
Department of Botany, Dr. Jagannath Mishra College, Muzaffarpur 842001, Bihar, India

Jia Feng and Shulian Xie
College of Life Science, Shanxi University, Taiyuan 030006, China

Priyanka Sati, Kusum Dhakar and Anita Pandey
Biotechnological Applications, G. B. Pant Institute of Himalayan Environment and Development, Kosi-Katarmal, Almora, Uttarakhand 263 643, India

Simon A. Black
Durrell Institute of Conservation and Ecology, School of Anthropology and Conservation, University of Kent, Canterbury, Kent CT2 7NZ, UK
Department of Human Resources, University of Kent, Canterbury, Kent, UK

Jim J. Groombridge
Durrell Institute of Conservation and Ecology, School of Anthropology and Conservation, University of Kent, Canterbury, Kent CT2 7NZ, UK

Carl G. Jones
Durrell Wildlife Conservation Trust, Trinity, Jersey, Channel Islands, UK
Mauritian Wildlife Foundation, Grannum Road, Vacoas, Mauritius

Marcos Karlin, Rodrigo Galán, Ana Contreras, Ricardo Zapata, Rubén Coirini and Eduardo Ruiz Posse
Facultad de Ciencias Agropecuarias, Universidad Nacional de Cordoba, Avenida Valparaıso s/n, Ciudad Universitaria, C.C. 509, 5000 Cordoba, Argentina

Mohammed Kasso and Mundanthra Balakrishnan
Department of Zoological Sciences, Addis Ababa University, P.O. Box 1176, Addis Ababa, Ethiopia